Rheology of Fluid and Semisolid Foods

Principles and Applications

Food Engineering Series
Series Editor
Gustavo V. Barbosa-Cánovas, Washington State University

J. M. Aguilera and D. W. Stanley, *Microstructural Principles of Food Processing and Engineering*, second edition
G. V. Barbosa-Cánovas and H. Vega-Mercado, *Dehydration of Foods*
P. Fito, E. Ortega-Rodríguez, and G. V. Barbosa-Cánovas, *Food Engineering 2000*
P. J. Fryer, D. L. Pyle, and C. D. Reilly, *Chemical Engineering for the Food Industry*
D. Holdsworth, *Thermal Processing of Packaged Foods*
M. J. Lewis and N. J. Heppell, *Continuous Thermal Processing of Foods*
R. Moreira, M. E. Castell-Perez, and M. A. Barrufet, *Deep-Fat Frying: Fundamentals and Applications*

Rheology of Fluid and Semisolid Foods

Principles and Applications

M. Anandha Rao, PhD
Department of Food Science and Technology
Cornell University
Geneva, New York

A Chapman & Hall Food Science Book

AN ASPEN PUBLICATION®
Aspen Publishers, Inc.
Gaithersburg, Maryland
1999

The author has made every effort to ensure the accuracy of the information herein. However, appropriate information sources should be consulted, especially for new or unfamiliar procedures. It is the responsibility of every practitioner to evaluate the appropriateness of a particular opinion in the context of actual clinical situations and with due considerations to new developments. The author, editors, and the publisher cannot be held responsible for any typographical or other errors found in this book.

Aspen Publishers, Inc., is not affiliated with the American Society of Parenteral and Enteral Nutrition.

Library of Congress Cataloging-in-Publication Data
Rao, M. A., 1937–
Rheology of fluid and semisolid foods :
principles and applications / M. Anandha Rao.
p. cm.
Includes bibliographical references and index.
ISBN 0-8342-1264-1
1. Food—Testing. 2. Rheology. I. Title.
TP372.5.R36 1999
664—dc21 99-20667
CIP

Orders: (800) 638-8437
Customer Service: (800) 234-1660

Editorial Services: Kathy Litzenberg
Library of Congress Catalog Card Number: 99-20667
ISBN: 0-8342-1264-1

Printed in the United States of America

1 2 3 4 5

To Jan
for her unlimited and unconditional love;
and Hari
for bringing much joy into my life.

Table of Contents

Contributors

José Antonio Lopes da Silva, PhD
Department of Chemistry
University of Aveiro
Aveiro, Portugal

Paul E. Okechukwu, PhD
Anambra State Polytechnic
Oko, Anambra State, Nigeria

M. Anandha Rao, PhD
Department of Food Science and Technology
Cornell University
Geneva, New York

Preface

Considering that the science of rheology of fluid and semisolid foods has advanced considerably during the past few decades, food professionals need to appreciate the basic principles of rheological behavior and proper measurement of rheological properties, as well as the influence of composition and structure on those properties. Fluid and semisolid foods encompass a wide range of composition and structure and exhibit rheological behavior, including simple Newtonian, shear-thinning, shear-thickening, time-dependent shear-thinning and shear-thickening, as well as viscoelastic behavior. In addition, many manufactured solid foods are often in a fluid state at some stage during their manufacture; thus, rheological changes during phase transitions are also important. A sound appreciation of food rheology should be helpful in quality control and product development of foods. This book is intended to be an introduction to the science of the rheology of fluid foods and its application to practical problems.

One goal of this book is to introduce to interested students and scientists the principles of rheological behavior (Chapter 1), rheological and functional models applicable to fluid foods (Chapter 2), and measurement of the viscous and viscoelastic rheological behavior of foods (Chapter 3). The first three chapters cover the basic principles necessary to understand food rheology, to conduct rheological experiments, and to interpret properly the results of the experiments. They also contain many functional relationships that are useful in understanding the rheological behavior of specific foods. The science of rheology is based on principles of physics of deformation and flow, and it requires a reasonable knowledge of mathematics. Readers are urged to refresh their mathematics and physics backgrounds to understand the importance of several functional relationships. The science of rheology has been touched on by many eminent scientists, including Newton, Maxwell, Kelvin, Reiner, Einstein (1921 physics Nobel laureate), Flory (1974 chemistry Nobel laureate), de Gennes (1991 physics Nobel laureate), and Chu (1997 physics Nobel laureate).

Rheological behavior of food gum and starch dispersions is covered in Chapter 4 and that of many processed foods is covered in Chapter 5, with emphasis on the important developments and relationships with respect to the role of composition and structure on rheological behavior. In Chapter 4, the rheological changes taking place during starch gelatinization as a result of changes in starch granule size are covered. In both Chapters 4 and 5, generally applicable results are presented. At the end of Chapter 5, an extensive compilation of literature values of the rheological properties of fluid foods is also provided, along with the experimental techniques and ranges of important variables covered in a specific study.

Novel and important applications of rheology to food gels and gelation phenomena are covered in Chapter 6. In this chapter, the pioneering studies of Paul Flory and Pierre de Gennes are introduced. Many foods are gel systems, so the theoretical and practical aspects of these subjects are important and useful in understanding the role of gelling polymers on rheological properties of homogeneous and phase-separated gel systems.

Every food professional should recognize that sensory assessment of foods is important. Chapter 7 covers the role of rheological behavior in the sensory assessment of viscosity and flavor of fluid foods. It should be noted that food rheologists have made many unique contributions to sensory assessment of viscosity.

Lest one ignore the important role of rheological behavior and properties of fluid foods in handling and processing foods, these are covered in Chapter 8, including applications under isothermal conditions (pressure drop and mixing) and under nonisothermal conditions (heat transfer: pasteurization and sterilization). In particular, the isothermal rheological and nonisothermal thermorheological models discussed in Chapter 3 and 4 are applied in Chapter 8.

Last but not least, I thank Paul Okechukwu and José Antonio Lopes da Silva for collaborating in Chapters 5, 6, and 7. I am very grateful to the many graduate students and visiting scientists who have worked with me and contributed to my better understanding of rheology of fluid and semisolid foods. Several references listed at the end of each chapter are studies conducted with my students, who have made valuable contributions. Herb Cooley has helped me with many figures used in this book and in the conduct of several studies in my laboratory. Many persons have helped me during my career, especially my professor at Ohio State, Bob Brodkey, and my colleagues, Alfredo A. Vitali, ITAL, Brazil, and Ashim Datta, Cornell. I thank the editors at Aspen Publishers, Inc., for their invaluable help. Any mistakes in this book are mine.

M. Anandha Rao, PhD
Department of Food Science and Technology
Cornell University
Geneva, New York

CHAPTER 1

Introduction

M. Anandha Rao

By definition, rheology is the study of the deformation and flow of matter. The science of rheology grew considerably due to research work done on synthetic polymers and their solutions in different solvents, which, in turn, was necessary due to the many uses of the polymers ("plastics") in day-to-day and industrial applications. Nevertheless, because of the biological nature of foods, food rheology offers many unique opportunities of study and a large body of food rheology literature exists. Rheological properties are based on flow and deformation responses of foods when subjected to stress. A thorough study of the rheology of fluid foods requires knowledge of tensors and the basic principles of fluid flow, such as the equation of continuity (conservation of mass) and the equation of motion (conservation of momentum). The necessary basic equations can be derived (Bird et al., 1960; Macosko, 1994) by conducting a balance of either mass or momentum on a stationary volume element of finite dimensions (e.g., Δx, Δy, Δz in rectangular coordinates) through which the fluid is flowing.

$$\text{Input} + \text{Generation} = \text{Output} + \text{Accumulation} \tag{1.1}$$

Because mass is not generated, in the derivation of the equation of continuity, the second term on the left-hand side of Equation 1.1 can be omitted. The applicable partial differential equations are obtained in the limit as the dimensions of the control volume tend to zero. The primary objective here is to point out to readers the origins of useful relationships and the assumptions made in deriving them. Appendix 1–A presents the equation of continuity and the equations of motion in Cartesian, cylindrical, and spherical coordinates; however, for the actual derivation of the transport equations, the aforementioned references should be consulted. The use of some of these equations in deriving equations for specific measurement geometries will be illustrated in Chapter 3 and in processing applications in Chapter 8. In vector and tensor form, the transport equations are quite concise (Appendix 1–A).

Foods can be classified in different manners, including as solids, gels, homogeneous liquids, suspensions of solids in liquids, and emulsions. Fluid foods are those that do not retain their shape but take the shape of their container. Fluid foods that contain significant amounts of dissolved high molecular weight compounds (polymers) and/or suspended solids exhibit non-Newtonian behavior. Many non-Newtonian foods also exhibit both viscous and elastic properties, that is, they exhibit viscoelastic behavior.

Fluid and semisolid foods exhibit a wide variety of rheological behavior ranging from Newtonian to time dependent and viscoelastic. Fluid foods containing relatively large amounts of dissolved low molecular weight compounds (e.g., sugars) and no significant amount of a polymer or insoluble solids can be expected to exhibit Newtonian behavior. A small amount (~1%) of a dissolved polymer can substantially increase the viscosity and also alter the flow characteristics from the Newtonian behavior of water to the non-Newtonian behavior of an aqueous dispersion. It is interesting to note that whereas the rheological properties are altered substantially, magnitudes of the thermal properties (e.g., density and thermal conductivity) of the dispersion remain relatively close to those of water. Further, with one or two exceptions, for example, food polymer dispersions, it is difficult to predict precise magnitudes of viscosity of fluid foods, mainly because foods are complex mixtures of biochemical compounds that show a wide variation in composition. Therefore, studies on rheological properties of foods are useful and important for applications that include handling and processing, quality control, and sensory assessment of foods. The latter is an important field of study to which food scientists have made significant contributions.

When studying the rheological behavior of a food, knowledge of the composition of the food, especially the important structuring components (e.g., dissolved polymers, suspended solids), the structure of the food itself (e.g., homogeneous or phase-separated gel, emulsion), and the processing and storage conditions would be helpful because they all often affect the behavior. In dispersions, however, the structure of the food particles plays a major role in defining the rheological behavior. Either large molecules or finely subdivided bulk matter could be considered to be colloidal matter; that is, the particles are in the range 10^{-9} to 10^{-6} m in dimension. Therefore, principles of colloidal science are also useful in understanding the rheological behavior. For example, colloids can be classified as lyophilic (solvent loving) or lyophobic (solvent fearing), and when the solvent is water they are called hydrophilic and hydrophobic, respectively.

The classification of rheological behavior and the measurement of rheological properties of fluid foods were reviewed by Sherman (1970), Ross-Murphy (1984), Rao (1992, 1995), Steffe (1996), and others. In addition to these references, one must consult books on synthetic polymer rheology for valuable information on viscoelastic behavior, measurement techniques, and the role of fundamental properties such as molecular weight on viscoelastic behavior. In particular, the texts by Bird et al. (1977a, 1977b), Ferry (1980), and Tschoegl (1989) have much useful information on viscoelasticity. Barnes et al. (1989) discussed rheology in a lucid manner. Techniques for measuring rheological properties, especially flow properties, were well covered by Van Wazer et al. (1963) and those of viscoelastic properties by Walters (1975), Whorlow (1980), Dealy (1982), and Macosko (1994). An examination of the early efforts would provide an education on the innovations in experimental techniques and developments in interpretation of rheological behavior. In this respect, a review (Markovitz, 1985) of the studies conducted soon after the Society of Rheology was formed provides a fascinating picture of the development of the science of rheology. While one should appreciate the ease with which many rheological measurements can be performed today, primarily due to the availability of powerful desktop computers, it is essential that food rheologists really understand the underlying principles in rheological measurements and interpretation of results. In addition, low-friction compressed-air bearings, optical deformation measurement systems, and other developments also have contributed to measurements on foods and other low viscosity materials.

STRESS AND STRAIN TENSORS

While scalar quantities have magnitudes, vectors are defined in terms of both direction and magnitude and have three components. Tensors are second-order vectors and have nine components. In a Cartesian system of coordinates, a stress tensor ($\tilde{\tilde{\tau}}$) of force imposed on unit surface area of a test material can be resolved in terms of nine components (σ_{ij}): three normal and six tangential stresses. In simple shearing (viscometric) flow that is encountered in the flow geometries (capillary, Coette, cone-and-plate, and parallel plate), $\sigma_{12} = \sigma_{21}$, while the other four tangential components, σ_{13}, σ_{23}, σ_{31}, and σ_{32}, are equal to zero, so the stress tensor may be written as

$$\tilde{\tilde{\tau}} = \begin{Vmatrix} \sigma_{11} & \sigma_{12} & 0 \\ \sigma_{21} & \sigma_{22} & 0 \\ 0 & 0 & \sigma_{33} \end{Vmatrix} \tag{1.2}$$

The stresses, σ_{11}, σ_{22}, and σ_{33} are normal stresses that are equal to zero for Newtonian fluids and may be of appreciable magnitudes for some foods, such as doughs. The components of the deformation tensor, $\tilde{\tilde{e}}$, in viscometric flows are

$$\tilde{\tilde{e}} = \begin{Vmatrix} 0 & \dot{\gamma} & 0 \\ \dot{\gamma} & 0 & 0 \\ 0 & 0 & 0 \end{Vmatrix} \tag{1.3}$$

where $\dot{\gamma}$ is the shear rate.

Rheological properties of food materials over a wide range of phase behavior can be expressed in terms of viscous (viscometric), elastic, and viscoelastic functions that relate some components of the stress tensor to specific components of the strain or shear rate response. In terms of fluid and solid phases, viscometric functions are generally used to relate stress with shear rate in liquid systems, while elastic functions are related to the appropriate stress function to strain in solids. Viscoelastic properties cover the combination where a material exhibits both viscous and elastic properties.

VISCOMETRIC PROPERTIES

Three material functions—the viscosity function, η, and the first and second normal stress coefficients, Ψ_1 and Ψ_2—can be evaluated from the stress tensor using Equations 1–4 through 1–6, which relate specific stress components σ_{ij} to the shear rate $\dot{\gamma}$:

$$\sigma_{21} = \eta(\dot{\gamma}) \tag{1.4}$$

$$\sigma_{11} - \sigma_{22} = \Psi_1(\dot{\gamma})^2 \tag{1.5}$$

$$\sigma_{22} - \sigma_{33} = \Psi_2(\dot{\gamma})^2 \tag{1.6}$$

Steady state experiments employing rotational, capillary, and tube flow viscometers are commonly used to generate data for evaluation of these material properties. Reported values of the viscometric coefficients for many materials, especially synthetic polymers, indicate that η and Ψ_1 are large and positive, whereas Ψ_2 is small and negative (Macosko, 1994). While the first normal stress difference is known to be responsible for the climbing fluid film phenomenon often referred to as the "Weissenberg effect," extrudate swell, and normal force pump, the second normal stress difference determines where the free surface of a fluid flowing down a trough would be convex (Barnes et al., 1989; Macosko, 1994). Typical magnitudes of Ψ_1 of foods are shown in Figure 1–1 (Dickie and Kokini, 1982).

Most rheological studies have concentrated on the viscosity function and dynamic viscoelastic properties and much less on normal stress differences (Rao, 1992). Since about the mid 1960s, there has been a gradual paradigm shift from normal stresses to dynamic rheological properties as measures of viscoelastic properties. One reason for the shift appears to be the difficulties in obtaining reliable magnitudes of normal stresses and, compared to dynamic rheological properties, their limited practical application. Another reason is that with the availability of automated rheometers, measurement of dynamic rheological parameters has become very popular and manageable. Therefore, one rarely encounters data on normal stress functions of foods.

The relationships between stress and strain, and the influence of time on them, are generally described by constitutive equations or rheological equations of state

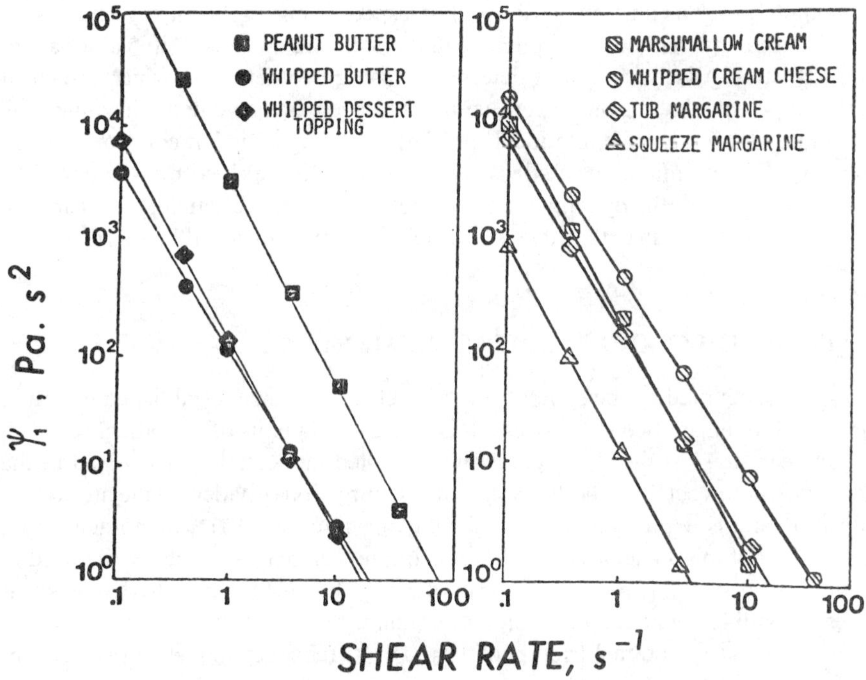

Figure 1–1 First Normal Stress Coefficient Data of Several Foods as a Function of Shear Rate. *Source*: Reprinted with permission from A.M. Dickie and J.L. Kokini, Use of the Bird-Leider Equation in Food Rheology, *Journal of Food Processing and Engineering*, Vol. 5, pp. 157–184. © 1981, Food and Nutrition Press.

(Ferry, 1980). When the strains are relatively small (i.e., in the linear range), the constitutive equations are relatively simple. Equations 1.5 and 1.6 are constitutive equations of viscoelastic behavior under continuous shear (nonlinear range). In the limit of low strains, relationships exist between the above material functions and those obtained from dynamic and creep experiments. Constitutive equations will be discussed further in Chapter 4.

SHEAR STRESS–SHEAR RATE RELATIONSHIPS

To study rheological properties of foods in a systematic manner, it is convenient to first learn the principles of rheological behavior of fluid and solid foods. Viscosity data must be obtained with well-designed viscometers and be expressed in fundamental units. Viscosity data in units that are specific to a particular brand viscometer, such as the time to flow out of a specific cup or the manufacturer's arbitrary viscosity, will not be applicable universally, and the viscosity values are of limited value. Viscosity is best determined using geometries in which the shear rate can be calculated from experimental data, such as the velocity or volumetric flow rate of a fluid or the rotational speed of a rotating cylinder, cone, or plate. In general, the determination of stress corresponding to the shear rates is not difficult. In addition to the dynamic viscosity determined in shear fields, one can also determine viscosity in extensional shear fields (Padmanabhan, 1995), as discussed later.

UNITS IN RHEOLOGICAL MEASUREMENT

Shear rate, denoted by the symbol $\dot{\gamma}$, is the velocity gradient established in a fluid as a result of an applied shear stress. It is expressed in units of reciprocal seconds, s^{-1}. Shear stress is the stress component applied tangentially. It is equal to the force vector (a vector has both magnitude and direction) divided by the area of application and is expressed in units of force per unit area (Pa). The nomenclature committee of the Society of Rheology recommends that the symbol σ be used to denote shear stress. However, the symbol τ, which was long used to denote shear stress, is still encountered in rheology literature.

Viscosity is the internal friction of a fluid or its tendency to resist flow. It is denoted by the symbol η for Newtonian fluids, whose viscosity does not depend on the shear rate, and for non-Newtonian fluids to indicate shear rate dependence by η_a. Depending on the flow system and choice of shear rate and shear stress, there are several equations to calculate. Here, it is defined by the equation

$$\eta_a = \frac{\text{Shear stress}}{\text{Shear rate}} = \frac{\sigma}{\dot{\gamma}} \tag{1.7}$$

The preferred units of viscosity are Pa s or mPa s. Some of the older units and their relation to the SI unit are given in Table 1–1. It is clear that the shear rate employed in the calculation must be specified when the magnitude of apparent viscosity is discussed. Apparent viscosity has many useful applications in characterizing a fluid food; in particular, in the characterization of shear-thinning fluids, the apparent viscosity at low shear rates, called zero-shear rate viscosity (η_0), is a useful parameter. Because much work was done with viscosity expressed in poise in sensory assessment of viscosity, the same tradition will be continued in Chapter 7 on sensory assessment of viscosity.

TYPES OF FLUID FLOW BEHAVIOR

The major types of fluid flow behavior can be described by means of basic shear diagrams of shear rate versus shear stress, such as Figures 1–2 and 1–3. In Figure

Table 1–1 Units and Conversion Factors for Viscosity, Kinematic Viscosity, and Activation Energy of Flow

The SI unit for viscosity is Pa s.
1 Pa s = 1000 cP = 1 N s m^{-2} = 1 kg m^{-1} s^{-1} = 0.67197 lbm ft^{-1} s^{-1}
1 poise (P) = 1 dyne s cm^{-2} = 0.1 Pa s
1 cP = 0.001 Pa s = 1 mPa s = 0.01 P
1 lbm ft^{-1} s^{-1} = 1.4882 kg m^{-1} s^{-1} = 1.4882 Pa s = 1488.2 cP
Kinematic viscosity (cSt) = Viscosity (cP) / Density (g cm^{-3})
1 St = 100 cSt = 0.0001 m^2 s^{-1}
1 calorie = 4.1868 joule (J)
Boltzmann constant, k = 1.38×10^{-23} N m K^{-1}

1–2, the shear stresses are plotted against the shear rates (independent variable), which is the conventional method. However, some authors plot shear rates against the shear stresses (independent variable), as shown Figure 1–3. With the introduction of controlled-stress rheometers, the use of shear stress as the independent variable is often desirable.

Newtonian Behavior

With Newtonian fluids, the shear rate is directly proportional to the shear stress, and the plot begins at the origin. Typical Newtonian foods are those containing compounds of low molecular weight (e.g., sugars) and that do not contain large concentrations of either dissolved polymers (e.g., pectins, proteins, starches) or insoluble solids. Examples of Newtonian foods include water, sugar syrups, most honeys, most carbonated beverages, edible oils, filtered juices, and milk.

All other types of fluid foods are non-Newtonian, which means either that the shear stress–shear rate plot is not linear and/or the plot does not begin at the origin or that the material exhibits time-dependent rheological behavior as a result of structural changes. Flow behavior may depend only on shear rate and not on the duration of shear (time independent) or may depend also on the duration of shear (time dependent). Several types of time-independent flow behavior of foods have been encountered.

Shear-Thinning Behavior

With shear-thinning fluids, the curve begins at the origin of the shear stress–shear rate plot but is concave upward; that is, an increasing shear rate gives a less than proportional increase in shear stress. Shear-thinning fluids are popularly called pseudoplastic. The expression *shear-thinning* is preferred compared to pseudo-

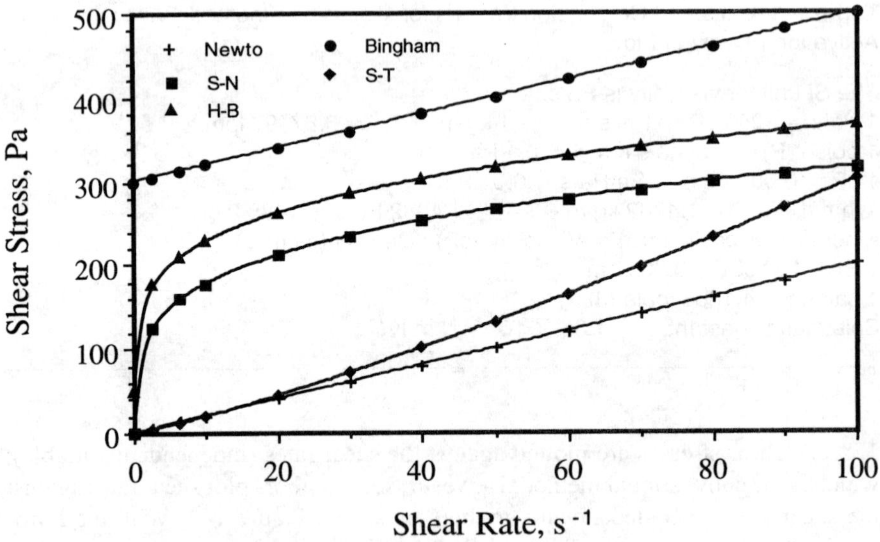

Figure 1–2 Basic Shear Diagram of Shear Rate versus Shear Stress for Classification of Time-Independent Flow Behavior of Fluid Foods: Newtonian (Newto), Shear-Thinning (S-N), and Shear Thickening (S-T). Some foods have yield stress that must be exceeded for flow to occur: Bingham and Herschel-Bulkley (H-B).

---

plastic because it is an accurate description of the shear rate–shear stress curve. Shear-thinning may be thought of being due to breakdown of structural units in a food due to the hydrodynamic forces generated during shear. Most non-Newtonian foods exhibit shear-thinning behavior, including many salad dressings and some concentrated fruit juices.

Yield Stress

The flow of some materials may not commence until a threshold value of stress, the yield stress (σ_0), is exceeded (see Figures 1–2 and 1–3). Although the concept of yield stress was questioned recently (Barnes and Walters, 1985), within the time scales of most food processes the concept of yield stress is useful in food process design, sensory assessment, and modeling. Shear-thinning with yield stress behavior is exhibited by foods such as tomato concentrates, tomato ketchup, mustard, and mayonnaise. In the event that the shear rate–shear stress data follow

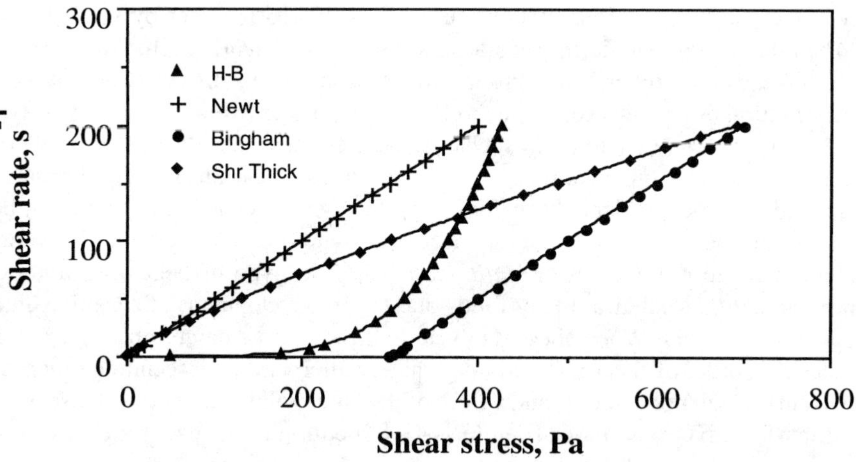

Figure 1–3 Basic Shear Diagram (Similar to Figure 1–2), with Shear Stress Plotted as the Independent Variable: Newtonian (Newt), Shear-Thickening (Shr Thick), Bingham, and Herschel-Bulkley (H-B). For clarity, shear-thinning behavior is not shown.

---

a straight line with a yield stress, the food is said to follow the Bingham plastic model.

Shear-Thickening Behavior

In shear-thickening behavior also, the curve begins at the origin of the shear stress–shear rate plot and is concave downward; that is, an increasing shear stress gives a less than proportional increase in shear rate. This type of flow has been encountered in partially gelatinized starch dispersions. The expression *dilatant* is popularly and incorrectly used to describe shear-thickening. However, because dilatancy implies an increase in the volume of the sample during the test, it is incorrect to use it to describe shear-thickening rheological behavior. Strictly speaking, shear-thickening should be due to increase in the size of the structural units as a result of shear.

Time-Dependent Behavior

Foods that exhibit time-dependent shear-thinning behavior are said to exhibit thixotropic flow behavior. Most of the foods that exhibit thixotropic behavior are heterogeneous systems containing a dispersed phase that is often very fine. At rest, the particles or molecules in the food are linked together by weak forces. When the hydrodynamic forces during shear are sufficiently high, the interparticle linkages are broken, resulting in reduction in the size of the structural units that, in turn, offer lower resistance to flow during shear (Mewis, 1979). This type of behavior is common to foods such as salad dressings and soft cheeses where the structural adjustments take place in the food due to shear until an equilibrium is reached. Time-dependent shear-thickening behavior is called antithixotropic behavior. Formerly, it was called rheopectic behavior, but the Society of Rheology recommended the expression *antithixotropic.* This type of behavior, although rare, is being detected in many foods due to the development of sensitive automated rheometers. When shear rate versus shear stress data are obtained first in ascending order of shear rate and immediately afterward in descending order, the two curves will not coincide and values of the latter will be lower than those of the former for thixotropic foods (Figure 1–4). Repetition of the experiments will result in a equilibrium hysteresis loop.

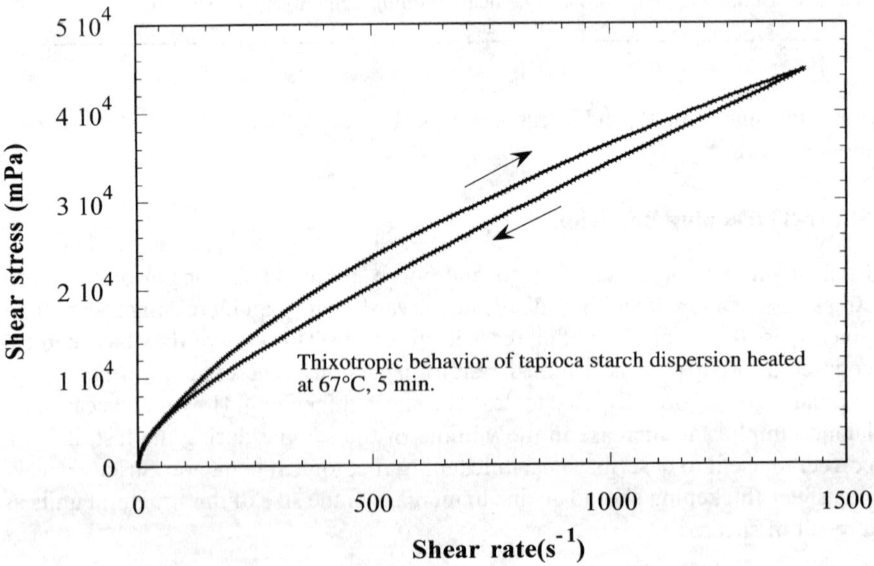

Figure 1–4 Time-Dependent Shear-Thinning (Thixotropic) Behavior of a Tapioca Starch Dispersion Heated at 67°C for 5 Minutes. *Source*: Reprinted with permission from J. Tattiyakul, *Studies on Granule Growth Kinetics and Characteristics of Tapioca Starch Dispersion during Gelatinization Using Particle Size Analysis and Rheological Methods.* M. S. Thesis, © 1997, Cornell University.

In antithixotropic behavior, the shear stress values in descending order of shear rates are higher than those in ascending order (Figure 1–5). The characterization of these types of flow is an important research problem at this time. The primary difficulty in obtaining reliable thixotropic or antithixotropic data is that loading of the test sample into a measurement geometry often induces structural changes that cannot be either controlled or expressed quantitatively. Therefore, considerable caution should be exercised in using the area of hysteresis loop as a measure of thixotropic behavior and a characteristic of the food sample because the magnitude of the area will depend on the shear to which the sample is subjected prior to and during loading of the sample into a viscometer.

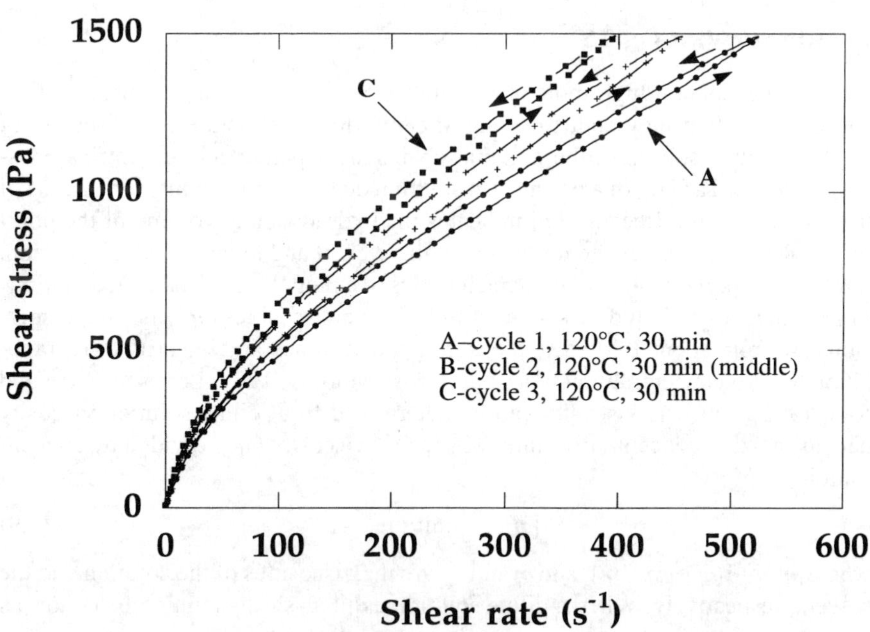

Figure 1–5 Time-Dependent Shear-Thickening (Antithixotropic) Behavior of a Cross-Linked Waxy Maize Starch Dispersion Heated at 120°C for 30 Minutes. Note that the decreasing shear curves of the samples had higher stress values than the increasing shear curves. *Source*: Reprinted with permission from E. K. Chamberlain, Characterization of Heated and Thermally Processed Cross-Linked Waxy Maize Starch Utilizing Particle Size Analysis, Microscopy and Rheology, M.S. Thesis. © 1996, Cornell University.

APPARENT VISCOSITY

Additional information can be obtained from a basic shear (shear stress–shear rate) diagram (see Figure 1–2) and the equations that describe the data in it. The yield stress (σ_0), if present, can be read off the diagram. One can calculate the apparent viscosity at any given shear rate. For example, the apparent viscosity at a shear rate of 50 s^{-1} is

$$\eta_{a,\,50} = \frac{\sigma_{50}}{\dot{\gamma}_{50}} \tag{1.8}$$

where σ_{50} is the shear stress corresponding to a shear rate of 50 s^{-1}. The zero-shear viscosity, η_0, is an important parameter in the study of shear-thinning fluids. It is the limiting viscosity at very low shear rates. The important role of η_0 in the study of biopolymer dispersions will be discussed in Chapters 2 and 4.

INTRINSIC VISCOSITY

Many foods contain high molecular weight polymers, such as proteins, pectins, and others. Often they contribute significantly to the structure and viscosity of foods. In dilute solutions, the polymer chains are separate, and the intrinsic viscosity, denoted as $[\eta]$, of a polymer in solution depends only on the dimensions of the polymer chain. Because $[\eta]$ indicates the hydrodynamic volume of the polymer molecule and is related to the molecular weight and to the radius of gyration, it reflects important molecular characteristics of a biopolymer. The concentrations of polymers used should be such that the relative viscosities (η/η_s) of the dispersions are from about 1.2 to 2.0 to ensure good accuracy and linearity of extrapolation to zero concentration (Morris and Ross-Murphy, 1981; Lopes da Silva and Rao, 1992). Intrinsic viscosity can be determined from dilute solution viscosity data as the zero concentration-limit of specific viscosity (η_{sp}) divided by concentration (c):

$$[\eta] = \operatorname*{limit}_{c \to 0} (\eta_{sp}/c) \tag{1.9}$$

where $\eta_{sp} = [(\eta - \eta_s)/\eta_s]$, and η and η_s are the viscosities of the solution and the solvent, respectively. When a dilute solution exhibits shear-thinning behavior, its zero shear viscosity, η_0, at very low shear rates ("vanishing shear rates") may be used in place of the Newtonian viscosity η. There are several ways of determining the intrinsic viscosity $[\eta]$ from experimental dilute solution viscosity data (Tanglertpaibul and Rao, 1987). The two equations commonly employed for determining $[\eta]$ of food gums are those of Huggins (Equation 1.10) and Kraemer (Equation 1.11).

$$\eta_{sp}/c = [\eta] + k_1\,[\eta]^2 c \tag{1.10}$$

$$\ln(\eta_r)/c = [\eta] + k_2 [\eta]^2 c \tag{1.11}$$

In Equation 1.11, η_r is the relative viscosity (η/η_s). Equations 1.10 and 1.11 imply that plots of η_{sp}/c versus c and ln (η_r)/c versus c would result in straight lines, respectively. The extrapolations are usually done for relative viscosity values between 1.2 and 2.0 when the corresponding specific viscosities are between about 0.2 and 1.0. The Huggins constant k_1 is considered to be an index of the polymer-polymer interaction, with a large number of the reported values being between 0.3 in good solvents and 1.0 in theta solvents; the higher values are attributed to the existence of association between the macromolecules. The Huggins and Kraemer constants (k_1 and k_2) are theoretically related by the equation

$$k_1 = k_2 + 0.5 \tag{1.12}$$

The Huggins equation has been used much more extensively. Figure 1–6 illustrates intrinsic viscosity determination of mesquite seed gum, a galactose-mannose polymer, based on the Huggins equation (Yoo et al., 1994).

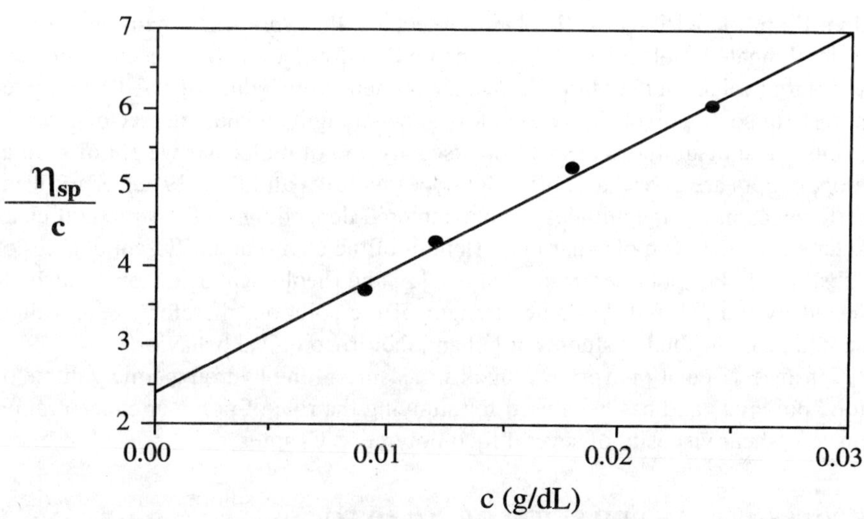

Figure 1–6 Intrinsic Viscosity Determination of Mesquite Seed Gum, a Galactose-Mannose Polymer, Based on the Huggins Equation. *Source*: B. Yoo, A.A. Figueiredo, and M.A. Rao, Rheological Properties of Mesquite Seed Gum in Steady and Dynamic Shear, *Lebensmittel Wissenschaft und Technologie*, Vol. 27, pp. 151–157, © 1994, Academic Press.

For polyelectrolytes (charged polymers), Tam and Tiu (1993) utilized the expression for specific viscosity in the equation of Fuoss-Strauss:

$$\frac{\eta_{sp}}{c} = \frac{[\eta]}{(1 + Bc^{1/2})} \quad (1.13)$$

When (c/η_{sp}) is plotted against $c^{1/2}$, a straight line is obtained with an intercept of $1/[\eta]$ and a slope of $B/[\eta]$.

The Mark-Houwink empirical equation relates intrinsic viscosity and the average molecular weight (M):

$$[\eta] = K\,(M)_a \quad (1.14)$$

where K and a are constants at a specific temperature for a given polymer-solvent system. The magnitude of the constant a is generally in the range 0.5 to 0.8 when conformation of the polymer is random coil (Launay et al., 1986); for polymers that assume compact conformations, the magnitudes of the exponent a are lower than 0.5. Magnitudes of a higher than 0.8 are encountered when hydrodynamic interactions are absent (free-draining random coil), that is, when the solvent can drain freely between the macromolecules, as is thc case when the conformations are elongated and rigid (rod-like polymers). A large number of models have been used to deduce intrinsic viscosity-molecular weight relationships (Launay et al., 1986), including the equivalent sphere model (Flory-Fox equation), the random flight model for flexible chain molecules, the worm-like chain model, and rigid elongated molecules. Rigid elongated conformations have been associated with large values of the Mark-Houwink exponent, with values of 1.7–1.8 being reported for some polyelectrolytes at low ionic strength and high degree of disassociation. The magnitudes of intrinsic viscosity and of molecular weight of several biopolymers are given in Table 1–2 (Lopes da Silva and Rao, 1992). Significant differences in the magnitudes are encountered depending on the structural characteristics, such as molecular mass (length of the chain) and different degrees of substitution. For polymers with charges, $[\eta]$ and rheological behavior are also affected by the pH and the ionic strength of the solution; therefore, their values should be controlled in studies of $[\eta]$ and their rheological behavior.

Intrinsic viscosity is very useful as a measure of the hydrodynamic volume of food polymers and has been used for studying the role of polymer concentration on zero-shear viscosity of several food polymers (Chapter 4).

STRESS-STRAIN BEHAVIOR OF SOLID FOODS

Equations 1.4 to 1.6 represent material functions under large deformations (e.g., continuous shear of a fluid). One may recall a simple experiment in an introductory physics course where a stress (σ) is applied to a rod of length L in a tension

Table 1–2 Examples of Magnitudes of Intrinsic Viscosity [η] and Average Molecular Weight (M) of Some Biopolymers

Biopolymer	*Solvent*	M	[η] *(dL/g)*	*Reference*
Apple pectin				
DE*—72.9%	NaCl 0.155 M	155 000	6.13	Michel, 1982
DE—60.5%		113 000	4.45	
Xanthan gum	Water		168	Launay et al.,
	NaCl 2%		36.7	1984
Sodium alginate	Ionic force			
	0.01	200 000	5.99	
		500 000	17.2	Smidsrød,
	1.0	200 000	3.72	1970
		500 000	8.26	
Locust bean gum	Water	1 600 000	11.2	Doublier,
		1 100 000	7.70	1975
Guar gum	Water	440 000	4.5	Robinson et al.,
		1 650 000	12.5	1982

*DE stands for degree of esterification.

mode and that results in a small deformation ΔL. The linear relationship between stress (σ) and strain (γ) (also relative deformation, $\gamma = \Delta L/L$) is used to define the Young's modulus of elasticity E (Pa):

$$\sigma = E\gamma \tag{1.15}$$

If the solid does not show time-dependent behavior, i.e., it deforms instantaneously, one has an ideal elastic body or a Hookean solid. The symbol E for the modulus is used when the applied strain is extension or compression, while the symbol G is used when the modulus is determined using shear strain. The conduct of the experiment such that a linear relationship is obtained between stress and strain should be noted. In addition, for an ideal Hookean solid, the deformation is instantaneous. In contrast, all real materials are either viscoplastic or viscoelastic in nature and, in particular, the latter exhibit time-dependent deformations. The rheological behavior of many foods may be described as viscoplastic, and the applicable equations are discussed in Chapter 2.

LINEAR VISCOELASTICITY

One convenient manner of studying viscoelasticity is by stress relaxation, where the time-dependent shear stress is studied for step increase in strain. In Figure 1–7,

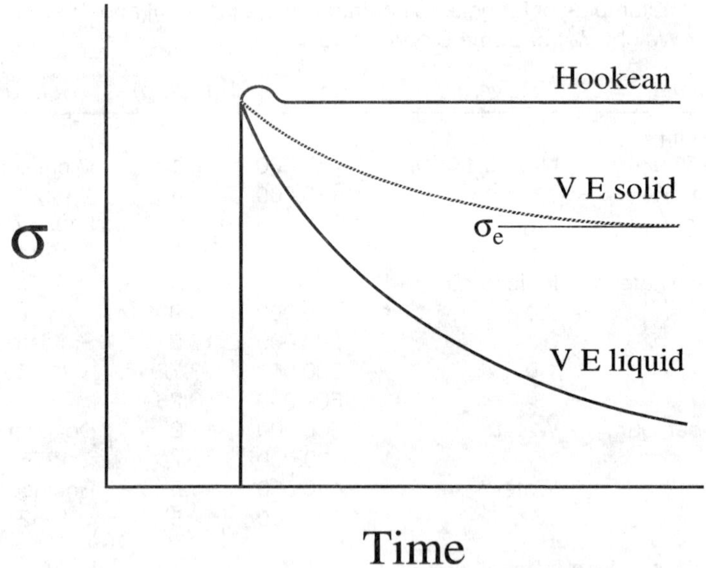

Figure 1–7 Stress Relaxation Response of a Fluid, an Ideal Elastic Solid, and a Viscoelastic Material. The experimental response of an ideal elastic solid is not shown separately because it is the same as that of the ideal elastic solid. *Source*: C.W. Macosko, *Rheology: Principles, Measurements, and Applications*, © 1994, John Wiley and Sons, Inc. Reprinted with permission of John Wiley and Sons, Inc.

the stress relaxation responses of a Hookean solid and a viscoelastic solid and liquid are shown when subjected to a strain instantaneously and held constant. The relaxation modulus can be calculated as

$$G(t, \gamma) = \frac{\sigma(t, \gamma)}{\gamma} \tag{1.16}$$

Because a Hookean solid deforms instantaneously, the imposed strain has a time-dependent profile similar to that of the stress. For relatively small strains, shear normal stresses and the type of deformation (e.g., linear or shear) will not be important. Further, the extensional and the shear relaxation moduli are related as

$$E(t) = 3G(t) \tag{1.17}$$

Boltzmann suggested that when the changes in stress and strain are small, one can express the change in stress as

$$\mathrm{d}\sigma = \gamma\,\mathrm{d}G \tag{1.18}$$

Defining the memory function as $M(t) = -\mathrm{d}G(t)/\mathrm{d}t$, one can express a large linear deformation as the sum of all the small linear deformations, which in turn can be written as the integral over all past times:

$$\sigma = -\int_{-\infty}^{t} M(t - t')\, \gamma(t')\, \mathrm{d}t' \qquad (1.19)$$

where t' is the past time variable in the range $-\infty$ to the present time t. The elapsed time is denoted as $s = (t - t')$, and Equation 1–19 can be written as

$$\sigma = -\int_{-\infty}^{t} M(s)\, \gamma(t - s)\, \mathrm{d}s \qquad (1.20)$$

The above equation is a one-dimensional model of linear viscoelastic behavior. It can be also written in terms of the relaxation modulus after noting that

$$\mathrm{d}\sigma = G\, \mathrm{d}\gamma = G \frac{\mathrm{d}\gamma}{\mathrm{d}t}\, \mathrm{d}t = G\dot{\gamma}\mathrm{d}t \qquad (1.21)$$

$$\sigma = -\int_{-\infty}^{t} G(t - t')\, \dot{\gamma}(t')\, \mathrm{d}t' \qquad (1.22)$$

Therefore, the stress is an integral over past time of the product of the relaxation modulus and the rate of strain. The relaxation modulus can be expressed in terms of a series of relaxation times, λ_n, and the constants, G_n:

$$G(t) = \sum_{n=1}^{N} G_n \exp(-t/\lambda_n) \qquad (1.23)$$

In the unlikely event of $G(t)$ being described by a single exponential term:

$$G(t) = G_0 \exp(-t/\lambda) \qquad (1.24)$$

From the above equation, one obtains the simple Maxwell model:

$$\sigma = -\int_{-\infty}^{t} G_0\, e^{-(t - t')/\lambda}\, \dot{\gamma}(t')\, \mathrm{d}t' \qquad (1.25)$$

One can express linear viscoelasticity using the relaxation spectrum $H(\lambda)$, that is, using the relaxation time λ. The relationship between the relaxation modulus and the spectra is

$$G(s) = \int_{0}^{\infty} \frac{H(\lambda)}{\lambda}\, e^{-s/\lambda}\, \mathrm{d}\lambda \qquad (1.26)$$

Linear Viscoelasticity in Differential Form

Three equations are basic to viscoelasticity: (1) Newton's law of viscosity, $\sigma = \eta\dot{\gamma}$; (2) Hooke's law of elasticity, Equation 1.15; and (3) Newton's second law of motion, $F = ma$, where m is the mass and a is the acceleration. One can combine the three equations to obtain a basic differential equation. In linear viscoelasticity, the conditions are such that the contributions of the viscous, elastic, and inertial elements are additive. The Maxwell model is

$$\sigma + \lambda \frac{d\sigma}{dt} = \eta\dot{\gamma} = \lambda G_0 \dot{\gamma} \tag{1.27}$$

A convenient manner of interpreting viscoelastic behavior of foods is in terms of a spring that has a modulus E and a dashpot that represents a Newtonian fluid with viscosity η that can be arranged either in series (Maxwell model) (Figure 1–8, left) or in parallel (Kelvin-Voigt model). In the Maxwell model, for slow motions the dashpot, that is, Newtonian behavior, dominates. For rapidly changing stresses, that is, at short times, the model approaches elastic behavior. The use of models such as the Maxwell and the Kelvin-Voigt (Figure 1–8, right) and their combinations is valid only when the experimental data are obtained within the linear viscoelastic range. Many studies on foods have reported measurements in the linear viscoelastic range, so the Maxwell and the Kelvin-Voigt models and their combinations were used in the interpretation of results of such studies (Chapter 3 and 5).

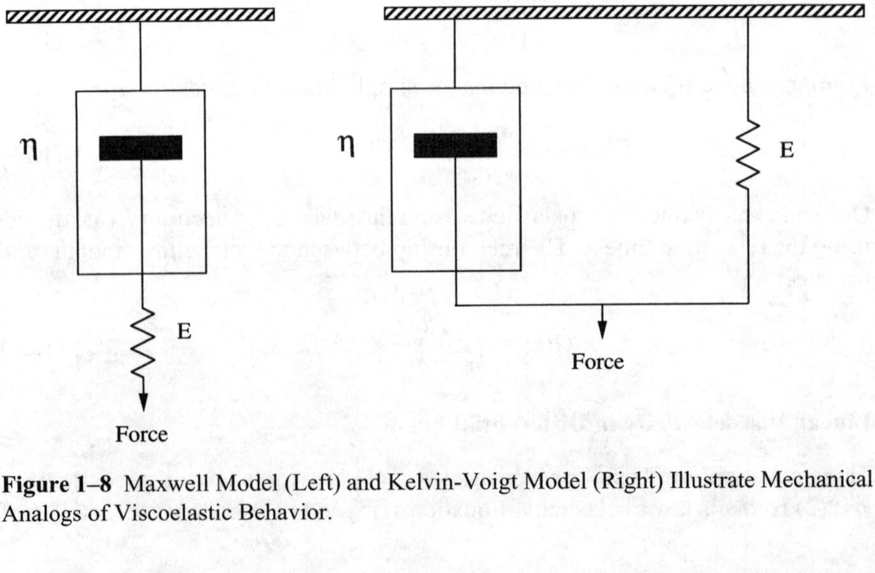

Figure 1–8 Maxwell Model (Left) and Kelvin-Voigt Model (Right) Illustrate Mechanical Analogs of Viscoelastic Behavior.

The relaxation time for a Maxwell element (Figure 1–8, left) is defined as

$$\tau_1 = \frac{\eta}{E} \tag{1.28}$$

From data in a stress relaxation experiment (Chapter 3), where the strain is constant and stress is measured as a function of time, $\sigma(t)$, the relaxation time may be estimated from the time necessary for $[\sigma(t)/\sigma(0)]$ to become $(1/e) = 0.368$. Typically, several Maxwell elements are used to fit experimental data, $\sigma(t)$. For the Kelvin-Voigt element (Figure 1–8, right) under stress, the equation is

$$\sigma = \eta \frac{d\gamma}{dt} + E\gamma \tag{1.29}$$

For constant stress, the above equation can be integrated to describe the deformation:

$$\gamma = \frac{\sigma}{E}[1 - \exp(-E/\eta)t] \tag{1.30}$$

In terms of the retardation time, τ_2, the above equation is

$$\gamma = \frac{\sigma}{E}[1 - \exp(-t/\tau_2)] \tag{1.31}$$

The Kelvin-Voigt elements are used to describe data from a creep experiment and the retardation time (τ_2) is the time required for the spring and the dashpot to deform to $(1 - 1/e)$, or 63.21% of the total creep. In contrast, the relaxation time is that required for the spring and dashpot to stress relax to $1/e$ or 0.368 of $\sigma(0)$ at constant strain. To a first approximation, both τ_1 and τ_2 indicate a measure of the time to complete about half of the physical or chemical phenomenon being investigated (Sperling, 1986).

REFERENCES

Barnes, H.A., and K. Walters. 1985. The yield stress myth? *Rheol Acta* 24: 323–326.

Barnes, H.A., J.F. Hutton, and K. Walters. 1989. *An introduction to rheology.* Amsterdam: Elsevier Science.

Bird, R.B., W.E. Stewart, and E.N. Lightfoot. 1960. *Transport phenomena.* New York: John Wiley & Sons.

Bird, R.B., R.C. Armstrong, and O. Hassager. 1977a. *Dynamics of polymeric liquids—Fluid mechanics.* New York: John Wiley & Sons.

Bird, R.B., O. Hassager, R.C. Armstrong, and C.F. Curtiss. 1977b. *Dynamics of polymeric liquids—Kinetic theory.* New York: John Wiley & Sons.

Dealy, J.M. 1982. *Rheometers for molten polymers.* New York: Van Nostrand Reinhold.

Dickie, A.M., and J.L. Kokini. 1982. Use of the Bird-Leider equation in food rheology. *J Food Process Eng* 5: 157–184.

Doublier, J.L. 1975. Propriétés rheologiques et caracteristiques macromoleculaires de solutions aqueuses de galactomannanes. Doctorate thesis, Universite Paris VI, France.

Ferry, J.D. 1980. *Viscoelastic properties of polymers.* New York: John Wiley & Sons.

Launay, B., G. Cuvelier, and S. Martinez-Reyes. 1984. Xanthan gum in various solvent conditions: intrinsic viscosity and flow properties. In *Gums and stabilisers for the food industry 2,* eds. G.O. Phillips, D.J. Wedlock, and P.A. Williams, 79–98. London: Pergamon Press.

Launay, B., J.L. Doublier, and G. Cuvelier. 1986. Flow properties of aqueous solutions and dispersions of polysaccharides. In *Functional properties of food macromolecules,* eds. J.R. Mitchell and D.A. Ledward, 1–78. London: Elsevier Applied Science Publishers.

Lopes da Silva, J.A.L., and M.A. Rao. 1992. Viscoelastic properties of food gum dispersions. In *Viscoelastic properties of food,* eds. M.A. Rao and J.F. Steffe, 285–316. London: Elsevier Applied Science Publishers.

Macosko, C.W. 1994. *Rheology: Principles, measurements, and applications.* New York: VCH Publishers.

Markovitz, H. 1985. Rheology: In the beginning. *J Rheol* 29: 777–798.

Mewis, J. 1979. Thixotropy—A general review. *J Non-Newtonian Fluid Mechanics* 6: 1–20.

Michel, F. 1982. Etude du comportement potentiometrique et viscosimetrique de pectines hautement methylees en presence de saccharose. Doctorate thesis, Universite de Dijon, France.

Morris, E.R., and B. Ross-Murphy. 1981. Chain flexibility of polysaccharides and glicoproteins from viscosity measurements. In *Techniques in carbohydrate metabolism,* ed. D.H. Northcote, B310, pp. 1–46. Amsterdam: Elsevier Science.

Padmanabhan, M. 1995. Measurement of extensional viscosity of viscoelastic liquid foods. *J Food Eng* 25: 311–327.

Rao, M.A. 1992. Measurement of viscoelastic properties of fluid and semisolid foods. In *Viscoelastic properties of foods,* eds. M.A. Rao and J.F. Steffe, 207–232. London: Elsevier Applied Science Publishers.

Rao, M.A. 1995. Rheological properties of fluid foods. In *Engineering properties of foods,* 2d ed., eds. M.A. Rao and S.S.H. Rizvi, 1–53. New York: Marcel Dekker, Inc.

Robinson, G., S.B. Ross-Murphy, and E.R. Morris. 1982. Viscosity-molecular weight relationships, intrinsic chain flexibility and dynamic solution properties of guar galactomannan. *Carbohydrate Research 107*: 17–32.

Ross-Murphy, S.B. 1984. Rheological methods. In *Biophysical methods in food research,* ed. H.W.S. Chan, 138–199. Oxford, England: Blackwell Scientific Publications.

Sherman, P. 1970. *Industrial rheology.* New York: Academic Press.

Smidsrød, O. 1970. Solution properties of alginate. *Carbohydrate Research 13*: 359.

Sperling, L.H. 1986. *Introduction to physical polymer science.* New York: John Wiley & Sons.

Steffe, J.F. 1996. *Rheological methods in food process engineering.* 2d ed. East Lansing, MI: Freeman Press.

Tam, K.C., and C. Tiu. 1993. Improved correlation for shear-dependent viscosity of polyelectrolyte solutions. *J Non-Newtonian Fluid Mechanics* 46: 275–288.

Tanglertpaibul, T., and M.A. Rao. 1987. Intrinsic viscosity of tomato serum as affected by methods of determination and methods of processing concentrates. *J Food Sci* 52: 1642–1645, 1688.

Tschoegl, N.W. 1989. *The phenomenological theory of linear viscoelastic behavior—An introduction.* New York: Springer-Verlag.

Van Wazer, J.R., J.W. Lyons, K.Y. Kim, and R.E. Colwell. 1963. *Viscosity and flow measurement. A handbook of rheology.* New York: Interscience.

Walters, K. 1975. *Rheometry.* London: Chapman and Hall.

Whorlow, R.W. 1980. *Rheological techniques.* New York: Halsted Press.

Yoo, B., A.A. Figueiredo, and M.A. Rao. 1994. Rheological properties of mesquite seed gum in steady and dynamic shear. *Lebens Wissen und Technol* 27: 151–157.

SUGGESTED READINGS

Barbosa-Cánovas, G.V., J.L. Kokini, L. Ma, and A. Ibarz. 1996. The rheology of semiliquid foods. In *Advances in food and nutrition research,* Vol. 39: 1–69.

Brodkey, R.S. 1967. *The phenomena of fluid motions.* Reading, MA: Addison-Wesley Publishing Co.

Rao, M.A. 1977a. Rheology of liquid foods—A review. *J Texture Studies* 8: 135–168.

Rao, M.A. 1977b. Measurement of flow properties of fluid foods—Developments, limitations, and interpretation of phenomena. *J Texture Studies* 8: 257–282.

Rao, M.A., and J.F. Steffe. 1992. *Viscoelastic properties of foods.* New York: Elsevier Applied Science Publishers.

Appendix 1–A

Momentum and Heat Transport Equations for Incompressible Fluids

Transport Equations in Vector Notation

The equations of continuity, motion, and energy in vector notation are written below:

$$\nabla \cdot \mathbf{V} = 0$$

$$\rho \frac{\mathbf{DV}}{\mathbf{DT}} = -\nabla p - [\nabla \cdot \tau] + \rho\, \mathbf{g}$$

$$k\, \nabla^2\, T = -(\nabla \cdot q) - (\tau : \nabla\, \mathbf{V})$$

The Equation of Continuity (Bird et al., 1960)

Cartesian coordinates (x, y, z):

$$\frac{\partial \rho}{\partial t} + \frac{\partial}{\partial x}(\rho v_x) + \frac{\partial}{\partial y}(\rho v_y) + \frac{\partial}{\partial z}(\rho v_z) = 0$$

Cylindrical coordinates (r, θ, z):

$$\frac{\partial \rho}{\partial t} + \frac{1}{r}\frac{\partial}{\partial r}(\rho r v_r) + \frac{1}{r}\frac{\partial}{\partial \theta}(\rho v_\theta) + \frac{\partial}{\partial z}(\rho v_z) = 0$$

Spherical coordinates (r, θ, ϕ):

$$\frac{\partial \rho}{\partial t} + \frac{1}{r^2}\frac{\partial}{\partial r}(\rho r^2 v_r) + \frac{1}{r \sin \theta}\frac{\partial}{\partial \theta}(\rho v_\theta \sin \theta) + \frac{1}{r \sin \theta}\frac{\partial}{\partial \phi}(\rho v_\phi) = 0$$

Source: R.B. Bird, W.E. Stewart, and E.N. Lightfoot, *Transport Phenomena,* © 1960, John Wiley and Sons, Inc. Reprinted with permission of John Wiley and Sons, Inc.

Equation of Motion in Rectangular Coordinates (*x*, *y*, *z*) in Terms of σ (Bird et al., 1960)

x-component:

$$\rho\left(\frac{\partial v_x}{\partial t} + v_x\frac{\partial v_x}{\partial x} + v_y\frac{\partial v_x}{\partial y} + v_z\frac{\partial v_x}{\partial z}\right) = -\frac{\partial p}{\partial x} + \left(\frac{\partial \sigma_{xx}}{\partial x} + \frac{\partial \sigma_{yx}}{\partial y} + \frac{\partial \sigma_{zx}}{\partial z}\right) + \rho g_x \tag{A}$$

y-component:

$$\rho\left(\frac{\partial v_y}{\partial t} + v_x\frac{\partial v_y}{\partial x} + v_y\frac{\partial v_y}{\partial y} + v_z\frac{\partial v_y}{\partial z}\right) = -\frac{\partial p}{\partial y} + \left(\frac{\partial \sigma_{xy}}{\partial x} + \frac{\partial \sigma_{yy}}{\partial y} + \frac{\partial \sigma_{zy}}{\partial z}\right) + \rho g_y \tag{B}$$

z-component:

$$\rho\left(\frac{\partial v_z}{\partial t} + v_x\frac{\partial v_z}{\partial x} + v_y\frac{\partial v_z}{\partial y} + v_z\frac{\partial v_z}{\partial z}\right) = -\frac{\partial p}{\partial z} + \left(\frac{\partial \sigma_{xz}}{\partial x} + \frac{\partial \sigma_{yz}}{\partial y} + \frac{\partial \sigma_{zz}}{\partial z}\right) + \rho g_y \tag{C}$$

Equation of Motion in Cylindrical Coordinates (*r*, θ, *z*) in Terms of σ (Bird et al., 1960)

r-component:

$$\rho\left(\frac{\partial v_r}{\partial t} + v_r\frac{\partial v_r}{\partial r} + \frac{v_\theta}{r}\frac{\partial v_r}{\partial \theta} - \frac{v_\theta^2}{r} + v_z\frac{\partial v_r}{\partial z}\right)$$
$$= -\frac{\partial p}{\partial r} + \left(\frac{1}{r}\frac{\partial}{\partial r}(r\sigma_{rr}) + \frac{1}{r}\frac{\partial \sigma_{r\theta}}{\partial \theta} - \frac{\sigma_{\theta\theta}}{r} + \frac{\partial \sigma_{rz}}{\partial z}\right) + \rho g_r \tag{A}$$

θ-component:

$$\rho\left(\frac{\partial v_\theta}{\partial t} + v_r\frac{\partial v_\theta}{\partial r} + \frac{v_\theta}{r}\frac{\partial v_\theta}{\partial \theta} + \frac{v_r v_\theta}{r} + v_z\frac{\partial v_\theta}{\partial z}\right)$$
$$= -\frac{1}{r}\frac{\partial p}{\partial \theta} + \left(\frac{1}{r^2}\frac{\partial}{\partial r}(r^2\sigma_{r\theta}) + \frac{1}{r}\frac{\partial \sigma_{\theta\theta}}{\partial \theta} + \frac{\partial \sigma_{\theta z}}{\partial z}\right) + \rho g_\theta \tag{B}$$

z-component:

$$\rho\left(\frac{\partial v_z}{\partial t} + v_r \frac{\partial v_z}{\partial r} + \frac{v_\theta}{r}\frac{\partial v_z}{\partial \theta} + v_z \frac{\partial v_z}{\partial z}\right)$$

$$= -\frac{\partial p}{\partial z} + \left(\frac{1}{r}\frac{\partial}{\partial r}(r\sigma_{rz}) + \frac{1}{r}\frac{\partial \sigma_{\theta z}}{\partial \theta} + \frac{\partial \sigma_{zz}}{\partial z}\right) + \rho g_z \quad \text{(C)}$$

The Equation of Motion in Spherical Coordinates in Terms of σ (Bird et al., 1960)

r-component:

$$\rho\left(\frac{\partial v_r}{\partial t} + v_r \frac{\partial v_r}{\partial r} + \frac{v_\theta}{r}\frac{\partial v_r}{\partial \theta} + \frac{v_\phi}{r \sin\theta}\frac{\partial v_r}{\partial \phi} - \frac{v_\theta^2 + v_\phi^2}{r}\right)$$

$$= -\frac{\partial p}{\partial r} + \left(\frac{1}{r^2}\frac{\partial}{\partial r}(r^2\,\sigma_{rr}) + \frac{1}{r\sin\theta}\frac{\partial}{\partial \theta}(\sigma_{r\theta}\sin\theta)\right.$$

$$\left. + \frac{1}{r\sin\theta}\frac{\partial \sigma_{r\phi}}{\partial \phi} - \frac{\sigma_{\theta\theta} + \sigma_{\phi\phi}}{r}\right) + \rho g_r \quad \text{(A)}$$

θ-component:

$$\rho\left(\frac{\partial v_\theta}{\partial t} + v_r \frac{\partial v_\theta}{\partial r} + \frac{v_\theta}{r}\frac{\partial v_\theta}{\partial \theta} + \frac{v_\phi}{r \sin\theta}\frac{\partial v_\theta}{\partial \phi} + \frac{v_r v_\theta}{r} - \frac{v_\phi^2 \cot\theta}{r}\right)$$

$$= -\frac{1}{r}\frac{\partial p}{\partial \theta} + \left(\frac{1}{r^2}\frac{\partial}{\partial r}(r^2\,\sigma_{r\theta}) + \frac{1}{r\sin\theta}\frac{\partial}{\partial \theta}(\sigma_{\theta\theta}\sin\theta)\right.$$

$$\left. + \frac{1}{r\sin\theta}\frac{\partial \sigma_{\theta\phi}}{\partial \phi} + \frac{\sigma_{r\theta}}{r} - \frac{\cot\theta}{r}\sigma_{\phi\phi}\right) + \rho g_\theta \quad \text{(B)}$$

ϕ-component:

$$\rho\left(\frac{\partial v_\phi}{\partial t} + v_r \frac{\partial v_\phi}{\partial r} + \frac{v_\theta}{r}\frac{\partial v_\phi}{\partial \theta} + \frac{v_\phi}{r \sin\theta}\frac{\partial v_\phi}{\partial \phi} + \frac{v_\phi v_r}{r} + \frac{v_\theta v_\phi}{r}\cot\theta\right)$$

$$= -\frac{1}{r\sin\theta}\frac{\partial p}{\partial \phi} + \left(\frac{1}{r^2}\frac{\partial}{\partial r}(r^2\,\sigma_{r\phi}) + \frac{1}{r}\frac{\partial \sigma_{\theta\phi}}{\partial \theta} + \frac{1}{r\sin\theta}\frac{\partial \sigma_{\phi\phi}}{\partial \phi}\right.$$

$$\left. + \frac{\sigma_{r\phi}}{r} + \frac{2\cot\theta}{r}\sigma_{\theta\phi}\right) + \rho g_\phi \quad \text{(C)}$$

CHAPTER 2

Flow and Functional Models for Rheological Properties of Fluid Foods

M. Anandha Rao

A flow model may be considered to be a mathematical equation that can describe rheological data, such as shear rate–shear stress data in a basic shear diagram, and it provides a convenient and concise manner of describing the data. Occasionally, such as for the viscosity versus temperature data during starch gelatinization, more than one equation may be necessary to describe the rheological data. In addition to mathematical convenience, it is important to quantify how magnitudes of model parameters are affected by state variables, such as temperature, and the effect of structure/composition (e.g., concentration of solids) of foods and establish widely applicable relationships that may be called functional models. Flow models have been used also for derivation of expressions for velocity profiles and volumetric flow rates in tube and channel flows, and in the analysis of heat transfer phenomena. Numerous flow models can be encountered in the rheology literature, and some from the food rheology literature are listed in Table 2–1. Models that have found extensive use in the analysis of the flow behavior of fluid foods are discussed herein. Models that account for yield stress are known as viscoplastic models (Bird et al., 1982). For convenience, the flow models can be divided into those for time-independent and those for time-dependent flow behavior.

TIME-INDEPENDENT FLOW BEHAVIOR

Newtonian Model

The model for a Newtonian fluid is described by the equation

$$\sigma = \eta \dot{\gamma} \qquad (2.1)$$

As per the definition of a Newtonian fluid, the shear stress and the shear rate are proportional to each other, and a single parameter η characterizes the data. For a

Table 2–1 Some Flow Models for Describing Shear Rate ($\dot{\gamma}$) versus Shear Stress (σ) Data

$\sigma = \eta\dot{\gamma}$	Newtonian model*
$\sigma = \dfrac{\dot{\gamma}}{\left[\dfrac{1}{\eta_0} + K_E(\sigma)^{(1/n_E)-1}\right]}$	Ellis model for low shear rate data containing η_0 (Brodkey, 1967)
$\sigma = [\eta_\infty\dot{\gamma} + K_S\dot{\gamma}^{n_S}]$	Sisko model for high shear rate data containing η_∞ (Brodkey, 1969)
$\eta_a = \eta_\infty + \dfrac{\eta_0 - \eta_\infty}{1 + (\alpha_c\dot{\gamma})^m}$	Cross model for data over a wide range of shear rates*
$\eta_a = \eta_\infty + \dfrac{\eta_0 - \eta_\infty}{[1 + (\lambda_c\dot{\gamma})^2]^N}$	Carreau model for data over a wide range of shear rates*
$\sigma = K\dot{\gamma}^n$	Power law model used extensively in handling applications*
$\sigma - \sigma_0 = \eta'\,\dot{\gamma}$	Bingham model*
$\sigma - \sigma_0 = K_H\,(\dot{\gamma})^{n_H}$	Herschel-Bulkley model*
$\sigma^{0.5} = K_{0c} + K_c\,(\dot{\gamma})^{0.5}$	Casson model used especially in treating data on chocolates*
$\sigma^{0.5} - \sigma_{0M} = K_M\dot{\gamma}^{n_M}$	Mizrahi and Berk (1972) model, a modification of the Casson model
$\sigma^{n_1} = \sigma_0^{n_1} + \eta_\infty\,(\dot{\gamma})^{n_2}$	Generalized model of Ofoli et al. (1987)*
$\sigma = [(\sigma_{0V})^{1/n_V} + K_V\,\dot{\gamma}]^{n_V}$	Vocadlo (Vocadlo and Moo Young, 1969) model

*Discussed in text.

Bingham plastic fluid that exhibits a yield stress σ_0, the model is

$$\sigma - \sigma_0 = \eta'\,\dot{\gamma} \tag{2.2}$$

where η' is called the Bingham plastic viscosity.

As shown in Figure 1–2, the Newtonian model and the Bingham plastic model can be illustrated by straight lines in terms of shear rate and shear stress, and the former can be described by one parameter, η, and the latter by two parameters, η' and σ_0, respectively. However, the shear rate–shear stress data of shear-thinning and shear-thickening fluids are curves that require more than one parameter to describe their data. Given that the equation of a straight line is simple, it is easy to understand attempts to transform shear rate–shear stress data into such lines. An additional advantage of a straight line is that it can be described by just two parameters: the slope and the intercept.

Power Law Model

Shear stress–shear rate plots of many fluids become linear when plotted on double logarithmic coordinates, and the power law model describes the data of shear-thinning and shear-thickening fluids.

$$\sigma = K\dot{\gamma}^n \tag{2.3}$$

where K, the consistency index (Pa s^n), is the shear stress at a shear rate of 1.0 s^{-1} and the exponent n, the flow behavior index, is dimensionless that also reflects the closeness to Newtonian flow. For the special case of a Newtonian fluid ($n = 1$), the consistency index K is identically equal to the viscosity of the fluid. When the magnitude of $n < 1$ the fluid is shear-thinning, and when $n > 1$ the fluid is shear-thickening in nature. Taking logarithms of both sides of Equation 2.3:

$$\log \sigma = \log K + n \log \dot{\gamma} \tag{2.4}$$

The parameters K and n are determined from a plot of log σ versus log $\dot{\gamma}$, and the resulting straight line's intercept is log K and the slope is n. If a large number of $\sigma - \dot{\gamma}$ data points, for example, > 15 (it is easy to obtain a large number of points with automated viscometers) is available, linear regression of log $\dot{\gamma}$ versus log σ will provide statistically best values of K and n. Nevertheless, a plot of experimental and predicted values of log $\dot{\gamma}$ and log σ is useful for observing trends in data and ability of the model to follow the data. Figure 2–1 illustrates applicability of the power law model to a 2.6% tapioca starch dispersion heated at 67°C for 5 minutes. Linear regression techniques also can be used for determination of the parameters of the Herschel-Bulkley (when the magnitude of the yield stress is known) and the Casson models discussed later in this chapter.

Because it contains only two parameters (K and n) that can describe shear rate–shear stress data, the power law model has been used extensively to characterize fluid foods. It is also the most used model in studies on handling of foods and heating/cooling of foods. Extensive compilations of the magnitudes of power law parameters can be found in studies by Holdsworth (1971, 1993). Because it is convenient to group foods into commodities, a compilation of magnitudes of power law parameters of food commodities is given in Chapter 5. In addition, the

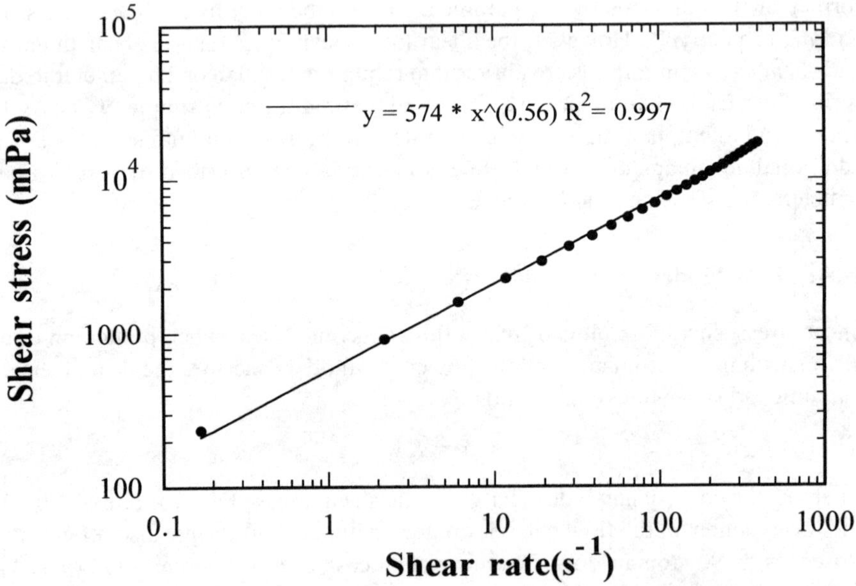

Figure 2–1 Plot of a Log Shear Rate ($\dot{\gamma}$) versus Log Shear Stress (σ) for a 2.6% Tapioca Starch Dispersion Heated at 67°C for 5 Minutes To Illustrate Applicability of the Power Law Model. *Source*: Reprinted with permission from J. Tattiyakul, *Studies on Granule Growth Kinetics and Characteristics of Tapioca Starch Dispersion during Gelatinization Using Particle Size Analysis and Rheological Methods.* M.S. Thesis, © 1997, Cornell University.

influence of temperature in quantitative terms of activation energies and the effect of concentration of soluble and insoluble solids on the consistency index are given.

Although the power law model is popular and useful, its empirical nature should be noted. One reason for its popularity appears to be its applicability over the shear rate range of 10^1 to 10^4 s^{-1} that can be obtained in many commercial viscometers. Often, the magnitudes of the consistency and the flow behavior indexes of a food sample depend on the specific shear rate range being used; so when comparing the properties of different samples, an attempt should be made to determine K and n over a specific range of shear rates. One drawback of the power law model is that it does not describe the low shear and high shear rate constant-viscosity data of shear-thinning foods.

Herschel-Bulkley Model

When yield stress of a food is measurable, it can be included in the power law model, and the model is then known as the Herschel-Bulkley model:

$$\sigma - \sigma_0 = K_{\mathrm{H}}(\dot{\gamma})^{n_{\mathrm{H}}} \tag{2.5}$$

where $\dot{\gamma}$ is shear rate (s^{-1}), σ is shear stress (Pa), n_{H} is the flow behavior index, K_{H} is the consistency index, and σ_0 is yield stress. The concept of yield stress has been challenged (Barnes and Walters, 1985) because a fluid may deform minutely at stress values lower than the yield stress. Nevertheless, yield stress may be considered to be an engineering reality that plays an important role in many food products.

If the yield stress of a sample is known from an independent experiment, K_{H} and n_{H} can be determined from linear regression of log $(\sigma - \sigma_0)$ versus log $\dot{\gamma}$ as the intercept and slope, respectively. Alternatively, nonlinear regression technique was used to estimate σ_0, K_{H}, and n_{H} (Rao and Cooley, 1983). However, estimated values of yield stress and other rheological parameters should be used only when experimentally determined values are not available.

Casson Model

The Casson model (Equation 2.6) is a structure-based model (Casson, 1959) that has been used for a number of food dispersions, although it was developed for characterizing printing inks originally.

$$\sigma^{0.5} = K_{0c} + K_c(\dot{\gamma})^{0.5} \tag{2.6}$$

For a food following the Casson model, a straight line results when the square root of shear rate, $(\dot{\gamma})^{0.5}$, is plotted against the square root of shear stress, $(\sigma)^{0.5}$, with slope K_c and intercept K_{0c} (Figure 2–2). The Casson yield stress is calculated as the square of the intercept, $\sigma_{0c} = (K_{0c})^2$ and the Casson plastic viscosity as the square of the slope, $\eta_{\mathrm{Ca}} = (K_c)^2$. The data in Figure 2–2 are of Steiner (1958) on a chocolate sample. The International Office of Cocoa and Chocolate has adopted the Casson model as the official method for interpretation of flow data on chocolates.

The Casson plastic viscosity can be used as the infinite shear viscosity (η_∞) (Metz et al., 1979) of dispersions by considering the limiting viscosity at infinite shear rate:

$$\left(\frac{d\sigma}{d\dot{\gamma}}\right)_{\dot{\gamma}\to\infty} = \left(\frac{d\sqrt{\sigma}}{d\dot{\gamma}}\frac{d\sigma}{d(\sqrt{\sigma})}\right)_{\dot{\gamma}\to\infty} \tag{2.7}$$

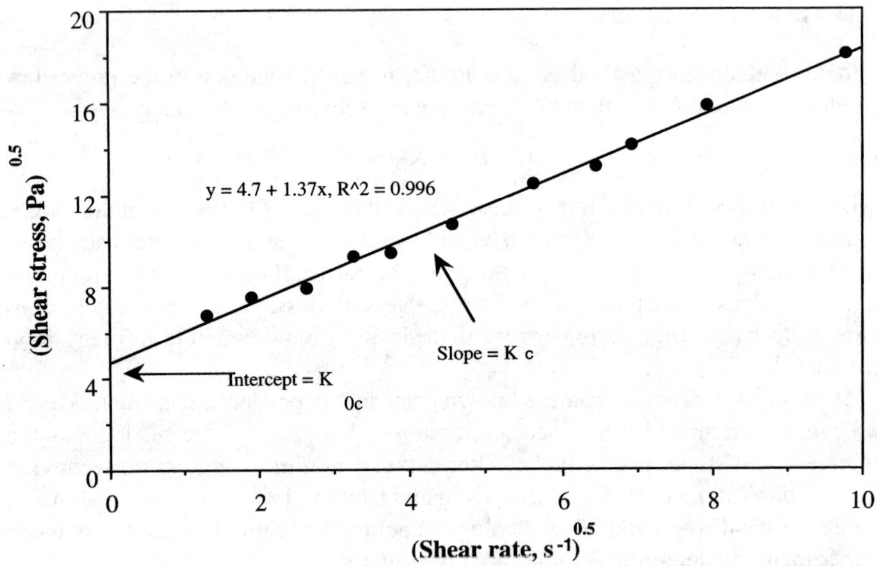

Figure 2–2 Plot of $(\dot{\gamma})^{0.5}$ versus $(\sigma)^{0.5}$ for a Food That Follows the Casson Model. The square of the intercept is the yield stress, and that of slope is the Casson plastic viscosity. Casson, 1959.

Using the Casson equation, the two terms in the right-side bracket are

$$\frac{d(\sqrt{\sigma})}{d\dot{\gamma}} = \frac{K_c}{2\sqrt{\dot{\gamma}}} \tag{2.8}$$

and

$$\frac{d\sigma}{d(\sqrt{\sigma})} = 2\sqrt{\sigma} \tag{2.9}$$

Combining the above two equations,

$$\eta_\infty = \eta_{\text{Ca}} = (K_c)^2 \tag{2.10}$$

Quemada Model

Quemada et al. (1985) proposed a viscosity equation for dispersed systems based on zero-shear (η_0) and infinite-shear (η_∞) viscosities, and a structural parameter

(λ) dependent on the shear rate, which may be written as:

$$\frac{\eta}{\eta_\infty} = \frac{1}{\left\{1 - \left[1 - \left(\frac{\eta_\infty}{\eta_0}\right)^{0.5}\right]\lambda\right\}^2} \tag{2.11}$$

where

$$\lambda = \frac{1}{[1 + (t_c\dot{\gamma})^{0.5}]} \tag{2.12}$$

The time constant t_c is related to the rate of aggregation of particles due to Brownian motion. For highly concentrated dispersed systems, η_∞ will be much lower than η_0, so that $(\eta_\infty/\eta_0) \ll 1$ and the dispersion may have a yield stress, and Equation 2.11 reduces to the Casson model (Equation 2.6) (Tiu et al., 1992) with the Casson yield stress, $\sigma_{0c} = (\eta_\infty/t_c)$. Thus, the Casson-Quemada models can be used to examine rheological behavior of dispersions that cover the range of shear-thinning alone to shear-thinning with yield stress. The Casson-Quemada models were used to study the role of cocoa solids and cocoa butter on cocoa dispersions (Fang et al., 1996, 1997), which is discussed in Chapter 5.

A general model for shear rate–stress data that, under specific assumptions, reduces to the Herschel-Bulkley, the Casson, and other models was presented by Ofoli et al. (1987):

$$\sigma^{n_1} = \sigma_0^{n_1} + \eta_\infty(\dot{\gamma})^{n_2} \tag{2.13}$$

where n_1 and n_2 are constants, and η_∞ is the infinite shear viscosity. It is important to note that one model may be applicable at low shear rates and another at high shear rates (Dervisoglu and Kokini, 1986). While applicability of the flow models themselves may be interesting, it is much more important to study the role of food composition on a model's parameters and apply the model to better understand the nature of foods.

APPARENT VISCOSITY—SHEAR RATE RELATIONSHIPS OF SHEAR-THINNING FOODS

At sufficiently high polymer concentrations, most shear-thinning biopolymer (also called a gum or a hydrocolloid) dispersions exhibit similar three-stage viscous response when sheared over a wide shear rate range (Figure 2–3): (1) At low shear rates, they show Newtonian properties with a constant zero-shear viscosity (η_0) over a limited shear range, which is followed by (2) a shear-thinning range where solution viscosity decreases in accordance with the power law relationship—the reciprocal of the shear rate at which the transition from Newtonian to pseudoplastic behavior occurs is the characteristic time or the time constant—and

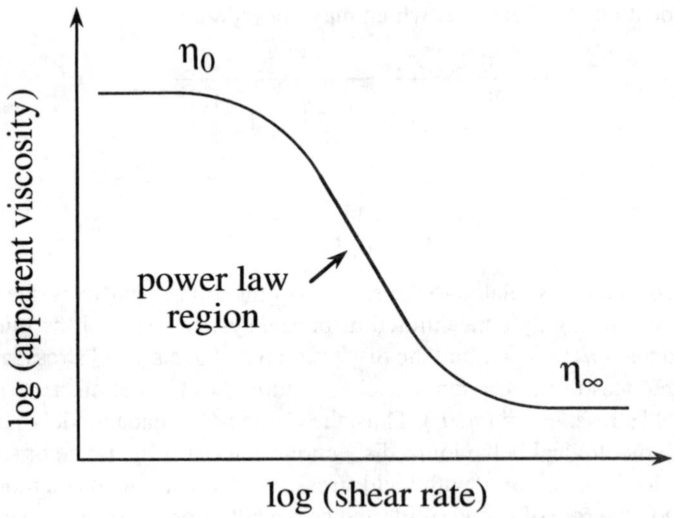

Figure 2–3 Plot of Shear Rate versus Apparent Viscosity for Shear-Thinning Foods Identifying Three Separate Regions: A Zero-Shear Viscosity at Low Shear Rates, a Power Law Region at Intermediate Shear Rates, and an Infinite-Shear Viscosity at High Shear Rates. Often, only data in the power law region are obtained. *Source*: Reprinted from P.E. Okechukwu and M.A. Rao, Rheology of Structured Polysaccharide Food Systems: Starch and Pectin, *Polysaccharide Association Structures in Food*, R.H. Walter, ed., pp. 289–328, by courtesy of Marcel Dekker, Inc.

(3) at high shear rates, they show a limiting and constant infinite-shear viscosity (η_∞). The three regions may be thought of being due to rearrangement in the conformation of the biopolymer molecules in the dispersion due to shearing. In stage 1 when the magnitude of $\dot{\gamma}$ is low, there is little rearrangement of the polymer chains, while in stage 2 the chains undergo gradual rearrangement with $\dot{\gamma}$ resulting in a power law behavior. In stage 3, the shear rate is sufficiently high that the polymer chains do not undergo much rearrangement.

Cross and Carreau Equations

The apparent viscosity (η_a) of the solution can be correlated with shear rate ($\dot{\gamma}$) using the Cross or Carreau equations, Equations 2.14 and 2.15, respectively.

$$\eta_a = \eta_\infty + \frac{\eta_0 - \eta_\infty}{1 + (\alpha_c \dot{\gamma})^m} \tag{2.14}$$

$$\eta_a = \eta_\infty + \frac{\eta_0 - \eta_\infty}{[1 + (\lambda_c \dot{\gamma})^2]^N} \tag{2.15}$$

where α_c and λ_c are time constants related to the relaxation times of the polymer in solution and m and N are dimensionless exponents. Because magnitudes of η_∞ of food polymer dispersions with concentrations of practical interest are usually very low in magnitude, they are difficult to determine experimentally. Therefore, to avoid consequent errors in estimation of the other rheological parameters in Equations 2.14 and 2.15, often η_∞ has been neglected (Abdel-Khalik et al., 1974; Lopes da Silva et al., 1992). The Cross and Carreau models described well the shear dependence of aqueous dispersions of high methoxyl pectins and locust bean gum (Lopes da Silva et al., 1992), konjac flour gum (Jacon et al., 1993), mesquite gum solution (Yoo et al., 1994), and other gums (Launay et al., 1986). In general, the model of Cross has been used in studies in Europe and that of Carreau in North America. In Chapter 4, the applicability of the Cross and Carreau models to locust bean gum dispersions will be discussed in more detail.

For small values of η_∞, the Cross exponent m tends to a value $(1 - n)$, where n is the power law flow behavior index (Launay et al., 1986; Giboreau et al., 1994). For the shear rate ($\dot{\gamma}_c$) where $\eta_{ap} = (\eta_0 + \eta_\infty)/2$, the Cross time constant $\alpha_c = 1/\dot{\gamma}_c$. Generally, $\dot{\gamma}_c$ gives an order of magnitude of the critical shear rate marking the end of the zero shear rate Newtonian plateau or the onset of the shear-thinning region. It is therefore important to recognize the shear rate dependence of the rheological behavior of polysaccharide polymers in solution and the difficulty involved in obtaining experimental data over the applicable shear rate range of 10^{-6}–10^4 s^{-1} (Barnes et al., 1989). The low shear rate region of about 10^{-3}–10^0 is often used for the characterization and differentiation of structures in polysaccharide systems through the use of stress-controlled creep and nondestructive oscillatory tests. The shear rate range of about 10^1–10^4 s^{-1} falls within the operational domain of most commercial rheometers, so the range of 10^{-3}–10^4 s^{-1} can sometimes be effectively covered by a combination of measuring procedures and instruments.

Both the Carreau and the Cross models can be modified to include a term due to yield stress. For example, the Carreau model with a yield term given in Equation 2.16 was employed in the study of the rheological behavior of glass-filled polymers (Poslinski et al., 1988):

$$\eta_a = \sigma_0 \dot{\gamma}^{-1} + \eta_p [1 + (\lambda_p \dot{\gamma})^2]^{-N} \tag{2.16}$$

where σ_0 is the yield stress, η_p is the plateau viscosity, and λ_p and N are constants to be determined from experimental data. Rayment et al. (1998) interpreted the rheological behavior of guar gum dispersions containing raw rice starch in terms of the Cross model with yield stress (Equation 2.17).

When yield stress is exhibited, the term *plateau viscosity* is used instead of *zero-shear viscosity*.

$$\eta_a = \sigma_0 \dot{\gamma}^{-1} + \eta_p[1 + (\alpha_c \dot{\gamma})]^{-m} \tag{2.17}$$

TIME-DEPENDENT FLOW BEHAVIOR

Considerable care should be exercised in determining reliable time-dependent rheological data because of the often unavoidable modification in structure due to sample handling and while loading the sample in a viscometer or rheometer measuring geometry. Nevertheless, with careful attention to details, such as allowing a sample to relax in the rheometer measuring geometry, rheological data can be obtained to characterize time-dependent rheological behavior.

Weltman Model

The Weltman (1943) model has been used to characterize thixotropic behavior (Paredes et al., 1988) and antithixotropic behavior (da Silva et al., 1997) of foods:

$$\sigma = A - B \log t \tag{2.18}$$

where σ is shear stress (Pa), t is time (s), and A (value of stress at t = 1s) and B are constants. A plot of σ versus log time should result in a straight line. In thixotropic behavior, B takes negative values, and in antithixotropic behavior it takes positive values. Table 2–2 shows typical magnitudes of the constants A and B for cross-linked waxy maize starch dispersions.

Tiu and Boger Model

A more elaborate model to study thixotropic behavior of foods exhibiting yield stress was devised by Tiu and Boger (1974), who studied the time-dependent rheological behavior of mayonnaise by means of a modified Herschel-Bulkley model:

$$\sigma = \lambda[\sigma_0 + K_H(\dot{\gamma})^{n_H}] \tag{2.19}$$

where σ is the shear stress (Pa), $\dot{\gamma}$ is the shear rate (s^{-1}), λ is a time-dependent structural parameter that ranges from an initial value of unity to an equilibrium value λ_e, σ_0 is the yield stress (Pa), K_H is the consistency index (Pa s^n), and n_H is the flow behavior index. The decay of the structural parameter with time was assumed to obey a second-order equation:

$$\frac{d\lambda}{dt} = -k_1 (\lambda - \lambda_e)^2 \tag{2.20}$$

where the constant k_1 is a function of shear rate to be determined experimentally.

Table 2–2 Weltman Equation Parameters for Cross-Linked Waxy Maize Gelatinized Starch Dispersions. Da Silva et al., 1997.

		Shear Rate (s^{-1})			
Weltman Parameter	*Conc. (%)*	*50*	*100*	*200*	*300*
A	3	6.97×10^{-2}	5.76×10^{-2}	4.22×10^{-2}	
	4	2.97×10^{-1}	2.08×10^{-1}	1.72×10^{-1}	9.54×10^{-2}
	5		5.39×10^{-1}	3.88×10^{-1}	3.55×10^{-1}
B	3	1.58×10^{-3}	5.71×10^{-4}	1.66×10^{-5}	
	4	3.15×10^{-3}	2.07×10^{-3}	5.29×10^{-3}	7.83×10^{-3}
	5		-8.30×10^{-5}	1.06×10^{-2}	8.79×10^{-3}
	3	1.00	1.00	1.00	
Correlation	4	1.00	1.00	1.00	1.00
coefficient	5		1.00	1.00	1.00

While the determination of σ_0, K_H, and n is straightforward, estimation of k_1 and λ_e requires the use of values of apparent viscosities (η_a) (Tiu and Boger, 1974):

$$\lambda = \frac{\eta_a \dot{\gamma}}{\sigma_0 + K_H \dot{\gamma}^{n_H}} \tag{2.21}$$

and

$$\frac{d\eta_a}{dt} = -a_1(\eta_a - \eta_e)^2 \tag{2.22}$$

where

$$a_1(\dot{\gamma}) = \frac{k_1 \dot{\gamma}}{\sigma_0 + K\dot{\gamma}^{n_H}} \tag{2.23}$$

It can be shown (Tiu and Boger, 1974) that a plot of $1/(\eta_a - \eta_e)$ versus time will yield a straight line of slope a_1, and repeating the procedure at other shear rates will establish the relationship between a_1 and $\dot{\gamma}$, and hence k_1 and $\dot{\gamma}$ from Equation 2.23. For a commercial mayonnaise sample, values of the different parameters were $\lambda_e = 0.63$, $\sigma_0 = 7.0$ Pa, $K = 28.5$ Pa s^n, $n = 0.32$, and k_1 was a weak function of shear rate, specifically, $k_1 = 0.012\ \dot{\gamma}^{0.13}$ (Tiu and Boger, 1974).

ROLE OF SOLIDS FRACTION IN RHEOLOGY OF DISPERSIONS

Following Einstein's work (Einstein, 1906, 1911) (Equation 2.24) on dilute rigid sphere dispersions, models for estimating viscosity of concentrated nonfood dispersions of solids are based on volume fraction (ϕ) of the suspended solids and the relative viscosity of the dispersion, $\eta_r = (\eta/\eta_s)$, where η is the viscosity of the dispersion and η_s is the viscosity of the continuous phase (Jinescu, 1974; Metzner, 1985).

$$\eta_r = 1 + 2.5\,\phi \tag{2.24}$$

Metzner (1985) pointed out that at high solids concentration levels a theoretical equation (Equation 2.25) of Frankel and Acrivos appears to do a good job of portraying experimental data of rigid solids dispersed in polymer melts.

$$\eta_r = \frac{9}{8}\left[\frac{(\phi/\phi_m)^{1/3}}{1-(\phi/\phi_m)^{1/3}}\right] \tag{2.25}$$

Wildemuth and Williams (1984) modeled the η_r of rigid sphere suspensions with a shear-dependent maximum volume fraction (ϕ_m). The applicability of a shear dependent ϕ_m to food dispersions has not been tested.

The Krieger-Dougherty relationship (Krieger and Dougherty, 1959) (Equation 2.26) is based on the assumption that an equilibrium exists between individual spherical particles and dumbbells that continuously form and dissociate:

$$\eta_r = \left(1 - \frac{\phi}{\phi_m}\right)^{-[\eta]\phi_m} \tag{2.26}$$

where $[\eta]$ and ϕ_m are the intrinsic viscosity and maximum packing fraction of solids. Theoretically, $[\eta]$ should be 2.5 for rigid spheres and ϕ_m should be about 0.62 if the spheres are of uniform diameter (Krieger, 1985), but Choi and Krieger (1986) found it necessary to use values of $[\eta]$ of 2.65–3.19 to fit Equation 2.26 to viscosity-volume fraction data on sterically stabilized polymethylmethacrylate spheres. For calculating values of η_r according to the Krieger-Dougherty equation, a value of $\phi_m = 0.62$ was assumed (Choi and Krieger, 1986). Both Equations 2.25 and 2.26 contain the ratio of volume fraction of solids in a dispersion to the maximum volume fraction, and the values of η_r predicted (Figure 2–4) are close to each other. In Figure 2–4 for the Krieger-Dougherty model, ϕ_m was taken to be 0.62 and $[\eta]$ to be 2.5. It is emphasized that the above equations were derived for rigid solids; because of the polydisperse and deformable nature of gelatinized starch dispersions, it is not surprising that attempts to predict their viscosity with Equation 2.26 (Noel et al., 1993; Ellis et al., 1989) were not successful.

Most food particles are not spherical in shape, so the empirical equation (Equation 2.27) that described well (Kitano et al., 1981; Metzner, 1985) the relative viscosity versus concentration behavior of suspensions of spheres and fibers with aspect (L/D) ratios ≤ 30 in polymer melts is of interest:

$$\eta_r = [1 - (\phi/A)]^{-2} \tag{2.27}$$

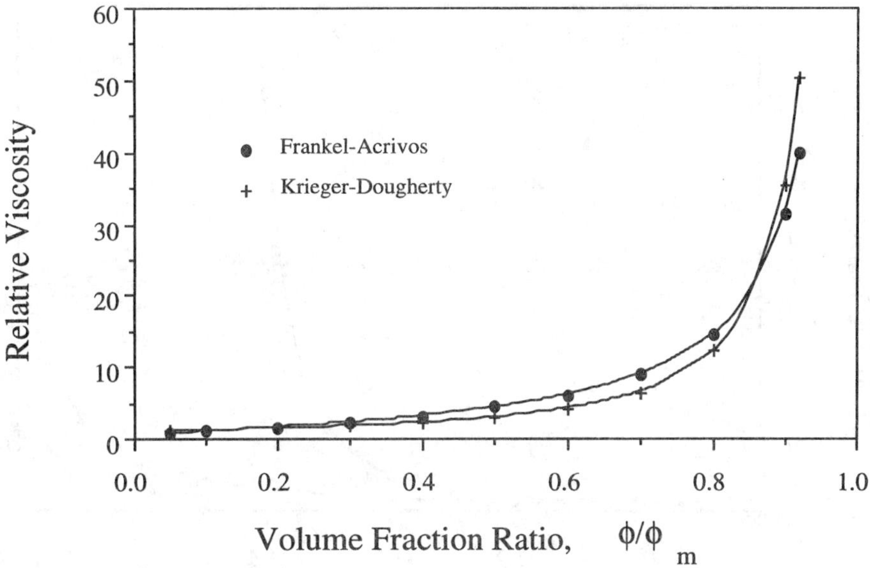

Figure 2–4 Relative Viscosity versus Volume Fraction Ratio (ϕ/ϕ_m) Predicted by Models of Frankel-Acrivos and Krieger-Dougherty ($\phi_m = 0.62$ and $[\eta] = 2.5$).

Equation 2.27 is an empirical equation that is a simple modification of the Maron-Pierce equation for dispersions of spherical rigid solids:

$$\eta_r = [1 - (\phi/\phi_m)]^{-2} \tag{2.28}$$

For fluids that obey the power law model (Equation 2.3), Metzner (1985) suggested that the viscosities of the suspension and of the continuous phase be evaluated at the same shearing stress. For rigid particles, the value of A decreases as the aspect ratio of suspended particles increases; for example, when the aspect ratio is 1.0 (smooth sphere) the magnitude of A is 0.68, and when the aspect ratio is 6 to 8 (rough crystal) A is 0.44 (Metzner, 1985). The shape of many food particles is not spherical and may be considered to be closer to a rough crystal. Figure 2–5 illustrates predictions of η_r by Equation 2.27 for dispersions of spherical and rough crystal-like rigid particles. For the dispersion of rough crystal-like rigid particles, high values of η_r are attained at solids loading much lower than for rigid spherical particles (Figure 2–5).

Because of the compressible nature of food dispersions, the direct determination of the magnitude of ϕ is difficult as it depends on the centrifugal force em-

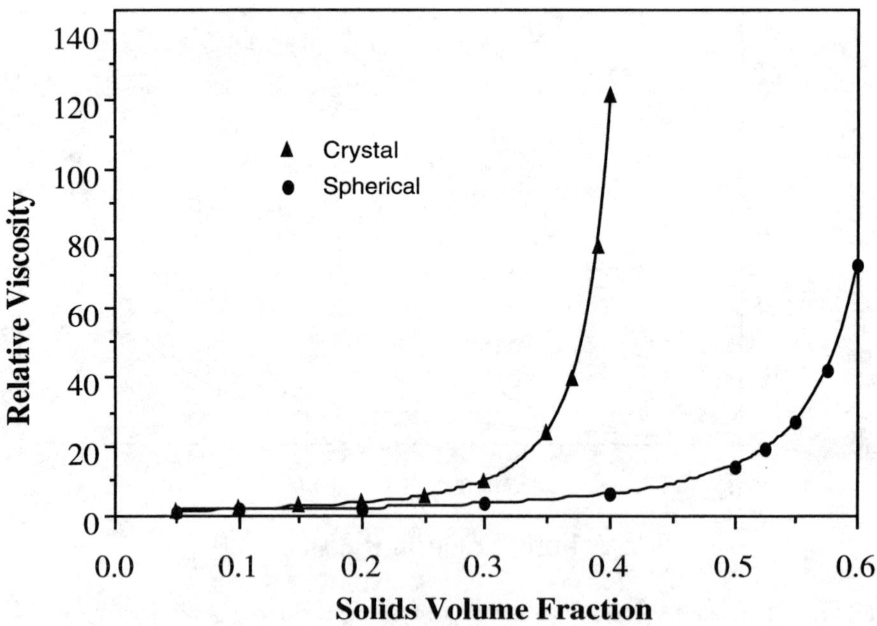

Figure 2–5 Relative Viscosity versus Volume Fraction of Solids Predicted by Model of Kitano et al. (1981) for Rigid Spheres (A = 0.68) and Crystal-Like Solids (A = 0.44). Kitano et al., 1991.

ployed in the separation of the phases. Therefore, rheological properties of plant food dispersions, such as tomato concentrates and concentrated orange juice, are based on the mass of pulp. Rheological properties of starch dispersions are based on the mass fraction of starch granules, denoted as cQ, as described in Chapter 4.

Applicability of Equation 2.27 for suspensions of tomato pulp of narrow size distribution was shown by Yoo and Rao (1994). Tomato pulp particles with average particle diameters 0.71 and 0.34 mm retained on the two sieves (sieve no. 40 and 60, respectively) were produced (Figure 2–6). The apparent average diameters were calculated as in Kimball and Kertesz (1952). Because the Casson viscosity was shown to be equal to infinite shear viscosity (Metz et al., 1979), it is less arbitrary and has a theoretical foundation. Considering the values of the single parameter A of tomato puree samples (Figure 2–7), the particle shape appears to be close to rough or irregular spherical shape (Yoo and Rao, 1994). Also, the magnitude of A of the TP6 sample was higher than that of the TP4 sample because of its lower aspect ratio.

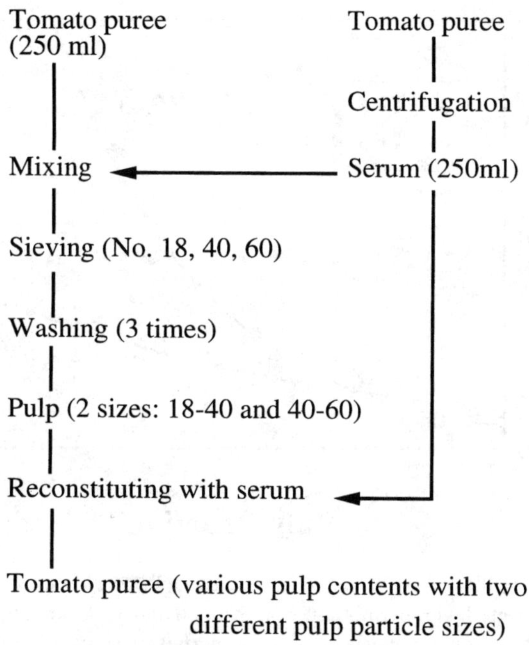

Figure 2–6 Procedure for Producing Tomato Pulp Particles of Narrow Particle Size Distribution. *Source*: Reprinted with permission from B. Yoo and M.A. Rao, Effect of Unimodal Particle Size and Pulp Content on Rheological Properties of Tomato Puree, *Journal of Texture Studies*, Vol. 25, pp. 421–436. © 1994, Food and Nutrition Press.

Quemada et al. (1985) proposed a model similar to that given by Equation 2.27, but with $A = 2.0$ and the structural parameter $k = 1.80$:

$$\eta_r = \left(1 - \frac{1}{2}k\,\phi\right)^{-2} \tag{2.29}$$

Quemada et al. (1985) reported values of k in the range 2.50 to 3.82 for dispersions of rigid solids, and 1.70 to 1.85 for red blood suspensions. For gelatinized 2.6% tapioca starch dispersions (STDs), the relative viscosity was calculated using Equation 2.30:

$$\eta_{r\infty} = \frac{\eta_\infty}{\eta_s} \tag{2.30}$$

where $\eta_{r\infty}$ is the relative viscosity based on the Casson viscosity, η_∞ (mPa s), at

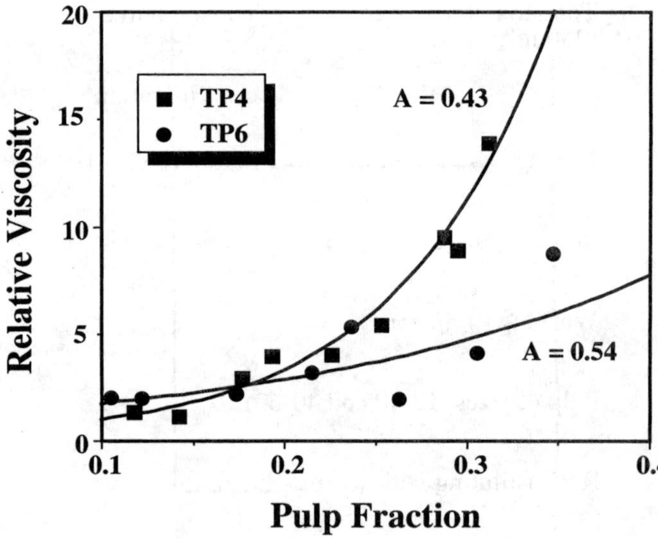

Figure 2–7 Applicability of the Model of Kitano et al. (1981) (Figure 2–5) to Tomato Particles. *Source*: Reprinted with permission from B. Yoo and M.A. Rao, Effect of Unimodal Particle Size and Pulp Content on Rheological Properties of Tomato Puree, *Journal of Texture Studies*, Vol. 25, pp. 421–436. © 1994, Food and Nutrition Press.

20°C, and η_s is the viscosity (mPa s) of the supernatant determined at 20°C. The model of Quemada et al. (1985) with the structural parameter $k = 1.80$ gave a satisfactory fit (Figure 2–8). Because heated tapioca starch granules are deformable, the value $k = 1.80$ of red blood cells was selected for the data on tapioca STDs. Given that the Krieger-Dougherty model predictions were substantially different than experimental η_r values of STDs (Ellis et al., 1989; Noel et al., 1993; Tattiyakul, 1997), the reasonable applicability of Equation 2.30 (Quemada et al., 1985) to tapioca STDs is noteworthy. However, the model of Quemada et al. (1985) is a phenomenological model, while that of Krieger-Dougherty is based on intrinsic viscosity of a single sphere. Saunders (1961) and Parkinson et al. (1970) found that the viscosity of a suspension increased with decrease in the average particle size. The particle size dependence can be explained by recognizing that as the particle size decreases, the number of particles in a given volume increases, resulting in a decrease in the mean distance between the particles. Another result of decreasing the

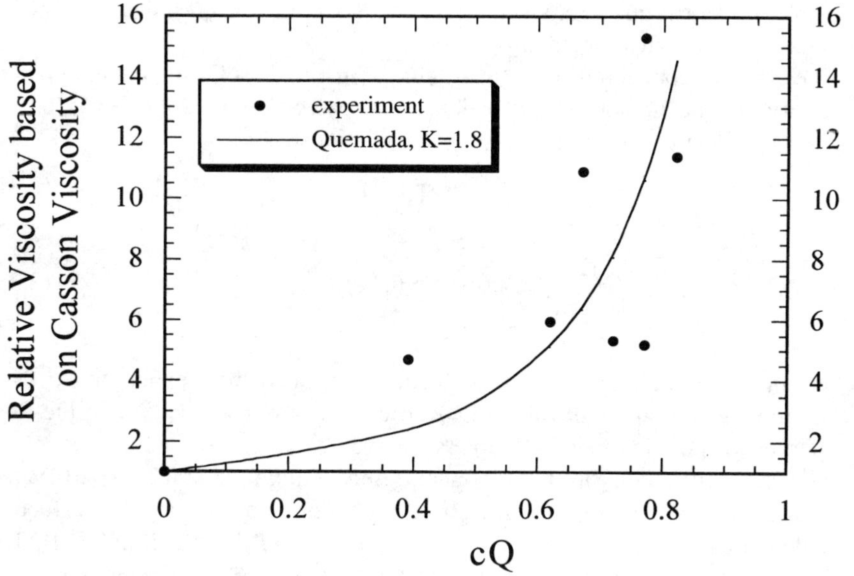

Figure 2–8 The Relative Viscosity (η_r) Values of Tapioca Starch Dispersions Strongly Depended on Their Volume Fraction. The line in the figure represents values predicted by the model of Quemada et al. (1985) with the structural parameter k = 1.80. *Source:* Reprinted from *Carbohydrate Polymers*, M.A. Rao and J. Tattiyakul, Granule Size and Rheological Behavior of Heated Tapioca Starch Dispersions, © 1999, with permission from Elsevier Science.

particle size is to increase the potential for particle-particle interaction (Agarwala et al., 1992).

EFFECT OF SOLUBLE AND INSOLUBLE SOLIDS CONCENTRATION ON APPARENT VISCOSITY OF FOODS

The effect of concentration on the zero-shear viscosity of biopolymer dispersions can be expressed in terms of the coil overlap parameter ($c[\eta]$) and the zero-shear specific viscosity, as described in Chapter 4 in connection with food gum dispersions.

Unlike biopolymer dispersions where the intrinsic viscosity is known and the concentration can be chosen a priori, for fluid foods the concentration of soluble (e.g., pectins in fruit juices) and insoluble solids often can be determined only posteriori, and the determination of their zero-shear viscosities is also difficult due to instrument limitation and the existence of yield stress. In most foods, it is often possible to identify the component(s), called key component(s), that play an important role in the rheological properties.

The effect of concentration (c) of soluble or insoluble solids on either apparent viscosity (η_a) or the consistency index of the power law model (K) can be described by either exponential or power law relationships:

$$\eta_a \propto \exp(ac) \tag{2.31}$$

$$\eta_a \propto c^b \tag{2.32}$$

$$K \propto \exp(a'c) \tag{2.33}$$

$$K \propto c^{b'} \tag{2.34}$$

where a, a', b, and b' are constants to be determined from experimental data. Other models that are applicable to specific foods are discussed under the flow properties of specific foods in Chapters 4 and 5.

The viscosity data on Pera concentrated orange juice (COJ) of Vitali and Rao (1984a, 1984b) illustrate the exponential model for the effect of soluble solids (°Brix) and insoluble solids (% pulp) (Tables 2–3 and 2–4). The influence of insoluble solids on apparent viscosity at a shear rate of 100 s^{-1} ($\eta_{a,100}$) and on K, shown in Figure 2–9, and that of soluble solids on the same rheological parameters, shown in Figure 2–10, can be described by exponential relationships.

The effect of concentration (c) of soluble solids (°Brix) and insoluble solids (pulp) on either apparent viscosity or the consistency index of the power law model of frozen concentrated orange juice (FCOJ) can be described by exponential relationships (Vitali and Rao, 1984a, 1984b). Equations 2.35 and 2.36 are applicable to the consistency index (K) of the power law model (Equation 2.3). In the case of FCOJ, it should be noted that insoluble solids are expressed in

Table 2–3 Effect of °Brix on Apparent Viscosity and Consistency Index of the Power Law Model, Pera Orange Juice, 5.7% Pulp, −10°C

°Brix	$\eta_{a,100}$ *Pa s*	*ln,* $\eta_{a,100}$	K, *Pa s*^{n}	*ln* K
50.8	0.29	−1.2379	0.89	−0.1165
56.5	0.65	−0.4308	2.18	0.7793
57.9	0.83	−0.1863	2.56	0.9400
61.6	1.62	0.4824	5.61	1.7246
65.3	3.38	1.2179	23.58	3.1604

Table 2–4 Effect of Pulp Content on Apparent Viscosity and Consistency Index of the Power Law Model, Pera Orange Juice, 65°Brix, −10°C

Pulp, %	$\eta_{a,100}$ Pa s	ln, $\eta_{a,100}$	K, Pa s^n	ln K
0	1.95	0.6678	3.74	1.3191
3.4	2.73	1.0043	7.11	1.9615
5.7	3.38	1.2179	8.63	2.1552
8.6	4.70	1.5476	15.77	2.7581
11.1	5.76	1.7509	23.91	3.1743

terms of pulp content determined on a 12°Brix sample by centrifugation for 10 minutes at 360 × *g*.

$$K = K^c \exp (B_K^c \text{ °Brix}) \tag{2.35}$$

$$K = K^P \exp (B_K^P \text{ pulp}) \tag{2.36}$$

where K^c, K^P, B_K^c, and B_K^P are constants.

The role of insoluble solids can be also studied in terms of the relative viscosity (η_r = apparent viscosity of FCOJ/apparent viscosity of serum) and pulp frac-

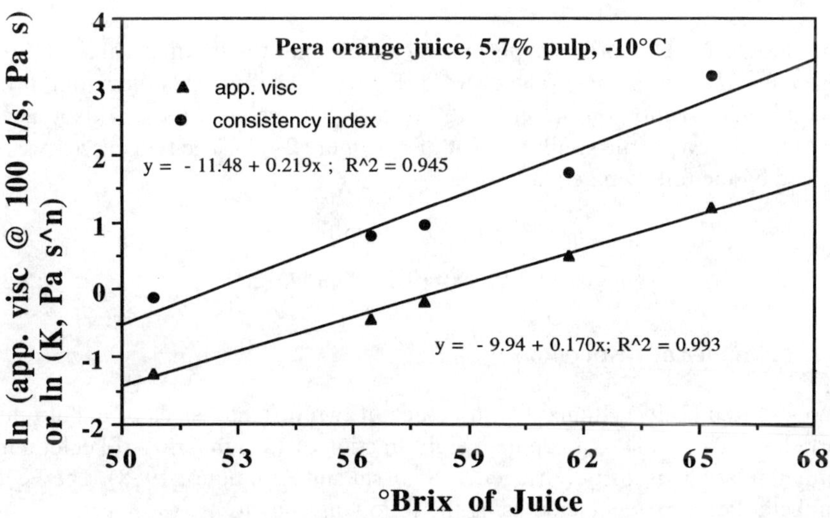

Figure 2–9 The Influence of Soluble Solids on Apparent Viscosity at a Shear Rate of 100 s^{-1} ($\eta_{a,\ 100}$) and on K of Pera Concentrated Orange Juice. Vitali and Rao, 1984a.

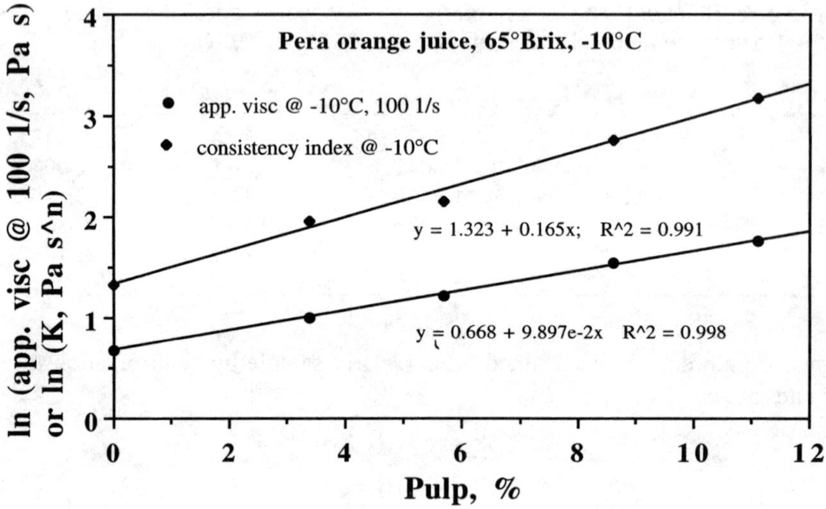

Figure 2–10 The Influence of Insoluble Solids on Apparent Viscosity at a Shear Rate of 100 s^{-1} ($\eta_{a,\,100}$) and on K of Pera Concentrated Orange Juice. Vitali and Rao, 1984b.

---

tion (Figure 2–11). And, as expected, such a plot has the limiting value of 1.0 at zero pulp fraction (serum). The curve in Figure 2–11 illustrates the strong influence of pulp fraction on the viscosity of COJ. The values of $\eta_{a,100}$ also, as expected, follow a profile similar to that of η_r (Figure 2–11). The two curves are described by the following equations:

$$\eta_r = \exp\,(9.90 \times \text{pulp fraction}) \tag{2.37}$$

$$\eta_{a,\,100} = 1.95 \exp\,(9.90 \times \text{pulp fraction}) \tag{2.38}$$

Peclet Number of Dispersions

Yoo and Rao (1994) studied the influence of two different sizes of tomato pulp particles (TP4 and TP6) at a pulp weight fraction of 17% in terms of Peclet number and relative viscosity (Krieger, 1985; Tsai and Zammouri, 1988). The Peclet numbers (Pe) were calculated using the following equation:

$$\text{Pe} = \frac{\eta_o\, r_o^3 \dot{\gamma}}{k\,T} \tag{2.39}$$

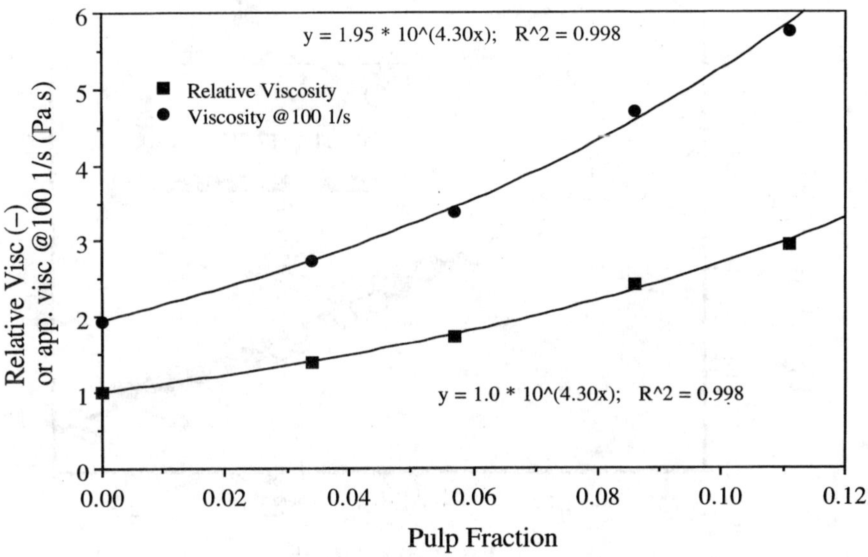

Figure 2–11 The Strong Influence of Pulp Fraction on the Viscosity and Relative Viscosity (η_r) of COJ.

---

where η_o is the viscosity of the suspending liquid (serum), r_o is the particle radius, k is the Boltzmann constant (1.38×10^{-23} N m/K), and T is the absolute temperature. The Peclet number compares the effect of imposed shear with the effect of diffusion of the particles. When Pe $\gg$ 1, hydrodynamic effects dominate, and a dispersion of spherical particles exhibits shear-thinning behavior. In contrast, when Pe $\ll$ 1, the distribution of particles is only slightly altered by the flow (Hiemenz and Rajagopalan, 1997). As shown in Figure 2–12, at equal values of pulp weight fraction, a TP6 sample with small diameter particles was more viscous than a TP4 sample with large diameter particles. A nearly linear relationship exists between the relative viscosity of TP sample and the Peclet number. A similar linear relationship was found for glass bead suspensions (Tsai and Zammouri, 1988). However, the slope of the linear relationship of TP6 sample is slightly higher than that of TP4 sample, indicating that at high shear rates, the TP6 aggregates with small particles are more sensitive to shear than those with large particles. The different aspect ratios of the TP4 and TP6 particles discussed above is another reason for the deviation from a single linear relationship. From these observations, it can be concluded that the effect of particle size on the relative vis-

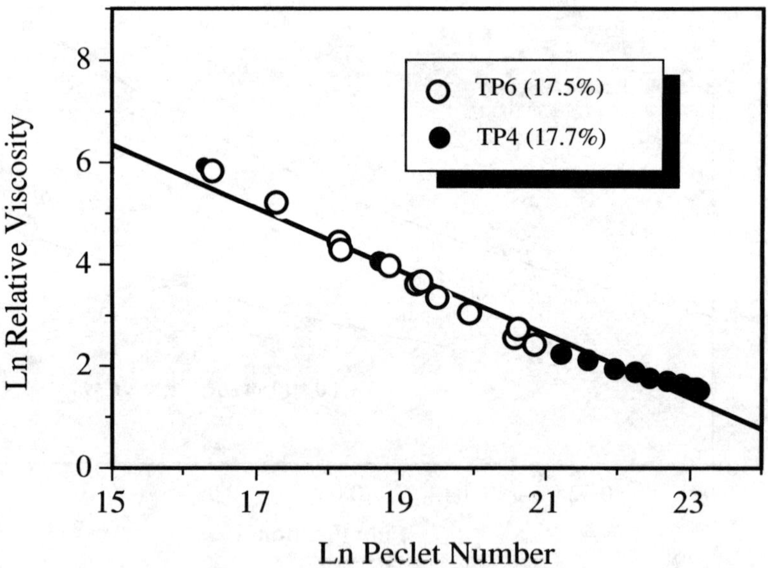

Figure 2–12 Peclet Number versus Relative Viscosity of Tomato Pulp Particles. At equal values of pulp weight fraction, a TP6 sample with small diameter particles was more viscous than a TP4 sample with large diameter particles. *Source*: Reprinted with permission from B. Yoo and M.A. Rao, Effect of Unimodal Particle Size and Pulp Content on Rheological Properties of Tomato Puree, *Journal of Texture Studies*, Vol. 25, pp. 421–436. © 1994, Food and Nutrition Press.

cosity of tomato puree sample can be correlated with the Peclet number of the particle.

Emulsions

Many foods are oil-in-water or water-in-oil emulsions (o/w), with dispersed particle size range of 0.01–10 μm (Rahalkar, 1992). Many of the equations discussed for food suspensions are also applicable to emulsions. In a dilute emulsion, the particles are far apart and the interparticle interactions are relatively weak. Skim milk is an example of a dilute emulsion with the concentration of fat droplets (dispersed phase) $<$ 1%. However, in a concentrated emulsion, the particles are close to each other and there are strong interparticle interactions. Mayonnaise has a dispersed phase (oil) concentration of 70–80%. As with foods containing insoluble solids, the volume fraction of the dispersed phase in emulsions is important.

The separation of phases in an emulsion occurs spontaneously in the direction of decreasing Gibbs free energy. Thus it can be said that there is more surface energy in an emulsion when the dispersed droplets are in a highly subdivided state than when they are in a coarser state of subdivision. To be stable, an o/w emulsion should exhibit yield stress, and the forces applied to the continuous phase by the dispersed phase due to the applicable forces (e.g., gravity and buoyancy) should be below the emulsion's yield stress.

The words *stable* and *unstable* are often used to describe emulsions that are better understood by examining the underlying processes. The coarsening of a thermodynamically unstable emulsion is called coalescence or aggregation: Coalescence is a process by which two or more particles fuse together to form a single larger particle; aggregation is a process by which small particles lump together to form aggregates (Hiemenz and Rajagopalan, 1997). The van der Waals force between particles in a dispersion is usually attractive and is strong at short distances between particles. Therefore, the emulsion will be unstable and coagulate unless there are repulsive interactions between particles. Two methods are used to overcome van der Waals attraction: (1) electrostatic stability, in which an electrostatic force results from overlapping electrical double layers of two particles, and (2) polymer-induced or steric stability, in which a suitable polymer that is adsorbed on the particle surfaces may be used. In food emulsions, polymer-induced stability is encountered commonly.

Very often, the microstructure and the macroscopic states of dispersions are determined by kinetic and thermodynamic considerations. While thermodynamics dictates what the equilibrium state will be, kinetics determines how fast that equilibrium state will be determined. While in thermodynamics the initial and final states must be determined, in kinetics the path and any energy barriers are important. The electrostatic and the electrical double-layer (the two charged portions of an interfacial region) play important roles in food emulsion stability. The Derjaguin-Landau-Verwey-Overbeek (DLVO) theory of colloidal stability has been used to examine the factors affecting colloidal stability.

An emulsifier may be thought of as a single chemical component or mixture of components that have the capacity to promote emulsion formation and stabilization by interfacial action; in contrast, a stabilizer confers long-term stability on an emulsion, often involving adsorption or another mechanism. Polymers, such as gum arabic, egg albumin, casein, and gelatin, have been used for a long time to stabilize food emulsions. Other stabilizers include xanthan gum, guar gum, and whey protein isolate. Because some of these are charged polymers (polyelectrolytes), their stabilizing influence is due to both electrostatic and polymeric effects, that is, electrosteric stabilization. Lecithins (e.g., from egg yolk and soybean) are the most important food emulsifiers derived from natural sources without chemical reaction (Dickinson, 1993). They consist of mixtures of many phospholipid components, with phosphatidylcholine (PC) and phosphatidylethanol-amine (PE) being present in large proportions.

A simplified view of some of the possible effects of polymer molecules on a dispersion (Hiemenz and Rajagopalan, 1997) include (1) at low polymer concentrations bridging flocculation, where a polymer chain forms bridges by adsorbing more than one particle; (2) at higher polymer concentrations, "brush-like" layers can form on the solid particles that can mask the influence of van der Waals attraction between the particles, thereby imparting stability, called steric stabilization; (3) at moderate to high polymer concentrations, the polymer chains may be excluded in the region between the particles, resulting in depletion flocculation; and (4) at high polymer concentrations, polymer-depleted regions may be created by demixing the polymer, resulting in depletion stabilization. Generally, polymers containing only one kind of repeat unit (homopolymers) are not good for steric stabilization. One requirement is that the polymer be adsorbed at the oil-water interface, and the polymer is considered to reside partially at surface sites and partially in loops or tails in the solution. Under the right circumstances, an adsorbed polymer layer stabilizes a dispersion against aggregation. It may be said that the approaching particles in an aggregation step experience an increase in free energy of repulsion, ΔG_R, that can be divided into enthalpic (ΔH_R) and entropic (ΔS_R) contributions:

$$\Delta G_R = \Delta H_R - T\,\Delta S_R \tag{2.40}$$

In Equation 2.40, the terms describe changes in enthalpy and entropy in the overlapping region of the adsorbed layers of two particles. Because ΔH_R and ΔS_R can be either positive or negative, it is possible for the value of ΔG_R to change sign with change in temperature, and the critical temperature for the onset of flocculation is known as the critical flocculation temperature (Hiemenz and Rajagopalan, 1997).

Because of the many factors affecting the rheology of food emulsions, the rheological properties cannot be predicted a priori, so experimental studies on model o/w and food emulsions are necessary. Whey proteins, especially β-lactoglobulin, possess good emulsification properties, and whey protein stabilized emulsions can be converted into emulsion gels by thermal treatment (Aguilera and Kessler, 1989; Dickinson and Yamamoto, 1996a). In general, the incorporation of protein-coated emulsion droplets into a heat-set globular protein network resulted in an increase in gel strength. Further, a self-supporting heat-set emulsion gel can be formed at a protein content much lower than that required under similar thermal processing conditions (Dickinson and Yamamoto, 1996b). Addition of pure egg yolk L-α-phosphatidylcholine after emulsification caused an increase in strength of heat-set β-lactoglobulin emulsion gels (Dickinson and Yamamoto, 1996b). The positive influence of pure egg lecithin, added after emulsification, on the elastic modulus of the whey protein concentrate emulsion gel was attributed to lecithin-protein complexation. Crude egg lecithin also gave a broadly similar increase in gel strength. However, pure soybean lecithin was not as effective in reinforcing the network, and crude soybean lecithin was ineffective (Dickinson and Yamamoto, 1996a). These results support the general hypothesis that filler particles (e.g., globular protein-coated droplets) that interact with a gel matrix (a network of aggregated denatured whey protein) tend to reinforce the network and increase the gel strength (Dickinson and Yamamoto, 1996b).

Other recent studies on physical properties of emulsions are those of McClements et al. (1993), Dickinson and Pawlowsky (1996), Dickinson et al. (1996), and Demetriades et al. (1997). The application of nuclear magnetic resonance (NMR) technique (Simoneau et al., 1993) and ultrasonic spectroscopy (Demetriades et al., 1996) to study the stability of emulsions has been discussed. The rheological behavior of salad dressings and mayonnaises that are emulsions will be discussed in Chapter 5.

EFFECT OF TEMPERATURE ON VISCOSITY

A wide range of temperatures are encountered during processing and storage of fluid foods, so the effect of temperature on rheological properties needs to be documented. The effect of temperature on either apparent viscosity at a specified shear rate (Equation 2.41) or the consistency index, K, of the power law model (Equation 2.42) of a fluid often can be described by the Arrhenius relationship. The effect of temperature on apparent viscosity can be described by the Arrhenius relationship:

$$\eta_a = \eta_{\infty A} \exp{(E_a/RT)} \tag{2.41}$$

where η_a is the apparent viscosity at a specific shear rate, $\eta_{\infty A}$ is the frequency factor, E_a is the activation energy (J mol^{-1}), R is the gas constant (J mol^{-1} K^{-1}), and T is temperature (K).

Although the name of Arrhenius is associated with Equation 2.41, Moore (1972) credited J. de Guzman Carrancio for first pointing out this relationship in 1913. The quantity E_a is the energy barrier that must be overcome before the elementary flow process can occur. The term exp $(-E_a/RT)$ may be explained as a Boltzmann factor that gives the fraction of the molecules having the requisite energy to surmount the barrier. Hence, E_a is the activation energy for viscous flow.

From a plot of ln η_a (ordinate) versus $(1/T)$ (abscissa), E_a equals (slope $\times$ R), and η_∞ is exponential of the intercept. The Arrhenius equation for the consistency index is

$$K = K_\infty \exp\,(E_{aK}/RT) \tag{2.42}$$

where K_∞ is the frequency factor, E_{aK} is the activation energy (J/mol), R is the gas constant, and T is temperature (K). A plot of ln K (ordinate) versus $(1/T)$ (abscissa) results in a straight line, and E_{aK} = (slope $\times$ R), and K_∞ is exponential of the intercept. The activation energy should be expressed in joules (J), but in the earlier literature it has been expressed in calories (1 calorie = 4.1868 joules). The applicability of the Arrhenius model to the apparent viscosity versus temperature data on a concentrated orange juice serum sample (Vitali and Rao, 1984a, 1984b) is shown in Figure 2–13.

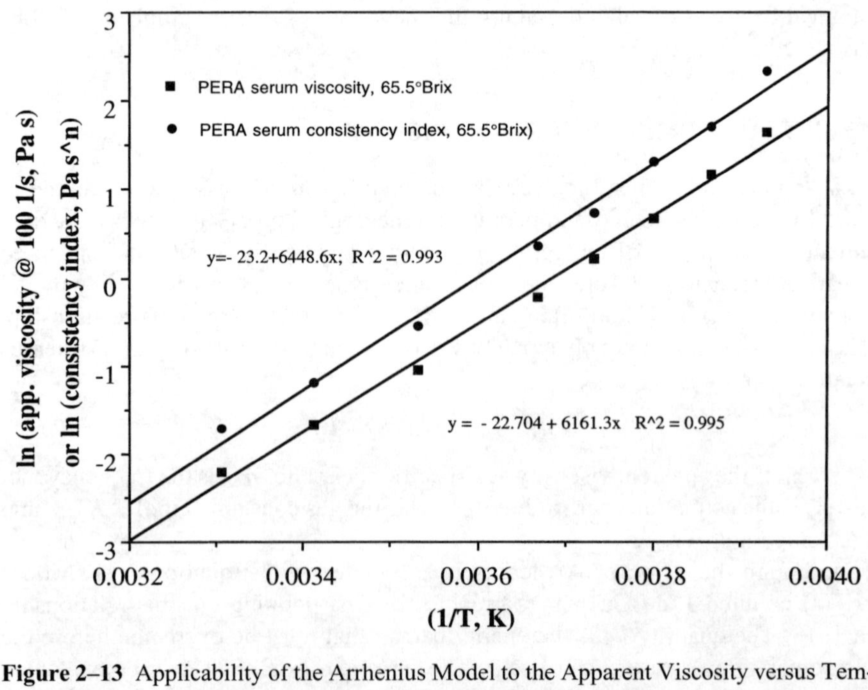

Figure 2–13 Applicability of the Arrhenius Model to the Apparent Viscosity versus Temperature Data on a Concentrated Orange Juice Sample. Vitali and Rao, 1984a.

Table 2–5 Magnitudes of the Parameters of the Arrhenius Equation for the Effect of Temperature on Concentrated Apple and Grape Juices. Rao et al., 1986.

Sample	η_∞	E_a, *kJ mol*$^{-1}$	*SSQ*	R^2
Apple 68.3° Brix	1.366E-15	79	0.10260	1.00
Apple 64.9° Brix	1.671E-15	76	0.02870	1.00
Apple 60.1° Brix	2.610E-13	62	0.00110	1.00
Apple 55.0° Brix	5.050E-11	48	0.00100	0.99
Grape 68.3° Brix	2.150E-14	73	0.31770	1.00
Grape 64.5° Brix	9.810E-13	63	0.03950	1.00
Grape 59.9° Brix	2.940E-12	58	0.00620	1.00
Grape 54.0° Brix	7.610E-11	49	0.00130	1.00

The Arrhenius equation did not describe very well the influence of temperature on viscosity data of apple and grape juices in the range 60–68°Brix (Rao et al., 1984, 1986). From nonlinear regression analysis, it was determined that the empirical Fulcher equation (see Ferry, 1980, p. 289; Soesanto and Williams, 1981) described the viscosity versus temperature data on those juice samples better than the Arrhenius model (Rao et al., 1986):

$$\log \eta = A' + \frac{B'}{T - T_\infty} \tag{2.43}$$

The magnitudes of the parameters of the Arrhenius and the Fulcher equations for the studied concentrated apple and grape juices are given in Table 2–5 and Table 2–6, respectively. The physical interpretation of the three constants in the Fulcher

Table 2–6 Magnitudes of the Parameters of the Fulcher Equation for the Effect of Temperature on Concentrated Apple and Grape Juices. Rao et al., 1986.

Sample	A′	B′	T_∞	*SSQ*	R^2
Apple 68.3° Brix	−7.26	1049	132.8	0.01356	1.00
Apple 64.9° Brix	−7.56	1035	130.9	0.01360	1.00
Apple 60.1° Brix	−6.87	911	124.3	0.00208	1.00
Apple 55.0° Brix	−6.07	896	94.0	0.00048	1.00
Grape 68.3° Brix	−6.75	1039	125.7	0.14110	1.00
Grape 64.5° Brix	−6.65	1041	115.1	0.02150	1.00
Grape 59.9° Brix	−6.81	1048	107.0	0.00080	1.00
Grape 54.0° Brix	−6.56	1085	84.5	0.00080	1.00

equation is ambiguous, but by translating them in terms of the Williams-Landel-Ferry (WLF) parameters, their significance can be clarified and it is functionally equivalent to the WLF equation (Ferry, 1980; Soesanto and Williams, 1981):

$$\log\left(\frac{\eta}{\rho T}\bigg/\frac{\eta_0}{\rho_0 T_0}\right) = -\frac{c_1^0\,(T - T_0)}{c_2^0 + (T - T_0)} \tag{2.44}$$

Specifically, T_∞ and B' are related to c_2^0, c_1^0, and a reference temperature T_0, often called the glass transition temperature, by the following two equations:

$$T_\infty = T_0 - c_2^0 \tag{2.45}$$

$$B' = \frac{c_1^0}{(T_0 - T_\infty)} \tag{2.46}$$

However, it should be pointed out that Soesanto and Williams (1981) also determined the values of the glass transition temperature of the very high concentration sugar solutions (91.9% to 97.6% by weight) by regression techniques and not experimentally. Because of the empirical nature of the Fulcher equation and the empirical origin of the WLF equation, their use with viscosity data of relatively dilute fruit juices serves mainly the objective of obtaining a useful correlation.

COMBINED EFFECT OF CONCENTRATION AND TEMPERATURE

The flow behavior index (n) is assumed to be relatively constant with temperature and concentration, and the combined effects of temperature and concentration on the power law consistency index, K, are described by

$$K = A\,c^{b'} \exp\,(E_a/RT) \tag{2.47}$$

Tomato concentrates and concentrated milk samples are examples of foods in this category. Alternatively, the combined effects of temperature and concentration on the power law consistency index, K, are described by

$$K = A' \exp\left(\frac{E_a}{RT} + b\,c\right) \tag{2.48}$$

MIXING RULES FOR TWO COMPONENT BLENDS

Earlier, the role of suspended particles in fluid media in increasing the viscosity of a suspension and different equations relating the volume fraction of solids were discussed. For general two-component or polymer composites, the Takayanagi models (Ross-Murphy, 1984; Sperling, 1986) provide means for calculating the upper (so-called isostrain) and lower (so-called isostress) limits of values for the shear modulus G_c of a composite formed from components x and y with shear moduli G_x and G_y, respectively. In the former, the polymers are arranged in parallel with respect to deformation, while in the latter they are arranged in series. In the parallel arrangement, the deformation of the weaker component is limited by the modulus of the stronger material, and both components are deformed to the same extent (isostrain). In the series arrangement, the strength of the weaker component limits the force transmitted to the stronger material; therefore, both are subjected to the same stress (isostress).

If ϕ_x and $\phi_y = 1 - \phi_x$ are the respective volume fractions of the two components, the equations for the upper and lower bound values are

$$G_c = \phi_x G_x + \phi_y G_y \qquad \text{(upper bound, isostrain)} \qquad (2.49)$$

$$(1/G_c) = (\phi_x/G_x) + (\phi_y/G_y) \qquad \text{(lower bound, isostress)} \qquad (2.50)$$

Values of G_c predicted by the above two models are shown in Figure 2–14. Implicit in the above models is that the experimental value of G_c lies between those

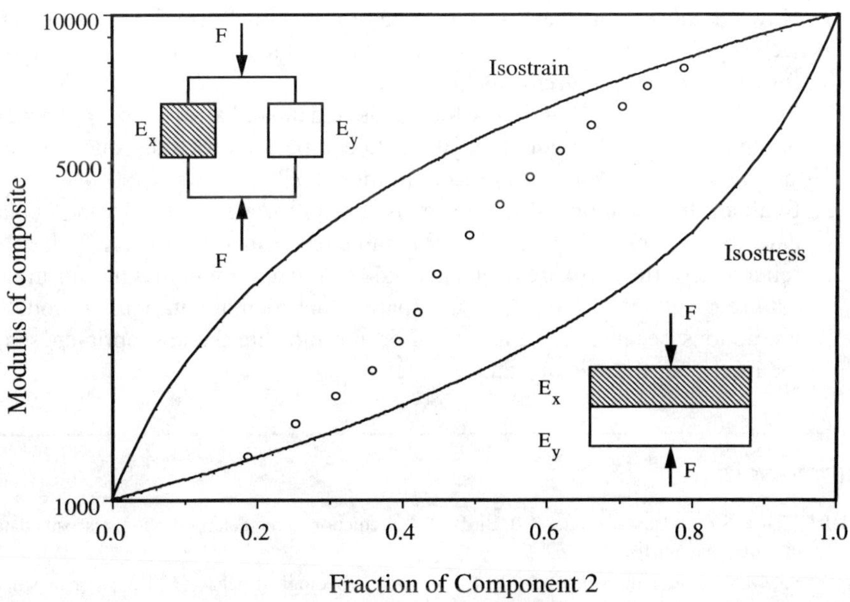

Figure 2–14 Modulus of the Composite Gel (G_c) Plotted against Volume Fraction of Component y ($G = 10{,}000$). When the weaker component (x, $G = 1{,}000$ Pa) dominates and becomes the continuous phase, G_c follows the lower bound isostress limit. With increasing fraction of y, there will be a phase inversion and G_c reaches the upper bound limit, path indicated by open circles.

of G_x and G_y. For a simple phase-separated system, the lower bound form (isostress) should predict G_c prior to phase inversion when the supporting phase (x) is the weaker one. It will then show a transition to the upper bound behavior. More complex models can be found in Sperling (1986). The Takayanagi models have been used to better understand the rheological behavior of a starch-soy protein system (Chapter 4), tomato paste (Chapter 5), and mixed gel systems (Chapter 6). However, given that most foods contain several major components ($\tilde{c}$), the number and distribution of phases (P) is much more complex, as seen from the phase rule:

$$P = (\tilde{c} - \tilde{f} + 2) \tag{2.51}$$

where $\tilde{f}$ is the number of degrees of freedom.

TREATMENT OF RHEOLOGICAL DATA USING MODELS

Prior to using a model for description of rheological data, it would be desirable to critically examine a number of issues, such as the following:

1. How reliable arc the data? Are the data reproducible? Are the measurement techniques reliable? Are the test samples reproducible, reliable, and suitable for rheological measurements?
2. Plot the data ($\sigma - \dot{\gamma}$) and look for trends in data and artifacts of rheometer. Sometimes most, but not all, of the data can be used for subsequent analysis. If the data are not reliable, why fit models?
3. Evaluate the rheological model: a. Is it appropriate for the experimental data? b. How reliable is the model parameter estimation software? Has the reliability of the software been checked? c. What are you looking for in the data (e.g., effect of temp.)? d. Compare experimental data with a model's predictions because R^2 values often do not indicate the appropriateness of the model for the selected data.

REFERENCES

Abdel-Khalik, S.I., O. Hassager, and R.B. Bird. 1974. Prediction of melt elasticity from viscosity data. *Polymer Eng and Sci* 14: 859–867.

Agarwala, M.K., B.R. Patterson, and P.E. Clark. 1992. Rheological behavior of powder injection molding model slurries. *J Rheol* 36: 319–334.

Aguilera, J.M., and H.G. Kessler. 1989. Properties of mixed and filled dairy gels. *J Food Science* 54: 1213–1217, 1221.

Barnes, H.A., J.F. Hutton, and K. Walters. 1989. *An introduction to rheology*. Amsterdam: Elsevier Science.

Barnes, H.A., and K. Walters. 1985. The yield stress myth? *Rheol Acta* 24: 323–326.

Bird, R.B., G.C. Dai, and B.J. Yarusso. 1982. The rheology and flow of viscoplastic materials. *Reviews Chem Eng* 1: 1–70.

Brodkey, R.S. 1967. *The phenomena of fluid motions.* Reading, MA: Addison-Wesley Publishing Co.

Casson, N. 1959. A flow equation for pigment-oil suspensions of the printing ink type. In *Rheology of disperse systems,* ed. C.C. Mill, 82–104. New York: Pergamon Press.

Choi, G.R., and I.M. Krieger. 1986. Rheological studies on sterically stabilized model dispersions of uniform colloidal spheres II. Steady-shear viscosity. *J Colloid Interface Sci* 113: 101–113.

Da Silva, P.M.S., J.C. Oliveira, and M.A. Rao. 1997. The effect of granule size distribution on the rheological behavior of heated modified and unmodified maize starch dispersions. *J Texture Studies* 28: 123–138.

Demetriades, K., J. Coupland, and D.J. McClements. 1996. Investigation of emulsion stability using ultrasonic spectroscopy. In *1996 IFT annual meeting book of abstracts,* 109–110. Chicago: Institute of Food Technologists.

Demetriades, K., J. Coupland, and D.J. McClements. 1997. Physical properties of whey protein stabilized emulsions as related to pH and NaCl. *J Food Sci* 62: 342–347.

Dervisoglu, M., and J.L. Kokini. 1986. Steady shear rheology and fluid mechanics of four semi-solid foods. *J Food Sci* 51: 541–546, 625.

Dickinson, E. 1993. Towards more natural emulsifiers. *Trends Food Sci Technol* 4: 330–334.

Dickinson, E., S-T. Hong, and Y. Yamamoto. 1996. Rheology of heat-set emulsion gels containing beta-lactoglobulin and small-molecule surfactants. *Netherlands Milk Dairy J* 50: 199–207.

Dickinson, E., and K. Pawlowsky. 1996. Effect of high-pressure treatment of protein on the rheology of flocculated emulsions containing protein and polysaccharide. *J Agric Food Chem* 44: 2992–3000.

Dickinson, E., and Y. Yamamoto. 1996a. Viscoelastic properties of heat-set whey protein-stabilized emulsion gels with added lecithin. *J Food Sci* 61: 811–816.

Dickinson, E., and Y. Yamamoto. 1996b. Effect of lecithin on the viscoelastic properties of β-lactoglobulin-stabilized emulsion gels. *Food Hydrocolloids* 10: 301–307.

Einstein, A. 1906. Eine neue bestimmung der molekuldimension. *Ann Physik* 19: 289–306.

Einstein, A. 1911. Berichtigung zu meiner arbeit: Eine neue bestimmung der molekuldimension. *Ann Physik* 34: 591–592.

Ellis, H.S., S.G. Ring, and M.A. Whittam. 1989. A comparison of the viscous behavior of wheat and maize starch pastes. *J Cereal Sci* 10: 33–44.

Fang, T.N., C. Tiu, X. Wu, and S. Dong. 1996. Rheological behaviour of cocoa dispersions. *J Texture Studies* 26: 203–215.

Fang, T., H. Zhang, T.T. Hsieh, and C. Tiu. 1997. Rheological behavior of cocoa dispersions with cocoa butter replacers. *J Texture Studies* 27: 11–26.

Ferry, J.D. 1980. *Viscoelastic properties of polymers.* New York: John Wiley & Sons.

Giboreau, A., G. Cuvelier, and B. Launay. 1994. Rheological behavior of three biopolymer/water systems with emphasis on yield stress and viscoelastic properties. *J Texture Studies* 25: 119–137.

Hiemenz, P.C., and R. Rajagopalan. 1997. *Principles of colloid and surface chemistry.* 3d ed. New York: Marcel Dekker.

Holdsworth, S.D. 1971. Applicability of rheological models to the interpretation of flow and processing behavior of fluid food products. *J Texture Studies* 4: 393–418.

Holdsworth, S.D. 1993. Rheological models used for the prediction of the flow properties of food products: A literature review. *Trans Inst Chem Engineers* 71, Part C: 139–179.

Jacon, S.A., M.A. Rao, H.J. Cooley, and R.H. Walter. 1993. The isolation and characterization of a water extract of konjac flour gum. *Carbohydrate Polymers* 20: 35–41.

Jinescu, V.V. 1974. The rheology of suspensions. *Int Chem Eng* 143: 397–420.

Kimball, L.B., and Z.I. Kertesz. 1952. Practical determination of size distribution of suspended particles in macerated tomato products. *Food Technol* 6: 68–71.

Kitano, T., T. Kataoka, and T. Shirota. 1981. An empirical equation of the relative viscosity of polymer melts filled with various inorganic fillers. *Rheologica Acta* 20: 207–209.

Krieger, I.J. 1985. Rheology of polymer colloids. In *Polymer colloids,* eds. R. Buscall, T. Corner, and J.F. Stageman, 219–246. New York: Elsevier Applied Science.

Krieger, I.M., and T.J. Dougherty. 1959. A mechanism for non-Newtonian flow in suspensions of rigid spheres. *Trans Soc Rheol* 3: 137–152.

Launay, B., J.L. Doublier, and G. Cuvelier. 1986. Flow properties of aqueous solutions and dispersions of polysaccharides. In *Functional properties of food macromolecules,* eds. J.R. Mitchell and D.A. Ledward, 1–78. London: Elsevier Applied Science.

Lopes da Silva, J.A.L., M.P. Gonçalves, and M.A. Rao. 1992. Rheological properties of high-methoxyl pectin and locust bean gum solutions in steady shear. *J Food Sci* 57: 443–448.

McClements, D.J., F.J. Monaham, and J.E. Kinsella. 1993. Effect of emulsion droplets on the rheology of whey protein isolate gels. *J Texture Studies* 24: 411–422.

Metz, B., N.W.F. Kossen, and J.C. van Suijdam. 1979. The rheology of mould suspensions. In *Advances in biochemical engineering.* Vol. 2, eds. T.K. Ghose, A. Fiechter, and N. Blakebrough, 103. New York: Springer Verlag.

Metzner, A.B. 1985. Rheology of suspensions in polymeric liquids. *J Rheol* 29: 739–775.

Mizrahi, S., and Z. Berk. 1972. Flow behaviour of concentrated orange juice: Mathematical treatment. *J Texture Studies* 3: 69–79.

Moore, W.J. 1972. *Physical Chemistry.* 4th ed. Englewood Cliffs, NJ: Prentice Hall, Inc.

Noel, T.R., S.G. Ring, and M.A. Whittam. 1993. Physical properties of starch products: Structure and function. In *Food colloids and polymers: Stability and mechanical properties,* eds. E. Dickinson and P. Wolstra, 126–137. Cambridge, England: Royal Society of Chemistry.

Ofoli, R.Y., R.G. Morgan, and J.F. Steffe. 1987. A generalized rheological model for inelastic fluid foods. *J Texture Studies* 18: 213–230.

Paredes, M.D.C., M.A. Rao, and M.C. Bourne. 1988. Rheological characterization of salad dressings. 1. Steady shear, thixotropy and effect of temperature. *J Texture Studies* 19: 247–258.

Parkinson, C., S. Matsumoto, and P. Sherman. 1970. The influence of particle-size distribution on the apparent viscosity of non-Newtonian dispersed system. *J Colloid Interface Sci* 33: 150–160.

Poslinski, A.J., M.E. Ryan, R.K. Gupta, S.G. Seshadri, and F.J. Frechette. 1988. Rheological behavior of filled polymeric systems I. Yield stress and shear-thinning effects. *J Rheol* 32: 703–735.

Quemada, D., P. Fland, and P.H. Jezequel. 1985. Rheological properties and flow of concentrated diperse media. *Chem Eng Comm* 32: 61–83.

Rahalkar, R.R. 1992. Viscoelastic properties of oil-water emulsions. In *Viscoelastic Properties of Foods,* eds. M.A. Rao and J.F. Steffe, 285–315. London: Elsevier Applied Science Publishers.

Rao, M.A., and H.J. Cooley. 1983. Applicability of flow models with yield for tomato concentrates. *J Food Process Eng* 6: 159–173.

Rao, M.A., H.J. Cooley, and A.A. Vitali. 1984. Flow properties of concentrated fruit juices at low temperatures. *Food Technology* 38(3): 113–119.

Rao, M.A., R.S. Shallenberger, and H.J. Cooley. 1986. Effect of temperature on viscosity of fluid foods with high sugar content. In *Engineering and food.* Vol. 1, eds. M. LeMaguer and P. Jelen, 23–31. New York: Elsevier Applied Science.

Rao, M.A., and J. Tattiyakul. 1999. Granule size and rheological behavior of heated tapioca starch dispersions. *Carbohydrate Polymers* 38: 123–132.

Rayment, P., S.B. Ross-Murphy, and P.R. Ellis. 1998. Rheological properties of guar galactomannan and rice starch mixtures. II. Creep measurements. *Carbohydrate Polymers* 35: 55–63.

Ross-Murphy, S.B. 1984. Rheological methods. In *Biophysical methods in food research,* ed. H.W. Chan, 138–199. London: Blackwell Scientific Publications.

Saunders, F.L. 1961. Rheological properties of monodisperse latex systems I. Concentration dependence of relative viscosity. *J Colloid Sci* 16: 13–22.

Simoneau, C., M.J. McCarthy, and J.B. German. 1993. Magnetic resonance imaging and spectroscopy for food systems. *Food Res Int* 26: 387–398.

Soesanto, T., and M.C. Williams. 1981. Volumetric interpretation of viscosity for concentrated and dilute sugar solutions. *J Phys Chem* 85: 3338–3341.

Sperling, L.H. 1986. *Introduction to physical polymer science.* New York: John Wiley & Sons.

Steiner, E.H. 1958. A new rheological relationship to express the flow properties of melted chocolate. *Rev Int Chocolatiere* 13: 290–295.

Tattiyakul, J. 1997. Studies on granule growth kinetics and characteristics of tapioca starch dispersion during gelatinization using particle size analysis and rheological methods. M.S. thesis, Cornell University, Ithaca, NY.

Tiu, C., and D.V. Boger. 1974. Complete rheological characterization of time-dependent food products. *J Texture Studies* 5: 329–338.

Tiu, C., A.K. Podolsak, T.N. Fang, and J.B. Watkins. 1992. Rheological behavior of water-creosote and creosote-water emulsions. *Rheol Acta* 31: 381–389.

Tsai, S.C., and K. Zammouri. 1988. Role of interparticular van der Waals force in rheology of concentrated suspensions. *J Rheol* 32: 737–750.

Vitali, A.A., and M.A. Rao. 1984a. Flow properties of low-pulp concentrated orange juice: Effect of temperature and concentration. *J Food Sci* 49: 882–888.

Vitali, A.A., and M.A. Rao. 1984b. Flow properties of low-pulp concentrated orange juice: Serum viscosity and effect of pulp content. *J Food Sci* 49: 876–881.

Vocadlo, J.J., and M. Moo Young. 1969. Rheological properties of some commercially available fats. *Can Inst Food Technol J* 2: 137–140.

Weltman, R.N. 1943. Breakdown of thixotropic structure as a function of time. *J Appl Phys* 14: 343–350.

Wildemuth, C.R., and M.C. Williams. 1984. Viscosity of suspensions modeled with a shear-dependent maximum packing fraction. *Rheol Acta* 23: 627–635.

Yoo, B., A.A. Figueiredo, and M.A. Rao. 1994. Rheological properties of mesquite seed gum in steady and dynamic shear. *Lebensmittel Wissenschaft und Technologie* 27: 151–157.

Yoo, B., and M.A. Rao. 1994. Effect of unimodal particle size and pulp content on rheological properties of tomato puree. *J Texture Studies* 25: 421–436.

Yoo, B., and M.A. Rao. 1996. Creep and dynamic rheological behavior of tomato concentrates: Effect of concentration and finisher screen size. *J Texture Studies* 27: 451–459.

CHAPTER 3

Measurement of Flow and Viscoelastic Properties

M. Anandha Rao

Techniques for measuring rheological properties of polymeric materials have been well described previously by others (e.g., Whorlow, 1980; Macosko, 1994). The text by Van Wazer et al. (1963) is still a valuable reference that explains in detail many facets of earlier attempts to measure rheological properties of polymeric materials. The unique nature of fluid foods prompted this author to review both the rheological properties of fluid foods and their measurement about 20 years ago (Rao, 1977a, 1977b). More recent efforts on rheology of foods include those of Rao (1992, 1995) and Steffe (1996).

The study of the Newtonian and non-Newtonian flow behavior of foods requires considerable care and adequate instrumentation. Poorly designed instruments can provide data that can be misleading and of little value. A viscometer must be capable of providing readings that can be converted to shear rate ($\dot{\gamma}$, s^{-1}) and shear stress (σ, Pa). Further, a well-designed instrument should allow for the recording of the readings so that time-dependent flow behavior can be studied. The shear stresses that need to be measured in the case of low-viscosity foods are low in magnitudes, so instruments that minimize friction by the use of gas bearings are very useful for such foods.

Ideally, for viscosity measurement the flow in the selected geometry should be steady, laminar, and fully developed, and the temperature of the test fluid should be maintained uniform. For obtaining data under isothermal conditions, the temperature of the sample should be maintained constant at the desired level. For foods that exhibit Newtonian behavior, viscometers that operate at a single shear rate (e.g., glass capillary) are acceptable. For foods that exhibit non-Newtonian behavior, data should be obtained at several shear rates, and the commonly used viscometric flow geometries in rheological studies on foods include (1) concentric cylinder, (2) plate and cone (cone-plate), (3) parallel disk (also called parallel plate), (4) capillary/tube/pipe, and (5) slit flow. The equations of continuity and motion can be applied to the aforementioned well-defined geometries. As shown later, a single cylinder rotating in an infinite fluid can also be used (Rao, 1995), if the experiments are conducted properly.

The concepts of laminar and fully developed flow are frequently used in studies on flow and heat transfer. Both are best explained using flow in a straight tube as an example. In steady fully developed flow, the radial velocity profile does not change along the length of the tube. Laminar flow means that all fluid elements are flowing parallel to each other and that there is no bulk mixing between adjacent elements. The criterion for laminar flow of a Newtonian fluid, with density ρ and viscosity η, and average velocity $\bar{v}$ in a tube is that the Reynolds number (Equation 3.1), named after Osborne Reynolds, is < 2000:

$$\frac{D\bar{v}\rho}{\eta} < 2000 \tag{3.1}$$

Viscosity measurements are done under laminar flow conditions. Under conditions of turbulent flow of Newtonian fluids, the measured viscosity will be higher. In general, because non-Newtonian fluid foods are highly viscous, usually laminar flows are encountered.

A priori estimation of applicable shear rates for food processing application is also not easy because they depend upon both the properties of the food and the shearing conditions of a specific operation. Often, order of magnitude estimates of the shear rates are made, assuming Newtonian flow behavior and that the process equipment is designed conservatively. Sestak et al. (1983) presented characteristic ranges of shear rates for a few operations that should be used with considerable caution (especially the very high shear rate values): coagulation tanks with agitators, 2–20 s^{-1}; helical channel extruder, 1,000–3,000 s^{-1}; and filling machine nozzle, 30–3,000 s^{-1}. For flow in pipes of circular cross section, assuming laminar flow far from fittings, the shear rates for Newtonian fluids can be estimated from the expression $32Q/\pi D^3$, where Q is the volumetric flow rate ($m^3\ s^{-1}$) of the food and D is the internal diameter of the tube (m). However, if the flow properties are being measured to correlate with human assessment of quality, Shama and Sherman (1973) showed that viscous foods are manipulated in the mouth at shear rates of about 10 to 30 s^{-1}, while low-viscosity foods are subjected to a shear stress of about 10 Pa that result in shear rates of about 100 to 200 s^{-1}.

The relationships for shear stress in capillary/tube and slit geometries can be deduced relatively easily from the pressure drop over a fixed length after fully developed flow has been achieved. Likewise, in rotational viscometer geometries (concentric cylinder, cone-plate, parallel disk), shear stress can be calculated from the measured torque and the dimensions of the test geometry being used. In contrast to shear stress, the derivation of expressions for the shear rate requires solution of the continuity and momentum equations with the applicable boundary conditions. The derivation of equations for the shear rates in capillary, concentric cylinder, and cone-plate geometries is outlined in Appendixes 3–A through 3–C. It will be shown that in a properly designed cone-plate geometry, the shear rate depends only on the rotational speed and not on the geometrical characteristics. In all other flow geometries, the dimensions of the measuring geometry (capillary, concentric cylinder, parallel disks) play important roles.

ROTATIONAL VISCOMETERS

In a rotational viscometer, the shear rate is derived from the rotational speed of a cylinder or a cone. Often, considerable effort is spent on obtaining low-shear viscosity data, such as zero-shear viscosity, that are difficult to obtain in view of the low shear rates needed (e.g., $< 0.01\ s^{-1}$). The maximum rotational speed that is used is left to the discretion of the operator. If the flow properties of the food are needed to design for processing operations, it is necessary to use shear rates that span the range expected to be used in the process, and these may be quite high.

Concentric Cylinder Viscometer

In this geometry, a cylinder (bob) is placed concentrically (coaxially) inside a cup containing a selected volume of the test fluid (Figure 3–1). It is easy to imagine that the process of loading a sample into the cup and insertion of the bob will either destroy or alter the structure of the test fluid. Also, it may require considerable effort to insert a bob into a cup containing a very high viscosity material, for example, a dough. The bob is rotated and the drag of the fluid on the bob is measured by means of a torque sensor. In the Coette geometry, the cup is rotated and the main advantage is that higher shear rates can be obtained prior to the onset of turbulence due to Taylor vortices. By changing the rotational speed (shear rate) and measuring the resulting shear stress, it is possible to obtain viscosity data over a wide range of shearing conditions. In automated viscometers, the bob is programmed for an increase-in-speed ramp up to a predetermined shear rate and a decrease-in-speed ramp back to zero speed. In controlled-stress instruments, the variable controlled is the stress (torque), and the rotational speed is the response measured. Today, many rotational viscometers are available commercially whose data collection function and calculation of shear rate–shear stress data and parameters of rheological models (e.g., power law model) are performed

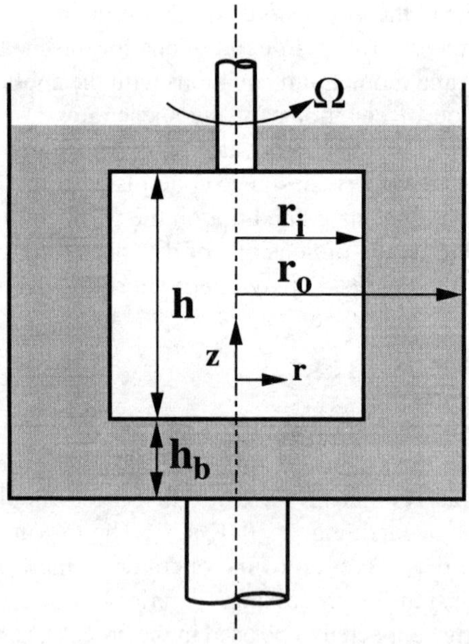

Figure 3–1 Schematic Diagram of a Concentric Cylinder Geometry.

by dedicated computers using software provided by the viscometer manufacturers. Nevertheless, it is instructive to know the applicable equations for the different geometries.

In a concentric cylinder geometry, the shear stress can be determined from the total torque (M):

$$\sigma = \frac{M}{2\pi r_i^2 h} \tag{3.2}$$

where r_i is the radius of the rotating bob.

Newtonian Shear Rate

The Newtonian shear rate in a concentric cylinder geometry can be calculated exactly from the following expression (see Appendix 3–A):

$$\dot{\gamma}_N = \frac{2\Omega r_o^2}{\left[r_o^2 - r_i^2\right]} = \frac{2\Omega}{\left[1 - \left(\frac{r_i}{r_o}\right)^2\right]} \tag{3.3}$$

where Ω is the angular velocity of the rotating bob, r_i is the radius of the bob, and r_o is the radius of the cup. Recalling that viscosity is obtained by dividing the magnitude of shear stress by the corresponding shear rate, the viscosity of a Newtonian fluid from concentric cylinder flow data is given by the Margules equation:

$$\eta = \left(\frac{M}{4\pi h\Omega}\right)\left(\frac{1}{r_i^2} - \frac{1}{r_o^2}\right) \tag{3.4}$$

where η is viscosity, M is the torque on the bob, Ω is the angular velocity of the rotating bob, h is the length of the bob in contact with the fluid, r_i is the radius of the bob, and r_o is the radius of the cup. For a given instrument with a given geometry this equation reduces to

$$\eta = C\,M/\Omega \tag{3.5}$$

where C is the instrument constant $= (1/4\pi h)\,(1/r_i^2 - 1/r_o^2)$. The instrument constant provided by manufacturers applies to Newtonian fluids and may not be correct for non-Newtonian fluids. Although viscometer manufacturers usually provide the torque values for each bob that accompanies their instrument, the validity of the values should be checked (e.g., by suspending weights from a bob positioned horizontally) periodically due to changes in response of torque transducers.

Krieger Method

In contrast to a cone-and-plate geometry, to be discussed next, the shear rate of non-Newtonian foods cannot be determined from a simple expression involving the angular velocity, and often one must use a suitable relationship between rotational speed and shear stress to correct for non-Newtonian behavior. More complex equations are needed to describe the flow of non-Newtonian fluids in concentric cylinder geometry. For example, for fluids that can be described by the power law model, an expression presented by Krieger and Elrod (Van Wazer et al., 1963) has been used extensively in the literature:

$$\dot{\gamma}_i = \frac{\Omega}{\ln\varepsilon}\left[1 + \ln\varepsilon\,\frac{d(\ln\Omega)}{d(\ln\sigma_i)} + \frac{(\ln\varepsilon)^2 d^2\Omega}{3\Omega d(\ln\sigma_i)^2}\right] \tag{3.6}$$

where Ω is the angular velocity of the rotating bob, $\dot{\gamma}_i$ is the shear rate at the bob, σ_i is the shear stress at the bob, and $\varepsilon =$ cup radius/bob radius.

Power Law Fluids

For fluids that can be described by the power law model, an expression that provides the influence of both the deviation from Newtonian behavior of the fluid and that of the cup to bob ratio is (McKelvey, 1962; Brodkey, 1967)

$$\dot{\gamma} = \frac{2\Omega}{\left[1 - \left(\frac{r_i}{r_o}\right)^2\right]} \left\{ \frac{1 - \left(\frac{r_i}{r_o}\right)^2}{n\left[1 - \left(\frac{r_i}{r_o}\right)^{2/n}\right]} \right\} \tag{3.7}$$

From Equation 3.7 it can be deduced (Table 3–1) that for non-Newtonian foods the correction term to Newtonian shear rate depends on the extent to which a fluid deviates from Newtonian behavior and the size of gap between the inner and outer cylinders. To minimize errors in the calculated shear rates, it would be preferable to employ concentric cylinders with a narrow gap between them. Therefore, some of the commercially available units have the ratio of the radii (r_i/r_o) of about 0.95. However, it is not obvious that the software provided by viscometer manufacturers contains corrections for non-Newtonian flow behavior, so, whenever possible, it is advisable to use narrow gap concentric cylinder systems for fluids that deviate considerably from Newtonian behavior.

For a single cylinder rotating in an infinite fluid (e.g., large container), assuming the shear rate far from a rotating cylinder ($\dot{\gamma}_o$) is zero in Equation 3A.12 (Ap-

Table 3–1 Values of Shear Rate Correction Factors for Concentric Cylinder Geometry

(r_i/r_o)	$XX = \left\{1 - \left(\frac{r_i}{r_o}\right)^2\right\}$	*Value of* n	$YY = n\left\{1 - \left(\frac{r_i}{r_o}\right)^{2/n}\right\}$	*Correction Factor* $\frac{XX}{YY}$
0.95	0.0975	0.20	0.0803	1.215
0.95	0.0975	0.50	0.0927	1.051
0.95	0.0975	1.00	0.0975	1.000
0.95	0.0975	1.20	0.0983	0.992
0.95	0.0975	2.00	0.1000	0.975
0.75	0.4375	0.20	0.1887	2.318
0.75	0.4375	0.50	0.3418	1.280
0.75	0.4375	1.00	0.4375	1.000
0.75	0.4375	1.20	0.4571	0.957
0.75	0.4375	2.00	0.5000	0.875
0.50	0.7500	0.20	0.1998	3.754
0.50	0.7500	0.50	0.4688	1.600
0.50	0.7500	1.00	0.7500	1.000
0.50	0.7500	1.20	0.8220	0.912
0.50	0.7500	2.00	1.0000	0.750

pendix 3–A), the expression for the shear rate is

$$2\sigma_i \frac{d\Omega_i}{d\sigma_i} = \dot{\gamma}_i \tag{3.8}$$

Equation 3.8 can be used when only a single cylinder system is available instead of a concentric cylinder system, but care should be exercised in making sure that the sample characteristics are not changed during collection of the experimental data, such as settling of solids of a dispersion. For a test sample, angular velocity versus shear stress data should be obtained, from which the magnitude of $(d\Omega_i/d\sigma_i)$ is calculated and substituted in Equation 3.8.

Other expressions for concentric cylinder geometry include that for Bingham plastic fluids where the yield stress must be taken into account, which leads to the Reiner-Riwlin equation:

$$\Omega = \frac{M}{4\pi h\eta'}\left(\frac{1}{r_i^2} - \frac{1}{r_o^2}\right) + \frac{\sigma_0}{\eta'}\, ln\, \frac{r_o}{r_i} \tag{3.9}$$

where σ_0 = yield stress and η' is the plastic viscosity. Equations for the shear rate at the rotating bob can be also derived for fluids that follow other rheological models.

Steiner's Method

For foods such as chocolates that can be described by the Casson model (Equation 2.6), Steiner (1958) chose not to develop an explicit expression for the non-Newtonian shear rate but related $\dot{\gamma}_N$ to the Casson model parameters K_{0c}, K_c, and shear stress. Steiner's approach is valid for values of $a = (r_i/r_o)$ between 0.5 and 0.9 and $\dot{\gamma}_N$ values greater than 0.1 s^{-1} when $a = 0.9$ and 0.01 s^{-1} when $a = 0.5$.

$$\{(1 + a)\sqrt{\sigma}\} = K_c\{(1 + a)\sqrt{\dot{\gamma}_N}\} + 2K_{0c} \tag{3.10}$$

The limited restrictions on the ranges of shear rate and the values of the ratio (r_i/r_o) for using Equation 3.10 are due to retaining only the first term in the series expansion of ln (r_i/r_o) and neglecting the higher order terms. From a plot $(1 + a)\sqrt{\dot{\gamma}_N}$ (abscissa) versus $(1 + a)\sqrt{\sigma}$ (ordinate) (Figure 3–2), where $a = (r_i/r_o)$, the Casson model parameters can be calculated from the slope (m) $K_c = m$ and intercept (b) $K_{0c} = b/2$. It will be recalled that the Casson yield stress is calculated as $\sigma_{0c} = (K_{0c})^2$ and the Casson plastic viscosity as $\eta_{Ca} = (K_c)^2$. Values of σ_{0c} and η_{Ca} (Table 3–2) of two chocolate samples whose viscosity data were obtained by Steiner (1958) with two concentric cylinder systems—MVI $(r_i/r_o) = 0.95$ and MVIII $(r_i/r_o) = 0.72$—were determined by plotting $(1 + a)\sqrt{\dot{\gamma}_N}$ versus $(1 + a)\sqrt{\sigma}$ (Figure 3–2). Magnitudes of the Casson parameters of the two chocolate

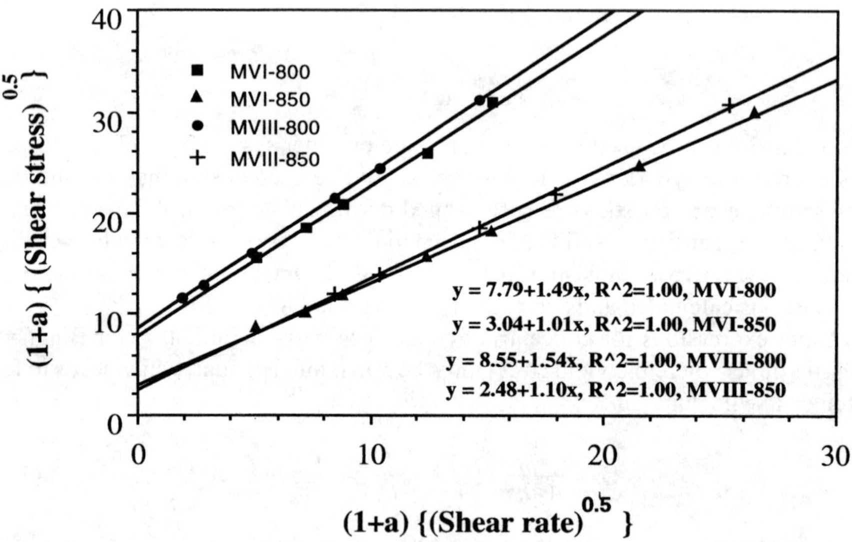

Figure 3–2 Steiner Plot for Flow Data on a Chocolate Sample. Data points shown were recalculated from original data of Steiner (1958). MVI and MVII indicate concentric cylinders with different dimensions, and 800 and 850 indicate different chocolate samples. Steiner, 1958.

Table 3–2 Casson Yield Stress ($\sigma_{0c} = (K_{0c})^2$) and Plastic Viscosity ($\eta_{Ca} = (K_c)^2$) Values of Chocolate Samples with Couvertures 800 and 850 Determined with Two Rotovisco Concentric Cylinder Systems: MVI (r_i/r_o) = 0.95 and MVIII (r_i/r_o) = 0.72. Steiner, 1958.

Measuring System and Sample	*Intercept* $= 2\,K_{0c}$	σ_{0c} *(Pa)*	*Slope* $= K_c$	η_{Ca} *Pa s*
MVI, 800	7.79	15.2	1.49	2.22
MVI, 850	3.04	2.3	1.01	1.02
MVIII, 800	8.55	18.3	1.54	2.37
MVIII, 850	2.48	1.5	1.10	1.21

samples determined with the two measuring systems are in reasonable agreement with each other. Steiner's approach to fluids that follow the Casson model is yet another example of the need to correct Newtonian shear rate data obtained with concentric cylinder systems.

A number of concentric cylinder design factors affect the rheological data obtained and these are discussed next.

End Effects in Concentric Cylinder Systems

In Equation 3.2, the measured torque should be due only to the lateral surface ($2\pi r_i h$) of a vertical cylinder. Therefore, any significant drag on the ends of the rotating cylinder should be either minimized by design, for example by providing an interface of air (Figure 3–3), or corrected for the influence of the end. The correction for the drag on the bottom as the equivalent height may be calculated by determination of the torque exerted by the rotating cylinder at different fill heights and extrapolation of the torque to zero height. This procedure is valid for a Newtonian fluid. Assuming torsional flow between parallel disks (Macosko, 1994), the extra torque (M_{bb}) contributed by the end can be estimated:

$$M_{bb} = \frac{2\pi r_i^3 (1/n)}{3+n} \left(\frac{\Omega_i}{h_b}\right)^n \tag{3.11}$$

where h_b is the gap between the bottom of the cylinder (Figure 3–1) and the cup, and n is the power law flow behavior index that can be obtained from a plot of log Ω versus log M. The total torque due to the lateral surface of the bob and the bob's end is $M + M_{bb}$.

Another approach of accounting for end effect, called the Mooney-Ewart approach (Whorlow, 1980), is to replace the end with a cone whose influence can be calculated (Figure 3–4). Ignoring the discontinuity where the conical and cylindrical surfaces meet, the total torque is given in terms of the arithmetic mean stress, σ_m:

$$M = 2\pi r_{eq}^2 \left(h + \frac{r_{eq}}{3 \sin \alpha}\right) \sigma_m \tag{3.12}$$

The stress σ_m occurs at a radius r_{eq}:

$$r_{eq}^2 = \frac{2\, r_i^2\, r_o^2}{r_i^2 + r_o^2} \tag{3.13}$$

Double Concentric Cylinder System

For low-viscosity fluids, one manner of obtaining the torque (stress) value is to increase the area of shear as shown in Figure 3–5. Compared to the geometry shown in Figure 3–1, in view of the thin wall of the inner compo-

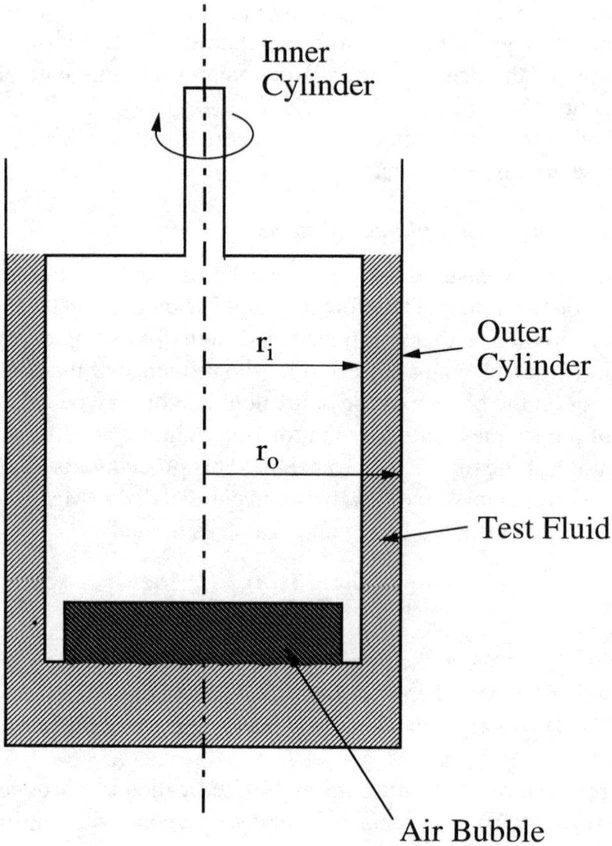

Figure 3–3 Schematic of a Concentric Cylinder Geometry Air Bubble To Minimize Shear at the Bottom of a Rotating Cylinder.

---

nent, the test sample is disturbed much less when the double concentric cylinder geometry is assembled for a rheological test, and the end effects are negligible. The design is such that $(r_1/r_2) = (r_3/r_4)$. For the geometry shown in Figure 3–5, the expression for the shear stress, analogous to Equation 3.2 is

$$\sigma = \frac{M}{2\pi h(r_2^2 + r_3^2)} \tag{3.14}$$

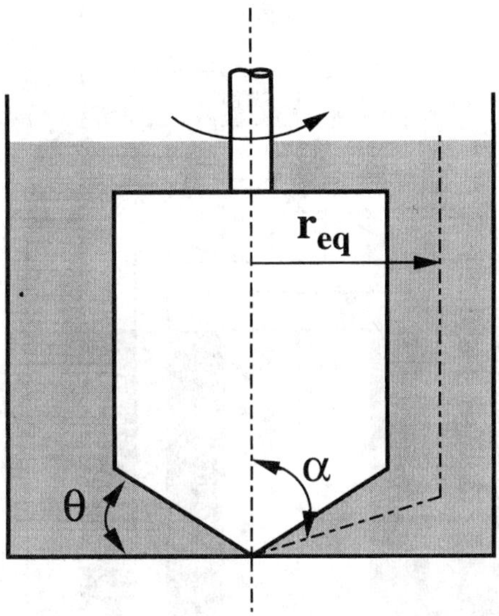

Figure 3–4 Mooney-Ewart Geometry To Account for End Effect in a Concentric Cylinder Geometry.

Recognizing that the fluid is sheared on the surfaces with radii r_2 and r_3, one can write an equation analogous to Equation 3.3:

$$\dot{\gamma}_N = 2\Omega\left[\frac{r_1}{1-\left(\frac{r_1}{r_2}\right)^2}+\frac{r_1}{1-\left(\frac{r_3}{r_4}\right)^2}\right] \tag{3.15}$$

Slip in Concentric Cylinder Systems

In some foods, a thin layer of low-viscosity fluid forms at the solid-fluid interface, which in turn contributes to lower viscosity values. The boundary condition that at the solid-fluid interface the fluid velocity is that of the wall is not satisfied. This phenomenon is known as slip effect. Mooney (1931) outlined the procedures for the quantitative determination of slip coefficients in capillary flow and in a Coette system. The development for the concentric cylinder system will be outlined here for the case of the bob rotating, and details of the derivation can be found in Mooney (1931).

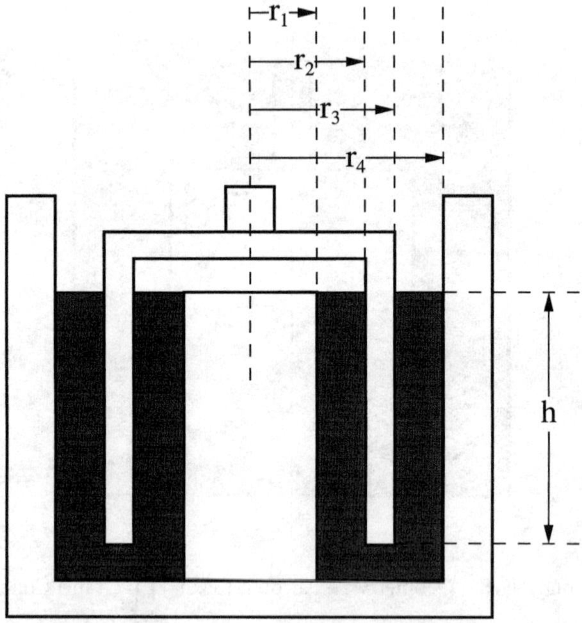

Figure 3–5 Double Concentric Cylinder Geometry for Low-Viscosity Fluids.

For a concentric cylinder system, the differential equation for angular velocity can be written as

$$\frac{d\omega}{dr} = \frac{T\varphi}{2\pi r^3} \tag{3.16}$$

Integration from the inner to the outer cylinder gives

$$\Omega = \frac{\beta(\sigma_1)T}{2\pi r_1^3} + \frac{\beta(\sigma_2)T}{2\pi r_2^3} + \int_{r_1}^{r_2} \left(\frac{T\varphi}{2\pi r^3}\right) dr \tag{3.17}$$

In the above equations, ω and Ω are the angular velocity of liquid and cylinder, respectively; β is the slip coefficient (defined as β = slip velocity/shearing stress); φ is the fluidity, and T is the torque per unit height ($M\,h^{-1}$).

In controlled-shear rate viscometers such as the Haake Rotovisco, T is determined as a function of Ω, while in constant stress rheometers, such as the Carri-Med, Ω is determined as a function of T. To determine surface slip in a coaxial cylinder viscometer, one must vary the ratio r_1/r_2 as well as T and Ω (Mooney, 1931). Specifically, one can determine Ω at constant T using three combinations of cups and bobs of radii r_2/r_1, r_3/r_2, and r_3/r_1. The coefficient of slip is given by

$$\beta = \frac{2\pi r_2^3}{T}(N_{12} + N_{23} - N_{13}) \tag{3.18}$$

where N is the rotational speed for the appropriate concentric cylinder system. When the bob is rotating, as is the case in many rheometers, the slip velocities for the cup (v'_{cup}) and bob (v'_{bob}) are given by the following equations:

$$v'_{cup} = \beta\left[\frac{T}{2\pi r_{cup}^2\, L}\right] \tag{3.19}$$

$$v'_{bob} = r_{bob}\,\Omega - \beta\left[\frac{T}{2\pi r_{bob}^2\, L}\right] \tag{3.20}$$

The velocity difference across the test material is $(v'_{bob} - v'_{cup})$ and the ratio of this velocity to the velocity difference without slip $[(v'_{bob} - v'_{cup})/v_{bob}]$ is the factor used to correct the shear rate. It must be noted that Mooney (1931) also suggested that one may use only two concentric cylinder systems, and this suggestion was interpreted in a quantitative manner by Yoshimura and Prud'homme (1988a), who derived the necessary equations for such systems.

To estimate slip velocities, the rotational speeds at specific values of torque for the three concentric cylinder units (Figure 3–6) specified in the theory of Mooney (1931) were determined (Qiu and Rao, 1989; Grikshtas and Rao, 1993). The magnitudes of the radii of the three concentric cylinder units r_1, r_2, and r_3 were 18.4, 20.04, and 21.0 mm, respectively. Experiments were conducted so that the rotational speeds N_{12}, N_{13}, and N_{23} were determined at the same magnitude of torque, where the subscripts 12, 13, and 23 denote the appropriate combination of cup and bob; for example, the subscript 12 denotes a bob of radius r_1 and a cup of radius r_2. The slip coefficient (β) and the slip velocity ratio (v-ratio) were calculated using Equations 3.18, 3.19, and 3.20. The shear rates corrected for slip were calculated as the product of the v-ratio and the uncorrected for slip shear rates calculated using Krieger and Elrod's equation (Equation 3.6).

Grikshtas and Rao (1993) showed that Kiljanski's method for estimating slip velocities using two concentric cylinder units gave results comparable to that of Mooney. Yoshimura and Prud'homme (1988a) described procedures for estimating slip velocities in a parallel plate geometry.

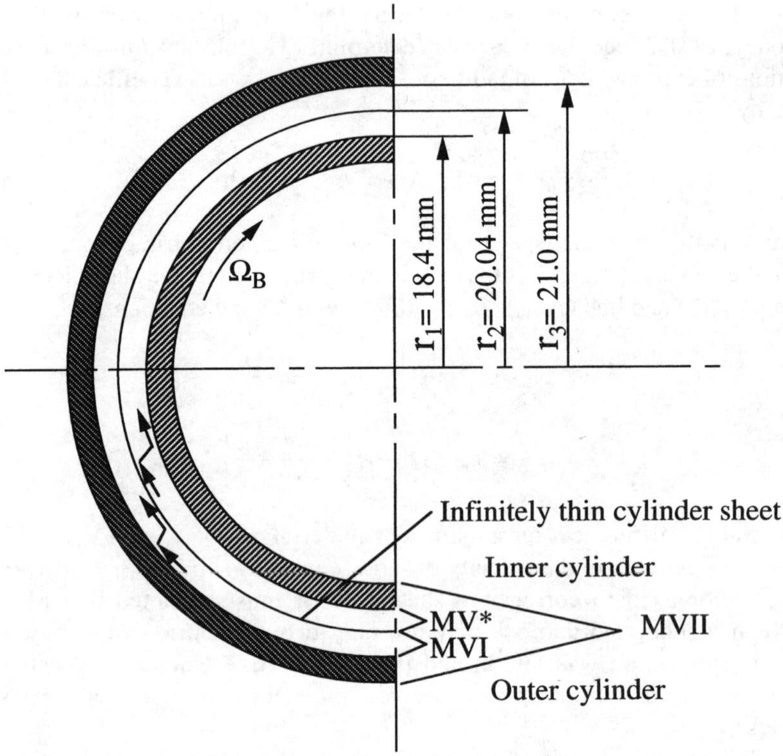

Figure 3–6 Diameters of Concentric Cylinders Used in a Study of Slip Velocities Using the Method of Mooney. *Source*: Reprinted with permission from C-G. Qiu and M.A. Rao, Effect of Dispersed Phase on the Slip Coefficient of Apple Sauce in a Concentric Cylinder Viscometer, *Journal of Texture Studies*, Vol. 20, pp. 57–70. © 1989, Food and Nutrition Press.

Turbulent Flow

As stated earlier, rheological measurements should be conducted under laminar flow conditions and turbulent flow conditions will result in higher viscosities. Thus, a sudden increase in viscosity with increase in rotational speed may be due to the onset of turbulent flow and may be erroneously attributed to shear-thickening behavior of the test fluid. In the design of concentric cylinder systems, one has the option of rotating either the inner or the outer cylinder. In this respect, G.I. Taylor showed that when the Reynolds number reaches a certain value, vortices appear whose axes are located along the circumference, and when the outer cylinder is rotated the vortices appear at higher rotational speeds (Schlichting, 1960). An equation for predicting the onset of vortex formation for Newtonian fluids, credited to Ludwig Prandtl, is (Schlichting, 1960)

$$\frac{v_i w}{\nu}\sqrt{\frac{w}{r_i}} > 41.3 \tag{3.21}$$

where w is the width of the gap, v_i is the peripheral velocity of the inner cylinder, r_i is the radius of the inner cylinder, and ν is the kinematic viscosity of the fluid ($\nu = \eta/\rho$).

Plate-and-Cone Geometry

The apex of a cone is brought into close proximity of a horizontal plate (Figure 3–7). Often the apex is truncated slightly to eliminate a sharp point. The minimum gap between the cone and plate is usually of the order of 50 μm, so this geometry may not be suitable for dispersions containing larger diameter solids. The test

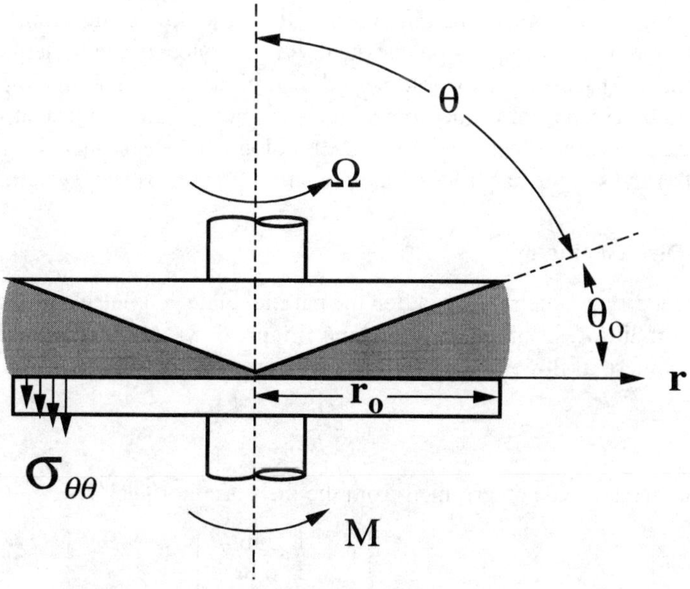

Figure 3–7 Schematic Diagram of a Cone-and-Plate Geometry.

fluid fills the gap between the cone and the plate, and because the gap is small, only a small volume (typically, 1–5 mL) of fluid is needed. The cone is rotated, and the torque is measured at various speeds of rotation. A cone-and-plate viscometer can be used to obtain shear stress–shear rate curves and shear stress versus time at constant shear rate curves as described above for concentric cylinder geometry. The shear rate and shear stress are given by the following equations:

$$\text{Shear stress, } \sigma_{\theta\phi} = \frac{3T_{\text{cn}}}{D} \tag{3.22}$$

$$\text{Shear rate, } \dot{\gamma} = \frac{\Omega}{\theta_{\text{o}}} \tag{3.23}$$

where T_{cn} is the torque per unit area, D is the diameter of the rotating cone or plate, Ω is the angular velocity, and θ_{o} is the cone angle in radians. In general θ_{o} is usually quite small (2°–4°). It is readily apparent from Equation 3.23 that the magnitude of $\dot{\gamma}$ depends only on the angular velocity and the cone angle, so it is uniform in a cone-plate geometry.

The concentric cylinder and the cone-plate geometries together with low-inertia rheometers are well suited for studying time-dependent viscous (e.g., thixotropic and antithixotropic) and viscoelastic (stress growth, creep, and dynamic) properties of foods. The alteration of the structure during sample loading is a concern in both geometries but especially if the concentric cylinder is inserted subsequently. At least a reasonable period of time for recovery of sample structure should be given prior to rheological measurements. With some samples (e.g., gelatinized starch and other food gels), the rheological measurements can be conducted after the sample has been created in situ in the measuring system.

Parallel Disk Geometry

The parallel disk geometry, also called the parallel plate geometry, consists of two disks with radius r_{o} separated by the gap h (Figure 3–8). Assuming steady, laminar, and isothermal flow, the expression for shear rate is

$$\dot{\gamma} = \frac{\Omega r_{\text{o}}}{h} \tag{3.24}$$

The shear stress can be determined from the measured torque, M:

$$\sigma = \frac{3M}{2\pi r_{\text{o}}^3}\left[1 + \frac{1}{3}\frac{\text{d}\ln M}{\text{d}\ln \dot{\gamma}}\right] \tag{3.25}$$

where Ω is the angular velocity, r_{o} is the disk radius, and h is the gap between the disks. Although the shear field is nonhomogeneous, the parallel disk geometry is

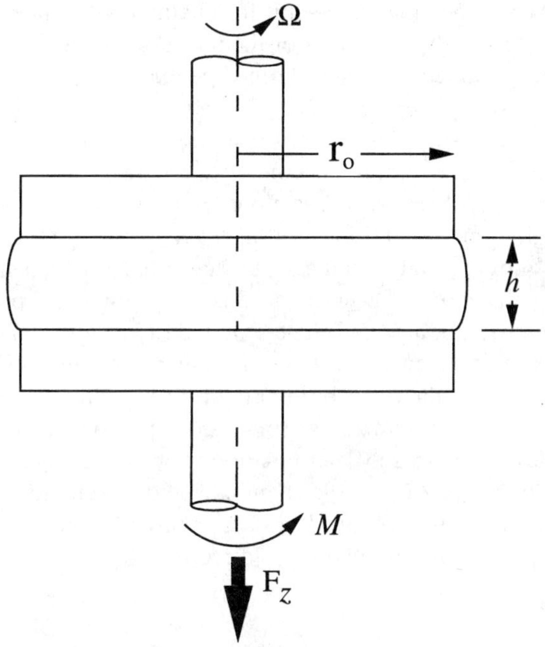

Figure 3–8 Schematic Diagram of a Parallel Disk Geometry.

useful in handling dispersions that contain relatively large size particles, such as gelatinized starch dispersions. Yang and Rao (1998) used a parallel disk geometry to obtain rheological data during gelatinization of a corn starch dispersion. They found that the paraffin oil placed at the edge of the starch sample did not penetrate into the starch. The gap, *h*, between the disks can be selected to accommodate foods containing large particles, and a rough guide for a suitable gap width is for $h \approx 10 \times$ particle diameter.

Mixer Viscometer

Some rotational viscometers employ a rotating disk, bar, paddle, or pin at a constant speed (or series of constant speeds). It is extremely difficult to obtain true shear stress, and the shear rate usually varies from point to point in the rotating member. In particular, the velocity field of a rotating disk geometry can be considerably distorted in viscoelastic fluids. Nevertheless, because they are simple to operate and give results easily, and their cost is low, they are widely used in the food industry. While they may be useful for quality control purposes, especially with Newtonian foods, the reliability of their values should be verified by comparison with data obtained with well-defined geometries (capillary/tube, concentric cylinder, and cone-plate).

Many foods are suspensions of solid matter in a continuous medium. In concentric cylinder and plate-and-cone viscometers, the solid matter may be larger than the gap between the stationary and the rotating parts, or it may settle due to agitation. In addition, a thin layer of the continuous phase may separate at the moving and stationary surfaces, and this can induce errors in the measured values of shear rate. Foods that can pose problems during measurement with rotational viscometers include mustard, tomato puree, apple sauce, and other suspensions. For these and other complex foods, approximate shear rate–shear stress data can be obtained by a technique using a rotating vane (Figure 3–9) that minimizes settling and separation of the product. The technique was developed for the determination of power consumption during mixing of fluids (Metzner and Otto, 1957) and adapted to studying rheology of fermentation broths (Bongenaar et al., 1973). It is based on the assumption that the average shear rate around the paddle is directly proportional to the rotational speed (N):

$$\dot{\gamma} = k_s N \tag{3.26}$$

where k_s is a constant that must be determined for each paddle (Rao, 1975; Rao and Cooley, 1984) and N is rotational speed (revolutions s^{-1}) of the paddle. The constant k_s can be determined from a semilogarithmic plot of $(1 - n)$ (abscissa) versus log $[p/(KN^{n+1}d^3)]$ (ordinate) (Rieger and Novak, 1973). For a given impeller, tests must be conducted such that the necessary data are obtained: p, the power (Nm s^{-1}); N, the rotational speed (s^{-1}); d, the diameter of the impeller (m); and the power law parameters (K and n) of several test fluids so that a wide range of values of $(1-n)$ can be obtained. It can also be determined by matching the torque with a Newtonian fluid of a known viscosity to that with a non-Newtonian fluid (Metzner and Otto, 1957; Rao and Cooley, 1984). Wood and Goff (1973) applied the matching viscosity method to determine the effective shear rate in a Brabender Viscograph. The latter method was called (Rao and Cooley, 1984) the Metzner-Otto-Wood-Goff (MOWG) method. Castell-Perez et al. (1987) and Briggs and Steffe (1996) employed the two techniques described above for small samples.

For foods that follow the simple power law model, the flow behavior index is determined as the slope of a double logarithmic plot of log N (abscissa) versus log torque (ordinate). In order to determine the consistency index, magnitudes of torque must be determined in the same container for the test fluid and for a standard fluid whose rheological characteristics are known. Recognizing that the ra-

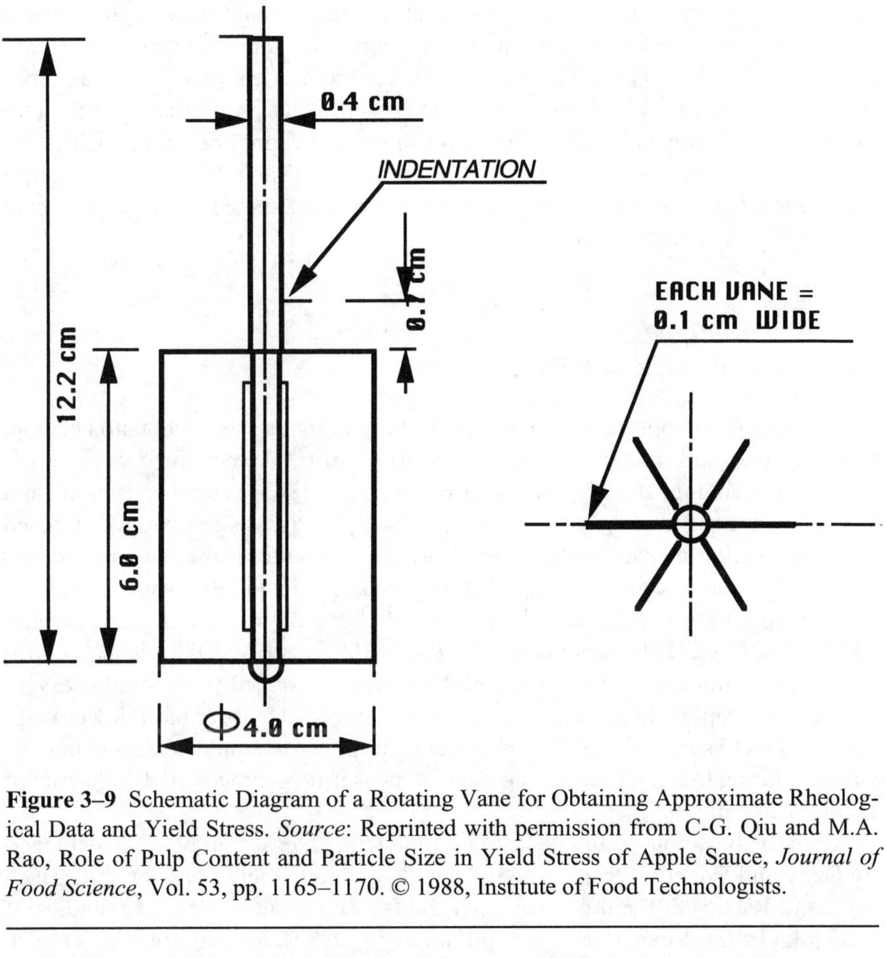

Figure 3–9 Schematic Diagram of a Rotating Vane for Obtaining Approximate Rheological Data and Yield Stress. *Source*: Reprinted with permission from C-G. Qiu and M.A. Rao, Role of Pulp Content and Particle Size in Yield Stress of Apple Sauce, *Journal of Food Science*, Vol. 53, pp. 1165–1170. © 1988, Institute of Food Technologists.

tios of the torques will be equal to the ratios of the shear stresses, one can write the algebraic expression for a test fluid (x) and the standard fluid (y):

$$\frac{M_x}{M_y} = \frac{\sigma_x}{\sigma_y} = \frac{K_x\,(k_s\,N_x)^{n_x}}{K_y\,(k_s\,N_y)^{n_y}} \tag{3.27}$$

where M_x and M_y are the magnitudes of torque measured with the test and the standard fluids, N_x and N_y are the rotational speeds employed with the test and standard fluids, K_x and K_y are the magnitudes of the consistency index of the test and the standard fluids, and n_x and n_y are the magnitudes of the flow behavior index of the test and the standard fluids. Because all the terms other than K_x, the consistency index of the test fluid, are known, it can be calculated. For the special case of a Newtonian standard fluid, $N_y = 1$. Studies are needed to verify the assumption that the average shear rate of the paddle is directly proportional to the rotational speed. However, in the interim, for foods that cannot be studied with conventional viscometers, the careful use of vanes will enable the determination of the power law parameters that in turn can be used to calculate shear rates and shear stresses from the relationship

$$\sigma = K\dot{\gamma}^n \tag{2.3}$$

Yield Stress of Foods with Vanes

Yield stress is an important rheological property in foods such as tomato ketchup, tomato sauce, and melted chocolate. Experimental measurement of yield by the relaxation technique described by Van Wazer et al. (1963) is time consuming and the values are affected by prior shearing of sample. Yield stress values obtained by extrapolation of shear stress–shear rate data depend on the model used and prior shearing of the sample, so different values of yield stress may be obtained for a given sample depending on the extent of alteration of structure. One manner of visualizing yield stress, shown in Figure 3–10 (Keentok, 1982), is that a sample with undisturbed structure has a high yield stress, called static yield stress. In contrast, a sample whose structure has been disturbed by shear has a lower magnitude of yield stress, called dynamic yield stress. For example, a sample may be sheared during loading of the sample into a measuring geometry and collection of rheological data.

A relatively simple technique using a vane with at least four blades, developed by Dzuy and Boger (1983), was employed to examine yield stress of applesauce (Qiu and Rao, 1988) and other pureed foods (Yoo et al., 1995). The studies of Keentok (1982), Yoshimura and Prud'homme (1988b), Yoshimura et al. (1987), and James et al. (1987) also contain useful discussion on yield stress. Figure 3–11 illustrates the experimental setup used by Qiu and Rao (1988). In this technique, the maximum torque reading is determined at a low rotational speed (e.g., 0.4 rpm) (Figure 3–12). The yield stress (σ_{0v}) is then calculated from the equation

$$T_m = \frac{\pi D_v^3}{2}\left(\frac{H}{D_v} + \frac{1}{3}\right)\sigma_{0_v} \tag{3.28}$$

where T_m is the maximum torque reading (N m), and D_v (m) is the diameter and H (m) the height of the paddle, respectively. The method is simple and consistent results were obtained for the studied applesauce samples (Qiu and Rao, 1988).

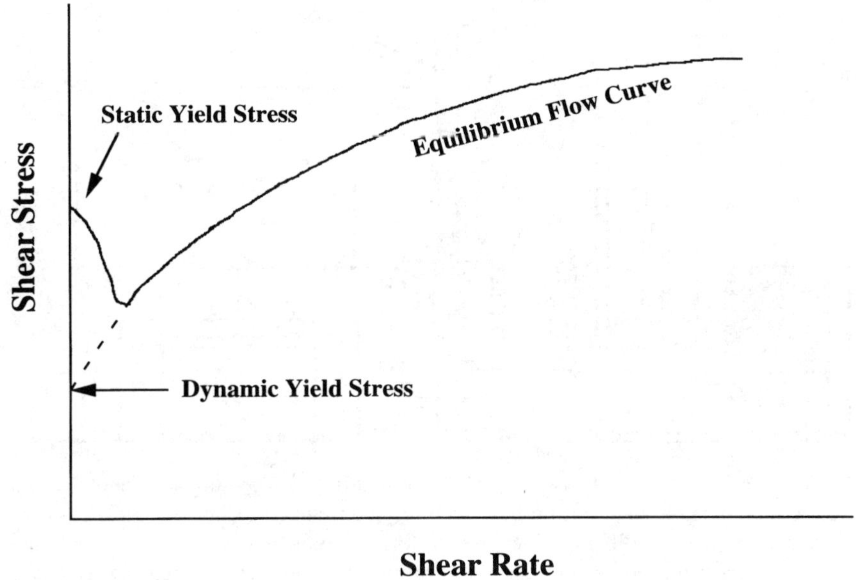

Figure 3–10 Illustration of Static and Dynamic Yield Stress. Keentok, 1982.

Further, in a study of the vane yield stress of tomato puree samples, a controlled shear rate viscometer and a controlled stress rheometer provided comparable values of yield stress (Yoo et al. 1995). However, accurate determination of the maximum torque with the controlled stress rheometer requires care because of the uncertainty in determining the representative value (Figure 3–13).

Helical Screw Rheometer

A helical screw enclosed in a tight-fitting barrel was suggested as an on-line viscometer for fluid food suspensions because it prevents settling or separation of particulates during measurement (Tamura et al., 1989). When the screw is rotated with the system closed, it generates a linearly increasing pressure differential along the axis of the screw, and the pressure is proportional to the viscosity of a Newtonian fluid. Comparable viscosity data were obtained with the helical screw rheometer (HSR) and a cone-and-plate viscometer on Newtonian and mildly non-Newtonian homogeneous fluids, but lower power law consistency index values were obtained for all suspensions. The time-dependent flow behavior of the foods

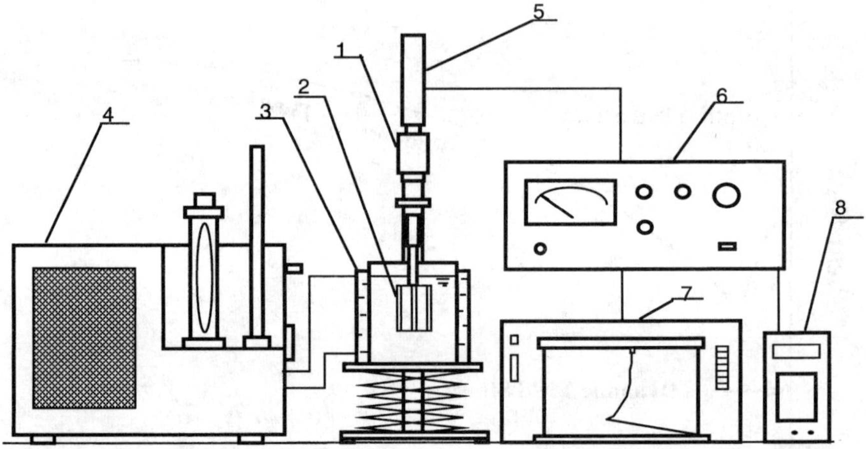

Figure 3–11 Vane Yield Stress Setup at Controlled Strain. The numbers denote separate pieces of equipment: 1—torque unit, 2—vane, 3—constant temperature vessel for sample, 4—constant temperature bath, 5—viscometer drive motor, 6—viscometer control panel, 7—chart recorder for torque output, and 8—voltmeter to monitor vane rpm signal. *Source*: Reprinted with permission from C-G. Qiu and M.A. Rao, Role of Pulp Content and Particle Size in Yield Stress of Apple Sauce, *Journal of Food Science*, Vol. 53, pp. 1165–1170. © 1988, Institute of Food Technologists.

studied and the greater structure breakdown in the HSR were suggested as the possible reasons for the discrepancy.

PRESSURE-DRIVEN FLOW VISCOMETERS

Capillary/Tube Viscometer

Viscosity measurements can also be conducted using nonrotational pressure-driven flow viscometers. For example, the glass capillary viscometer in which flow is induced by gravitational force is used extensively for obtaining data on low-viscosity Newtonian fluids. In addition, small and large diameter tube viscometers can be constructed so that one can obtain shear rate–shear stress data of non-Newtonian foods (Figure 3–14). The viscometers must be designed such that the pressure drop, Δp (Pa), over a known length of tube, L (m), and the corresponding volumetric flow rate, Q ($m^3\ s^{-1}$), of the food in laminar flow can be measured. Care must be exercised in designing the viscometer so that the Δp of food

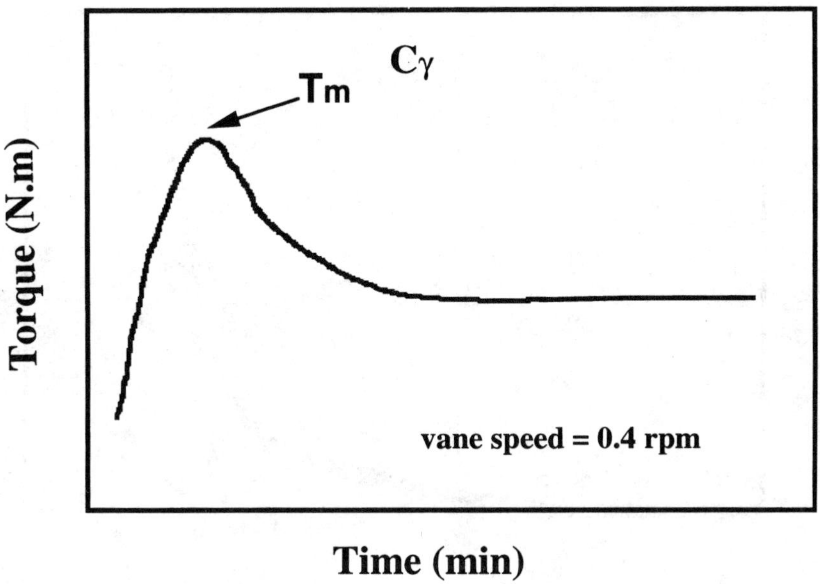

Figure 3–12 Torque versus Time Plot with a Vane in Controlled Strain Operation Where the Maximum Torque Is Used To Calculate the Yield Stress. *Source*: Reprinted with permission from Yoo et al., Yield Stress of Food Suspensions with the Vane Method at Controlled Shear Rate and Shear Stress, *Journal of Texture Studies*, Vol. 26, pp. 1–10. © 1995, Food and Nutrition Press.

flow is measured over a section L of the tube free from the influence of fittings and the disturbances at the entrance and exit of the tube; the latter are known as end effects and are due to the higher pressure drop in flows that are not fully developed.

In the study of Vitali and Rao (1982), an entrance length of $100 \times D$ and an exit length of $40 \times D$ were used, and the pressure drop was measured in fully developed flow; therefore corrections for end (entrance and exit) effects were avoided. When it is not possible to measure Δp over a length of fully developed tube flow, the excess pressure drop due to end effects can be corrected using a Bagley plot obtained by using tubes with different (L/D) ratios and plotting pressure drop values at constant shear rate against those of (L/D) (Figure 3–15). The correction (ζ) to (L/D) is obtained by extrapolating to zero pressure drop so that the effective length at a specific shear rate for the specific food material is $(L + \zeta D)$. Values of ζ from Figure 3–15 were 1.10, 2.14, 2.59, 2.69, and 2.87 at the shear rates 40, 60,

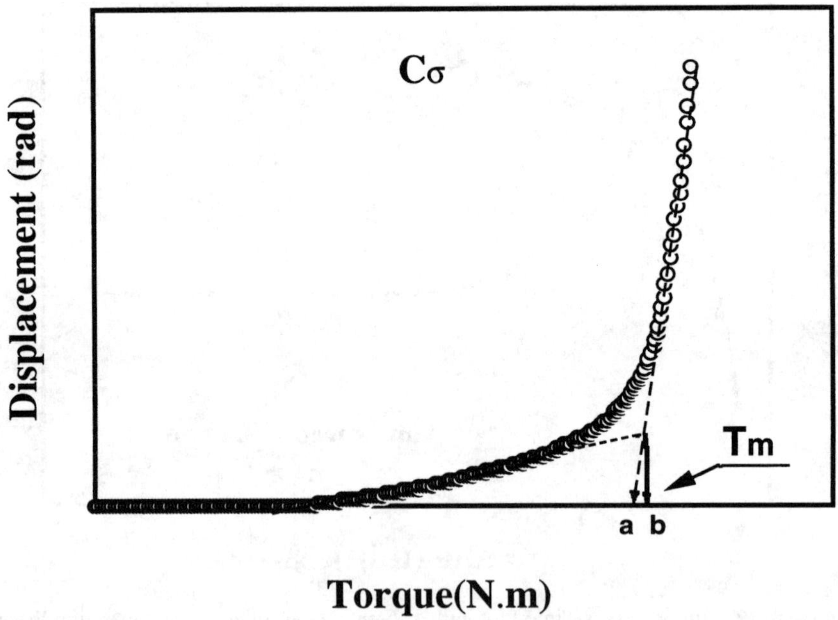

Figure 3–13 Torque versus Deformation Plot with a Vane in Controlled Stress Operation. *Source*: Reprinted with permission from Yoo et al., Yield Stress of Food Suspensions with the Vane Method at Controlled Shear Rate and Shear Stress, *Journal of Texture Studies*, Vol. 26, pp. 1–10. © 1995, Food and Nutrition Press.

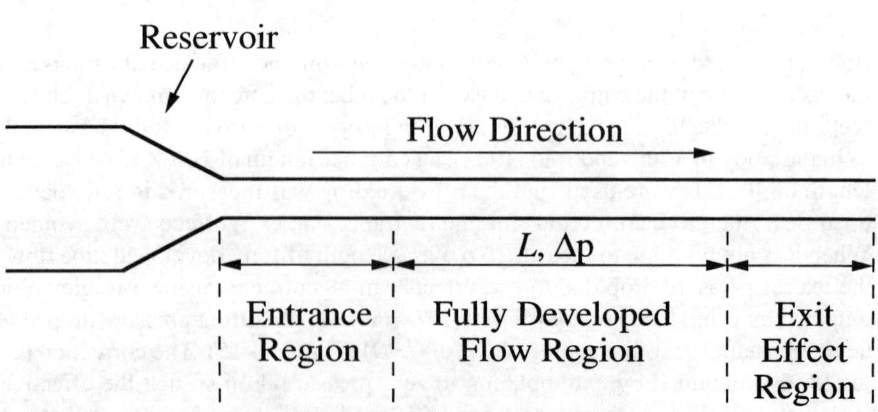

Figure 3–14 Tube Viscometer Design Considerations. Vitali and Rao, 1982.

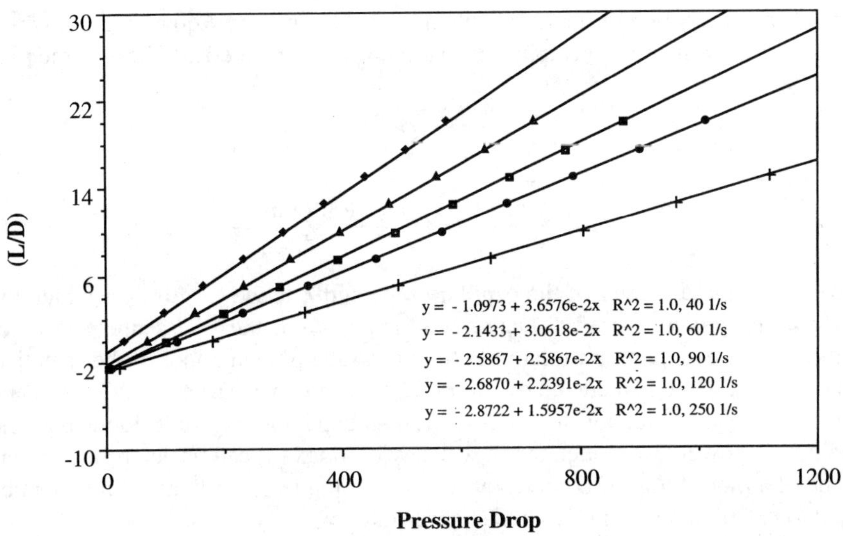

Figure 3–15 Illustration of Bagley End Correction Technique. Pressure drop units in pounds per square inch as in Brodkey (1967). *Source:* From *The Phenomena of Fluid Motions*, by R.S. Brodkey. Copyright © 1967, by Addison-Wesley Publishing Company, Inc. Reprinted by permission.

90, 120, and 250 s^{-1}. Another use for the data shown in Figure 3–15 is to estimate the entrance pressure drop, Δp_e.

Therefore, when pressure drop data over fully developed flow length cannot be obtained, the steps involved in obtaining shear rate–shear stress data can be summarized as follows: (1) Values of $(32\ Q/\pi D^3)$ are plotted against the corresponding Δp for each tube. (2) At selected values of $(32\ Q/\pi D^3)$, the corresponding pressures are obtained and plotted versus the (L/D) ratio of the tubes as in Figure 3–15. (3) The correction factor ζ is obtained from Figure 3–15 and the effective length at a specific shear rate is calculated as $(L + \zeta D)$. (4) From the corrected shear stress and $(32Q/\pi D^3)$, the shear rate–shear stress data can be obtained as discussed above. In the study of Jao et al. (1978) on soy dough rheology, the Bagley correction to end effects was used. Senouci and Smith (1988) found entrance pressure losses to be less than 10% of the total pressure drop in a slit die viscometer for maize grits; in contrast, for potato flour it was as large as 58%.

An outline of the steps involved to obtain the equations for shear rate and shear stress for fully developed flow in a tube is given in Appendix 3–B. The shear stress (σ_w) is given by Equation 3.29 and the shear rate by Equation 3.30, where the subscript w is used to emphasize that the values obtained are those at the pipe wall.

$$\sigma_w = \frac{D\,\Delta p}{4L} \tag{3.29}$$

$$\dot{\gamma}_w = \frac{3}{4}\left(\frac{32\,Q}{\pi D^3}\right) + \frac{\sigma_w}{4}\,\frac{d(32Q/\pi D^3)}{d\sigma_w} \tag{3.30}$$

where D is the diameter of the capillary/tube and L is the length over which the pressure drop is measured. Equation 3.30 is known as the Weissenberg-Rabinowitsch-Mooney equation for laminar flow in a tube/pipe in honor of the scientists who derived it and is made up of a term for Newtonian shear rate ($32Q/\pi D^3$) and a term that is a correction for non-Newtonian behavior. For fluid foods obeying the simple power law model, the flow behavior index (n) can be determined as the slope of a plot of log ($32Q/\pi D^3$) (abscissa) and log (σ_w) (ordinate), and it can be used to calculate the non-Newtonian shear rate ($\dot{\gamma}$):

$$\dot{\gamma} = \frac{3n+1}{4n}\left(\frac{32Q}{\pi D^3}\right) \tag{3.31}$$

One can now construct diagrams of shear rate versus shear stress for non-Newtonian foods. Vitali and Rao (1982) obtained volumetric flow versus pressure drop data given in Table 3–3, which were then converted to shear rate and shear stress values shown in Figure 3–16 and also listed in the table.

Capillary and tube viscometers are relatively inexpensive. However, jackets for temperature control and large reservoirs for holding food samples can increase their cost, and care must be exercised in analyzing the data obtained with them. As in the case of rotational viscometers, when food suspensions are being studied, the errors (slip effects) in the shear rates due to the formation of a thin layer of the continuous phase at the walls of the tube can be large. Much effort was spent on analyzing the potential errors in tube viscometers, and at least the reviews (e.g., Van Wazer et al., 1963; Rao, 1977a), if not the original references, should be consulted to appreciate the experiences of earlier workers. Capillary viscometers are not well suited for studying time-dependent rheological behavior because of the difficulty in measuring reliable values of time-dependent wall stresses.

Yield Stress

If one can determine accurately the minimum pressure required (Δp) to cause flow in a horizontal tube, then the yield stress of the material can be estimated from

$$\sigma_0 = \frac{D\,\Delta p_{\min}}{4L} \tag{3.32}$$

Table 3–3 Tube Viscometer Flow Data on Guava Puree 13.6° Brix, 58°C

Test No.	*Mercury Column cm*	*Q cm³ s⁻¹*	*Shear Stress Pa*	*Pseudo Shear Rate s⁻¹*	*True Shear Rate s⁻¹*
1	11.3	3.31	29.1	285	394
2	16.4	7.95	42.3	684	945
3	20.2	12.58	52.1	1080	1500
4	23.0	20.28	59.3	1750	2410
5	27.2	28.58	70.2	2460	3400
6	30.1	38.04	77.6	3270	4520
7	33.0	47.39	85.1	4080	5630
8	34.8	55.54	89.8	4780	6600

Sample Calculation, Test No. 5

The following steps illustrate calculation of shear stress and shear rate. Note that centimetergram-second units were used as in the original work (Vitali and Rao, 1982), but the final values are in SI units.

1. The pressure drop in cm of Hg was converted to Pa:

$$\Delta p = \rho\, g\, dh = 13.6 \text{ g cm}^{-3} \times 981 \text{ cm s}^{-2} \times 27.2 \text{ cm} = 3.63 \times 10^5 \text{ dyne cm}^{-2}$$

2. The shear stress at the wall,

$$\sigma_W = \frac{D\,\Delta p}{4L} = \frac{0.491 \times 3.63 \times 10^5}{4 \times 63.5} = 702 \text{ dyne cm}^{-2} = 70.2 \text{ Pa}$$

3. The pseudo shear rate was calculated using $\dot{\gamma}_N = (32Q/\pi D^3)$:

$$\dot{\gamma}_N = \frac{32 \times 28.58}{3.142 \times (0.491)^3} = 2.46 \times 10^3 \text{ s}^{-1}$$

4. The same procedures were used to calculate shear stresses and pseudo shear rates for all the data. A plot of log $\dot{\gamma}_N$ versus log σ_W shown in Figure 3–16 and from linear regression of the data the slope (n') and the intercept (K') were calculated to be 0.40 and 3.17 Pa s^n. Because $n' = n$, the power law flow behavior index, the consistency index K was evaluated from

$$K = \frac{K'}{(3n + 1/4n)^n} = [3.17/(1.38)^{0.4}] = 2.78 \text{ Pa s}^n$$

The structure of the food sample would be disturbed considerably during the determination of Δp_{min}, so the measured yield stress would be closer to the dynamic yield stress than the static yield stress (Figure 3–10). In contrast, in the vane method for determination of yield stress both the static and dynamic yield stresses can be determined.

Die Swell or Jet Expansion

In a highly elastic material with a high first normal stress difference ($\sigma_{11} - \sigma_{22}$), the material expands to a larger diameter, known as die swell or jet expansion. The diameter of the expanded jet (D_1) may be estimated from an equation developed

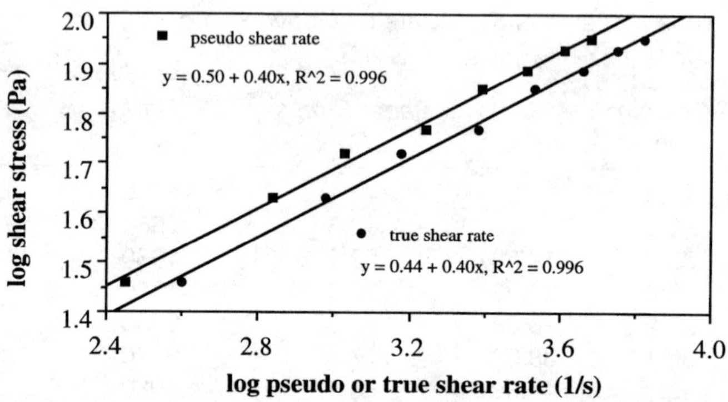

Figure 3–16 Shear Rate–Shear Stress Data on Guava Puree Obtained with a Tube Viscometer. Both pseudo and true rates are shown. Vitali and Rao, 1982.

by Tanner (1988):

$$\frac{D_1}{D} = \left(1 + \frac{1}{8}\left\{\frac{\sigma_{11} - \sigma_{22}}{\sigma_w}\right\}^2\right)^{1/6} \tag{3.33}$$

The above equation provides approximate values because the effects of extensional viscosity and moisture evaporation are not considered.

Glass Capillary Viscometer

For low-viscosity Newtonian foods that can flow under normal gravitational force, glass capillary viscometers can be used. Recalling that the magnitude of viscosity can be obtained by dividing the equation for shear stress by that for shear rate, one can derive from the Hagen-Poiseuille equation:

$$\eta = (\pi\, \Delta p\, r^4\, t/8\, V L) \tag{3.34}$$

where η is the viscosity; Δp, the driving pressure for flow (usually product of fluid density × gravity × mean height of fluid column, h, during flow $\Delta p = \rho g h$); r, the capillary radius; t, the time of flow; V, the volume of flow (note that the volumetric flow rate $Q = V/t$); and L, the length of the capillary. In a given instrument Δp, r, V, and L are usually held constant, so the viscosity is directly proportional to the time of flow, t. Therefore, the time for a standard volume of fluid to pass through a length of narrow bore glass tube is measured for a standard viscosity (η_{st}) fluid (t_{st}) and a test fluid (t). Because the magnitude of viscosity is directly proportional to the time of flow, the viscosity of a fluid can be calculated as

$$\eta = (t/t_{st}) \times \eta_{st} \tag{3.35}$$

It is important that magnitudes of t and t_{st} are determined with care at a specific temperature and that magnitude of viscosity of the standard fluid η_{st} is reliable. Glass capillary viscometers are not suitable for liquids that deviate substantially from Newtonian flow or contain large size particles or a high concentration of suspended solids.

Ubbelohde Viscometer

As stated in Chapter 1, for the determination of intrinsic viscosity, $[\eta]$, of a polymer, viscosity values of several dilute solutions, when the relative viscosities (η/η_s) of the dispersions are from about 1.2 to 2.0, are determined. To facilitate such measurements, the so-called Ubbelohde glass capillary viscometer is used, which has a large reservoir to permit several successive dilutions of a polymer solution (Figure 3–17). Because intrinsic viscosity measurement is important, the test procedure for using the Ubbelohde viscometer is outlined here in brief (Cannon Instrument Co., 1982).

(1) A measured volume of sample is charged directly from a pipette through tube G into the lower reservoir J. (2) The viscometer is placed in a constant temperature bath in a vertical position. (3) Tube B is closed with a finger and suction is applied to tube C until the test fluid reaches the center of bulb C and suction on tube is removed. The finger from tube B is removed and immediately placed over tube A. (4) After the finger is removed from tube A, the efflux time for the liquid meniscus to pass from mark D to mark F is measured to the nearest 0.1 s. Steps 3 and 4 should be repeated to ensure that reproducible results can be obtained. The viscosity of the sample can be calculated from the efflux time by multiplying it with the viscometer constant. Alternatively, the values of times for flow of a standard volume of fluid of different polymer concentrations and of the solvent can be used to calculate the intrinsic viscosity. (5) The sample in reservoir is diluted by pipetting a measured volume of solvent and the viscosity of the diluted sample is determined as described in steps 3 and 4. The intrinsic viscosity is determined by means of either the Higgins or Kraemer equation for non-polyelectrolytes, that is, plots of η_{sp}/c versus c or ln $(\eta_r)/c$ versus c, respectively. For a polyelectrolyte, (c/η_{sp}) is plotted against $c^{1/2}$ from which a straight line is obtained with an intercept of $1/[\eta]$ and a slope of $B/[\eta]$, where B is a constant (see Equation 1.13). Further, determination of the time of flow of a polymer solution can be automated by detecting the position of the fluid's meniscus by optical means.

In glass capillary viscometers, both entrance and exit effects depend on the kinetic energy of the fluid stream in the capillary. Based on extensive experimental data, the kinematic viscosity data $(\nu = \eta/\rho)$ with a correction term for kinetic energy is expressed as follows:

$$\nu = kt - K/t^m \tag{3.36}$$

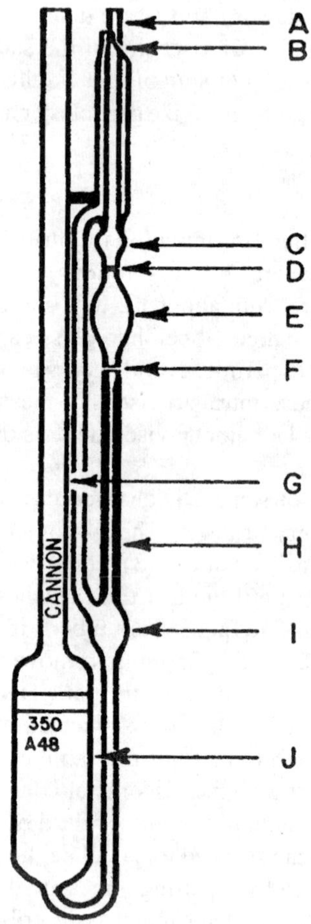

Figure 3–17 Schematic Diagram of an Ubbelohde Viscometer. Courtesy of the Cannon Instrument Company, State College, Pennsylvania.

where $k = \pi r^4 hg/8LQ$, and K and m are constants. The constant $m = 2$ for capillaries ending in trumpet-shaped ends, and $m = 2/3$ for square ends. Typically, when $m = 2$, the value of K is constant for a specific viscometer; for an Ubbelohde viscometer, K was 60 (Van Wazer et al., 1963). More importantly, when viscosity measurements are conducted such that the efflux times are very long, the kinetic energy correction factor K/t^m may be omitted completely.

Slit (Channel) Rheometer

Slit or channel geometries have been used for studying primarily the rheological behavior of products processed in extruders, such as cooked cereal doughs. The applicable equations for shear rate and shear stress have been derived by Dealy (1982), Steffe (1996), and others, and their derivation is analogous to the corresponding equations in tube flow. For the fully developed flow of a Newtonian fluid in a slit of length L, width w, and height (thickness) h (Figure 3–18), the relationship between the shear stress (σ) and the gradient ($\Delta p/L$) is given by

$$\sigma = \left(\frac{\Delta p}{L}\right)\left(\frac{h}{2}\right) \tag{3.37}$$

When the ratio of width to height should be greater than 10, edge effects can be neglected. The velocity profile for a Newtonian fluid is given by

$$v = \frac{3Q}{2hw}\left[1 - 4\left(\frac{y}{h}\right)^2\right] \tag{3.38}$$

where y is the coordinate in the direction of height. At the wall ($y = h/2$), the shear stress and the shear rate of a Newtonian fluid are given by Equations 3.37 and 3.39, respectively.

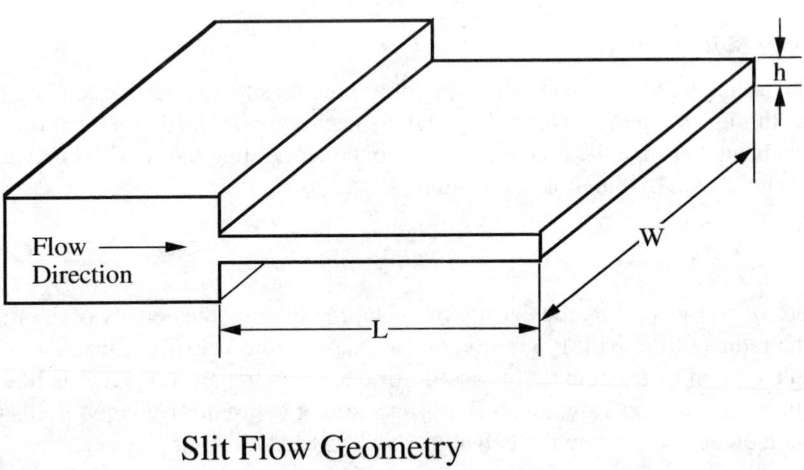

Figure 3–18 Slit Rheometer Geometry. *Source*: Reprinted from M.A. Rao, Rheological Properties of Fluid Foods, *Engineering Properties of Foods*, M.A. Rao and S.S.H. Rizvi, eds., pp. 1–53, by courtesy of Marcel Dekker, Inc.

$$\dot{\gamma}_{\mathrm{N}} = \frac{6Q}{h^2 w} \tag{3.39}$$

The shear rate for a non-Newtonian fluid can be obtained from

$$\dot{\gamma} = \left(\frac{6Q}{h^2 \mathrm{W}}\right)\left(\frac{2+\beta}{3}\right) \tag{3.40}$$

where β equals

$$\beta = \frac{\mathrm{d}\ln\left(\dfrac{6Q}{wh^2}\right)}{\mathrm{d}\ln\left[\dfrac{h}{2}\left(\dfrac{\Delta p}{L}\right)\right]} = \frac{\mathrm{d}\ln\dot{\gamma}_{\mathrm{N}}}{\mathrm{d}\ln\sigma} \tag{3.41}$$

Equation 3.41 is analogous to the Weissenberg-Rabinowitsch-Mooney equation for laminar flow in a tube/pipe. As in pipe flow, the shear stress and the Newtonian shear rate are calculated first using Equations 3.37 and 3.39, respectively. Using the slope (β) of a plot of ln σ and ln $\dot{\gamma}_{\mathrm{N}}$ in Equation 3.40, the non-Newtonian shear rates ($\dot{\gamma}$) can be calculated. It is of interest to note that the slit rheometer has been used also for determining normal stress differences for use in understanding extrudate swell. The study by Senouci and Smith (1988) is an example of the use of the slit rheometer for characterizing the rheological properties of foods.

Miscellaneous Viscometers

Falling Ball Viscometer

In the falling ball viscometer, the time for a standard sphere to fall a standard distance through a liquid at its limiting velocity is measured. The diameter of the ball is much smaller than the diameter of the container holding the fluid. This type of test is described by the Stokes equation:

$$\eta = \frac{(2/9)\,(\rho_{\mathrm{s}} - \rho_{\mathrm{e}})\,g\,R^2}{V} \tag{3.42}$$

where η = viscosity; ρ_{s}, the density of the falling ball; ρ_{e}, the density of the fluid; R, the radius of the ball; g, gravity; and V, the limiting velocity. This is another simple type of instrument that is useful for clear Newtonian fluids. It has limited applicability for non-Newtonian fluids and cannot be used for opaque fluids unless a method for locating the ball at any time is part of the instrument.

Orifice Viscometer

The time for a standard volume of fluid to flow through an orifice is measured. The orifice may be regarded as a very short capillary. The viscometer is made up of a cup or bowl of standard dimensions with an accurately designed diameter orifice in the bottom. This type of viscometer is often used for quality control of fluid foods in food processing plants where a rapid, inexpensive method is required and extreme accuracy is not needed.

Other types of viscometers, such as an oscillation viscometer, that are useful for characterizing Newtonian foods are also available. However, their use for characterizing non-Newtonian fluid food can be justified only if the complex flow fields can be analyzed and expressions are derived for the shear rate and shear stress.

VISCOSITY MEASUREMENT AT HIGH TEMPERATURES

Many fluid foods are subjected to temperatures greater than 90°C during pasteurization and thermal processing. Therefore, flow properties of fluid foods at these temperatures are useful in these processing applications. Relatively few studies have been conducted on rheological behavior of foods at high temperatures (~95°C). Dail and Steffe (1990) used a pressurized tube (12.7 mm dia) viscometer to study the flow behavior of 1.82% and 2.72% waxy maize (WM) starch dispersions at 121° to 143°C. This system required a large mass (~160 kg) of test sample and, as with many tube/capillary viscometers, time-dependent rheological behavior at a fixed shear rate could not be detected. Abdelrahim et al. (1995) studied the flow behavior of WM starch dispersions with an indirect-drive (magnetic coupling) pressurized (D100/300, Haake) concentric cylinder viscometer system. The uncertainty in rotational speeds due to slip of the magnetic coupling, friction in the mechanical bearings of the rotating cylinder, and the effect of the clutch mechanism were of concern. Reliable rheological data, in terms of the basic units of shear rate (s^{-1}) and shear stress (Pa), can be obtained on a specific food product at process temperatures using any one of the flow geometries (e.g., capillary, concentric cylinder, cone-plate).

A pressurized direct-drive concentric cylinder viscometer system (Figure 3–19) was used to obtain shear rate versus shear stress data on a tomato puree at several fixed temperatures between 76° and 120°C (Figure 3–20), and temperature versus apparent viscosity data at several shear rates on a 4% waxy rice (WR) starch dispersion during gelatinization over the temperature range 30° to 110°C (Figure 3–21) (Rao et al., 1999). The drive motor, torque unit, and concentric cylinder unit and temperature control vessel of a Haake RV2 viscometer system (Haake Inc.) were placed in a chamber (PRC) that could be pressurized to 0.2 MPa (two atmospheres). The temperature control vessel was insulated to minimize heat loss. A copper-constantan (36 gauge wires) thermocouple placed in the well of the inner concentric cylinder measured the temperature of the test sample.

The control unit of the viscometer was kept outside the PRC so that the desired rotational speeds could be selected. The rotational speed and the corresponding

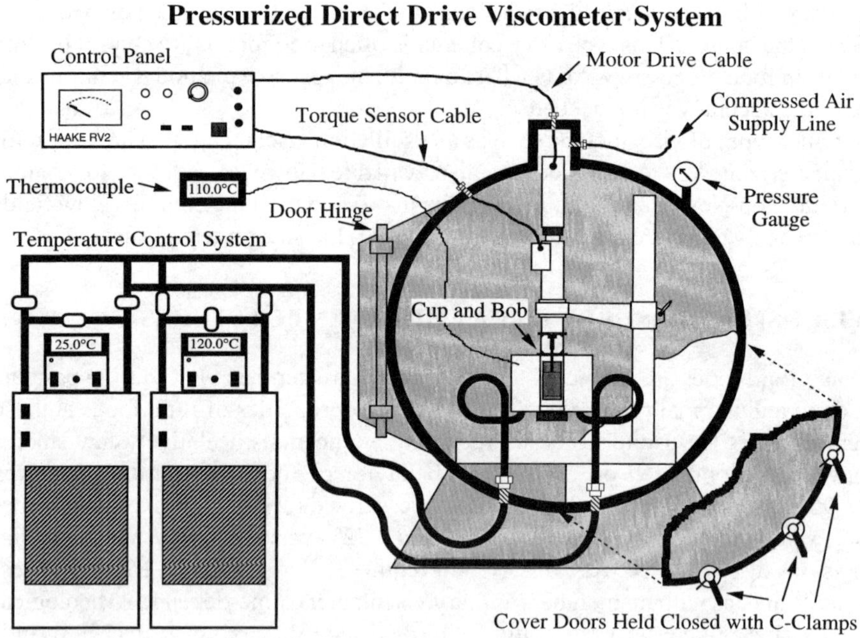

Figure 3–19 Pressurized Viscometer System for Obtaining Data at High Temperatures. *Source*: Reprinted with permission from H-J. Liao, *Simulation of Continuous Sterilization of Fluid Food Products: The Role of Thermorheological Behavior of Starch Dispersion and Process*, PhD Thesis. © 1998, Cornell University.

torque data, as well as the temperatures sensed by the thermocouples, were recorded on a lap-top computer (Toshiba, T1200 XE) after analog to digital conversion via a data logger (Digistrip II, Kaye Instruments). The electrical lines to the drive motor and the torque unit inside the PRC were passed through air-tight Swagelok fittings.

A high-temperature bath (Model 9500, Fisher Scientific) and a low-temperature bath (Model MH, Julabo, Germany), both containing silicone oil (Aldrich Chemical Co.) as the heating/cooling medium, were used to heat and cool the test sample, respectively. Insulated copper tubing was used for flow of the silicone oil between the oil baths and the viscometer temperature control vessel; the rigid tubing also helped in holding the viscometer firmly in the PRC. A set of three valves was used to circulate the hot oil outside the PRC and keep it well agitated in the high-temperature bath or through the viscometer temperature control vessel. Another set of three valves was used with the low-temperature oil bath.

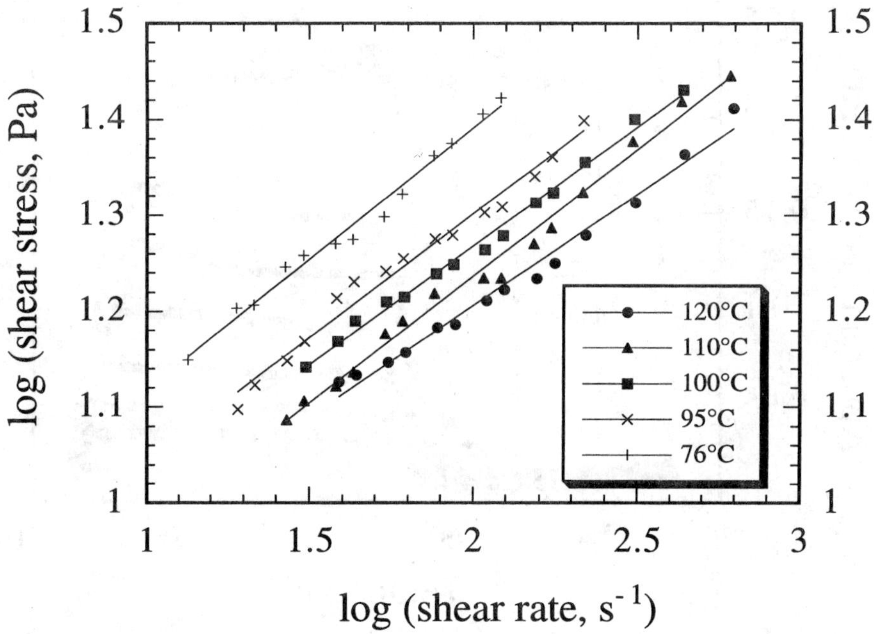

Figure 3–20 Flow Behavior (Power Law) Data on Tomato Concentrates at 76°–120°C. *Source*: Reprinted with permission from H-J. Liao, *Simulation of Continuous Sterilization of Fluid Food Products: The Role of Thermorheological Behavior of Starch Dispersion and Process*, PhD Thesis, © 1998, Cornell University.

In a typical experiment, the temperatures of the high- and low-temperature baths were first set and allowed to reach equilibrium values while the silicone oil was circulated in the lines outside the PRC. Because of energy loss in the high-temperature lines, there was a substantial difference between the set temperature of the high-temperature bath and the sample temperature. Therefore, the temperature of the high-temperature bath was set about 20°C higher than the desired sample temperature. The concentric cylinder system (MVI, r_i = 20 mm, r_o = 21 mm, H = 60 mm) was carefully filled (~120 ml) with either tomato puree or starch dispersion and secured in the temperature control vessel. To keep evaporation losses at a minimum, total experiment time at the high temperatures (95 to 120°C) was kept close to 20 minutes. A Teflon® cover at the top of the concentric cylinder system and a piece of wet cloth wrapped above the concentric cylinder system helped considerably in reducing moisture loss from both the tomato puree and the starch dispersion samples.

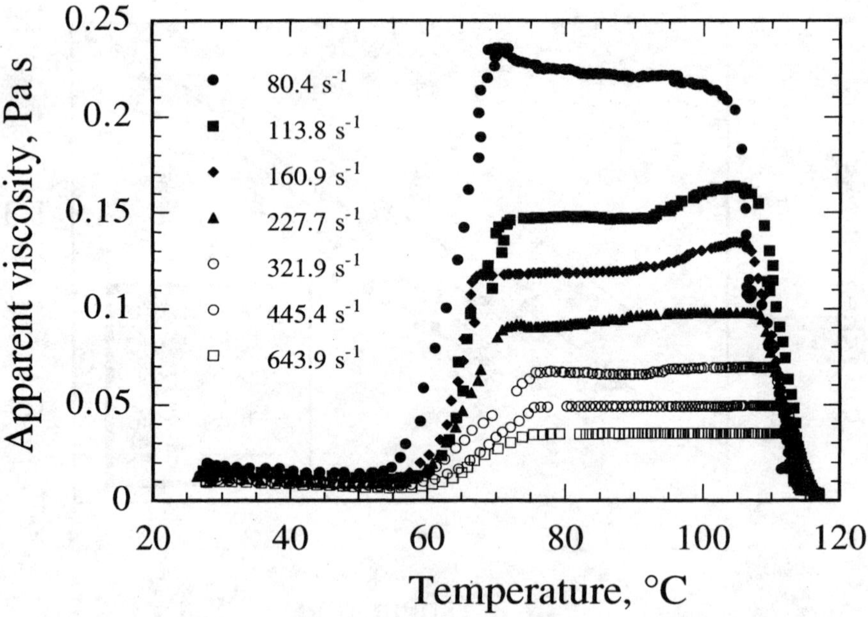

Figure 3–21 Apparent Viscosity versus Temperature Data on a Starch Dispersion as a Function of Shear Rate (80–640 s^{-1}) and Temperature (35–118°C). *Source*: Reprinted with permission from H-J. Liao, *Simulation of Continuous Sterilization of Fluid Food Products: The Role of Thermorheological Behavior of Starch Dispersion and Process*, PhD Thesis. © 1998, Cornell University.

EXTENSIONAL FLOW VISCOMETRY

Extensional or elongational flow is another type of deformation that can be used to obtain information regarding the rheological behavior of foods, and it is different than the shear flows discussed previously. Any abrupt change in the flow geometry (e.g., sudden contraction, flow across an orifice) will generate a flow with an extensional shear component. In elongational flows, a velocity gradient along the direction of flow can stretch polymers far from equilibrium. In turn, extended polymers exert a force back on the solvent that leads to non-Newtonian behavior: viscosity enhancement and turbulent drag reduction. Applications of extensional flows to foods include atomization, swallowing, and dough sheeting (Padmanabhan, 1995). Extensional flow data are very important because they differ considerably from shear data, and the resistance to flow in extension can exceed that in shear by orders of magnitude (Sridhar et al., 1991). Three types of extensional flows have found wide acceptance: (1) uniaxial, (2) biaxial, and (3) planar. For any extensional flow, one can define the "stressing viscosities" η_1^+ and η_2^+:

$$\eta_1^+ = \frac{\sigma_{11} - \sigma_{33}}{2(2+m)\dot{\varepsilon}} \tag{3.43}$$

$$\eta_2^+ = \frac{\sigma_{22} - \sigma_{33}}{2(1+2m)\dot{\varepsilon}} \tag{3.44}$$

For uniaxial extension $m = -1/2$ and only η_1^+ is defined, and for biaxial extension $m = 1$ and $\eta_1^+ = \eta_2^+$. However, for planar extension $m = 0$, and η_1^+ and η_2^+ are not equal (Macosko, 1994).

Uniaxial Extension

In uniaxial extension the material is stretched in one direction and compressed equally in the other two. For uniaxial extension or simple extension of a liquid sample of length L one meaningful measure of strain is (Dealy, 1982)

$$\mathrm{d}\varepsilon = \mathrm{d}L/L \tag{3.45}$$

The strain rate is given by

$$\dot{\varepsilon} = \frac{\mathrm{d}\varepsilon}{\mathrm{d}t} = \frac{1}{L}\frac{\mathrm{d}L}{\mathrm{d}t} = \frac{\mathrm{d}\ln L}{\mathrm{d}t} \tag{3.46}$$

Because $(\mathrm{d}L/\mathrm{d}t)$ is velocity (v), the above equation can be written as

$$\dot{\varepsilon} = \frac{v}{L} \tag{3.47}$$

When the strain rate ($\dot{\varepsilon}$) is maintained constant, the deformation obtained is called steady simple extension or steady uniaxial extension (Dealy, 1982) and the extensional viscosity (η_E) is related to the normal stress difference:

$$\eta_E(\dot{\varepsilon}) = \frac{\sigma_{11} - \sigma_{22}}{\dot{\varepsilon}} \tag{3.48}$$

One application of the above equation is to be able to estimate magnitudes of the normal stress difference ($\sigma_{11} - \sigma_{22}$) as the product of $\dot{\varepsilon}$ and η_E. However, to obtain magnitudes of η_E from the above equation, one must have the means to measure both the normal stress difference and the strain rate ($\dot{\varepsilon}$). A fluid is said to be tension-thickening when η_E increases with increasing $\dot{\varepsilon}$, and tension-thinning when η_E decreases with increasing $\dot{\varepsilon}$. For the special case of a Newtonian fluid with viscosity η:

$$\eta_E = 3\,\eta \tag{3.49}$$

For a shear-thinning fluid with a zero shear viscosity (η_0):

$$\lim_{\dot{\varepsilon} \to 0} [\eta_E (\dot{\varepsilon})] = 3\eta_0 \qquad (3.50)$$

The Trouton ratio, T_R, is defined as the ratio of the extensional viscosity to the shear viscosity:

$$T_R = \frac{\eta_E(\dot{\varepsilon})}{\eta(\dot{\gamma})} \qquad (3.51)$$

Often, it is not possible to reach a steady state in extension, and it is convenient to define a transient extensional viscosity, $\bar{\eta}_E$, that is a function of time (t) and the extensional strain rate ($\dot{\varepsilon}$) (Barnes et al., 1989).

Even with solid foods, with which extensional viscosity measurement is relatively easy, the main difficulty in extensional viscosity measurement is in achieving steady state with respect to the applied rate of strain or stress. For example, Barnes et al. (1989) state that in the experiments conducted by Meissner, the Hencky strain ($\ln L/L_o$, where L_o is the original length and L is the length in deformation) of 7 corresponded to stretching the sample to 1100 times its original length. Therefore, the data often obtained may be on the transient function, $\bar{\eta}_E (t, \dot{\varepsilon})$. Extensional flow viscosity measurement poses a special challenge in low-molecular-weight polymers and the main problem is to achieve a continuous extensional flow field (Barnes et al., 1989) because the residence time (t_{res}) of polymer molecules in the velocity gradient is limited and because of the polydispersity of the molecules (Perkins et al., 1997). Although several techniques for measuring extensional stresses in polymer solutions have been proposed, rigorous interpretation of the results from these tests was only possible through the use of a rheological constitutive equation (Sridhar et al., 1991). This is because one generates a flow that is dominated by extension and the data obtained are then interpreted in terms of rheologically meaningful material functions. Two such flows are called a spin-line or fiber-spin rheometer (Figure 3–22) and the open-siphon technique (Figure 3–23) (Barnes et al., 1989). These two techniques, often suitably modified, have been employed by various authors for fluids and semisolids.

Filament Stretching

Sridhar et al. (1991) presented a filament stretching technique for measuring the extensional viscosity of polymer solutions (0.185% polyisobutylene in kerosene with zero-shear viscosity 36 Pa s at 20°C) at a constant stretch rate. The sample to be tested was held between two Teflon® disks whose diameter ranged between 1 and 7 mm. The stationary top disk was attached to a microbalance system with a precision of 10^{-3} N. The output of the microbalance was recorded and converted

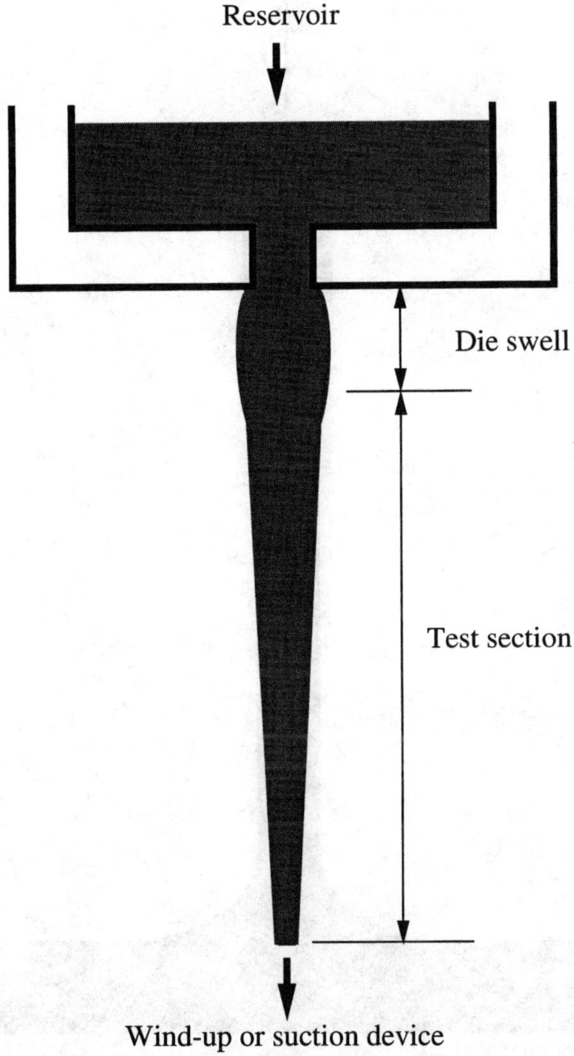

Figure 3–22 A Spin-Line or Fiber-Spin Rheometer for Extensional Viscosity Measurement. *Source*: Reprinted from Barnes et al., *An Introduction to Rheology*. © 1989, pp. 85–89, with permission from Elsevier Science.

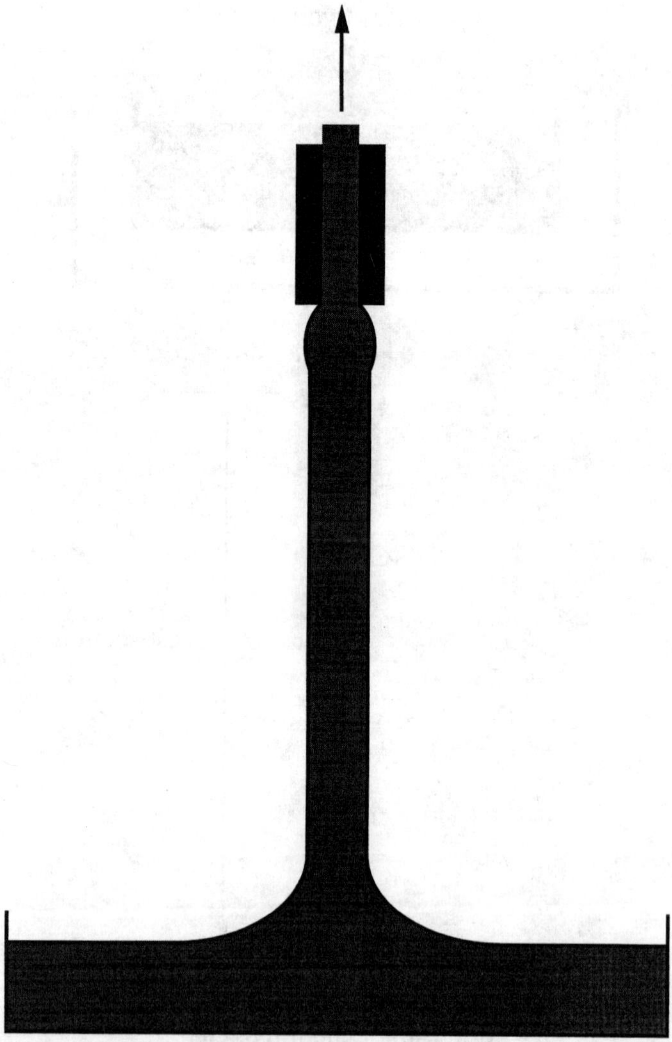

Figure 3–23 Schematic Diagram of the Open-Siphon Technique for Extensional Viscosity Measurement. *Source*: Reprinted from Barnes et al., *An Introduction to Rheology*. © 1989, pp. 85–89, with permission from Elsevier Science.

to force on the filament. The bottom disk was attached to a central shaft of a material testing unit whose movement was controlled. A small amount of the test fluid in the vicinity of the disks remained undeformed. Therefore, the stretch rate was calculated from the decrease of filament radius with time, instead of increase in filament length with time. The diameter of the filament within the homogeneous portion of the filament during the stretching experiment was determined with an optics-based diameter measuring device. Several experiments were conducted to obtain a range of diameters as a function of time. As an alternative technique for measurement of the diameter, the authors suggested photographing the filament at up to four frames per second during the experiment and determination of the filament dimensions from the negatives. Two distinct stretching regions were observed: (1) The first local extension rate, $\dot{\varepsilon}_1$, was significantly greater than the overall extensional rate and corresponded to significant necking of the fluid column when subjected to sudden stretching. This nonuniform stretching lasted up to half the time it took to stretch the fluid to the maximum length of about 160 mm. (2) The second extension rate region, $\dot{\varepsilon}_2$, corresponded to the homogeneous deformation of the fluid element when only a small amount of fluid remained next to the disks. Typical magnitudes of $\dot{\varepsilon}_1$ and $\dot{\varepsilon}_2$ obtained were 0.72 s^{-1} and 0.33 s^{-1}, and 0.92 s^{-1} and 0.57 s^{-1}, respectively. Initially, when the extensional viscosity was independent of the stretch rate, the Trouton ratio increased slowly up to 3. Beyond this linear viscoelastic region, the polymer molecules extended considerably and the extensional viscosity increased dramatically to Trouton ratios $> 10^3$. The flow rate and change in cross-sectional area of a fluid filament are used to determine the strain rate (Khagram et al., 1985), and such a device is called a fiber-spinning instrument.

Stagnation Point Method

In addition, stagnation point (a point at which the net magnitude of velocity is zero, such as at the midpoint of a cylinder or a sphere in cross flow) devices, such as the dough roll mill, cross slot flow, and opposing jet devices can be used in which an area of intense extensional flow is created without the need to sustain a continuous filament. The latter technique is illustrated in Figure 3–24. Clark (1997) obtained extensional viscosity data on syrups and food gum dispersions using creation of a stagnation point technique. The ratio of extensional to shear viscosity was about 3.8 for a sugar syrup in reasonable agreement of 3.0 predicted by Trouton rule. The Trouton ratio for a 0.1% xanthan gum dispersion in 0.1% NaCl was about 4 at low and 16 at high strain rates, reflecting the limited and extensive extension of the molecule by the extensional flow field, respectively.

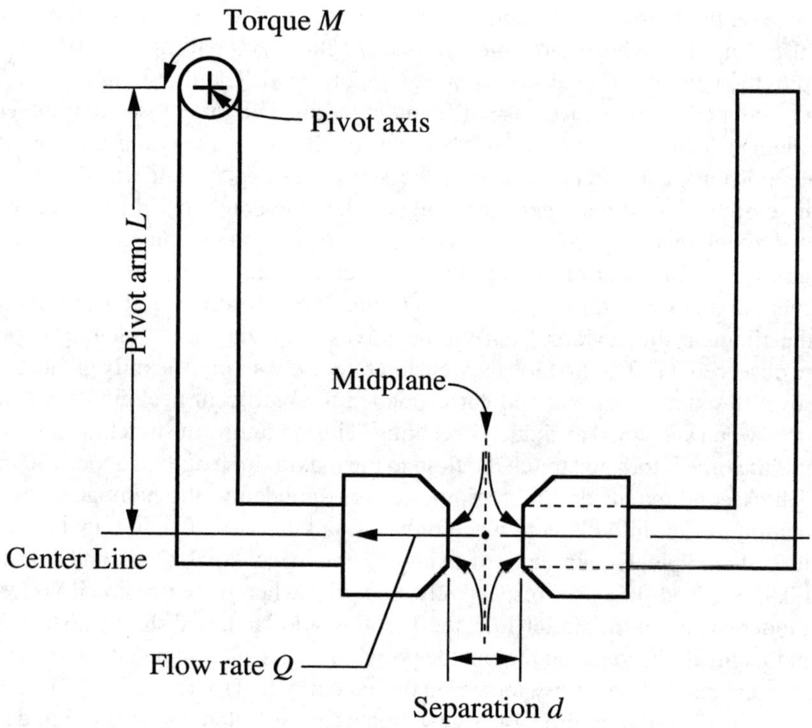

Figure 3–24 Opposing Jet Devices Can Be Used in Which an Area of Intense Extensional Flow Is Created. *Source*: C.W. Macosko, *Rheology: Principles, Measurements, and Applications,* © 1994, John Wiley and Sons, Inc. Reprinted with permission of John Wiley and Sons, Inc.

Biaxial Extension

Uniaxial extensional viscosity data are scarce, in part because of difficulties in conducting uniaxial extension measurements in a reliable manner. In contrast, biaxial extension studies are relatively easy to conduct, and the availability of universal testing machines in many food science and agricultural engineering departments that can be adapted for conducting biaxial extension measurements should help popularize them. In equal biaxial extension, the material is stretched equally in two directions and compressed in the other. It is convenient to work with circular disks that can be subjected to biaxial extension (Figure 3–25) together with a universal testing machine (Chatraei et al., 1981). This technique, called squeezing flow, was adapted to measurements on foods

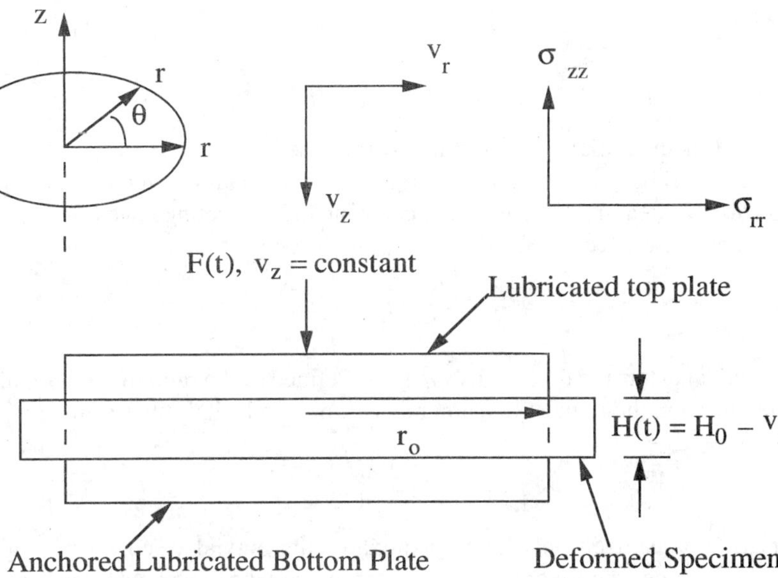

Figure 3–25 Biaxial Extensional Flow Test Setup. Campanella et al., 1987.

using lubrication between the food sample and the metal platens. Squeezing flow between two disks with lubricated surfaces can generate a homogeneous compression or equal biaxial extension in a high viscosity polymer and other materials, such as cheeses and peanut butter (Chatraei et al., 1981; Casiraghi et al., 1985; Campanella et al., 1987). The extensional viscosity can be calculated assuming a homogeneous deformation, which implies that there is perfect slip at the wall. The requirement of perfect slip at the walls can be an asset in foods that tend to slip at walls in conventional viscometric flows, such as melted cheese and other foods. The pertinent equations for obtaining biaxial extensional viscosity are obtained assuming an incompressible material undergoing homogeneous deformation; the latter implies perfect slip at the wall (Chatraei et al., 1981; Campanella et al., 1987). For the system illustrated in Figure 3–25, the velocity field is

$$v_z = \dot{\varepsilon}_T\, H(t); \qquad v_r = -\dot{\varepsilon}_T \frac{r}{2}; \qquad v_\theta = 0 \tag{3.52}$$

where v_z, v_r, and v_θ are the vertical, radial, and angular velocity components; $H(t)$ is the instantaneous height at time t, and r is the radial distance. In a Universal Testing Machine operated at a constant cross head speed, the strain rate is given by

$$\dot{\varepsilon}_{\mathrm{T}} = \frac{v_z}{H_0 - v_z t} \tag{3.53}$$

where H_0 is the initial height of the test specimen.

The experiment can be carried out at either constant or time-dependent force, and with the area of the plates being constant and neglecting edge effects, the normal stress difference is related to the applied force:

$$\sigma_{\mathrm{zz}} - \sigma_{\mathrm{rr}} = \frac{F(t)}{\pi r_{\mathrm{o}}^2} \tag{3.54}$$

The biaxial extensional viscosity (η_{b}) was defined as the ratio of the normal stress difference and the radial extension rate assumed to be half the strain rate:

$$\eta_{\mathrm{b}} = \frac{2F(t)\,(H_0 - v_z t)}{\pi r_{\mathrm{o}}^2 v_z} \tag{3.55}$$

For an unvulcanized polydimethylsiloxane, the biaxial viscosity was approximately six times the shear viscosity over the biaxial extensional rates from 0.003 to 1.0 s^{-1} (Chatraei et al., 1981), a result expected for Newtonian fluids; that is, the relationship between the limiting value of biaxial extensional viscosity (η_{b}) at zero strain rate and the steady zero-shear viscosity (η_0) of a non-Newtonian food is

$$\lim_{\varepsilon_{\mathrm{b}} \to 0} [\eta_{\mathrm{b}}(\dot{\varepsilon})] = 6\eta_0 \tag{3.56}$$

The lubricated squeezing flow technique was applied to melting American cheese where a Universal Testing Machine with lubricated platens was employed (Campanella et al., 1987) to obtain experimental data. This technique is particularly well suited for highly viscous semisolid foods prone to either slip or require considerable effort in pressure-driven methods, such as melted cheese, cheese spreads, corn massa, tomato pastes, mayonnaise, and peanut butter, and it provides data at low values of strain rate of the order of about 0.01 s^{-1}. Conversely, it is not easy to obtain high strain rates with Universal Testing Machines. In addition, it should be ensured that the lubricant is not depleted and is in place during a biaxial strain experiment.

Planar Extension

In planar extension, the material is stretched in one direction, held to the same dimension in a second, and compressed in the third. The work of Perkins et al. (1997) (coauthor Steven Chu is a 1997 physics Nobel laureate for studies on atomic physics) is an excellent example of the considerations in generating planar elongational flow for studying extensional viscosity of fluorescently labeled DNA molecules. They tracked individual molecules in an imaging area 100 μm by 94 μm and measured the extension and the residence time of each DNA molecule. They classified the DNA molecular conformations as dumbbell, kinked, half-dumbbell, folded, uniform, coiled, or extended; the first three types were dominant at $\dot{\varepsilon} = 0.86\ \mathrm{s}^{-1}$ that were significantly greater than the inverse relaxation time ($\tau_r = 3.89$ s) determined at strains < 0.3. For a single polymer coil, theory suggests that the polymer stretching occurs at a critical extensional strain rate, $\dot{\varepsilon}_c$:

$$\dot{\varepsilon}_c \approx \frac{0.5}{\tau_1} \tag{3.57}$$

where τ_1 is the longest relaxation time of the polymer. For $\dot{\varepsilon} < \dot{\varepsilon}_c$, the molecules are in a coiled state and as $\dot{\varepsilon}$ is increased, the hydrodynamic force exerted across the polymer just exceeds the linear portion of the polymer's entropic elasticity (Perkins et al., 1997).

Cogswell's Treatment of Extensional Flow

One technique for measurement of extensional flow that has been used to study various doughs is that of Cogswell (1972, 1978) for entrance flows. The analysis is simple and is based on several assumptions (Padmanabhan and Bhattacharya, 1993): (1) The flow is isothermal and creeping (negligible inertial effects); (2) the fluid is incompressible and has a pressure-independent viscosity; (3) the shear viscosity follows the power law model, $\eta_a = K\dot{\gamma}^{n-1}$; (4) there is no slip at the edge of the converging profile; and (5) the entrance pressure drop (Δp_e) in converging flow from a circular barrel into a circular die can be considered to be made up of that due to shear ($\Delta p_{e,S}$) and extensional flow ($\Delta p_{e,E}$):

$$\Delta p_e = \Delta p_{e,E} + \Delta p_{e,S} \tag{3.58}$$

By applying force balances over a cone or wedge-shaped element, Cogswell (1972) developed relationships between pressure drop and stress. For a sudden planar contraction ($\theta = 90°$ in Figure 3–26), the apparent planar extensional viscosity is (Padmanabhan and Bhattacharya, 1993)

$$\eta_{p,E} = \frac{3(n+1)^2 (\Delta p_{e,E})^2}{4\dot{\gamma}^2 \eta_a} \tag{3.59}$$

and the apparent extension rate is

$$\dot{\varepsilon}_E = \frac{2\dot{\gamma}^2 \eta_a}{3(n+1)\,\Delta p_{e,E}} \tag{3.60}$$

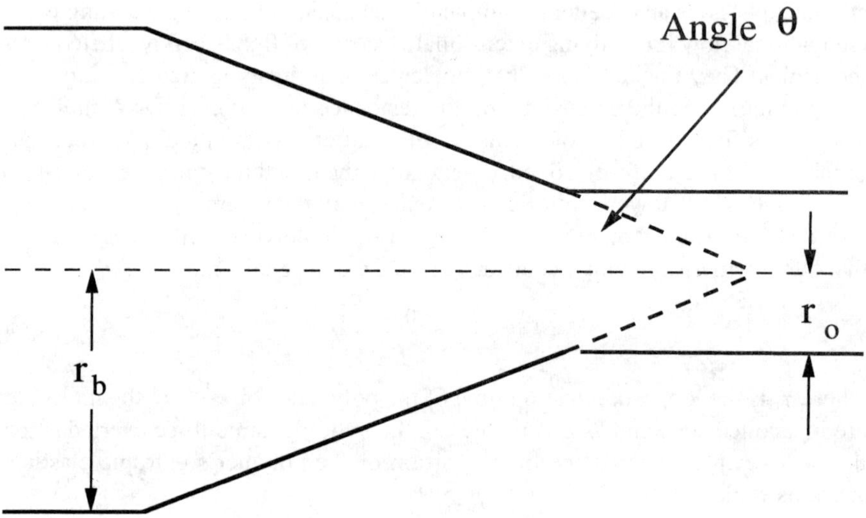

Figure 3–26 Schematic Diagram of Extensional Flow Analysis Geometry. Cogswell, 1972.

The magnitude of $\Delta p_{e,E}$ is best obtained directly from experimental studies or application of Bagley's analysis. If it cannot be obtained from measurements as in Padmanabhan and Bhattacharya (1993), it can be estimated by (1) first estimating the Δp_e using the procedure of Bagley discussed earlier and (2) calculating $\Delta p_{e,E}$ from Equation 3.58 after estimating $\Delta p_{e,S}$ from (Steffe, 1996)

$$\Delta p_{e,S} = \left(\frac{4Q}{\pi r_o^3}\right)^n \left(\frac{3n+1}{4n}\right)^n \frac{2K}{3n \tan\theta}\left(1 - \left(\frac{r_o}{r_b}\right)^{3n}\right) \tag{3.61}$$

The required general equations for calculating extensional viscosity can be derived assuming that the average stress, σ_E, and the average extension rate, $\dot{\varepsilon}_E$, are first related by a power law relationship (Steffe, 1996)

$$\sigma_E = K_E(\dot{\varepsilon}_E)^{n_E} \tag{3.62}$$

$$\Delta p_{e,E} = \left(\frac{4Q}{\pi r_o^3}\right)^{n_E} \left(\frac{2K_E}{3n_E}\right)\left(\frac{\tan\theta}{2}\right)^{n_E}\left(1 - \left(\frac{r_o}{r_b}\right)^{3n_E}\right) \tag{3.63}$$

$$\dot{\varepsilon}_E = \left[\left(\frac{4Q}{\pi r_o^3}\right)\tan\theta\right]/2 \tag{3.64}$$

It should be noted that the above analysis requires knowledge of the actual flow angle θ of the test fluid, which may not be the same as the one in the physical geometry.

MEASUREMENT OF VISCOELASTIC BEHAVIOR OF FLUID FOODS

While the normal stress and transient flow experiments provide useful data, particularly with respect to evaluation of constitutive equations, viscoelastic behavior of many foods has been studied by means of dynamic shear, creep-compliance, and stress relaxation techniques. The paradigm shift from emphasis on normal stress measurements to dynamic rheological technique took place gradually beginning in the late 1950s. The application of each of these techniques to foods is considered in the following sections. It is customary to employ different symbols for the various rheological parameters in different types of deformation: shear, bulk, or simple extension. Following Ferry (1980), the symbols employed for the three types of deformation are given in Table 3–4.

Oscillatory Shear Flow

Small amplitude oscillatory shear (SAOS), also called dynamic rheological experiment, can be used to determine viscoelastic properties of foods. Here, both the expressions SAOS and dynamic rheological experiment will be used. The experiments can be conducted with the Coette, plate-and-cone, and parallel plate geometries. As noted earlier, uniform shear is achieved in the gap of a cone-plate geometry. However, the narrow gap may not be suitable for studying foods containing solids with dimension greater than about 100 μm. In a SAOS experiment, a sinusoidal oscillating stress or strain with a frequency ω is applied to the material, and the phase difference between the oscillating stress and strain, as well

Table 3–4 Viscoelastic Parameters from Shear, Simple Extension, and Bulk Compression

Parameter	*Shear*	*Simple Extension*	*Bulk Compression*
Stress relaxation modulus	$G(t)$	$E(t)$	$K(t)$
Creep compliance	$J(t)$	$D(t)$	$B(t)$
Storage modulus	$G'(\omega)$	$E'(\omega)$	$K'(\omega)$
Loss modulus	$G''(\omega)$	$E''(\omega)$	$K''(\omega)$
Complex modulus	$G^*(\omega)$	$E^*(\omega)$	$K^*(\omega)$
Dynamic viscosity	$\eta'(\omega)$	$\eta'_e(\omega)$	$\eta'_v(\omega)$
Complex viscosity	$\eta^*(\omega)$	$\eta^*_e(\omega)$	$\eta^*_v(\omega)$

as the amplitude ratio, is measured. The information obtained should be equivalent to data from a transient experiment at time $t = \omega^{-1}$.

In the SAOS tests, a food sample is subjected to a small sinusoidally oscillating strain or deformation $\gamma(t)$ at time t according to Equation 3.65.

$$\gamma(t) = \gamma_0 \sin(\omega t) \tag{3.65}$$

where γ_0 is the strain amplitude and ω the angular frequency. The applied strain generates two stress components in the viscoelastic material: an elastic component in line with the strain and a 90° out of phase viscous component. Differentiation of Equation 3.65 yields Equation 3.66, which shows the strain rate $\dot{\gamma}(t)$ for evaluating the viscous component to be $\pi/2$ radians out of phase with the strain.

$$\dot{\gamma}(t) = \gamma_0\, \omega \cos(\omega t) \tag{3.66}$$

For deformation within the linear viscoelastic range, Equation 3.67 expresses the generated stress (σ_0) in terms of an elastic or storage modulus G' and a viscous or loss modulus G''.

$$\sigma_0 = G'\, \gamma_0 \sin(\omega t) + G''\, \gamma_0 \cos(\omega t) \tag{3.67}$$

For a viscoelastic material the resultant stress is also sinusoidal but shows a phase lag of δ radians when compared with the strain. The phase angle δ covers the range of 0 to $\pi/2$ as the viscous component increases. Equation 3.68 also expresses the sinusoidal variation of the resultant stress.

$$\sigma(t) = \sigma_0 \sin(\omega t + \delta) \tag{3.68}$$

The following expressions that define viscoelastic behavior can be derived from Equations 3.67 and 3.68:

$$G' = \left[\frac{\sigma_0}{\gamma_0}\right] \cos\delta \tag{3.69}$$

$$G'' = \left[\frac{\sigma_0}{\gamma_0}\right] \sin\delta \tag{3.70}$$

$$\tan\delta = \frac{G''}{G'} \tag{3.71}$$

where G' (Pa) is the storage modulus, G'' (Pa) is the loss modulus, and $\tan\delta$ is the loss tangent.

The storage modulus G' expresses the magnitude of the energy that is stored in the material or recoverable per cycle of deformation. G'' is a measure of the energy that is lost as viscous dissipation per cycle of deformation. Therefore, for a perfectly elastic solid, all the energy is stored; that is, G'' is zero and the stress and the strain will be in phase (Figure 3–27). In contrast, for a liquid with no elastic prop-

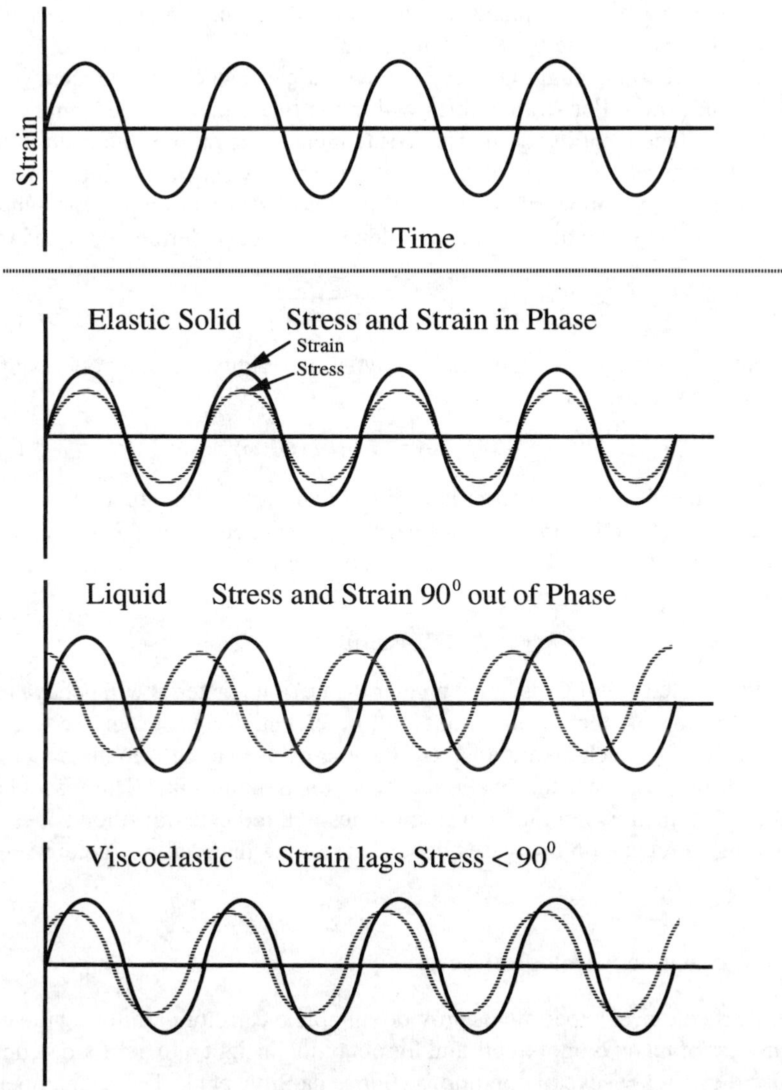

Figure 3–27 Stress versus Strain Response of a Newtonian Liquid and a Perfectly Elastic Solid in Dynamic Tests. Rao, 1992.

erties, all the energy is dissipated as heat; that is G' is zero and the stress and the strain will be out of phase by 90° (Figure 3–27).

For a specific food, magnitudes of G' and G'' are influenced by frequency, temperature, and strain. For strain values within the linear range of deformation, G' and G'' are independent of strain. The loss tangent is the ratio of the energy dissipated to that stored per cycle of deformation. These viscoelastic functions have been found to play important roles in the rheology of structured polysaccharides. One can also employ notation using complex variables and define a complex modulus $G^*(\omega)$:

$$|G^*| = \sqrt{(G')^2 + (G'')^2} \tag{3.72}$$

Note that the dynamic viscosity and the dynamic rigidity are components of the complex viscosity, η^*:

$$\eta^* = (G^*/\omega) = \eta' - i\,(G'/\omega) \tag{3.73}$$

where ω is the frequency of oscillation; the real part of the complex viscosity η' is equal to (G''/ω), and the imaginary part η'' is equal to (G'/ω). One can also determine the loss modulus, G'', from oscillatory shear data using the expression

$$G'' = \omega\,\eta' \tag{3.74}$$

It should be noted that if G' is much greater than G'', the material will behave more like a solid; that is, the deformations will be essentially elastic or recoverable. However, if G'' is much greater than G', the energy used to deform the material is dissipated viscously and the material's behavior is liquid-like. The viscoelastic moduli determined over a range of frequencies will indicate transition zones corresponding to relaxation processes dependent on the material's structure (Ferry, 1980).

Types of Dynamic Rheological Tests

Dynamic rheological experiments provide suitable means for monitoring the gelation process of many biopolymers and for obtaining insight into gel/food structure because they satisfy several conditions (Lopes da Silva et al., 1998; Champenois et al., 1998): (1) They are nondestructive and do not interfere with either gel formation or softening of a structure, (2) the time involved in the measurements is short relative to the characteristic times of the gelation and softening processes, and (3) the results are expressible in fundamental terms and therefore relate to the structure of the network (e.g., Durand et al., 1990). For the same reasons, they were used to follow the changes induced on potato cells by cellulase (Shomer et al., 1993) and α-amylase (Champenois et al., 1998) on wheat starch, and they can be used in studying other phase transitions in foods.

Three types of dynamic rheological tests can be conducted to obtain useful properties of viscoelastic foods, such as gels, and of gelation and melting. They will be described briefly here and their results discussed extensively in Chapter 4 on food biopolymer dispersions and Chapter 6 on food gels. The three types of tests are as follows:

1. Frequency sweep studies in which G' and G'' are determined as a function of frequency (ω) at a fixed temperature. Figure 3–28 illustrates such data obtained on high-methoxyl pectin gels with a plate-cone geometry (Rao and Cooley, 1993). Based on frequency sweep data, one can designate "true gels" when the molecular rearrangements within the network are very reduced over the time scales analyzed, such that G' is higher than G'' throughout the frequency range and is almost independent of frequency (ω) (Clark and Ross-Murphy, 1987). In contrast, for "weak gels," there is higher dependence on frequency for the dynamic moduli, suggesting the existence of relaxation processes occurring even at short time scales, and lower difference between moduli values. Additionally (Ross-Murphy, 1984), with a di-

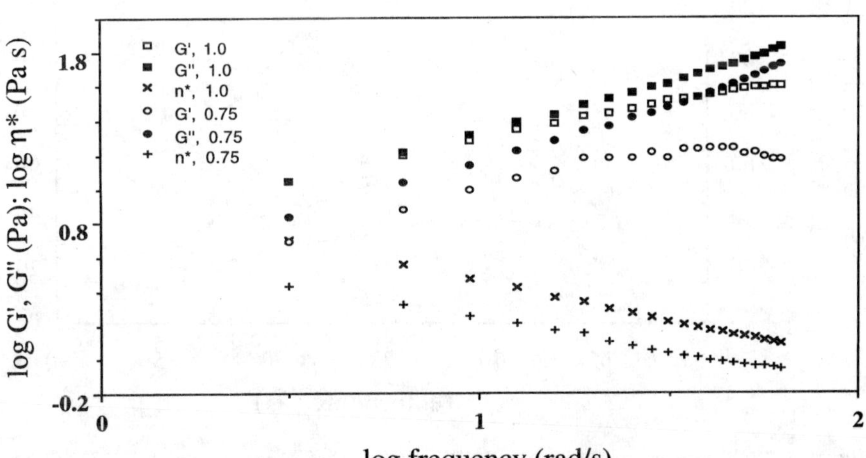

Figure 3–28 Frequency Sweep Studies in Which G' and G'' Are Determined as a Function of Frequency (ω) at a Fixed Temperature. Data obtained on high-methoxyl pectin gels. Rao and Cooley, 1993.

lute biopolymer solution, G'' is larger than G' over the entire frequency range, but at high frequencies they approach each other; at the lowest frequency, G' is proportional to ω^2 and G'' is proportional to ω. For a concentrated biopolymer solution, G'' is larger than G' at lower frequencies, but the G' and G'' curves intersect at a crossover frequency; at frequencies higher than the crossover frequency, values of G' are higher than those of G''.

2. Temperature sweep studies in which G' and G'' are determined as a function of temperature at fixed ω. This test is well suited for studying gel formation during cooling of a heated dispersion (Rao and Cooley, 1993), gelatinization of a starch dispersion during heating (Figure 3–29) (Yang and Rao, 1998), and gel formation of proteins (Owen et al., 1992). In Figure 3–29, the gelatinization of starch was studied with a parallel plate (disk) geometry, and considerable care was exercised to minimize moisture loss from the dispersions. The existence of the linear viscoelastic range can be verified in pre-

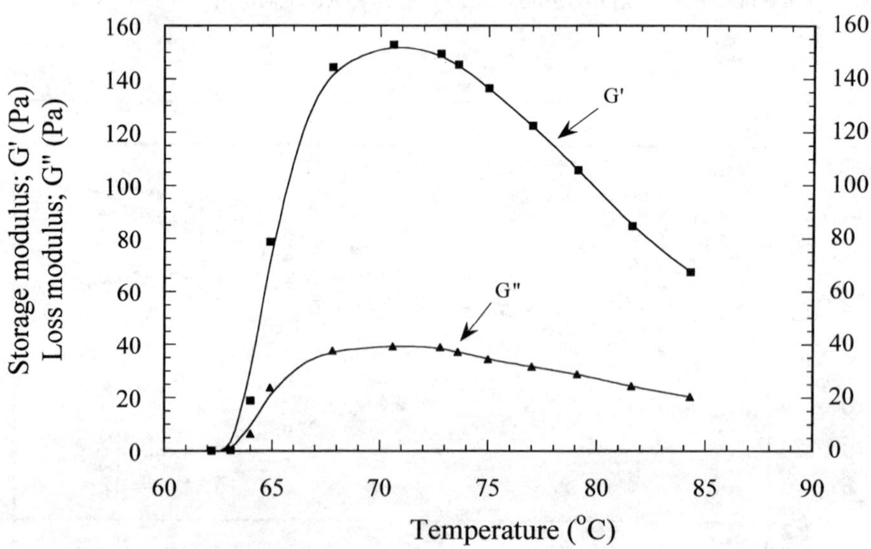

Figure 3–29 Temperature Sweep Data in Which G' and G'' Are at 6.28 rad s^{-1}, 3% Strain. Data obtained on an 8% tapioca starch dispersion heated from 60–85°C at 4°C min^{-1}. *Source*: Reprinted with permission from J. Tattiyakul, *Studies on Granule Growth Kinetics and Characteristics of Tapioca Starch Dispersion during Gelatinization Using Particle Size Analysis and Rheological Methods.* M.S. Thesis, © 1997, Cornell University.

liminary experiments conducted on a given product at several temperature steps (Owen et al., 1992).

3. Time sweep study in which G' and G'' are determined as a function of time at fixed ω and temperature. This type of test, often called a gel cure experiment, is well suited for studying structure development in physical gels. Figure 3–30 is an example of a gel cure experiment on low-methoxyl pectin + Ca^{2+} gels containing sucrose with a cone-plate geometry (Grosso and Rao, 1998). Additional insight into gelation and melting phenomena can be obtained from the temperature and time sweep data by calculation of structure development rates (dG'/dt or $d\eta^*/dt$) or structure loss rates ($-dG'/dt$ or $-d\eta^*/dt$) (Rao and Cooley, 1993; Lopes da Silva et al., 1998). From the time-sweep data, because G' values of gels were recorded at the same time intervals, the influence of sugars and calcium concentrations on them could be examined quantitatively in terms of differences ($\Delta G'$) in G' values; Figure 3–31 shows values of $\Delta G'$ of low-methoxyl pectin + Ca^{2+} gels containing 20% sucrose due to increasing the calcium content from 0.10% to 0.15% (Grosso and Rao, 1998).

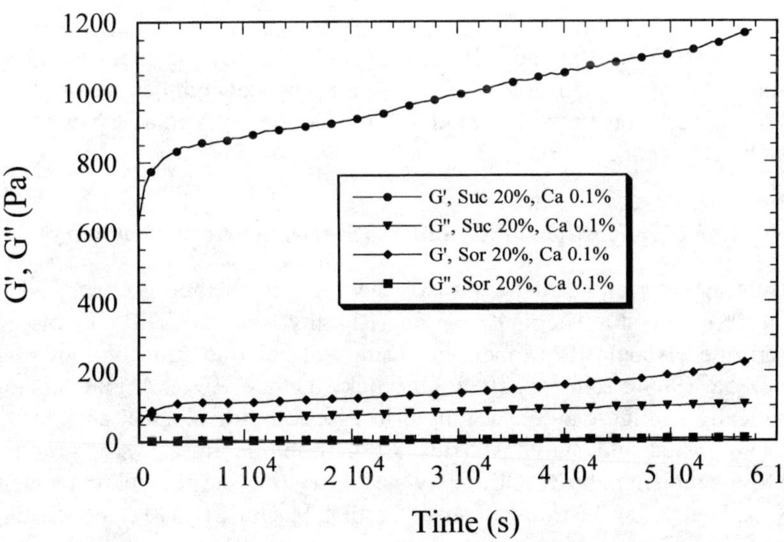

Figure 3–30 Values of G' and G'' during a Gel Cure Experiment on Low-Methoxyl Pectin + Ca^{2+} Gels Containing Sucrose and Sorbitol. Grosso and Rao, 1998.

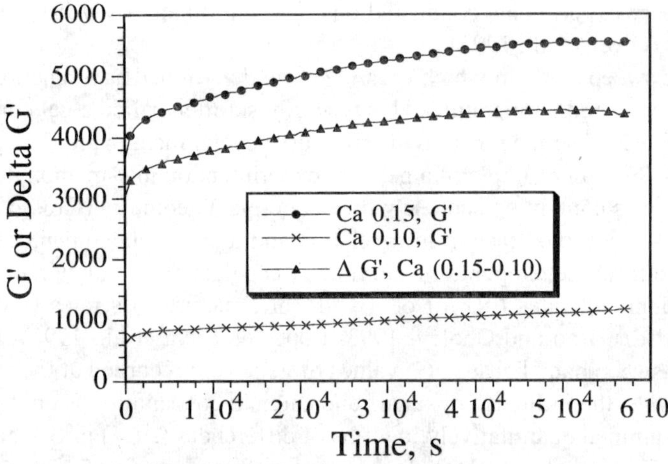

Figure 3–31 Values of G' of 20% Sucrose-Low-Methoxyl Pectin Gels with 0.10% and 0.15% Calcium Content, and of $\Delta G'$ Due to Increasing the Calcium Content. Grosso and Rao, 1998.

Dynamic rheological studies also can be used to study the characteristics of solid foods. Often it is convenient to impose a sinusoidal compressive or elongational strain, so storage and loss moduli obtained are denoted as E' and E'', respectively (Rao et al., 1995).

Axial Motion between Coaxial Cylinders (Segel-Pochettino Geometry)

Axial motion between coaxial cylinders, the Segel-Pochettino geometry (Ferry, 1980, p. 98), named a Thermal Scanning Rigidity Monitor (TSRM), was used to determine viscoelastic properties during sol-gel transition of foods (Wu et al., 1985a, 1985b; Hamann, 1987; Okechukwu et al., 1991). A TSRM, shown schematically in Figure 3–32, was used to assess the rheological changes that occur in cowpea and starch slurries as their temperatures were gradually raised from ambient to 90°C (Okechukwu et al., 1991). The plunger was attached to a B load cell of an Instron Universal Testing Machine (UTM) (Instron Engineering Corp., Canton, MA) and the UTM was operated in an oscillatory mode with a cross head speed of 0.05 cm min^{-1}, plunger displacement of 0.1 mm (Okechukwu et al., 1991). At the end of the test, the length of the plunger covered

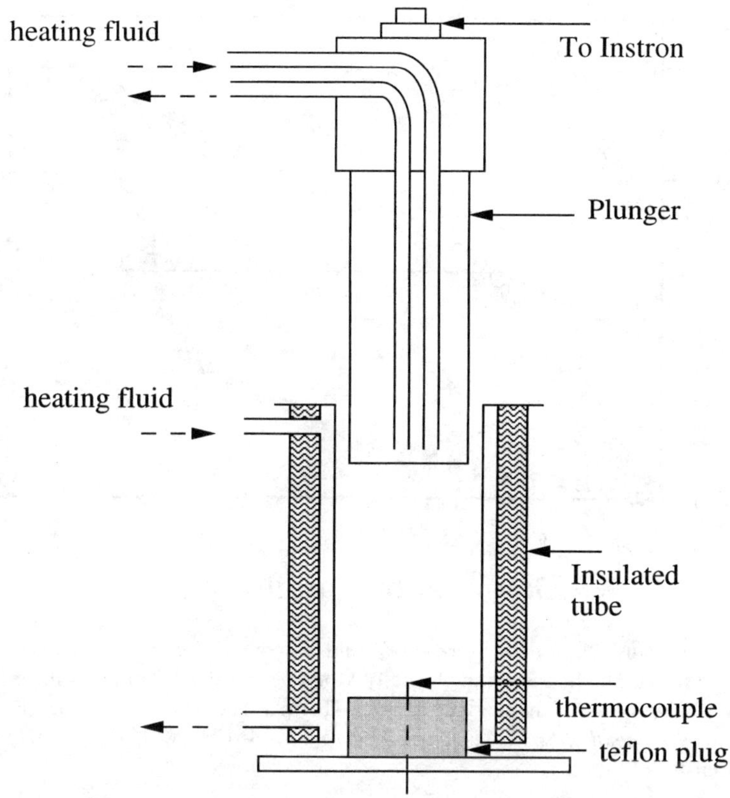

Figure 3–32 Schematic Diagram of a Thermal Scanning Rigidity Monitor. *Source*: Reprinted with permission from Okechukwu et al., Rheology of Sol-gel Thermal Transition in Cowpea Flour and Starch Slurry, *Journal of Food Science*, Vol. 56, pp. 1744–1748. © 1991, Institute of Food Technologists.

by the slurry was measured and used in estimating the apparent storage rigidity modulus, G'.

Values of force were plotted against plunger displacement and delineated for estimates of the storage (F_s) and viscous force (F_V) components of the peak force (Figure 3–33) that are considered typical components of a Kelvin-Voigt (Figure 1–8) viscoelastic model. The areas under the force-displacement curves were numerically estimated using a subroutine (IMSL, Houston, TX). The storage force

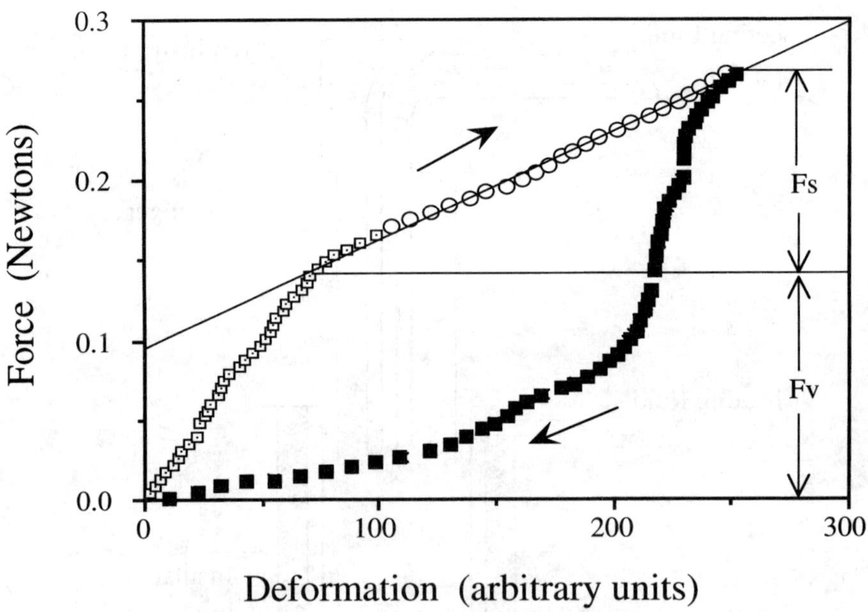

Figure 3–33 Estimation of the Storage (F_s) and Viscous Force (F_v) Components of the Peak Force with a Thermal Scanning Rigidity Monitor. *Source*: Reprinted with permission from Okechukwu et al., Rheology of Sol-gel Thermal Transition in Cowpea Flour and Starch Slurry, *Journal of Food Science*, Vol. 56, pp. 1744–1748. © 1991, Institute of Food Technologists.

---

and the calculated areas were used to evaluate the storage rigidity modulus, G', and the energy loss, respectively (Wu et al., 1985a; Hamann, 1987). G' was evaluated for a thin section of gel of length L, at radial distance r, within the annular region of the TSRM by Equation 3.75:

$$\frac{F_s}{2\pi L r} = -G' \frac{dD}{dr} \tag{3.75}$$

where D is the plunger displacement and L is the length of the plunger covered by the slurry. Equation 3.75 can be solved for the physical boundaries of shear: at $r = r_1$, deformation (D) $= D_1$ and at $r = r_2$, $D = 0$, where r_1 and r_2 are the radii of the plunger and tube, respectively, to give Equation 3.76.

$$G' = \frac{F_s}{2\pi L D} \ln \frac{r_2}{r_1} \tag{3.76}$$

To compensate for the additional compressive force experienced by the plunger at the flat surface, an additional term is needed:

$$G' = \frac{F_s}{2\pi L D_1}\left(\ln \frac{r_2}{r_1} - \frac{r_2^2 - r_1^2}{r_2^2 + r_1^2}\right) \tag{3.77}$$

The % energy loss for each cycle ($\%E_{loss}$) was evaluated using Equation 3.78:

$$E_{loss} = \frac{\text{Area}_1 - \text{Area}_2}{\text{Area}_1} \times 100 \tag{3.78}$$

where Area_1 and Area_2 are the areas under the force deformation plots for the positive and negative displacements of the plunger for a cycle, respectively. Because the TSRM can be used in conjunction with a UTM, viscoelastic behavior can be studied at several frequencies and deformations.

Vibrating Sphere Rheometer

A sphere vibrating at a specific frequency can be used to obtain magnitudes of G' and G'' at that frequency. Such an instrument was used to follow sol-gel transition of 5%, 7%, and 10% starch dispersions (Hansen et al., 1990). However, such instruments seem to have a limited range of oscillation frequencies (e.g., 676 to 680 Hz). In addition, the reliability of the data obtained in comparison to data from dynamic rheological tests in which cone-plate, parallel plate, and concentric cylinder geometries have been used needs to be established.

Relationship between Dynamic Rheological Properties and Normal Stress Functions

When properly conducted, frequency sweep tests provide data over a wide range of frequencies. However, if fundamental parameters are required, each test must be restricted to linear viscoelastic behavior. This can be verified with a given sample by increasing the amplitude of oscillation and noting the magnitude of phase lag and the amplitude ratio. The limit of linearity can be detected when dynamic rheological properties (e.g., G' and G'') change rapidly from their almost constant values (Davis, 1973). The relationship between G' obtained from oscillatory tests and the zero shear limiting value (when $\omega = \dot{\gamma}$) of the first normal stress coefficient is (Dealy, 1982)

$$\lim_{\omega \to 0,\, \dot{\gamma} \to 0} \left| \frac{2G'(\omega)}{\omega^2} = \Psi_1(\dot{\gamma}) = \frac{N_1(\dot{\gamma})}{\dot{\gamma}^2}, \text{ at } \omega = \dot{\gamma} \right| \tag{3.79}$$

The above equation implies that a plot of log G' versus log ω should approach a straight line with a slope of 2 at low frequencies and that the limiting values of dynamic rigidity and the first normal stress difference are related. It may not be possible to demonstrate such relationships due to the inability to employ very low frequencies and shear rates.

Relationships between Steady Shear and Oscillatory Shear Parameters

It would be of interest to determine the relationship between the rheological parameters determined from oscillatory tests such as η^* and η' on one hand and the apparent viscosity (from shear stress and shear rate) on the other. The superimposition of the shear rate dependence of steady shear viscosity (η_a) and of the frequency dependence of the complex viscosity (η^*), at equal values of frequency and shear rate, was first reported by Cox and Merz (1958) for polystyrene samples and is known as the Cox-Merz rule.

$$\eta^* (\omega) = \eta_a (\omega) \big|_{\omega=\dot{\gamma}} \tag{3.80}$$

This empirical correlation has been confirmed experimentally for several synthetic polymers (Ferry, 1980) and for several solutions of random-coil polysaccharides (Morris, 1981). However, many exceptions to the Cox-Merz rule have also been found for synthetic polymers (Matsumoto et al., 1975; Kulicke and Porter, 1980) as well as for biopolymer systems (Morris et al., 1981; Mills and Kokini, 1984; Yang and Rao, 1998) and foods (Bistany and Kokini, 1983; Rao and Cooley, 1992). Relationships between the parameters were determined for equal magnitudes of the shear rate, $\dot{\gamma}$, and the frequency, ω (Bistany and Kokini, 1983). Based on data on 10 fluid and semisolid foods, power type relationships were found to be satisfactory between the parameters.

$$\eta^* (\omega) = C[\eta(\dot{\gamma})]^{\alpha} \big|_{\dot{\gamma}=\omega} \tag{3.81}$$

$$\eta' (\omega) = C'[\eta(\dot{\gamma})]^{\alpha'} \big|_{\dot{\gamma}=\omega} \tag{3.82}$$

Similarly, a power type equation was found for G'/ω^2 and the first normal stress coefficient, Ψ_1:

$$G'/\omega^2 (\omega) = C^* [\Psi_1(\dot{\gamma})]^{\alpha'} \tag{3.83}$$

Deviation from the Cox-Merz rule appears to be an indication of structural heterogeneities in a food. For example, significant deviations from the Cox-Merz rule were found in dispersed systems, such as tomato concentrates (Rao and Cooley, 1992) and cross-linked waxy maize starch dispersions (da Silva et al., 1997). In contrast to the observation on several foods of dispersed nature, the Cox-Merz rule was found to be applicable to fluids with homogeneous structure, such as dispersions of guar gum (Mills and Kokini, 1984) and locust bean gum (Lopes da Silva et al., 1993). Doraiswamy et al. (1991) presented theoretical treatment for data on suspensions of synthetic polymers. They suggested that by using effective shear rates, the Cox-Merz rule can be applied to products exhibiting yield stresses. The shift factors discussed above can be used to calculate effective shear rates.

When departures from the Cox-Merz rule are attributed to structure decay in the case of steady shear, the complex viscosity is usually larger than the steady viscosity (Mills and Kokini, 1984). Notwithstanding this feature, the relation between magnitudes of η_a and η^* can be dependent on the strain amplitude used (Lopes da Silva et al., 1993).

The results of oscillatory shear data can be expressed in terms of mechanical models (Davis, 1973). Elliott and Ganz (1977) employed a modified Bingham body to interpret results of dynamic rheological tests on salad dressings. The model shown in Figure 3–34 consists of an elastic element with modulus, *E*; a friction element having a rest yield stress $\sigma_{0,r}$ and a sliding yield stress $\sigma_{0,s}$; and a viscous element having a viscosity η_m. This model was developed to deal with viscoelastic materials with yield stress. In dynamic rheological testing of these materials, when the amplitude of the imposed strain is sufficiently low so that the yield stress is not exceeded, the stress response will be sinusoidal and in phase

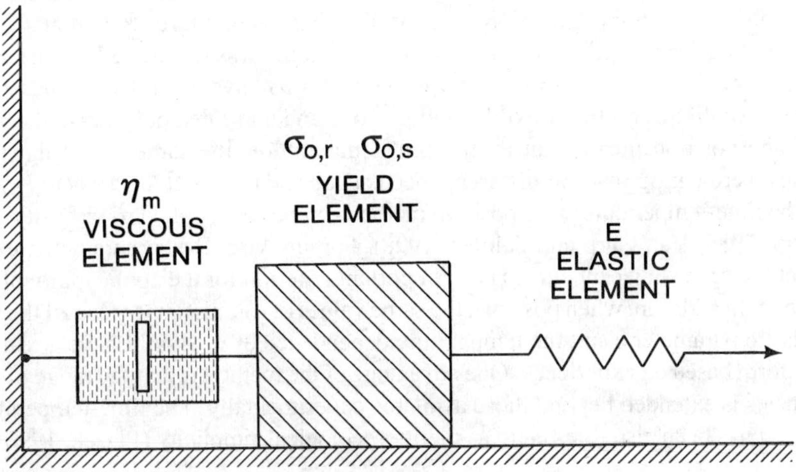

Figure 3–34 Modified Bingham Body To Interpret Results of Dynamic Rheological Tests on Salad Dressings. *Source*: Reprinted with permission from J.H. Elliott and A.J. Ganz, Salad Dressings—Preliminary Rheological Characterization, *Journal of Texture Studies*, Vol. 8, pp. 359–371. © 1977, Food and Nutrition Press.

with the strain. For strains that result in stresses greater than the yield stress, the response will not be sinusoidal but somewhat square. In addition to salad dressings, this model was employed also for materials judged organoleptically to be unctuous (Elliott and Ganz, 1971).

Time-Temperature Superposition

Methodology

The method of reduced variables or time-temperature superposition (Ferry, 1980, pp. 266–280) is based on the assumption that all relaxation times belonging to a given relaxation process have the same temperature dependence and that all contributions to a modulus (and compliance) should be proportional to ρT. Therefore, for example, changing the temperature from T to a reference temperature T_o would cause a change in G' and G'' equivalent to multiplying them by the temperature-density correction factor, due to the effect of thermal expansion and to the entropy-spring nature of the stored elastic energy in the flexible chain theory ($T_0\, \rho_0 / T\rho$). It was not possible to distinguish between $T_0\, \rho_0 / T\, \rho$ and T_0/T because they differ by only about 20% for synthetic polymers (Ferry, 1980). The difference between ($T_0\, \rho_0 / T\, \rho$) and T_0/T would be even less for aqueous food systems. In addition, multiplying the frequency values (ω) by the empirically determined shift factor a_T, that is, using ωa_T for the ordinate, has the physical significance of the ratio of maximum relaxation times at different temperatures to that at T_o. Although the time-temperature superposition technique was developed empirically, there is much theoretical support. Expressions for a_T have been derived based on theories of dilute solution flexible random coils and undiluted polymers. Successful superposition means that at different temperatures the same molecular processes were taking place at different speeds (Lopes da Silva et al., 1994).

The time-temperature superposition method can be also applied to viscosity data (Ferry, 1980; Van Gurp and Palmen, 1998). For any viscoelastic parameter, exact matching of the adjacent curves is an important criterion for the applicability of the method. In addition, when possible, the same values of a_T must superpose all the viscoelastic parameters, and the temperature dependence of a_T should have a reasonable form based on experience. One advantage of the method is that the range of frequencies is extended beyond those available experimentally. The time-temperature method has been also referred to as thermorheological simplicity (Plazek, 1996).

Time-Temperature Superposition of LB Gum and Pectin Dispersions

Lopes da Silva et al. (1994) found that the frequency-temperature superposition, analogous to time-temperature superposition in transient rheological experiments, was applicable to a 1% locust bean (LB) gum dispersion so that master curves at

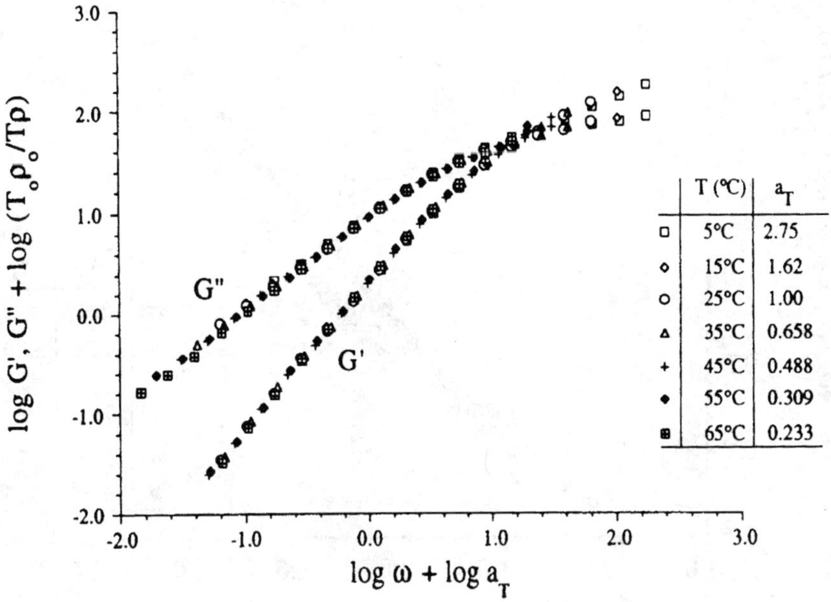

Figure 3–35 Smooth Master Curves at T_o = 25°C Were Obtained for G' and G'' of a 1% Locust Bean Gum Dispersion by Time-Temperature Superposition. *Source*: Reprinted from *Carbohydrate Polymers*, Vol. 23, Lopes da Silva et al., Influence of Temperature on Dynamic and Steady Shear Rheology of Pectin Dispersions, pp. 77–87, © 1994, with permission from Elsevier Science.

T_o = 25°C were obtained for G' and G'' (Figure 3–35). In contrast, smooth master curves could not be obtained for G' and G'' values of 3.5% high-methoxyl pectin dispersions either separately or for both simultaneously (Figure 3–36). The discrepancies were higher for a 3.5% low-methoxyl pectin dispersion. It was concluded that the time-temperature superposition technique was not applicable to the pectin dispersions due to their aggregated structure. For the studied samples, the vertical shift factor for the moduli ($T_o\ \rho_o/T\rho$) had a small effect on the master curve (Lopes da Silva et al., 1994).

Creep-Compliance Studies

Creep-compliance studies conducted in the linear viscoelastic range also provide valuable information on the viscoelastic behavior of foods (Sherman, 1970; Rao, 1992). The existence of linear viscoelastic range may be determined from

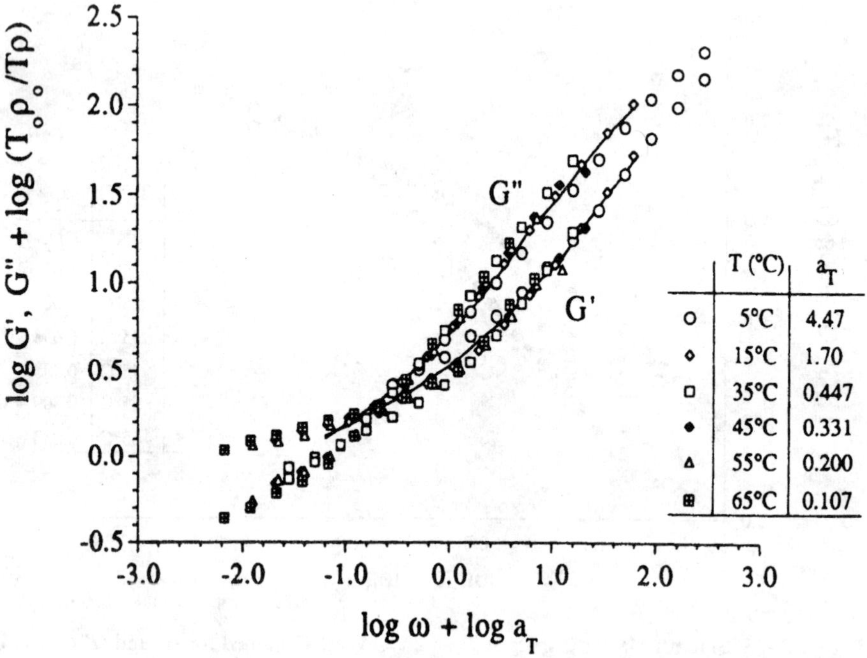

Figure 3–36 Smooth Master Curves Could Not Be Obtained for *G′* and *G″* Values of 3.5% High-Methoxyl Pectin Dispersions by Time-Dependent Superposition. *Source*: Reprinted from *Carbohydrate Polymers,* Vol. 23, Lopes da Silva et al., Influence of Temperature on Dynamic and Steady Shear Rheology of Pectin Dispersions, pp. 77–87, © 1994, with permission from Elsevier Science.

---

torque-sweep dynamic rheological experiments. The creep-compliance curves obtained at all values of applied stresses should superimpose on each other. In a creep experiment, an undeformed sample is suddenly subjected to a constant shearing stress, σ_c. As shown in Figure 3–37, the strain (γ) will increase with time and approach a steady state where the strain rate is constant. The data are analyzed in terms of creep-compliance, defined by the relations

$$J(t, \sigma_c) = \gamma/\sigma_c \tag{3.84}$$

A typical creep-compliance curve can be divided into three principal regions (Sherman, 1970), as described below.

1. A region (A-B) of instantaneous compliance J_0 in which the bonds between the different structural units are stretched elastically. In this region, if the stress

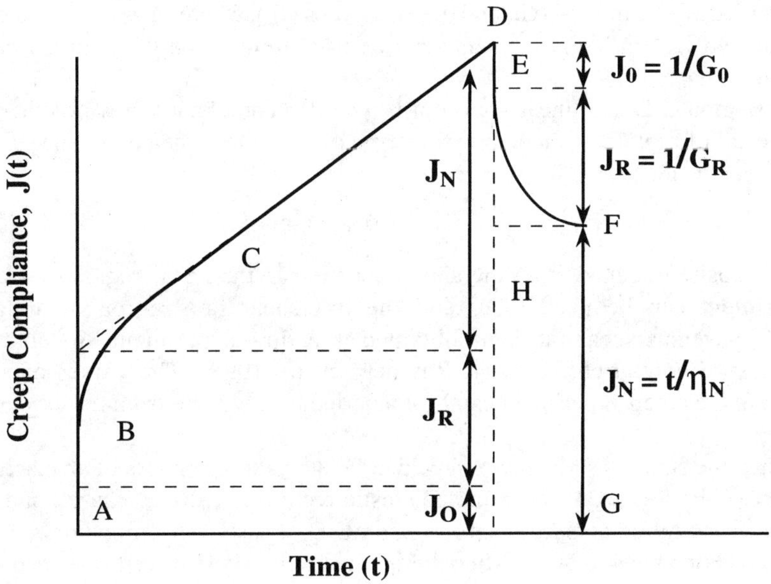

Figure 3–37 Illustration of a Creep-Compliance Curve.

---

was removed, the structure of the sample would recover completely. The compliance is related to the corresponding modulus as

$$J_0 = 1/G_0 \tag{3.85}$$

where G_0 is the instantaneous elastic modulus. Magnitudes of instantaneous elastic modulus, G_0, and of storage modulus G' at 1 rad s^{-1} were found to be of the same order of magnitude for a starch paste (Giboreau et al., 1994) and tomato concentrates (Yoo and Rao, 1996).

2. Region B-C corresponds to a time dependent retarded elastic region with a compliance J_R. In this region the bonds break and reform, but all of them do not break and reform at the same rate. The equation for this part, using mean values for the parameters, is

$$J_R = J_m\,[1 - \exp(-t/\tau_m)] \tag{3.86}$$

In Equation 3.86, J_m is the mean compliance of all the bonds, and τ_m is the mean retardation time; τ_m equals $J_m\ \eta_m$, where η_m is the mean viscosity associated with elasticity. One can replace the mean quantities with the spectrum of retarded elastic moduli (G_i) and the viscosities (η_i), where, $J_i = 1/G_i$. Typically, one or two Kelvin-Voigt elements can be used to describe the retarded elastic region.

3. Region C-D is a linear region of Newtonian compliance in which the units as a result of rupture of the bonds flow past one another. The compliance J_N and the viscosity are related by

$$J_N = t/\eta_N \tag{3.87}$$

The viscosity obtained from the above equation in the linear region of a creep experiment can be used to extend the low-shear rate region of apparent viscosity versus shear rate data obtained in a flow experiment by about two decades (Giboreau et al., 1994; Rayment et al., 1998). The values of strain (γ) from the creep experiment can be converted to shear rate from the expression $\dot{\gamma} = \gamma(t)/t$.

When the stress is suddenly removed at D, the pattern consists of an elastic recovery (D-E) followed by a retarded elastic recovery (E-F). Because bonds between structural units are broken in region C-D, a part of the structure is not recovered. For values of time after the applied stress is removed, the recovered strain approaches a maximum value, S_r, called recoverable shear (Dealy, 1982):

$$S_r = \sigma_c \, J_e^0 \tag{3.88}$$

Where σ_c is the applied stress and J_e^0 is the value of the straight line portion of the creep-compliance curve extrapolated to zero time. A shear modulus, G_r, can be defined as the reciprocal of the steady state compliance, J_e^0:

$$\sigma_c = G_r \, S_r \tag{3.89}$$

This equation is called "Hooke's law of shear."

The creep test is a simple and inexpensive test for viscoelastic foods that provides valuable information on the rheological parameters. Davis (1973) pointed out that indeed too much information can be obtained from the creep test. For example, an eight-element rheological model defined by 11 parameters was required for shortening and lard. One drawback with creep studies using concentric cylinder systems is that the material's structure is disturbed when the sample is being loaded.

The creep-compliance technique has been used extensively by Sherman and coworkers for the study of ice cream, model emulsions, margarine, and butter (Sherman, 1966; Vernon Carter and Sherman, 1980; Sherman and Benton, 1980; Shama and Sherman, 1969). In these studies, the methodology employed was similar to that for ice cream; that is, the creep-compliance data on a sample were described in terms of mechanical models, usually containing four or six elements. Attempts were made to relate the parameters of the models to the structure of the samples studied. However, with increased emphasis on dynamic rheological tests and interpretation of results in terms of composition and structure, the use of mechanical models to interpret results of rheological tests has declined steadily.

Sherman (1966) studied the viscoelastic properties of ice cream mix and melted ice cream. The creep-compliance curve for the ice cream mix was described by the relation

$$J(t) = J_0 + J_1\,(1 - e^{-t/\tau_1}) + (t/\eta_N) \tag{3.90}$$

The term J_0 is the instantaneous compliance, and J_1 and J_2 are the compliances associated with retarded elastic behavior; $t_1 = \eta_1/G_1$ and $t_2 = \eta_1/G_2$ are the retardation times associated with retarded elasticity; η_1 and η_N are the viscosities associated with Newtonian flow.

For melted ice cream, the equation for creep-compliance data was

$$J(t) = J_0 + J_1\,(1 - e^{-t/\tau_1}) + J_2\,(1 - e^{-t/\tau_2}) + (t/\eta_N) \tag{3.91}$$

The mechanical analogs corresponding to Equations 3.90 and 3.91 are shown in Figure 3–38. It is important to note that an association was established between the components of the ice cream and the components of the rheological model. This in turn permitted the justification of a six-element model for the mix and a four-element model for the melted ice cream.

The creep test can be also used for solid foods, and, as in the dynamic tests, a compressive or elongational stress can be employed (Rao et al., 1995). However, when a compressive stress is imposed, the sample–metal platen interface should be either lubricated or bonded to minimize frictional effects.

Transient Viscoelastic Flow

One of the characteristics of viscoelastic foods is that when a shear rate is suddenly imposed on them, the shear stress displays an overshoot and eventually reaches a steady state value. Kokini and Dickie (1981) and Dickie and Kokini (1982) presented stress overshoot data for the foods indicated in the previous section. Figure 3–39 illustrates stress overshoot as a function of shear rate.

The stress overshoot data can be modeled by means of equations that contain rheological parameters related to the stresses (normal and shear) and shear rate. One such equation is that of Leider and Bird (1974):

$$\sigma = -K(\dot{\gamma})^n[1 + (b\dot{\gamma}t - 1)\exp(-t/a\,n\,\lambda)] \tag{3.92}$$

Where σ is the shear stress, K and n are power law parameters relating the shear stress and the shear rate, $\dot{\gamma}$ is the imposed sudden shear rate, t is the time, λ is a time constant, and a and b are adjustable parameters. The time constant is related

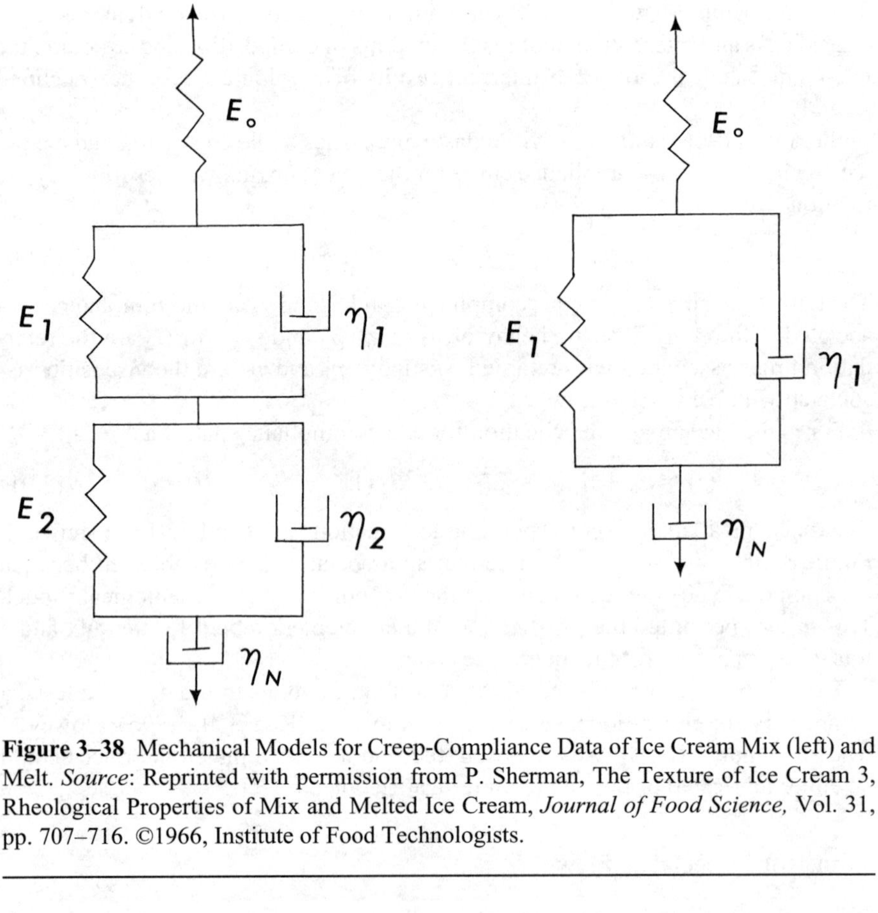

Figure 3–38 Mechanical Models for Creep-Compliance Data of Ice Cream Mix (left) and Melt. *Source*: Reprinted with permission from P. Sherman, The Texture of Ice Cream 3, Rheological Properties of Mix and Melted Ice Cream, *Journal of Food Science,* Vol. 31, pp. 707–716. ©1966, Institute of Food Technologists.

to the power law parameters relating shear stress and shear rate and those relating the primary normal stress difference and shear rate:

$$\lambda = (m'/2K)^{1/n' - n} \tag{3.93}$$

The Bird-Leider model predicted well the maximum and the steady state stresses as well as the times at which they occurred, but its prediction of the shear stress decay was not satisfactory (Dickie and Kokini, 1982).

A number of models have been developed to describe transient viscoelastic behavior. However, one must have at hand carefully obtained rheological data (e.g., obtained using low-inertia measurement systems) in order to test the applicability of the models. Another example of the applicability of models to viscoelastic data

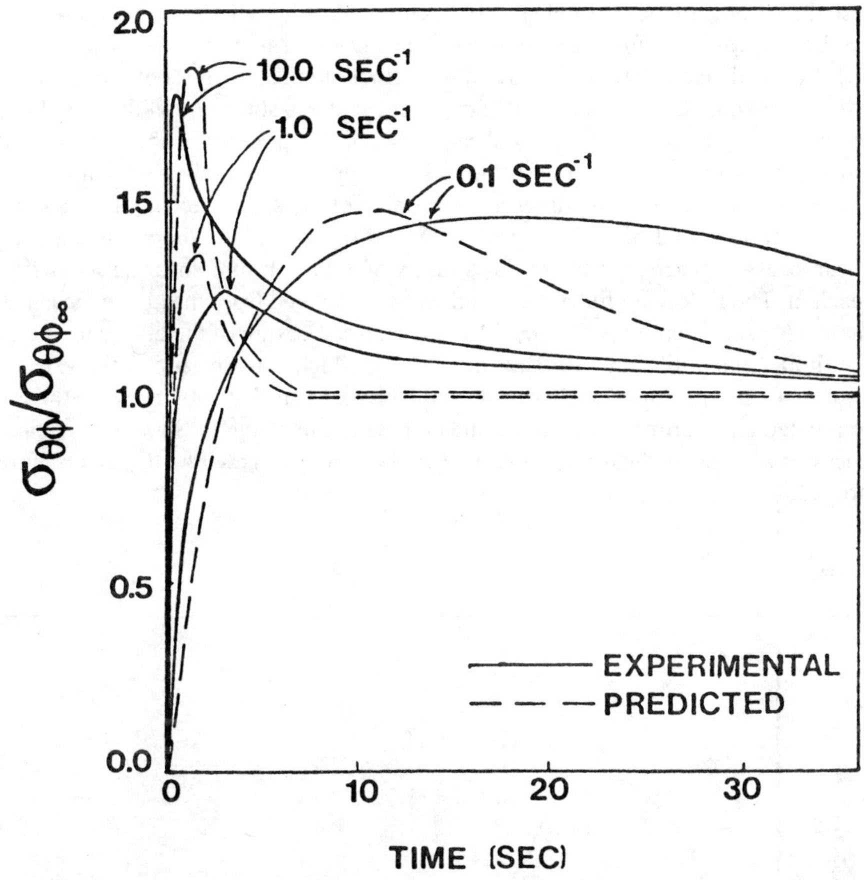

Figure 3–39 Stress Overshoot Data of Several Foods as a Function of Shear Rate and Values Predicted by Leider and Bird Model. *Source*: Reprinted with permission from A.M. Dickie and J.L. Kokini, Use of the Bird-Leider Equation in Food Rheology, *Journal of Food Process Engineering*, Vol. 5, pp. 157–184. © 1982, Food and Nutrition Press.

is the study of Leppard and Christiansen (1975) in which the models proposed by Bogue and Chen, Carreau, and Spriggs were evaluated. In the case of foods, Mason et al. (1982) developed an empirical model to describe the transient data on stick butter, tub margarine, and canned frosting (Kokini and Dickie, 1981; Dickie and Kokini, 1982). For mayonnaises, a model proposed by Larson (1985) was tested by Campanella and Peleg (1987), as described in Chapter 5 in the section on rheological behavior of mayonnaises and salad dressings.

Elliott and Ganz (1977) proposed a phenomenological analysis of the recorded stress-overshoot data (Figure 3–40): σ_A = shear stress at the end of the initial linear portion of the plot, σ_P = peak shear stress, σ_∞ = equilibrium shear stress, and σ_R = residual shear stress after rotation has stopped. The total strain that has occurred at different shearing times is the product of the shear rate and time, so that $\dot{\gamma}\,t_A$ = strain at end of initial linear portion of the plot, $\dot{\gamma}\,t_P$ = strain when peak shear stress is reached, and $\dot{\gamma}\,t_\infty$ = strain when equilibrium shear stress is first reached. The following functions can then be calculated from the above measurements: (1) shear modulus $G = \sigma_A/\gamma\,t_A$, which is a measure of the elastic nature of the fluid; (2) equilibrium viscosity $\eta_\infty = \sigma_\infty/\dot{\gamma}$, which is a measure of the viscous nature of the fluid; and (3) work of structure breakdown W = area between the extrapolated equilibrium shear stress line and the curve above it. Stress growth and decay at a constant shear rate were analyzed as the excess work (W) of structure breakdown.

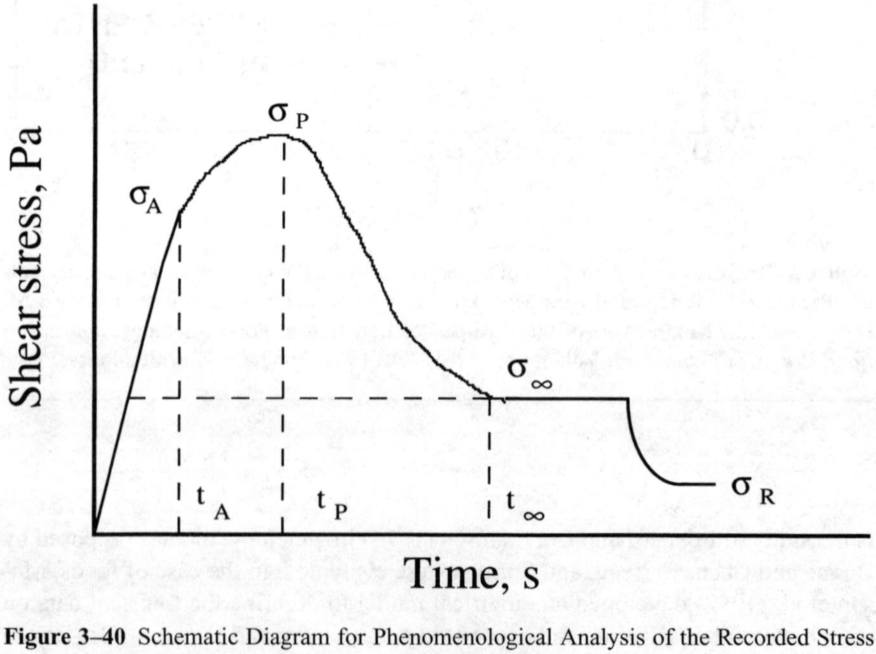

Figure 3–40 Schematic Diagram for Phenomenological Analysis of the Recorded Stress Overshoot Data at a Constant Shear Rate. Elliott and Ganz, 1977.

$$W = \dot{\gamma} \int_{A}^{B} (\sigma - \sigma_{\infty})\, dt \tag{3.94}$$

where $\dot{\gamma}$ is the constant shear rate, σ_{∞} is the equilibrium shear stress value, and A and B are the times at which the stress curve overshot the equilibrium stress value and the time at which the equilibrium value was reached, respectively. The analysis of Elliott and Ganz (1977) is simple and can provide useful information.

Stress Relaxation

In a stress-relaxation experiment, the decay of stress as a function of time is recorded at a fixed magnitude of deformation. The experiments can be performed with an instrument such as the Instron Universal Testing machine. However, to minimize errors due to friction between the test sample and the surfaces of the platens that are in contact with the sample, the sample must be either bonded to the platens or the contact surfaces must be well lubricated. In many studies, lubrication of the contact surfaces was employed, as it is easier than bonding the sample to the platens. The salient features of the stress-relaxation technique and the type of results that can be obtained are illustrated by a study of the rheological behavior of pectin sugar gels made from a commercial slow set pectin as a function of concentration, pH, and sugar content, and aged for 48 hours, by means of the stress-relaxation technique (Comby et al., 1986). A typical stress versus time curve is presented in Figure 3–41. During the loading period ($t < t^*$), a linear relation between force and displacement was observed, and a compression modulus (E_0) was calculated from the relation

$$E_0 = \sigma^*/\varepsilon^* = (L_0/A_0)(F^*/\Delta l) \tag{3.95}$$

where F^* is the force corresponding to the deformation ε^* and A_0 is the initial cross-section of the gel; $\varepsilon^* = (\Delta l/L_0)$, where Δl is the displacement, and L_0 is the initial height of the gel sample; $F^*/\Delta l$ is the slope of the loading curve. From the relaxation curves, the Young's modulus $E(t)$ can be obtained after taking into consideration the relaxation phenomenon during the loading period. One such procedure is the Zappas approximation reported by Lin and Aklonis (1980):

$$E(t - t^*/2) = \sigma(t)/\varepsilon^* \tag{3.96}$$

This method can be used to estimate $E(t)$ very accurately using data from $t = 2t^*$; but from $t = 1.3t^*$ the approximation is good. The relaxation curves were characterized by an inflection point between 20 and 60 seconds. Assuming that the in-

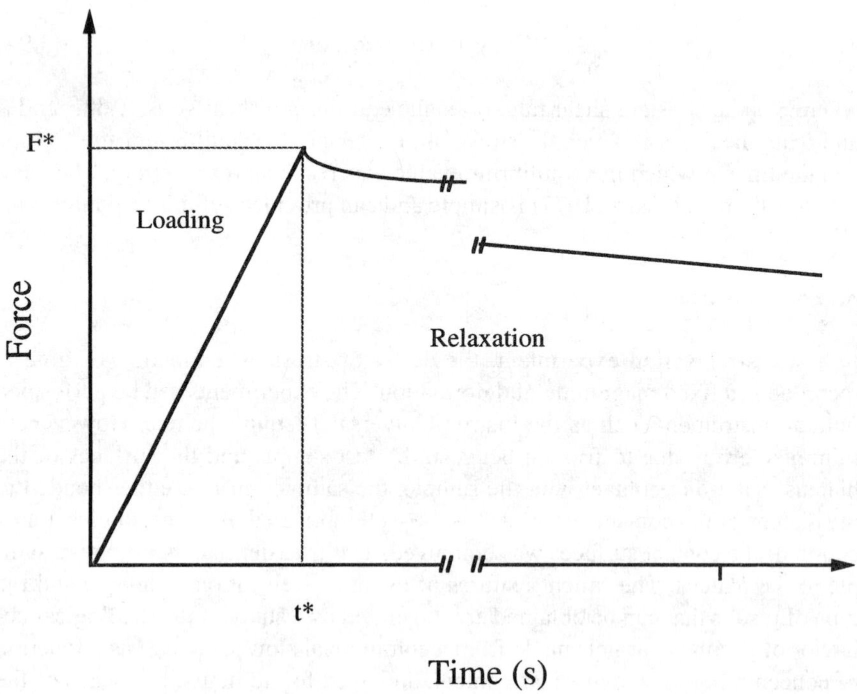

Figure 3–41 Typical Stress Time versus Time Curve Obtained from a Stress-Relaxation Experiment. Rao, 1992.

flection point corresponds to the classical plateau zone, the value of the modulus at this point (E_p) may be viewed as a pseudo-elastic modulus and interpreted according to theory of gel elasticity that was proposed for gels with low concentrations ($c < 0.4\%$) (Oakenfull, 1984). One important determination was that the molecular weight of the junction zones was about one-fourth of the number average molecular weight; that is, each molecule can be involved in only two or three junction zones.

In addition, Comby et al. (1986) calculated the parameters of a two-element Maxwell model:

$$E(t) = \Sigma E_i \exp(-t/\tau_i) \tag{3.97}$$

where τ_i is the relaxation time. Magnitudes of E_0, E_p, and E_1 increased with increase in pectin concentration. However, E_2 and τ_2 were not much affected up to pectin concentration (c) of 1.7%, suggesting that in turn the relaxation processes of chain segments between cross-linking zones were not affected; however, when the polymer concentration (c) was increased from 1.7% to 1.8%, there was a sharp increase in τ_1, reflecting a decrease in chain mobility. The effect of pH on stress-relaxation curves was examined for 1.7% pectin gels: E_p and E_0 increased steadily when pH decreased from 3.2 to 2.6 but remained constant below a pH of 2.6. Increasing sucrose content of 1.7% pectin gels from 60.3% to 72.3% resulted in increase in the elastic modulus from 6750 to 28,300 Pa.

The stress-relaxation test can be used for characterizing solid foods after taking precautions to minimize friction at the platen-food interface (Rao et al., 1995). For deformations in the nonlinear range, the empirical data analysis technique of Peleg (1980) has been used in several studies (e.g., Nussinovitch et al., 1990):

$$\frac{F_0 t}{F_0 - F(t)} = k_1 + k_2 t \tag{3.98}$$

where F_0 and $F(t)$ are the initial force and where, at any instant, and k_1 and k_2 are constants.

Static Measurement of Modulus

Saunders and Ward (1954) showed that a sample of gel placed in a U-tube together with a capillary containing an indexing fluid (Figure 3–42) can be used to determine the rigidity modulus of weak gels. Assuming that the gel is entirely elastic, the shear modulus, E, can be calculated from the following equation:

$$E = Pr_o^4/8a^2hL \tag{3.99}$$

where P is the effective pressure (i.e., the applied pressure corrected for the back pressure of the indexing liquid); r_o and a are the radii of the wide and narrow tubes, respectively; h is the displacement of the index liquid; and L is the length of the gel. Because the strain varies from a maximum value of $4a^2h/r_o^3$ at the tube wall to zero at the tube center, the volume average strain is obtained as $8a^2h/3r^3$ $8a^2h/3r_o^3$. The stress is usually applied for 30 seconds, and the displacement of the indexing liquid is recorded as a function of time. However, many food gels show a time-dependent strain (Stainsby et al., 1984). For gels that exhibit syneresis, the test sample as a whole slips along the tube when pressure is applied. Attempts to minimize slip of the sample include introducing dimples into the wall of the wide

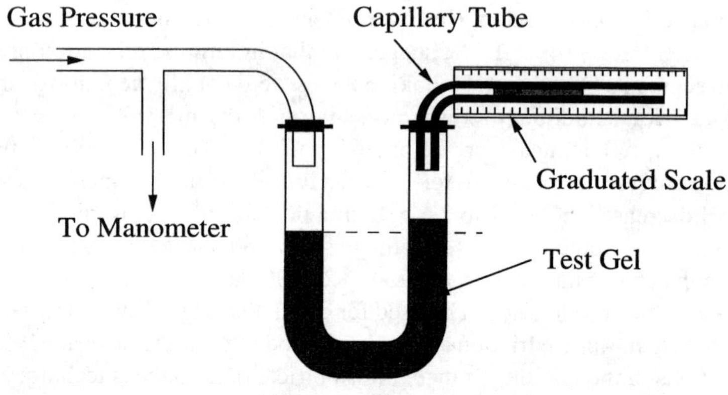

Figure 3–42 U-Tube Apparatus for Determination of Gel Modulus. Rao, 1992.

tube (Komatsu and Sherman, 1974) and coating the tube with a single layer of antibumping granules about 1 mm in diameter (Stainsby et al., 1984). The latter technique, using tubes 6.25 mm and 0.478 mm in radii, was found to be suitable for determining the rigidity moduli of starch, amylose, κ-carrageenan, and egg white gels. Of further interest here is that with the modified apparatus, time-dependent strain was not observed for the studied gels so that it may not be possible to obtain creep data with this technique.

Oakenfull et al. (1989) presented a method for determining the absolute shear modulus (E) of gels from compression tests in which the force, F, and the strain or relative deformation (δ/L) are measured with a cylindrical plunger with radius r on samples in cylindrical containers of radius R, as illustrated in Figure 3–43. Assuming that the gel is an incompressible elastic solid, the following relationships were derived:

$$E = f\, Y_e \tag{3.100}$$

where f is a parameter whose magnitude depends on the geometrical ratios r/R and R/L, as shown in Table 3–5, and

$$Y_e = (F/\pi r^2)/(\delta/L) \tag{3.101}$$

A correction must be applied to the measured force by subtracting the buoyancy effect, F_h:

$$F_h = \pi\, q\, g\, \delta\, r^2\, R^2/(R^2 - r^2) \tag{3.102}$$

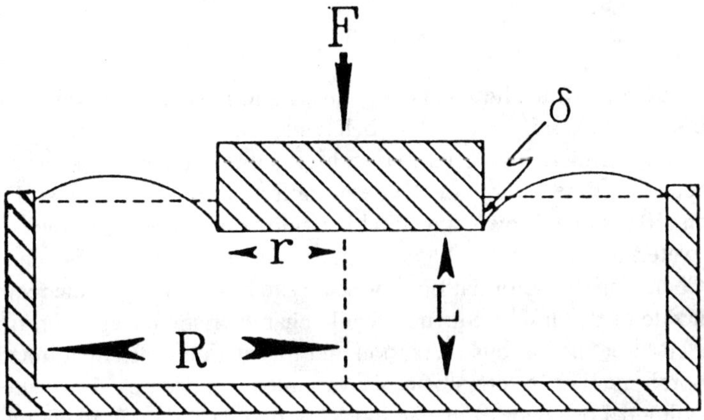

Figure 3–43 Plunger-Cup Geometry for Determination of Modulus of Weak Gels. *Source*: Reprinted with permission from Oakenfull et al., Method for Determining Absolute Shear Modulus of Gels from Compression Tests, *Journal of Texture Studies*, Vol. 19, pp. 407–417. ©1989, Food and Nutrition Press.

Although the technique was illustrated for gelatin gels, it should be applicable to other gels also. Note that because magnitudes of the force, *F*, were very small, they were measured with a sensitive electronic balance.

Deborah Number

A unique dimensionless number in rheology is the Deborah (De) number (named after the song of Deborah, Book of Judges, the Old Testament) that compares the time scale of the deformation process (t_D) to the observation time (t_O):

Table 3–5 Factors for Converting Apparent Young's Modulus into Absolute Shear Modulus for Different Sizes of Dish and Probe. Oakenfull et al., 1989

r/R	L/R = *0.1*	L/R = *0.2*	L/R = *0.5*	L/R = *1.0*
0.05	0.1570	0.0752	0.0381	0.0186
0.10	0.2080	0.104	0.0634	0.0331
0.20	0.1850	0.102	0.0912	0.0522
0.40	0.0549	0.0632	0.0930	0.0574
0.50	0.0264	0.0490	0.0803	0.0485
0.60	0.0138	0.0384	0.0622	0.0359

$$\mathrm{De} = \frac{t_\mathrm{D}}{t_\mathrm{o}} \tag{3.103}$$

typically, the time t_D for liquids is very small and is very large for solids. Small Deborah numbers signify liquid-like behavior, and large numbers signify solid-like behavior. Further, a material may exhibit solid-like character because either its characteristic time is large or the deformation process is very fast. For example, when a flat stone thrown at a suitable angle bounces several times on a calm body of water, because of the fast deformation process, the water is behaving like a solid. Characteristic deformation time scale can be estimated as the reciprocal of the shear rate or dynamic frequency, while characteristic observation times may be assumed to be the various relaxation and retardation times mentioned earlier. The Deborah number in terms of other variables has been used to better understand rheological behavior of dispersions (Chapter 4) and other phenomena.

Commercial Rheometers

Many of the earlier commercial viscometers have been of the constant shear rate design where the speed of rotation was the controlled variable and the resulting shear stress was measured. However, viscometers and rheometers in which the shear stress is controlled and resulting shear rate or strain can be measured are available now. Because they provide an opportunity to conduct studies related to yield stress, creep-compliance, stress relaxation, and rate of breakdown of weak structures, it seems likely that their use will become more widespread in the future. In addition, because with most fluid foods the stresses to be measured are of relatively low magnitudes, a few low-shear rheometers that are capable of measuring them have been introduced.

Lists of available models of rheometers and their manufacturers were given by Whorlow (1980), Mitchell (1984), and Ma and Barbosa-Cánovas (1995). However, given that numerous models of a viscometer or rheometer may be available, accurate description of all commercially available viscometers and rheometers is a futile effort. It is much more important to focus on the proper design of a measurement geometry (e.g., cone-plate, concentric cylinder), precision in measurement of strain and/or shear rate, and inertia of a measuring system and correction for it, as well as to verify that the assumptions made in deriving the applicable equations of shear rate have been satisfied and to ensure that the results provided by the manufacturer are indeed correct.

Some Precautions with Foods

In addition to using a well-designed viscometer or a rheometer, many precautions should be taken to ensure that the rheological data obtained on a food sample are reliable. The first step is to choose a measurement geometry that is amenable to analysis of the flow in the geometry so that a mathematical expression for the shear rate can be derived. A specific food and a measuring geometry can pose their unique set of problems during rheological experiments. Often, these problems can be identified by careful examination of the steps involved in the measurements and careful observation during the experiment; they should be analyzed and steps taken to ensure that they are overcome. Some of the factors to be considered in using the capillary, cone-plate, parallel plate, and concentric cylinder geometries were mentioned under each geometry in this chapter.

One serious problem in working with foods is dehydration of the test sample, such as from the edge of a parallel plate or a cone-plate geometry, or the exposed annular ring of a concentric cylinder geometry. Moisture loss from a test sample can be a significant factor (1) during a long-duration low-temperature experiment, (2) during a short-duration high-temperature rheological experiment, or (3) when the humidity of the surrounding atmosphere is low. In particular, the exposed sample surface area is large in the cone-plate and the parallel plate geometries. In dynamic rheological tests, silicone or mineral oil can be placed at the sample edge to prevent moisture loss, but care must be taken that the oil does not penetrate into the food or dissolve a food component (e.g., lipids by a mineral oil). In the concentric cylinder geometry, only the annular ring-shaped food surface at the top is exposed.

Attempts to minimize moisture loss from a food sample during a shear rate versus shear stress test include placing a solvent trap above the cone-plate and parallel plate geometries, placing an annular ring/disk above the exposed surface of a concentric cylinder geometry, and saturating the surrounding atmosphere with moisture. It is emphasized that sometimes with rotating viscometer geometries cone-plate, parallel plate, and concentric cylinder—one may at best minimize moisture loss but may not be able to eliminate it. For example, Rayment et al. (1998) found moisture loss from guar gum solutions containing rice starch was very severe at concentrations greater than 33% of the latter and could not obtain reliable data. The converse problem, that is, moisture gain due to condensation at low measurement temperatures, is of concern because the moisture can dilute the test sample significantly. With a tube/capillary viscometer, a closed system can be designed to prevent both moisture loss and gain.

Another common problem with two-phase foods (e.g., mustard) and melted cheese is the formation of a thin water or oil layer next to the walls of the measuring geometry. Because this layer of water or oil has a finite velocity during the experiment (slip), the zero-velocity boundary condition (no slip) at the interphase between the food and the measuring geometry that is used in deriving the equations for shear rate is not satisfied.

It is tempting to study time-dependent rheological behavior of a food and apply different flow models to the data collected. However, during loading of the sample into a viscometer geometry, the structure of the food is altered, and the degree to which the structure is altered depends on the shearing during sample loading. Therefore, the results obtained will depend on the extent of shearing the sample undergoes prior to data collection. Often, the remedy chosen is to allow the sample to rest for a period of time during which the structure recovers. Obviously, another option, when possible, is to create the test sample in situ in the test geometry and conduct the rheological tests without going through the steps of loading the sample into the measuring geometry.

REFERENCES

Abdelrahim, K.A., H.S. Ramaswamy, and F.R. Van de Voort. 1995. Rheological properties of starch solutions under aseptic processing temperatures. *Food Res Int* 28: 473–480.

Barnes, H.A., J.F. Hutton, and K. Walters. 1989. *An introduction to rheology.* Amsterdam: Elsevier Science.

Bird, R.B., W.E. Stewart, and E.N. Lightfoot. 1960. *Transport phenomena.* New York: John Wiley and Sons.

Bistany, K.L., and J.L. Kokini. 1983. Dynamic viscoelastic properties of foods in texture control. *J Rheol* 27: 605–620.

Bongenaar, J.J.T., N.W.F. Kossen, B. Metz, and F.W. Meijboom. 1973. A method for characterizing the rheological properties of viscous fermentation broths. *Biotech Bioeng* 15: 201–206.

Briggs, J.L., and J.F. Steffe. 1996. Mixer viscometer constant (k′) for the Brookfield small sample adapter and flag impeller. *J Texture Studies* 27: 671–677.

Brodkey, R.S. 1967. *The phenomena of fluid motions.* Reading, MA: Addison-Wesley Publishing Co.

Campanella, O.H., and M. Peleg. 1987. Analysis of the transient flow of mayonnaise in a coaxial cylinder viscometer. *J Rheol* 31: 439–452.

Campanella, O.H., L.M. Popplewell, J.R. Rosenau, and M. Peleg. 1987. Elongational viscosity measurements of melting American process cheese. *J Food Sci* 52: 1249–1251.

Cannon Instrument Co. 1982. *Instructions for the use of the Cannon-Ubbelohde dilution viscometer.* State College, PA.

Casiraghi, E.M., E.B. Bagley, and D.D. Christianson. 1985. Behavior of mozzarella, cheddar and processed cheese spread in lubricated and bonded uniaxial compression. *J Texture Studies* 16: 281–301.

Castell-Perez, M.E., J.F. Steffe, and R.G. Morgan. 1987. Adaptation of a Brookfield (HBTD) viscometer for mixer viscometer studies. *J Texture Studies* 18: 359–365.

Champenois, Y.C., M.A. Rao, and L.P. Walker. 1998. Influence of α-amylase on the viscoelastic properties of starch-gluten pastes and gels. *J Sci Food Agric* 127–133.

Chatraei, S.H., C.W. Macosko, and H.H. Winter. 1981. A new biaxial extensional rheometer. *J Rheol* 25: 433–443.

Clark, A.H., and S.B. Ross-Murphy. 1987. Structural and mechanical properties of biopolymer gels. *Adv Polymer Sci* 83: 57–192.

Clark, R. 1997. Evaluating syrups using extensional viscosity. *Food Technol* 511: 49–52.

Cogswell, F.N. 1972. Converging flow of polymer melts in extrusion dies. *Poly Eng Sci* 12: 64–73.

Cogswell, F.N. 1978. Converging flow and stretching flow: A compilation. *J Non-Newtonian Fluid Mech* 4: 23–38.

Comby, S., J.L. Doublier, and J. Lefebvre. 1986. Stress-relaxation study of high-methoxyl pectin gels. In *Gums and stabilisers for the food industry 3,* eds. G.O. Phillips, D.J. Wedlock, P.A. Williams, 203–212. New York: Elsevier Science.

Cox, W.P., and E.H. Merz. 1958. Correlation of dynamic and steady flow viscosities. *J Polymer Sci* 28: 619–622.

Dail, R.V., and J.F. Steffe. 1990. Rheological characterization of crosslinked waxy maize starch solutions under low acid aseptic processing conditions using tube viscometry techniques. *J Food Sci* 55: 1660–1665.

da Silva, P.M.S., J.C. Oliveira, and M.A. Rao. 1997. The effect of granule size distribution on the rheological behavior of heated modified and unmodified maize starch dispersions. *J Texture Studies* 28: 123–138.

Davis, S.S. 1973. Rheological properties of semi-solid foodstuffs. *J Texture Studies* 4: 15–40.

Dealy, J.M. 1982. *Rheometers for molten polymers. A practical guide to testing and property measurement.* New York: Van Nostrand Reinhold.

Dickie, A.M., and J.L. Kokini. 1982. Use of the Bird-Leider equation in food rheology. *J Food Process Eng* 5: 157–174.

Doraiswamy, D., A.N. Mujumdar, I. Tsao, A.N. Beris, S.C. Danforth, and A.B. Metzner. 1991. The Cox-Merz rule extended: A rheological model for concentrated suspensions and other materials with a yield stress. *J Rheol* 35: 647–685.

Durand, D., C. Bertrand, J.P. Busnel, J. Emery, M.A.V. Axelos, J.F. Thibault, J. Lefebvre, J.L. Doublier, A.H. Clark, and A. Lips. 1990. Physical gelation induced by ion complexation: Pectin-calcium systems. In *Physical networks,* eds. W. Burchard and S.B. Ross-Murphy, 283–300. London: Elsevier Applied Science.

Dzuy, N.Q., and D.V. Boger. 1983. Yield stress measurement for concentrated suspensions. *J Rheol* 27: 321–349.

Elliott, J.H., and A.J. Ganz. 1971. Modification of food characteristics with cellulose hydrocolloids, I. Rheological characterization of an organoleptic property. *J Texture Studies* 2: 220–229.

Elliott, J.H., and A.J. Ganz. 1977. Salad dressings—Preliminary rheological characterization. *J Texture Studies* 8: 359–371.

Ferry, J.D. 1980. Viscoelastic properties of polymers. New York: John Wiley & Sons.

Geankoplis, C.J. 1983. *Transport processes and unit operations.* 2d ed. London: Allyn Bacon, Inc.

Giboreau, A., G. Cuvelier, and B. Launay. 1994. Rheological behavior of three biopolymer/water systems with emphasis on yield stress and viscoelastic properties. *J Texture Studies* 25: 119–137.

Grikshtas, R., and M.A. Rao. 1993. Determination of slip velocities in a concentric cylinder viscometer with Mooney and Kiljanski methods. *J Texture Studies* 24: 173–184.

Grosso, C.R.F., and M.A. Rao. 1998. Dynamic rheology of structure development in low-methoxyl pectin+Ca^{2+}+sugar gels. *Food Hydrocolloids* 12: 357–363.

Hamann, D.D. 1987. Methods for measurement of rheological changes during thermally induced gelation of proteins. *Food Technol 41,* no. 3: 100, 102–108.

Hansen, L.M., R.C. Hoseney, and J.M. Faubion. 1990. Oscillatory probe rheometry as a tool for determining the rheological properties of starch-water systems. *J Texture Studies* 21: 213–224.

James, A.E., D.J.A. Williams, and P.R. Williams. 1987. Direct measurement of static yield properties of cohesive suspensions. *Rheol Acta* 26: 437–446.

Jao, Y.C., A.H. Chen, D. Lewandowski, and W.E. Irwin. 1978. Engineering analysis of soy dough rheology in extrusion. *J Food Process Eng* 2: 97–112.

Keentok, M. 1982. The measurement of the yield stress of liquids. *Rheol Acta* 21: 325–332.

Khagram, M., R.K. Gupta, and T. Sridhar. 1985. Extensional viscosity of xanthan gum solutions. *J Rheol* 29: 191–207.

Kokini, J.L., and A. Dickie. 1981. An attempt to identify and model transient viscoelastic flow in foods. *J Texture Studies* 12: 539–557.

Komatsu, H., and P. Sherman. 1974. A modified rigidity modulus technique for studying the rheological properties of w/o emulsions containing microcrystalline wax. *J Texture Studies* 5: 97–104.

Kulicke, W.M., and R.S. Porter. 1980. Relation between steady shear flow and dynamic rheology. *Rheol Acta* 19: 601–605.

Larson, R.G. 1985. Constitutive relationships for polymeric materials with power-law distributions of relaxation times. *Rheol Acta* 24: 327–334.

Leider., P.J., and R.B. Bird. 1974. Squeezing flow between parallel disks—I. Theoretical analysis. *Ind Eng Chem Fundamentals* 13: 336–341.

Leppard, W.R., and E.B. Christiansen. 1975. Transient viscoelastic flow of polymer solutions. *Am Inst Chem Engrs J* 21: 999–1006.

Lin, K.S.C., and J.J. Aklonis. 1980. Evaluation of the stress-relaxation modulus of materials with rapid relaxation rates. *J Appl Phys* 51: 5125–5130.

Lopes da Silva, J.A.L., M.P. Gonçalves, and M.A. Rao. 1993. Viscoelastic behavior of mixtures of locust bean gum and pectin dispersions. *J Food Eng* 18: 211–228.

Lopes da Silva, J.A.L., M.P. Gonçalves, and M.A. Rao. 1994. Influence of temperature on dynamic and steady shear rheology of pectin dispersions. *Carbohydrate Polymers* 23: 77–87.

Lopes da Silva, J.A., M.A. Rao, and J-T. Fu. 1998. Rheology of structure development and loss during gelation and melting. In *Phase/state transitions in foods: Chemical, rheological and structural changes,* eds. M.A. Rao and R.W. Hartel, 111–156. New York: Marcel Dekker.

Ma, L., and G.V. Barbosa-Cánovas. 1995. Instrumentation for the rheological characterization of foods. *Food Sci Technol Int* 1: 3–17.

Macosko, C.W. 1994. *Rheology: Principles, measurements and applications.* New York: VCH Publishers.

Mason, P.L., K.L. Bistany, M.G. Puoti, and J.L. Kokini. 1982. A new empirical model to simulate transient shear stress growth in semi-solid foods. *J Food Process Eng* 6: 219–233.

Matsumoto, T., C. Hitomi, and S. Onogi. 1975. Rheological properties of disperse systems of spherical particles in polystyrene solution at long time-scales. *Trans Soc Rheol* 194: 541.

McKelvey, J.N. 1962. *Polymer processing.* New York: John Wiley & Sons.

Metzner, A.B., and R.E. Otto. 1957. Agitation of non-Newtonian fluids. *Am Inst Chem Engrs J* 3: 3–10.

Mills, P., and J.L. Kokini. 1984. Comparison of steady shear and dynamic viscoelastic properties of guar and karaya gums. *J Food Sci* 49: 1–4, 9.

Mitchell, J.R. 1984. Rheological techniques. In *Food analysis: Principles and techniques,* eds. D.W. Gruenwedel and J.R. Whitaker, 151–220. New York: Marcel Dekker.

Mooney, M. 1931. Explicit formulas for slip and fluidity. *J Rheol* 2: 210–222.

Morris, E.R. 1981. Rheology of hydrocolloids. In *Gums and stabilisers for the food industry 2,* eds. G.O. Philips, D.J. Wedlock, and P.A. Williams, 57–78. Oxford, England: Pergamon Press.

Morris, E.R., A.N. Cutler, S.B. Ross-Murphy, and D.A. Rees. 1981. Concentration and shear rate dependence of viscosity in random coil polysaccharide solutions. *Carbohydrate Polymers* 1: 5–21.

Nussinovitch, A., G. Kaletunc, M.D. Normand, and M. Peleg. 1990. Recoverable work versus asymptotic relaxation modulus in agar, carrageenan and gellan gels. *J Texture Studies* 21: 427–438.

Oakenfull, D. 1984. A method for using measurements of shear modulus to estimate the size and thermodynamic stability of junction zones in non-covalently cross-linked gels. *J Food Sci* 49: 1103–1104, 1110.

Oakenfull, D.G., N.S. Parker, and R.I. Tanner. 1989. Method for determining absolute shear modulus of gels from compression tests. *J Texture Studies* 19: 407–417.

Okechukwu, P.E., M.A. Rao, P.O. Ngoddy, and K.H. McWatters. 1991. Rheology of sol-gel thermal transition in cowpea flour and starch slurry. *J Food Sci* 56: 1311–1315.

Owen, S.R., M.A. Tung, and A.T. Paulson. 1992. Thermorheological studies of food polymer dispersions. *J Food Eng* 16: 39–53.

Padmanabhan, M. 1995. Measurement of extensional viscosity of viscoelastic liquid foods. *J Food Eng* 25: 311–327.

Padmanabhan, M., and M. Bhattacharya. 1993. Planar extensional viscosity of corn meal dough. *J Food Eng* 18: 389–411.

Peleg, M. 1980. Linearization of relaxation and creep curves of solid biological materials. *J Rheol* 24: 451–463.

Perkins, T.T., D.E. Smith, and S. Chu. 1997. Single polymer dynamics in an elongational flow. *Science* 276: 2016–2021.

Plazek, D.J. 1996. 1995 Bingham medal address: Oh, thermorheological simplicity, wherefore art thou? *J Rheol* 40: 987–1014.

Qiu, C-G., and M.A. Rao. 1988. Role of pulp content and particle size in yield stress of apple sauce. *J Food Sci* 53: 1165–1170.

Qiu, C-G., and M.A. Rao. 1989. Effect of dispersed phase on the slip coefficient of apple sauce in a concentric cylinder viscometer. *J Texture Studies* 20: 57–70.

Rao, M.A. 1975. Measurement of flow properties of food suspensions with a mixer. *J Texture Studies* 6: 533–539.

Rao, M.A. 1977a. Rheology of liquid foods—A review. *J Texture Studies* 8: 135–168.

Rao, M.A. 1977b. Measurement of flow properties of fluid foods—Developments, limitations, and interpretation of phenomena. *J Texture Studies* 8: 257–282.

Rao, M.A. 1992. Measurement of viscoelastic properties of fluid and semisolid foods. In *Viscoelastic properties of food,* eds. M.A. Rao and J.F. Steffe, 207–232. London: Elsevier Applied Science.

Rao, M.A. 1995. Rheological properties of fluid foods. In *Engineering properties of foods,* 2d ed., eds. M.A. Rao and S.S.H. Rizvi, 1–53. New York: Marcel Dekker.

Rao, M.A., and H.J. Cooley. 1984. Determination of effective shear rates of complex geometries. *J Texture Studies* 15: 327–335.

Rao, M.A., and H.J. Cooley. 1992. Rheology of tomato pastes in steady and dynamic shear. *J Texture Studies* 23: 415–425.

Rao, M.A., and H.J. Cooley. 1993. Dynamic rheological measurement of structure development in high-methoxyl pectin/fructose gels. *J Food Sci* 58: 876–879.

Rao, M.A., H.J. Cooley, and H-J. Liao. 1999. Flow properties of tomato puree and starch dispersion at high temperatures with a direct-drive concentric cylinder viscometer. *J Food Process Eng* (In press).

Rao, V.N.M., R.A.M. Delaney, and G.E. Skinner. 1995. Rheological properties of solid foods. In *Engineering properties of foods,* 2d ed., eds. M.A. Rao and S.S.H. Rizvi, 55–97. New York: Marcel Dekker.

Rayment, P., S.B. Ross-Murphy, and P.R. Ellis. 1998. Rheological properties of guar galactomannan and rice starch mixtures. II. Creep measurements. *Carbohydrate Polymers* 35: 55–63.

Rieger, F., and V. Novak. 1973. Power consumption of agitators in highly viscous non-Newtonian liquids. *Trans Inst Chem Engrs* 51: 105–111.

Saunders, P.R., and A.G. Ward. 1954. An absolute method for the rigidity modulus of gelatine gel. In *Proceedings of the second international congress on rheology,* ed. V.G.W. Harrison, 284–290. New York: Academic Press.

Schlichting, H. 1960. *Boundary layer theory.* New York: McGraw-Hill.

Senouci, A., and A.C. Smith. 1988. An experimental study of food melt rheology. I. Shear viscosity using a slit die viscometer and a capillary rheometer. *Rheological Acta* 27: 546–554.

Sestak, J., R. Zitny, and M. Houska. 1983. Simple rheological models of food liquids for process design and quality assessment. *J Food Eng* 2: 35–49.

Shama, F., and P. Sherman. 1969. The influence of work softening on the viscoelastic properties of butter and margarine. *J Texture Studies* 1: 196–205.

Shama, F., and P. Sherman. 1973. Identification of stimuli controlling the sensory evaluation of viscosity. II. Oral methods. *J Texture Studies* 4: 111–118.

Sherman, P. 1966. The texture of ice cream 3. Rheological properties of mix and melted ice cream. *J Food Sci* 31: 707–716.

Sherman, P. 1970. *Industrial rheology.* New York: Academic Press.

Sherman, P., and M. Benton. 1980. Influence of skim milk powder/recodan R S ratio on the viscoelasticity of groundnut oil-in-water imitation milks. *J Texture Studies* 11: 1–13.

Shomer, I., M.A. Rao, M.C. Bourne, and D. Levy. 1993. Rheological behavior of potato tuber cell suspensions during temperature fluctuations and cellulase treatments. *J Sci Food Agric* 63: 245–250.

Smith, T.L., J.D. Ferry, and F.W. Schremp. 1949. Measurement of the mechanical properties of polymer solutions by electromagnetic transducers. *J App Phys* 20: 144.

Sridhar, T., V. Tirtaatmadja, D.A. Nguyen, and R.K. Gupta. 1991. Measurement of extensional viscosity of polymer solutions. *J Non-Newtonian Fluid Mechanics* 40: 271–280.

Stainsby, G., S.G. Ring, and G.R. Chilvers. 1984. A static method for determining the absolute shear modulus of a syneresing gel. *J Texture Studies* 15: 23–32.

Steffe, J.F. 1996. *Rheological methods in food process engineering.* East Lansing, MI: Freeman Press.

Steiner, E.H. 1958. A new rheological relationship to express the flow properties of melted chocolate. *Revue Int Chocolaterie* 13: 290–295.

Tamura, M.S., J.M. Henderson, R.L. Powell, and C.F. Shoemaker. 1989. Evaluation of the helical screw rheometer as an on-line viscometer. *J Food Sci* 54: 483–484.

Tanner, R.I. 1988. Recoverable elastic strain and swelling ratio. In *Rheological measurements,* eds. A.A. Collyer and D.W. Clegg, 93–118. New York: Elsevier Applied Science.

Van Gurp, M., and J. Palmen. 1998. Time-temperature superposition for polymer blends. *Rheol Bull 67,* no. 1: 5–8.

Van Wazer, J.R., J.W. Lyons, K.Y. Kim, and R.E. Colwell. 1963. *Viscosity and flow measurement.* New York: Interscience Publishers.

Vernon Carter, E.J., and P. Sherman. 1980. Rheological properties and applications of mesquite tree Prosopis juliflora gum 2. Rheological properties and stability of o/w emulsions containing mesquite gum. *J Texture Studies* 11: 351–365.

Vitali, A.A., and M.A. Rao. 1982. Flow behavior of guava puree as a function of temperature and concentration. *J Texture Studies* 13: 275–289.

Whorlow, R.W. 1980. *Rheological techniques.* New York. Halsted Press.

Wood, F.W., and T.C. Goff. 1973. The determination of the effective shear rate in the Brabender Viscograph and in other systems of complex geometry. *Die Starke* 25: 89–91.

Wu, M.C., T.C. Lanier, and D.D. Hamann. 1985a. Rigidity and viscosity changes of croacker actomyosin during thermal gelation. *J Food Sci* 50: 14–19.

Wu, M.C., T.C. Lanier, and D.D. Hamann. 1985b. Thermal transitions of admixed starch/fish protein systems during heating. *J Food Sci* 50: 20–25.

Yang, W.H., and M.A. Rao. 1998. Complex viscosity-temperature master curve of cornstarch dispersion during gelatinization. *J Food Process Eng* 21: 191–207.

Yoo, B., and M.A. Rao. 1996. Creep and dynamic rheological behavior of tomato concentrates: Effect of concentration and finisher screen size. *J Texture Studies* 27: 451–459.

Yoo, B., M.A. Rao, and J.F. Steffe. 1995. Yield stress of food suspensions with the vane method at controlled shear rate and shear stress. *J Texture Studies* 26: 1–10.

Yoshimura, A., R.K. Prud'homme, H.M. Princen, and A.D. Kiss. 1987. A comparison of techniques for measuring yield stresses. *J Rheol* 31: 699–710.

Yoshimura, A., and R.K. Prud'homme. 1988a. Wall slip corrections for Couette and parallel disk viscometers. *J Rheol* 32: 53–67.

Yoshimura, A., and R.K. Prud'homme. 1988b. Wall slip effects on dynamic oscillatory measurements. *J Rheol* 32: 575–584.

Appendix 3–A
Analysis of Flow in a Concentric Cylinder Geometry

One can derive applicable equations for shear rate starting from either the general equations of conservation of mass (continuity) and conservation of momentum (motion) or by conducting balances of mass or momentum on a differential shell. Numerous examples of using the conservation equations or shell balances can be found in Bird et al. (1960) and other texts on transport phenomena. For the concentric cylinder geometry shown in Figure 3–1, it will be assumed that the inner cylinder is rotating and the outer cylinder is stationary. Interested readers can derive the applicable equations for the case of outer cylinder rotating using appropriate boundary conditions.

First, from a simple force balance, the total torque (M) on the inner cylinder is:

$$\sigma_{r\theta} \times 2\pi r_i h \times r_i = -M \tag{3A.1}$$

In terms of torque per unit height, $T = \dfrac{M}{h}$, Equation 3A.1 becomes

$$\sigma_{r\theta} = -\frac{T}{2\pi r_i^2} \tag{3A.2}$$

To describe the flow between the inner and outer cylinders, cylindrical coordinates ($r,\ \theta,\ z$) are used. Note that there is no flow in the radial ($v_r = 0$) and axial direction ($v_z = 0$) and due to symmetry $\dfrac{\partial}{\partial\theta} = 0$. The r-component of the equation of motion reduces to

$$-\rho\frac{v_\theta^2}{r} = \frac{1}{r}\frac{\partial(r\,\sigma_{rr})}{\partial r} - \frac{\sigma_{\theta\theta}}{r} \tag{3A.3}$$

Equation 3A.3 deals with the normal stresses σ_{rr} and $\sigma_{\theta\theta}$. For the purpose of illustration we consider the θ-component of the equation of motion in detail and note in particular that $v_\theta = f(r)$:

$$\rho\left(\frac{\partial v_\theta}{\partial t} + v_r\frac{\partial v_\theta}{\partial_r} + \frac{v_\theta}{r}\frac{\partial v_\theta}{\partial_\theta} + \frac{v_r\, v_\theta}{r} + v_z\frac{\partial v_\theta}{\partial_z}\right)$$

$$= \frac{1}{r}\frac{\partial p}{\partial r} - \left(\frac{1}{r^2}\frac{\partial}{\partial r}(r^2\sigma_{r\theta}) + \frac{1}{r}\frac{\partial \sigma_{\theta\theta}}{\partial \theta} + \frac{\partial \sigma_{\theta z}}{\partial z}\right) + \rho g_\theta \quad \text{(3A.4)}$$

Under steady flow condition $\frac{\partial v_\theta}{\partial t} = 0$, $v_r = 0$, $\frac{\partial v_\theta}{\partial \theta} = 0$, and $\frac{\partial v_\theta}{\partial z} = 0$, the left-hand side of Equation 3A.4 reduces to zero. On the right-hand side, $\frac{\partial p}{\partial r} = 0$ because there is no pressure gradient in the r direction and $\rho g_\theta = 0$ because there is no θ-component of gravity, and $\frac{\partial \sigma_{\theta\theta}}{\partial \theta}$ and $\frac{\partial \sigma_{\theta z}}{\partial z} = 0$ because there are no shear gradients in the θ and z directions. We thus have for the θ-component:

$$\frac{\partial(r^2\sigma_{r\theta})}{\partial r} = 0 \quad \text{(3A.5)}$$

The z-component reduces to

$$0 = -\frac{\partial p}{\partial z} + \rho g_z \quad \text{(3A.6)}$$

Equation 3A.6 describes the hydrostatic pressure in the gap between the two cylinders. Differentiation of Equation 3A.5 results in

$$r^2\, \mathrm{d}\sigma_{r\theta} + 2r\sigma_{r\theta}\, \mathrm{d}r = 0 \quad \text{(3A.7)}$$

Equation 3A.7 can be rearranged to

$$\mathrm{d}r = -\frac{r}{2}\frac{\mathrm{d}\sigma_{r\theta}}{\sigma_{r\theta}} \quad \text{(3A.8)}$$

The boundary condition at the inner cylinder rotating at an angular velocity Ω is $v_\theta = r_\mathrm{i}\Omega$ at $r = r_1$, and at the stationary outer cylinder is $v_\theta = 0$ at $r = r_\mathrm{o}$; both expressions are based again on the assumption of no-slip condition at a solid-fluid interface. The velocity distribution is obtained from

$$\frac{v_\theta}{r} = \int_{r_\mathrm{i}}^{r} \frac{\mathrm{d}(v_\theta/r)}{\mathrm{d}r}\, \mathrm{d}r \quad \text{(3A.9)}$$

Substituting for dr in 3A.9 and using the appropriate integration limits:

$$\frac{v_\theta}{r} = -\int_{\sigma_\mathrm{i}}^{\sigma_{r\theta}} \left(\frac{r\mathrm{d}(v_\theta/r)}{\mathrm{d}r}\right)\frac{\mathrm{d}\sigma_{r\theta}}{2\sigma_{r\theta}} = \int_{\sigma_{r\theta}}^{\sigma_\mathrm{i}} \left(\frac{r\,\mathrm{d}(v_\theta/r)}{\mathrm{d}r}\right)\frac{1}{2\sigma_{r\theta}}\,\mathrm{d}\sigma_{r\theta} \quad \text{(3A.10)}$$

By using the Leibnitz rule and noting to differentiate with respect to $\sigma_{r\theta}$, and with the boundary conditions $\sigma_{r\theta} = \sigma_\mathrm{i}$ at $r = r_\mathrm{i}$ and $\sigma_{r\theta} = \sigma_\mathrm{o}$ at $r = r_\mathrm{o}$:

$$\frac{\mathrm{d}(v_\theta/r)}{\mathrm{d}\sigma_\mathrm{i}} = \frac{1}{2\sigma_\mathrm{i}}\left[\left(r\frac{\mathrm{d}(v_\theta/r)}{\mathrm{d}r}\right)_\mathrm{i} - \left(r\frac{\mathrm{d}(v_\theta/r)}{\mathrm{d}r}\right)_\mathrm{o}\right] \quad \text{(3A.11)}$$

Noting that $\Omega = v_\theta/r$ and $\dot{\gamma} = \left(r\frac{d(v_\theta/r)}{dr}\right)$, Equation 3A.11 can be written as

$$2\sigma_i \frac{d\Omega_i}{d\sigma_i} = \dot{\gamma}_i - \dot{\gamma}_o \tag{3A.12}$$

Therefore, while one can calculate easily the shear stress at the surface of the rotating cylinder from Equation 3A.2, one can only derive an expression for the difference in shear rates at the surfaces of the inner and outer cylinders from the basic equations of flow. Additional work is required to calculate the corresponding shear rate $\dot{\gamma}_i$, and there have been several approaches to determination of $\dot{\gamma}_i$. One approach has been to apply infinite series solution to the differential equation in 3A.12.

A popular method of calculating the shear rate at the surface of the rotating cylinder is to assume that the test fluid follows the simple power law model:

$$\sigma = -K\left[r\frac{d(v_\theta/r)}{dr}\right]^n \tag{3A.13}$$

Equation 3A.12 can be solved after recalling Equation 3A.2 for shear stress:

$$\frac{v_\theta}{r} = -\left(\frac{T}{2nK}\right)^{1/n}\frac{n}{2}r^{-2/n} + c \tag{3A.14}$$

Using the boundary conditions $v_\theta = 0$ at $r = r_o$ and $v_\theta = r_i\Omega$ at $r = r_i$, one can obtain Equation 3.7:

$$\dot{\gamma}_i = \frac{2\Omega}{\left[1-\left(\frac{r_i}{r_o}\right)^2\right]}\left\{\frac{1-\left(\frac{r_i}{r_o}\right)^2}{n\left[1-\left(\frac{r_i}{r_o}\right)^{2/n}\right]}\right\} \tag{3.7}$$

When $n = 1$ (Newtonian fluid), the Margules equation for shear rate is obtained:

$$\dot{\gamma}_i = \frac{2\Omega_i}{\left[1-\left(\frac{r_i}{r_o}\right)^2\right]} \tag{3.3}$$

Examining the limiting form of the second part of Equation 3.7 as $r_i \to r_o$:

$$\text{Limit}\left\{\frac{1-\left(\frac{r_i}{r_o}\right)^2}{n\left[1-\left(\frac{r_i}{r_o}\right)^{2/n}\right]}\right\} = \lim_{r_i \to r_o}\left[\frac{-2(r_i/r_o)}{-n(2/n)\,(r_i/r_o)^{(2/n)-1}}\right] = 1 \tag{3A.15}$$

Therefore, when $r_i \to r_o$, the shear rate of a non-Newtonian fluid tends to that of a Newtonian fluid. In Table 3–1, values of the correction factor in parentheses in

Equation 3.7 are given for several values of the flow behavior index and the ratio (r_i/r_o) of the concentric cylinders, and they confirm that (1) they are small when $r_i \rightarrow r_o$, and (2) they may be large when the fluids are substantially non-Newtonian and when $(r_i/r_o) < 0.95$.

While the concentric cylinder geometry is relatively easy to use in rheological studies, some of its limitations, as shown in Figure 3A–1, should be recognized.

Thixotropic/Antithixotropic Behavior Expected? Yes → CCG may be suitable if sample is prepared in situ or time is allowed for sample to recover and measuring system inertia is low.

↓

No, and Steady Shear Data are Desired → Perhaps a CCG can be used if end effects are minimized.

↓

Is the fluid a suspension in which the solids tend to settle (e.g., ungelatinized starch granules in water) at any time in the CCG?

↓ Yes → Do not use CCG.

No

↓

Is the gap between cylinders very small? Yes → Perhaps one can use Newtonian $\dot{\gamma}$ as the true shear rate.

↓

No

↓

Correct Newtonian $\dot{\gamma}$ using Krieger or power law method.

Figure 3A–1 Decision Tree for Concentric Cylinder Geometry (CCG).

Appendix 3–B
Analysis of Steady Laminar Fully Developed Flow in a Pipe

The shell balance method will be used to examine steady laminar flow of a fluid in a pipe. For the geometrical system illustrated in Figure 3B–1 of steady laminar fully developed flow of a fluid, a shell momentum balance can be conducted (Bird et al., 1960; Geankoplis, 1983) using the cylindrical coordinates, r, θ, and z. The momentum balance is conducted on a control volume shell at a radius r with dimensions Δr and Δz.

1. The balance of two opposing forces is:

$$p\,A|_z - p\,A|_{z+\Delta z} = p(2\pi r\,\Delta r)|_z - p(2\pi r\,\Delta r)|_{z+\Delta z} \tag{3B.1}$$

2. The shear or drag force at radius $r = \sigma_{rz}\,(2\pi r\,\Delta z)$, which can be considered to be the momentum flow into the cylindrical surface of the shell. The net efflux of momentum is the difference between the magnitudes of the momentum out and momentum in:

$$\sigma_{rz}\,2\pi r\Delta z|_{r+\Delta r}\,(\text{OUT}) - \sigma_{rz}\,2\pi r\Delta z|_r\,(\text{IN}) \tag{3B.2}$$

3. The momentum flux across the annular surface at z and $z + \Delta z$ is zero because the axial velocity v_z is of the same magnitude at z and $z + \Delta z$.

4. Recognizing that the sum of forces acting on control volume = rate of momentum out of control volume − rate of momentum into control volume:

$$p(2\pi r\,\Delta r)|_z - p(2\pi r\,\Delta r)|_{z+\Delta z} = \sigma_{rz}\,2\pi r\Delta z|_{r+\Delta r}\,(\text{OUT}) - \sigma_{rz}\,2\pi r\Delta z|_r\,(\text{IN}) \tag{3B.3}$$

The above equation can be simplified to

$$r\,\Delta r\,(p|_z - p|_{z+\Delta z}) = \Delta z\,[(r\sigma_{rz})|_{r+\Delta r} - (r\sigma_{rz})|_r] \tag{3B.4}$$

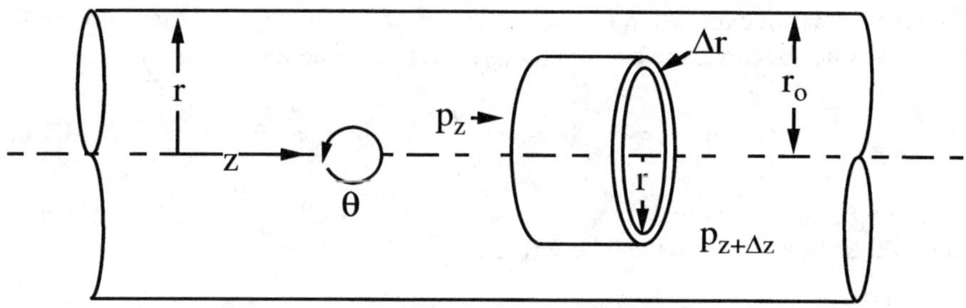

Figure 3B–1 Schematic Diagram for Force Balance in Laminar Capillary Flow.

Dividing both sides by $(\Delta r\ \Delta z)$ results in

$$\frac{r\,(p|_z - p|_{z+\Delta z})}{\Delta z} = \frac{(r\,\sigma_{rz})|_{r+\Delta r} - (r\,\sigma_{rz})|_r}{\Delta r} \tag{3B.5}$$

As Δr and $\Delta z \to 0$, the above equation becomes

$$r\,\frac{\mathrm{d}p}{\mathrm{d}z} = -\frac{\mathrm{d}}{\mathrm{d}r}(r\,\sigma_{rz}) \tag{3B.6}$$

After integration one gets

$$r\,\sigma_{rz} = -\frac{r^2}{2}\frac{\mathrm{d}p}{\mathrm{d}z} + c_1 \tag{3B.7}$$

which can be rearranged to

$$\sigma_{rz} = -\frac{r}{2}\frac{\mathrm{d}p}{\mathrm{d}z} + \frac{c_1}{r} \tag{3B.8}$$

By using the condition $\sigma_{rz} = 0$ at $r = 0$ (center line), the integration constant $c_1 = 0$, so that the stress distribution across the radius of the pipe is

$$\sigma_{rz} = -\frac{r}{2}\frac{\mathrm{d}p}{\mathrm{d}z} \tag{3B.9}$$

that for the wall of the pipe becomes $\sigma_w = -\left(\frac{r_o}{2}\frac{\mathrm{d}p}{\mathrm{d}z}\right)$, because at $r = r_o$, $\sigma_{rz} = \sigma_w$. When the pressure drop, Δp, is measured for fully developed flow over a length L the equation for the shear stress at the wall is

$$\sigma_w = \frac{D\,\Delta p}{4L} \tag{3B.10}$$

We will derive an expression for the volumetric flow rate, Q, and then another for the shear rate. Because the axial velocity, v_z, is dependent on the radial position:

$$Q = \int_0^{r_o} 2\pi r\, v_z\, dr = \pi \int_0^{r_o} v_z\, 2r\, dr + \pi \int_{v_z \text{ at } r=0}^{v_z \text{ at } r=r_o} r^2\, dv_z \tag{3B.11}$$

Assuming a no-slip boundary condition, that is, $v_z = 0$ at $r = r_o$, the right-hand side of the above equation can be written as

$$\pi \int_0^{r_o} v_z\, 2r\, dr - \pi \int_0^{v_z} r^2\, dv_z \tag{3B.12}$$

The above equation can be integrated and, noting that $v_z = 0$ at $r = r_o$, $-dv_z = -\left(\frac{dv_z}{dr}\right) dr$ and $r = \left(\frac{r_o}{\sigma_w}\right) \sigma_{rz}$, one can obtain the expression

$$Q = \left(\frac{\pi r_o^3}{\sigma_w^3}\right) \int_0^{\sigma_w} \sigma_{rz}^2 \left(-\frac{dv_z}{dr}\right) d\sigma_{rz} \tag{3B.13}$$

Multiplying both sides of the above equation by 4 and rearranging:

$$\frac{4Q}{\pi r_o^3} = \frac{4}{\sigma_w^3} \int_0^{\sigma_w} \sigma_{rz}^2 \left(-\frac{dv_z}{dr}\right) d\sigma_{rz} \tag{3B.14}$$

The above equation can be used for deriving a general solution for tube flow and specific expressions for the volumetric flow rates of fluids exhibiting different rheological behaviors. For a Newtonian fluid, noting that the shear rate $-\left(\frac{dv_z}{dr}\right) = \frac{\sigma_{rz}}{\eta}$, it can be shown that

$$\frac{4Q}{\pi r_o^3} = \frac{\sigma_w}{\eta} = \frac{1}{\eta} \frac{D\, \Delta p}{4L} \tag{3B.15}$$

or that the viscosity of a Newtonian fluid

$$\eta = \frac{(D\, \Delta p/4L)}{(8\bar{v}_z/D)};\ \eta \sim \frac{1}{\bar{v}_z} \sim \text{flow time} \tag{3B.16}$$

Equation 3B.16 is the basis for calculation of viscosity of a Newtonian fluid using a glass capillary viscometer. It should be recognized that $(4Q/\pi r_o^3) = (32\, Q/\pi D^3)$ gives the shear rate for Newtonian fluids but not for non-Newtonian fluids and is called pseudo shear rate. Additional steps are required to obtain an expression for the true shear rate.

Differentiating both sides of Equation 3B.14 and setting the limits

$$\frac{d[\sigma_w^3 (4Q/\pi r_o^3)]}{d\sigma_w} = 4\sigma_w^2 \left(-\frac{dv_z}{dr}\right) \tag{3B.17}$$

Further differentiating the left-hand side and rearranging, one obtains the general solution for the true shear rate in tube flow:

$$\left(-\frac{dv_z}{dr}\right) = \left(\frac{3}{4}\right)\frac{4Q}{\pi r_0^3} + \frac{\sigma_w}{4}\frac{d(4Q/\pi r_o^3)}{d\sigma_w} \tag{3B.18}$$

Equation 3B.18 is known as the Weissenberg-Rabinowitsch-Mooney (WRM) equation in honor of the three rheologists who have worked on this problem. An alternate equation can be derived for fluids obeying the power law model between shear stress and the pseudo shear rate:

$$\sigma = K'(4Q/\pi r_o^3)^{n'} \tag{3B.19}$$

$$\left(-\frac{dv_z}{dr}\right) = \frac{3n'+1}{4n'}\left(\frac{4Q}{\pi r_o^3}\right) \tag{3B.20}$$

To use the WRM equation, the steps involved are as follows:

1. Using a tube/pipe/capillary of known diameter (D) and length (L), several sets of volumetric flow rate (Q) versus pressure drop (Δp) data are obtained under isothermal fully developed and no slip at the pipe wall conditions.
2. The quantities ($4Q/\pi r_o^3$) and ($D\,\Delta p/4L$) are calculated, plotted, and a smooth curve fitted.
3. At each value of ($4Q/\pi r_o^3$), the value of $d(4Q/\pi r_o^3)/d\sigma_w$ is determined and the true shear rate calculated using Equation 3B.18. Alternatively, for many foods, log-log plots of the quantities ($4Q/\pi r_o^3$) and ($D\,\Delta p/4L$) are straight lines with slopes n' that can be used in Equation 3B.20 to obtain the true shear rate.
4. Generally, for a given fluid, values of n' and of n of the power law model $\sigma = K\dot{\gamma}^n$ will be the same. However, the value of K' in $\sigma = K'(4Q/\pi r_o^3)^{n'}$ is related to the consistency index K by the following equation:

$$K' = K\left(\frac{3n+1}{4n}\right)^n \tag{3B.21}$$

As an example, for $n = 0.3$, $K' = 1.15K$.

From Equation 3B.19, an expression can be derived for a pipe-flow apparent viscosity (η_{ap}) based on the diameter D and the average axial velocity in the tube ($\overline{v}_z$):

$$\eta_{ap} = K'\left(\frac{8\,\overline{v}_z}{D}\right)^{n'-1} \tag{3B.22}$$

When $n = n'$, together with Equations 3B.21 and 3B.22, one can obtain the generalized Reynolds number (GRe):

$$\mathrm{GRe} = \frac{D^n \, \overline{v}_z^{\,2-n} \, \rho}{8^{(n-1)} K} \left(\frac{4n}{3n + 1} \right)^n \qquad (3B.23)$$

Some of the considerations in selecting a capillary/tube viscometer are shown in Figure 3B–2.

The test fluid is low viscosity Newtonian? Yes → Use glass capillary.

↓

No, fluid is non-Newtonian. → Perhaps a capillary/tube viscometer can be used.

↓

Is the fluid a suspension in which the solids tend to settle (e.g., ungelatinized starch granules in water) in the reservoir and tube?

↓ Yes → Do not use tube viscometer.

No

↓

Can ΔP be measured over fully developed flow length? Yes → A single tube can be used.

↓

No → Use several tubes and Bagley's end correction method.

Figure 3B–2 Decision Tree for Capillary/Tube Viscometer

Appendix 3–C
Analysis of Flow in a Cone-Plate Geometry

We assume that the cone of radius r_o and cone angle θ_o is on top over the plate rotating with an angular velocity Ω while the bottom plate is stationary, as shown in Figure 3–7. We consider the equation of motion in spherical coordinates r, θ, ϕ. For steady fully developed flow, the velocity components $v_\theta = 0$ and $v_r = 0$, and v_ϕ is a function of r and θ. In addition we consider pressure variations (body forces) to be negligible and that the value of θ_o is very small < 0.1 rad. The r-component reduces to

$$\frac{\rho v_\phi^2}{r} = \frac{1}{r^2}\frac{\partial}{\partial r}(r^2\,\sigma_{rr}) - \frac{\sigma_{\theta\theta} + \sigma_{\phi\phi}}{r} \tag{3C.1}$$

The θ-component reduces to

$$\frac{1}{r\sin\theta}\frac{\partial(\sigma_{\theta\theta}\sin\theta)}{\partial\theta} - \frac{\cot\theta\,\sigma_{\theta\theta}}{r} = 0 \tag{3C.2}$$

The ϕ-component of the equation of motion reduces to

$$\frac{1}{r}\frac{\partial\sigma_{\theta\phi}}{\partial\theta} + \frac{2}{r}\cot\theta\,\sigma_{\theta\phi} = 0 \tag{3C.3}$$

We note that the ϕ-component contains the shear stress $\sigma_{\theta\phi}$. The applicable boundary conditions are $v_\phi = 0$ at $\theta = \frac{\pi}{2}$ because the plate is stationary and $v_\phi = r\Omega\cos\theta_o$ at $\theta = (\pi/2) - \theta_o$. First, we derive an expression for the shear stress by conducting a simple torque balance on the plate:

$$M = \int_0^{2\pi}\int_0^{r_o} r^2\,\sigma_{\theta\phi}\,dr\,d\phi = \frac{2\pi r_o^3}{3}\,\sigma_{\theta\phi}\big|_{\pi/2} \tag{3C.4}$$

Integration of Equation 3C.3 results in the expression: $\sigma_{\theta\phi}|_{(\pi/2)} = \sigma_{\theta\phi}(\theta)\sin^2\theta$. Therefore,

$$\sigma_{\theta\phi}(\theta) = \frac{3M}{2\pi r_o^3 \sin^2\theta} \tag{3C.5}$$

When $\theta_o < 0.1$ rad, $\sin^2\left(\frac{\pi}{2} - \theta_o\right) \sim 1$, so that:

$$\sigma_{\theta\phi} = \frac{3M}{2\pi r_o^3} \tag{3C.6}$$

Defining T_{cn} as the torque per unit area, Equation 3C.6 can be written in terms of the cone diameter D:

$$\sigma_{\theta\phi} = \frac{3T_{cn}}{D} \tag{3C.7}$$

From Equations 3C.4 and 3C.5, it follows that the shear stress in a cone-plate geometry is essentially uniform.

To obtain an expression for shear rate, we can simply say that for low values of the cone angle θ_o, we need not distinguish between $\sin\theta_o$ and θ_o (Whorlow, 1980) when the shear rate at radius r will be

$$\dot{\gamma}_r = \frac{r\Omega}{r\theta_o} = \frac{\Omega}{\theta_o} \tag{3C.8}$$

Alternatively, we can consider the velocity distribution (Brodkey, 1967) for v_ϕ:

$$\frac{v_\phi}{r\sin\theta} = \int_{\theta=\frac{\pi}{2}}^{\theta} \frac{d(v_\phi/r\sin\theta)}{d\theta}\, d\theta \tag{3C.9}$$

We can also substitute for $d\theta$ from Equation 3C.3: $d\theta = -\frac{1}{2}\frac{\sin\theta}{\cos\theta}\frac{1}{\sigma_{\theta\phi}}d\sigma_{\theta\phi}$, and we can also switch limits: $\sigma_{\theta\phi}$ at $\theta = 0$ and σ_{plate} at $\theta = \pi/2$. For a Newtonian fluid, we can substitute in Equation 3C.9 the relationship between shear rate and shear stress and get 3C.11

$$\sigma_{\theta\phi} = \eta\left[\sin\theta\frac{d(v_\phi/r\sin\theta)}{d\theta}\right] \tag{3C.10}$$

$$\frac{v_\phi}{r\sin\theta} = \int_{\theta}^{\pi/2} \frac{1}{\eta}\frac{\sigma_{\theta\phi}}{\sin\theta}\, d\theta \tag{3C.11}$$

Substituting for $\sigma_{\theta\phi}$ in Equation 3C.11 and performing the integration results in

$$\frac{v_\phi}{r} = \frac{3T}{4\eta r_o}\left[\cot\theta_o - \ln\left(\tan\frac{\theta}{2}\right)\sin\theta\right] \tag{3C.12}$$

Equation 3C.12 can be written for the boundary condition at $\theta = \frac{\pi}{2} - \theta_o$:

$$\Omega \cos \theta_o = \frac{3T}{4\eta r_o}\left[\tan \theta_o - \cos \theta_o \ln\left(\frac{1 - \tan \frac{\theta_o}{2}}{1 + \tan \frac{\theta_o}{2}}\right)\right] \tag{3C.13}$$

For low values of the cone angle θ_o, the above equation becomes

$$\Omega = \frac{3T}{4\eta r_o}(\theta_o + \theta_o) = \frac{3T\theta_o}{\eta D} \tag{3C.14}$$

Therefore,

$$\sigma_{\theta\phi} \cong \eta \frac{\Omega}{\theta_o} \tag{3C.15}$$

CHAPTER 4

Rheology of Food Gum and Starch Dispersions

M. Anandha Rao

Gums and starches are used extensively as thickening and gelling agents in foods. Therefore, understanding their rheological characteristics is of considerable interest. Because many food gums in dispersions have random coil configuration and starch dispersions have granules, it is better to study their rheological behavior separately.

RHEOLOGY OF FOOD GUM DISPERSIONS

Several water-soluble gums, also called hydrocolloids, are being used in foods, and their chemical composition and applications have been well documented (Glicksman, 1969; Davidson, 1980), and some of their characteristics are summarized in Table 4–1. Imparting viscosity, gelation, and film formation are the main functionalities of food gums. The predominant interest in the food industry is in dispersions of relatively low gum concentrations because of their use in practice. It is emphasized that considerable care should be exercised in preparation of the gum/starch dispersions, and the references mentioned in this chapter should be consulted for suitable preparation techniques. In particular, a reasonable period of hydration should be allowed for all polymer dispersions, and to control the conformation of charged polymers, the pH and the ionic strength of the dispersion should be controlled.

Rheological Models for Apparent Viscosity-Shear Rate Data

Apparent viscosity-shear rate data of food polymer dispersions have been reviewed by Launay et al. (1986) and Lopes da Silva and Rao (1992). The general log η_a − log $\dot{\gamma}$ curve, discussed in Chapter 2, has been used to characterize food polymer dispersions. For example, Lopes da Silva et al. (1992) found that both the

Table 4–1 Principal Biopolymers Used in Foods and Their Classification Based on Origin

Natural	*Semisynthetic*	*Synthetic*
Plant Exudates	**Cellulose Derivatives**	Polyvinylpyrolidone (PVP)
Gum arabic	Sodium carboxymethylcellulose	Polymers from polyethylene oxide (Polyox)
Adraganta gum	Methylcellulose	
Karaya gum	Hydroxyethylcellulose	
Gatti gum	Hydroxypropylmethylcellulose	
Algal Extracts	**Other Derivatives**	
Agar	Modified starches	
Alginates	Low-methoxyl pectins	
Carrageenans	Propylene glycoalginates	
Furcellaran		
Gums from Seeds		
Guar		
Locust bean gum		
Mesquite		
Gums from Cereals		
Starches		
Plant Extracts		
Pectins		
Gums from Fermentation		
Xanthan		
Dextran		
Gums of Animal Origin		
Gelatin		
Casseinates		

modified Carreau and the Cross models, wherein the infinite shear viscosity was considered to be negligible, described the apparent viscosity-shear rate data of locust bean (LB) gum (Figure 4–1) and pectin (Figure 4–2) dispersions well, especially when there were experimental data in the plateau region at low shear rates.

$$\eta_a = \frac{\eta_0}{1 + (\alpha_c \dot{\gamma})^m} \tag{4.1}$$

$$\eta_a = \frac{\eta_0}{[1 + (\lambda_c \dot{\gamma})^2]^N} \tag{4.2}$$

where α_c and λ_c are time constants related to the relaxation times of the polymer in solution and m and N are dimensionless exponents. However, when experimental data were scarce in the plateau region, as was the case with low-concentration pectin solutions of pH 7.0, 0.1M, the Carreau model was adept at predicting the zero-shear rate viscosity. In general, the deviations between predicted

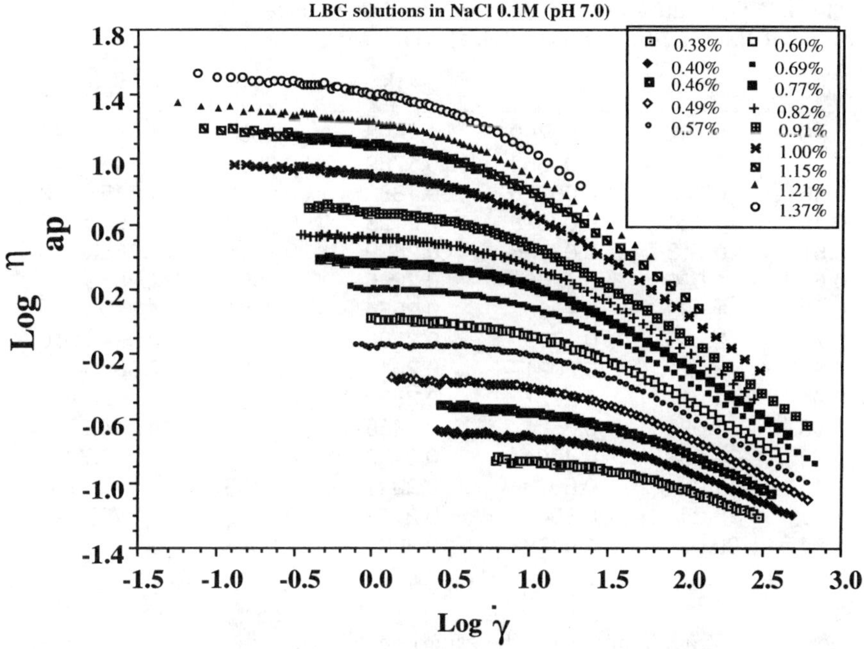

Figure 4–1 Apparent Viscosity (η_a) versus Shear Rate ($\dot{\gamma}$) of Locust Bean Gum Dispersions (0.1M, pH 7) of Different Concentrations.

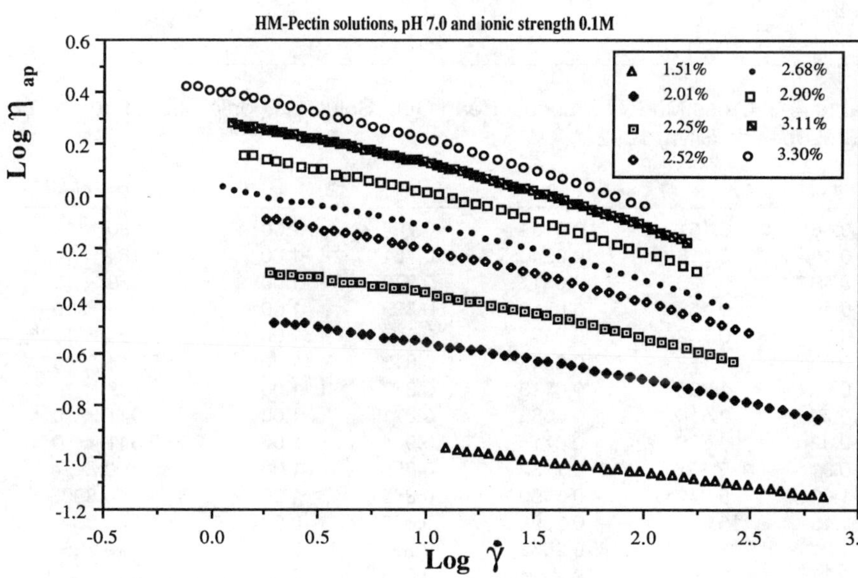

Figure 4–2 Apparent Viscosity (η_a) versus Shear Rate ($\dot{\gamma}$) of High-Methoxyl Pectin (0.1M, pH 7) Dispersions with Different Concentrations.

Table 4–2 Carreau Model Parameters for Locust Bean Gum Solutions (Ionic Strength 0.1M, pH 7.0). Lopes da Silva, 1992.

Conc. (%)	η_o	λ	N	R^2	*Residual*
0.38	0.139	0.03799	0.1593	1.000	1.844×10^{-4}
0.40	0.199	0.04552	0.1769	0.9999	1.147×10^{-3}
0.46	0.300	0.06076	0.1882	0.9999	1.645×10^{-3}
0.49	0.424	0.06618	0.2125	0.9998	4.851×10^{-3}
0.57	0.713	0.07798	0.2399	0.9998	7.762×10^{-3}
0.60	1.020	0.1200	0.2265	0.9995	2.886×10^{-2}
0.69	1.589	0.1250	0.2554	0.9995	4.819×10^{-2}
0.77	2.345	0.1755	0.2504	0.9985	2.094×10^{-1}
0.82	3.321	0.2091	0.2515	0.9986	2.988×10^{-1}
0.91	4.917	0.2727	0.2590	0.9954	1.193
0.98	7.086	0.3584	0.2459	0.9896	4.592
1.00	8.681	0.4064	0.2456	0.9886	5.769
1.15	14.00	0.5973	0.2341	0.9738	20.20
1.21	19.24	0.8044	0.2152	0.9606	39.72
1.37	30.63	1.0739	0.2006	0.9638	52.24

values by the Carreau model and the experimental data were at high shear rates. It is emphasized that, as illustrated by the high magnitudes of R^2 in Table 4–2 for Carreau model and in Table 4–3 for Cross model with data on LB gum dispersions, pH 7.0, ionic strength 0.1M, both models followed the experimental data well.

Table 4–3 Cross Model for Locust Bean Gum Solutions (Ionic Strength 0.1M, pH 7.0). Lopes da Silva, 1992.

Conc. (%)	η_o	α	m	R^2	*Residual*
0.38	0.151	0.0194	0.764	1.00	1.180×10^{-4}
0.40	0.216	0.0240	0.761	1.00	5.184×10^{-4}
0.46	0.328	0.0317	0.773	1.00	4.279×10^{-4}
0.49	0.456	0.0312	0.809	1.00	2.998×10^{-3}
0.57	0.752	0.0277	0.897	1.00	1.733×10^{-2}
0.60	1.113	0.0557	0.813	1.00	5.443×10^{-3}
0.69	1.708	0.0513	0.877	1.00	2.889×10^{-2}
0.77	2.516	0.0685	0.871	1.00	6.941×10^{-2}
0.82	3.526	0.0715	0.904	1.00	9.541×10^{-2}
0.91	5.361	0.1132	0.856	1.00	0.2124
1.00	9.340	0.1459	0.853	1.00	0.8833
1.15	15.23	0.2098	0.822	0.994	4.362
1.21	21.00	0.2552	0.797	0.990	9.725
1.37	33.25	0.2894	0.816	0.992	11.46

Yoo et al. (1994) also found that the Cross and Carreau models were satisfactory for the most part in describing the data that covered the zero-shear and the power law regions of mesquite seed gum. Also in agreement with Lopes da Silva et al. (1992), the Carreau model described well the viscosity data shown in Figure 4–3 over a wide range of shear rates except in the power law region at high shear rates in concentrations > 1.4 g $(100 \text{ mL})^{-1}$.

Time Constants of Viscoelastic Foods

Time constants are related to the relaxation times and can be found in equations based on mechanical models (phenomenological approaches) and in constitutive equations (empirical or semiempirical) for viscoelastic fluids that are based on either molecular theories or continuum mechanics. Equations based on mechanical models are covered in later sections, particularly in the treatment of creep-compliance studies, while the Bird-Leider relationship is an example of an empirical relationship for viscoelastic fluids.

Time constants based on molecular theories have been derived for rod-like and bead-spring models (Ferry, 1980; Bird et al., 1977). For example, Whitcomb and Macosko (1978) showed that the conformation of xanthan gum in solution is rod-like with some flexibility. The bead-spring model has found extensive use in the literature on polymers. The development of molecular theory for dilute solutions

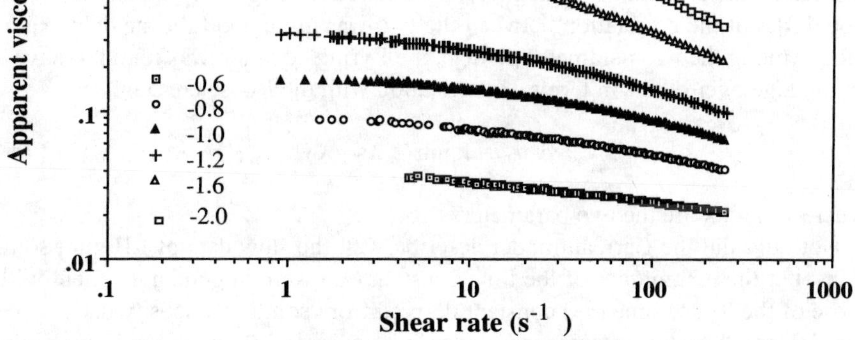

Figure 4–3 Apparent Viscosity (η_a) versus Shear Rate ($\dot{\gamma}$) of Mesquite Seed Gum Dispersions with Different Concentrations. Yoo et al., 1994.

of linear polymers credited to Rouse relates the time constant to molecular weight and other basic parameters. The theory assumed that a polymer molecule is made up of beads joined together by flexible spring segments. The equation for the first spring in terms of the zero-shear rate viscosity of the polymer solution (η_0) and the viscosity of the solvent (η_s) is

$$\tau_R = \frac{6(\eta_0 - \eta_s)M}{\pi^2 cRT} \tag{4.3}$$

where τ_R is the Rouse time constant, M is the molecular weight, c is the concentration, R is the gas constant, and T is the absolute temperature. Equations for the time constant in terms of intrinsic viscosity of a polymer solution (Ross-Murphy, 1984) and zero-shear viscosity of a polymer melt (Elbirli and Shaw, 1978) can be found; in the case of polydisperse polymers, it was suggested that a weight average molecular weight can be used in the equation. A time constant due to Zimm derived a few years after Rouse can also be used.

Because shear rate–shear stress data can be obtained relatively easily, a number of rheologists explored the possibility of determining the time constants from these data. Most shear-thinning (pseudoplastic) fluids behave as Newtonian fluids at low shear rates, and at a particular shear rate they begin to show their pseudoplastic behavior; the reciprocal of the shear rate at which the transition from Newtonian to pseudoplastic behavior occurs is the characteristic time or the time constant.

Data must be obtained at very low shear rates in order to observe the aforementioned transition, and because obtaining such data may not be an easy task, a number of models have been proposed and 14 of these were evaluated by Elbirli and Shaw (1978) for experimental data on polymer melts. Some of the models containing two parameters were found to fit the data and also predict characteristic times as well as some of the models containing three and four parameters. The predicted time constants were higher than those based on Rouse's equation (Equation 4.3), but the correlation between the two was quite good for the data examined. Among the two parameter models, the Eyring equation was found to have a slight edge over others in terms of correlation with the Rouse constant:

$$\eta/\eta_0 = [\sinh^{-1} \lambda\dot{\gamma}]/\lambda\dot{\gamma} \tag{4.4}$$

where η_0 and λ are the two parameters.

Not only did the Carreau model describe well the flow data of LB gum solutions, but the magnitudes of the time constant (λ_c) were in good agreement with those of the Rouse time (τ_R) constant derived from solution viscosity data, while the Cross (α_c) time constants were lower in magnitudes. Both the Carreau and the Cross time constants followed power relationships with respect to the concentration (c) of the solutions (Lopes da Silva et al., 1992):

$$\lambda_c = 0.672\, c^{2.55} \qquad R^2 = 0.981 \tag{4.5}$$

$$\alpha_c = 0.425\, c^{2.12} \qquad R^2 = 0.958 \tag{4.6}$$

The lower magnitudes of the empirical constants in Equation 4.6 reflect the lower magnitudes of the Cross time constant discussed above.

Rheological Properties of Binary Mixtures of Equal Concentration

Equation 4.7 (Kaletunc-Gencer and Peleg, 1986) and Equation 4.8 (Miller and Mann, 1944) have been used to calculate the "expected" viscosity of the mixtures of equal concentration and to find a blend equation to predict the viscosity properties of LB gum and pectin mixtures (Lopes da Silva et al., 1992):

$$\eta_{\text{mix}} = X_A\, \eta_A + X_B\, \eta_B \tag{4.7}$$

$$\eta_{\text{mix}} = \eta_A^{X_A} \cdot \eta_B^{X_B} \tag{4.8}$$

where X_A and X_B are the weight fractions of gum A and gum B, respectively, and η_A and η_B are the apparent viscosities (fitted values) of the pure solutions of gums A and B at the same concentration. Equation 4.7 does best when the difference between molecular weights of the components is not too great (Bird et al., 1977). This can be one of the possible reasons for the divergence of the predicted values relative to the experimental values.

Equation 4.7 does indicate the absence of synergistic effect, but it does not predict the viscosity of the pectin-LB gum polymer blends with concentrations of 1.5%. Equation 4.8 was first used by Miller and Mann (1944) in calculations of power requirements for agitation of mixtures of immiscible liquids. Harris (1970) has used the logarithmic form of this equation to relate the steady zero-shear viscosity of two-component blends of polystyrene solutions with the steady zero-shear viscosity of the separate components and for CMC/guar gum blends with satisfactory results (Plutchok and Kokini, 1986). Equation 4.8 does best when the difference between molecular weights of the components is not too great (Bird et al., 1977), and it predicted a geometric mean viscosity that provided better approximation to the experimental values for the viscosity of pectin-LB gum solutions (Lopes da Silva et al., 1992).

A possible and most likely explanation for the low value of viscosity for the mixtures based on Equation 4.7 is the very different viscosity dependence on concentration for the pectin and LB gum solutions, and not a true antagonism. To work with these gum solutions at equal concentration and to detect some possible synergism or antagonism between them, one must take into account the viscosity-concentration dependence of dilute gum solutions over the range of concentrations of each gum in the mixture.

Morris (1981) suggested that when the apparent viscosities of two gum dispersions at equal concentrations are substantially different, to account for the different viscosity-concentration relationships for the two gums, the apparent viscosity of mixtures of solutions of two gums could be studied by mixing solutions of the two biopolymers with concentrations adjusted to give similar viscosities. The steps in this procedure for pectin and LB gum dispersions were (Lopes da Silva et al., 1992) as follows: (1) The shear rate–apparent viscosity data of solutions of pectin and LB gum with different concentrations were obtained, and (2) the concentrations of pectin and LB gum solutions that had nearly equal zero-shear viscosities and that permitted gathering of apparent viscosity-shear rate data over a wide range of shear rates were selected.

Effect of Biopolymer Concentration on Zero-Shear Viscosity

In earlier studies on solutions of synthetic polymers (Ferry, 1980), the zero-shear viscosity was found to be related to the molecular weight of the polymers. Plots of log η versus log M often resulted in two straight lines with the lower M section having a slope of about 1 and the upper M section having a slope of about 3.4. Because the apparent viscosity also increases with concentration of a specific polymer, the roles of both molecular size and concentration of polymer need to be understood. In polymer dispersions of moderate concentration, the viscosity is controlled primarily by the extent to which the polymer chains interpenetrate, which is characterized by the coil overlap parameter $c[\eta]$ (Graessley, 1980). Determination of intrinsic viscosity $[\eta]$ and its relation to molecular weight were discussed in Chapter 1. The product $c[\eta]$ is dimensionless and indicates the volume occupied by the polymer molecule in the solution.

The various concentration regimes of importance following de Gennes and coworkers (Tirrell, 1994) are dilute ($c < c^*$), semidilute ($c^* < c < c^\ddagger$), and concentrated ($c > c^\ddagger$). At low concentrations ($c < c^*$), the polymer chains are not in contact with each other, the polymer coils have infinite dilution radii, and the viscosity is relatively low (Figure 4–4). A good rule of thumb for determining that concentration effects will become important is when the magnitude of $c[\eta]$ is about unity. At the overlap threshold concentration ($c = c^*$), the coils begin to overlap and there is no contraction. It can be shown that the functional dependence of c^* on M is $c^* \sim M^{-4/5}$. In the semidilute region, the coils contract (Figure 4–4), but the shrinking does not continue indefinitely and the polymer chain reaches minimum (θ) dimensions at a concentration $c^\ddagger$ that is independent of molecular weight.

In contrast to $c[\eta]$, the quantity $M[\eta]$ is termed the hydrodynamic volume and is directly proportional to the volume occupied by each molecule, as can be de-

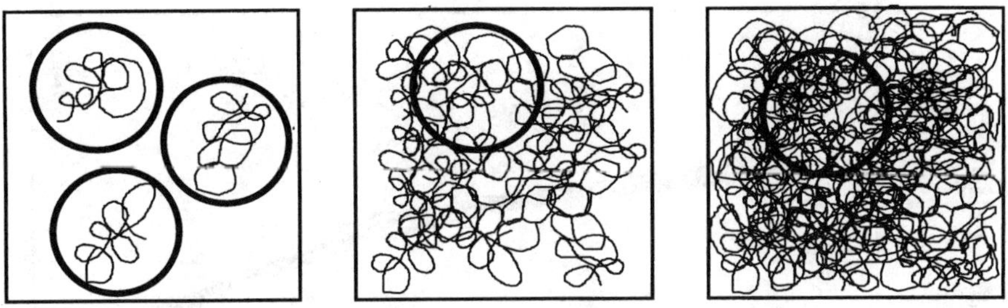

Figure 4–4 Polymer Chain Entanglement in Dispersions. At low concentrations, the polymer chains are not in contact with each other. At high concentrations, the polymer chains are in contact with each other, contributing to a large increase in viscosity. *Source*: C.W. Macosko, *Rheology: Principles, Measurements, and Applications*, © 1994, John Wiley and Sons, Inc. Reprinted with permission of John Wiley and Sons, Inc.

duced from the equation to rotate a polymer molecule in a complex spherical shape:

$$[\eta] = \Phi \frac{(\overline{r^2})^{3/2}}{M} \tag{4.9}$$

where Φ is a universal constant (Rodriguez, 1989).

Graessley (1967) introduced the use of reduced variables, η/η_0 and $\dot{\gamma}/\dot{\gamma}_0$, to represent viscosity data at several concentrations and temperatures. Tam and Tiu (1993) determined $\dot{\gamma}_0$ as the shear rate where $\eta_a = 0.9\ \eta_0$, while Morris et al. (1981) determined $\dot{\gamma}_0$ when $\eta_a = 0.1\ \eta_0$. Figure 4–5 illustrates applicability of the reduced variables technique to data on mesquite seed gum dispersions, where data shown in Figure 4–3 were converted using the following values of η_0 and $\dot{\gamma}_0$: 0.088 Pa s and 7.63 s^{-1} for the 0.80% dispersion, 0.33 Pa s and 4.29 s^{-1} for the 1.2% dispersion, 1.56 Pa s and 0.89 s^{-1} for the 1.6% dispersion, and 4.50 Pa s and 0.45 s^{-1} for the 2% dispersion, respectively.

A versatile reduced parameter approach of Graessley (1974) accounted for not only the shear rate but also polymer concentration and temperature:

$$\frac{\eta - \eta_s}{\eta_0 - \eta_s} = \left[\frac{(\eta_0 - \eta_s)M}{cRT}\right](\dot{\gamma}) \tag{4.10}$$

where η_s is the solvent viscosity, M the molecular weight, c the polymer concentration, R the gas constant, and T the temperature. The molecular relaxation time,

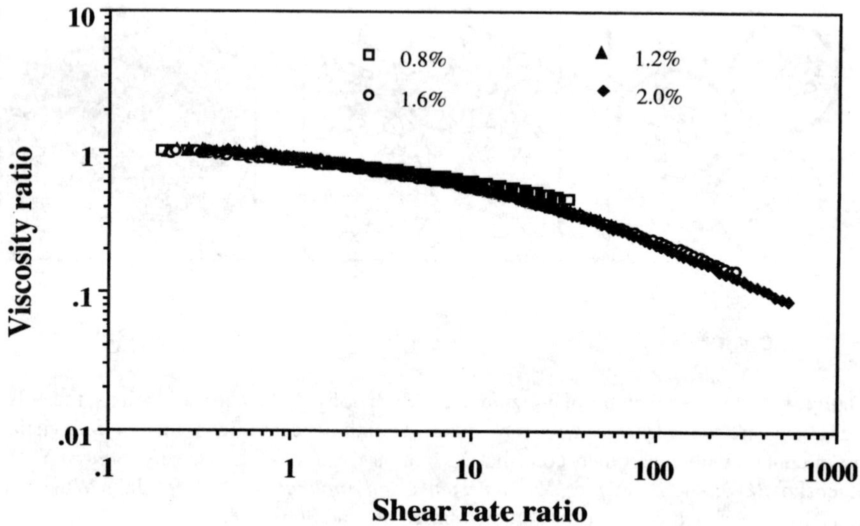

Figure 4–5 Plot of the Reduced Variables η/η_0 versus $\dot{\gamma}/\dot{\gamma}_0$ with Mesquite Seed Gum Dispersions ($\dot{\gamma}_0$ is the shear rate when $\eta_a = 0.9\ \eta_0$). Yoo et al., 1994.

τ_M, is defined by the terms in the bracket. When $\eta_0 \gg \eta_\infty$, the above equation can be simplified to

$$\frac{\eta}{\eta_0} = \left[\frac{(\eta_0 - \eta_s)M}{cRT}\right](\dot{\gamma}) \tag{4.11}$$

For polyelectrolytes (charged polymers), Tam and Tiu (1993) utilized the expression for specific viscosity in the equation of Fuoss-Strauss:

$$\frac{\eta_{sp}}{c} = \frac{[\eta]}{(1 + Bc^{1/2})} \tag{4.12}$$

Tam and Tiu (1993) derived the expression:

$$\frac{\eta}{\eta_s} = (1 + Bc^{1/2})\left[\frac{(\eta_0 - \eta_s)M}{cRT}\right](\dot{\gamma}) \tag{4.13}$$

To determine the value of B, (c/η_{sp}) is plotted against $c^{1/2}$, resulting in an intercept of $1/[\eta]$ and a slope of $B/[\eta]$.

A plot of log $c[\eta]$ against log $[(\eta_0 - \eta_S)/\eta_S]$ of several food gum solutions at 25°C, adapted from Morris et al. (1981), is shown in Figure 4–6. There are two re-

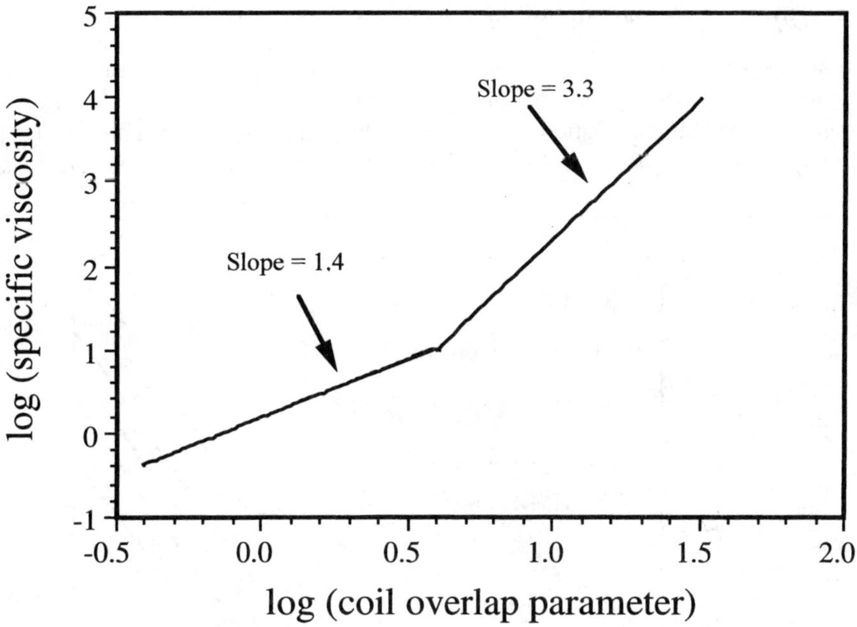

Figure 4–6 Plot of Log $c[\eta]$ (Coil Overlap Parameter) against Log $\eta_{sp} = [(\eta_0 - \eta_s)/\eta_s]$ (η_{sp} = specific viscosity). Morris et al., 1981.

gions: (1) a region of dilute dispersions where the viscosity dependence on concentration follows a 1.4 power, and (2) a region of concentrated dispersions where the viscosity dependence on concentration follows a 3.3 power. The transition from the dilute to the concentrated region occurred at a value of $c[\eta] = 4$. However, data on a few biopolymers such as guar gum and LB gum were found to deviate from the above observations. First, the region of concentrated solution behavior began at lower values of the coil overlap parameter $c[\eta] = 2.5$. Second, the viscosity showed a higher dependence on concentration with a slope of 5.1 instead of about 3.3. These deviations were attributed to specific intermolecular associations (hyperentanglements) between regular and rigid chain sequences in addition to the simple process of interpretation. Launay et al. (1986) suggested that there could be two transitions, instead of one transition shown in Figure 4–6, before the onset of high concentration–viscosity behavior. Because intrinsic viscosity of a biopolymer can be determined with relative ease, Figure 4–6 can be used to estimate the zero-shear viscosity of any food polymer at a specific polymer concentration at 25°C.

Concentration Dependence of the Zero-Shear Viscosity of Mixtures

Figure 4–7 shows viscosity data of high-methoxyl (HM) pectin and LB gum dispersions, and blends of dispersions with similar viscosity, at pH 7.0 and ionic strength 0.1M. All the solutions studied were above the critical concentration (c^*), corresponding to $c[\eta] \approx 4.5$ for random-coil polysaccharides (Morris and Ross-

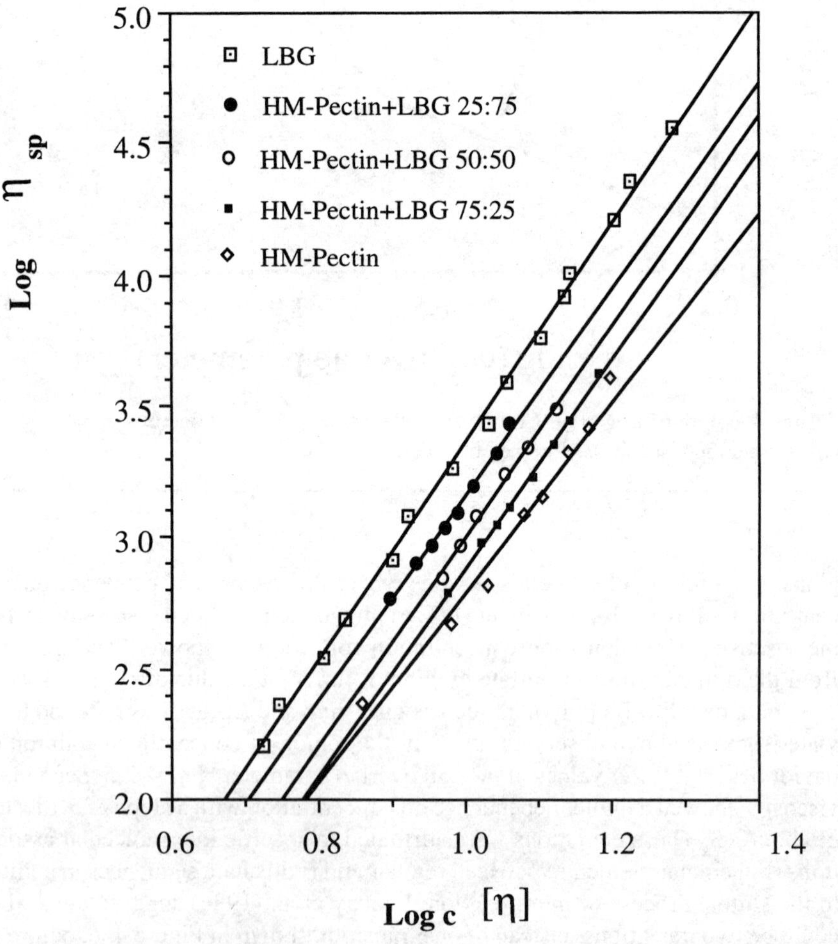

Figure 4–7 Plot of Log $c[\eta]$ against Log $\eta_{sp} = [(\eta_0 - \eta_s)/\eta_s]$ for High-Methoxyl Pectin and Locust Bean Gum, and Blends of Solutions with Similar Viscosity, at pH 7.0 and Ionic Strength 0.1M. *Source:* Reprinted with permission from J.A. Lopes da Silva, M.P. Gonçalves, and M.A. Rao, Rheological Properties of High-Methoxyl Pectin and Locust Bean Gum Solutions in Steady Shear, *Journal of Food Science*, Vol. 57, pp. 443–448, © 1992, Institute of Food Technologists.

Murphy, 1981). For this concentration regime, relationships with different exponents for the dependence of the specific viscosity on the coil overlap parameter ($c[\eta]$) were obtained:

for LB gum solutions: $$\eta_{sp} \propto c[\eta]^{4.2} \quad (4.14)$$

for LB gum 38:HM pectin 62 blend solutions: $$\eta_{sp} \propto c[\eta]^{3.9} \quad (4.15)$$

for LB gum 18:HM pectin 82 blend solutions: $$\eta_{sp} \propto c[\eta]^{4.0} \quad (4.16)$$

for LB gum 6:HM pectin 94 blend solutions: $$\eta_{sp} \propto c[\eta]^{4.0} \quad (4.17)$$

for HM pectin solutions: $$\eta_{sp} \propto c[\eta]^{3.6} \quad (4.18)$$

The slope 3.6 is near the expected value for disordered random-coil polysaccharides (Morris et al., 1981) and consistent with the value of 3.3 obtained for citrus pectin samples (Chou and Kokini, 1987; Axelos et al., 1989). For the LB gum solutions, a higher coil overlap dependence was obtained (slope $\approx$ 4.2) in good agreement with the data for galactomannan samples (Morris et al., 1981; Robinson et al., 1982), which can be attributed to the presence of more specific polymer-polymer interactions. Intermediate values of exponents were obtained for the mixtures, suggesting that the hydrodynamic volume of each polymer was not significantly affected by the presence of the other, even in concentrated solutions. However, the slopes obtained for the mixtures were slightly greater than expected by simple additivity, which can be due to the errors in the η_0 or to the very important presence of galactomannan entanglements.

Viscoelastic Behavior of Food Gum Dispersions

The magnitudes of G' and G'' obtained for LB gum solutions of 0.75%, 1.02%, and 1.22% in distilled water are shown in Figure 4–8; similar behavior was observed with 2.19% and 3.97% HM pectin solutions (Lopes da Silva et al., 1993). For both LB gum and HM pectin, the magnitudes of G' were lower than those of G'' over the range of gum concentrations and over a large range of the oscillatory frequencies studied. Log G' and log G'' vs. log ω plots followed either one or two straight lines. The frequency at which G' became equal to G'' will be called crossover frequency, while, in the case of data that followed two straight lines, the frequency at which the break occurred will be denoted ω^*.

The crossover frequency for LB gum and low-methoxyl (LM) pectin decreased with increase in concentration: for LB gum from 27.4 to 12.0 rad s^{-1} as the con-

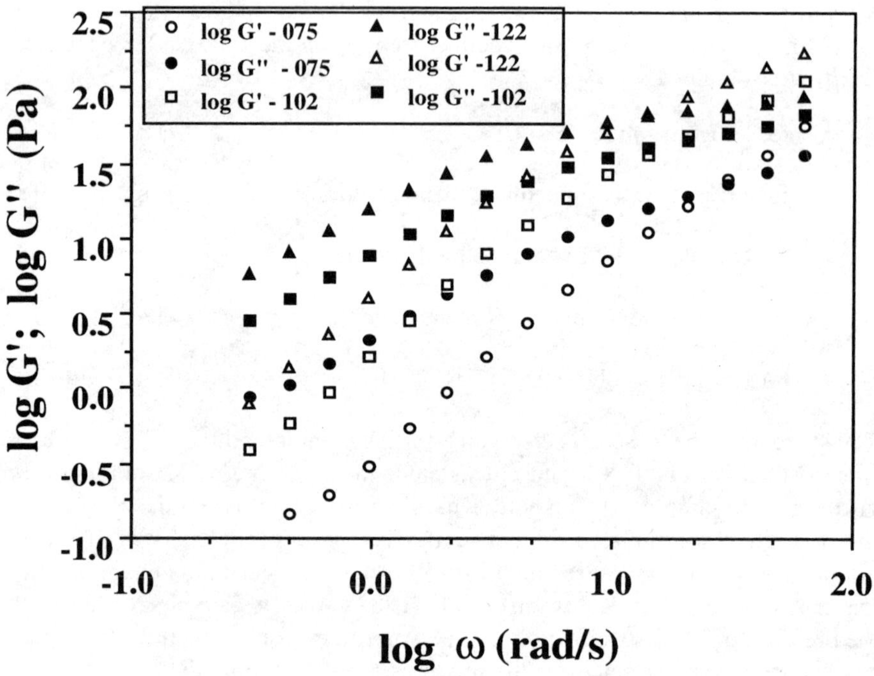

Figure 4–8 Magnitudes of G' and G'' for Locust Bean Gum Solutions of 0.75%, 1.02%, and 1.22% in Distilled Water. Lopes da Silva, 1993.

centration was increased from 0.75% to 1.22% and for LM pectin from 64.4 to 31.9 rad s^{-1} as the concentration was increased from 2.78% to 3.98% (Table 4–4). In contrast, the crossover frequency for HM-pectin solutions increased from 49.2 to 231 rad s^{-1} when the concentration was increased from 2.19% to 3.93% (Table 4–5).

In the study of mixtures of LB gum and HM-pectin solutions, the single gum solutions were selected so that the apparent viscosities were equal at a shear rate of about 10 s^{-1}. The LB gum and HM-pectin concentrations of the solutions with pH 7.0, ionic strength 0.1M, were 0.72% and 3.38%, while those dissolved in distilled water were 0.96% and 3.51%, respectively; the proportion of LB gum in the mixtures was 25%, 50%, and 75%. The crossover frequency for the mixtures decreased as the proportion of LB gum in the mixture was increased (Lopes da Silva et al., 1993).

Table 4–4 Power Law Correlation between G' and G'' and Frequency (rad s^{-1}) and the Crossover Frequency for Solutions of LB Gum (Natural pH and Ionic Strength)

Sample	*A′*	*B′*	n_2	*A″*	*B″*	R_2	*Crossover Frequency*	ω^* G′	ω^* G″
LBG 0.75% ($\omega < \omega^*$)	0.316	1.46	1.00	2.15	0.907	1.00		6.6	5.2
LBG 0.75% ($\omega > \omega^*$)	0.668	1.07	0.999	4.05	0.522	0.995	27.4		
LBG 1.02% ($\omega < \omega^*$)	1.72	1.42	1.00	7.85	0.867	1.00		5.0	4.4
LBG 1.02% ($\omega > \omega^*$)	5.05	0.751	0.999	18.3	0.298	0.997	17.3		
LBG 1.22% ($\omega < \omega^*$)	4.05	1.42	1.00	15.4	0.852	1.00		4.5	4.0
LBG 1.22% ($\omega > \omega^*$)	13.9	0.605	0.999	38.0	0.199	0.989	12.0		

Cox-Merz Rule for Biopolymer Dispersions

The Cox-Merz rule (Cox and Merz, 1958) is useful in predicting steady shear viscosity from complex viscosity and vice versa:

$$\eta^* (\omega) = \eta_a (\omega) \big|_{\omega = \dot{\gamma}} \tag{4.19}$$

Lopes da Silva et al. (1993) obtained the flow data by increasing the shear stress from zero to a maximum value, which depended on the concentration of the sample, and reducing it over the same range in a continuous manner. The time for the up and down curves was standardized. For examining the correlation between dynamic and steady shear properties, the down flow curves (decreasing shear stress) and the dynamic data obtained also by decreasing the oscillatory frequency (from 10 to 0.01 Hz) were used.

Table 4–5 Power Law Correlation between G' and G'' and Frequency (rad/s) Solutions of HM Pectin (Natural pH and Isoionic Dilutions)

Sample	*A′*	*B′*	R_2	*A″*	*B″*	R_2	*Crossover Frequency*
HMP 2.19%	0.0349	1.61	0.998	0.758	0.824	0.997	49.2
HMP 2.62%	0.199	1.20	0.995	1.33	0.844	1.00	197
HMP 2.76%	0.186	1.28	0.996	1.66	0.821	0.999	116
HMP 3.07%	0.386	1.17	0.997	2.55	0.813	1.00	211
HMP 3.48%	0.878	1.07	0.999	4.66	0.773	1.00	261
HMP 3.79%	1.91	0.934	0.997	6.53	0.741	1.00	584
HMP 3.93%	2.04	0.960	0.998	3.12	0.554	0.998	231
				1.66	0.912	0.999	(ω^* = 7.5 rad/s)

A common feature observed was the departure between the steady shear viscosity (η_a) and the real component of the dynamic viscosity (η') at large values of shear rate and frequency, with the expected more rapid decrease of η' with frequency than η_a decreases with shear rate (Bird et al., 1977), which can be attributed to the very different molecular motions involved in the dynamic and steady shear at high ω and $\dot{\gamma}$ (Ferry, 1980). Because of the relatively high value of strain amplitude used in the tests (36%) and the two-phase nature of the HM dispersions, the observations with respect to η^* and η are in agreement with those of Matsumoto et al. (1975).

High-Methoxyl Pectin Dispersions

For high-methoxyl pectin dispersions, magnitudes of η^* (ω) were almost always lower than those of η_a ($\dot{\gamma}$), only converging at higher values of frequency and shear rate (Figure 4–9). Such behavior is not very common in biopolymeric systems and was reported for semidilute solutions of xanthan gum in 0.5% NaCl (Rochefort and Middleman, 1987) and aqueous solutions of hydroxyethyl guar gum (Lapasin et al., 1991). In addition, Kulicke and Porter (1980) reported that for clustered synthetic ionomers, magnitudes of η^* (ω) were < η_a ($\dot{\gamma}$). However, most

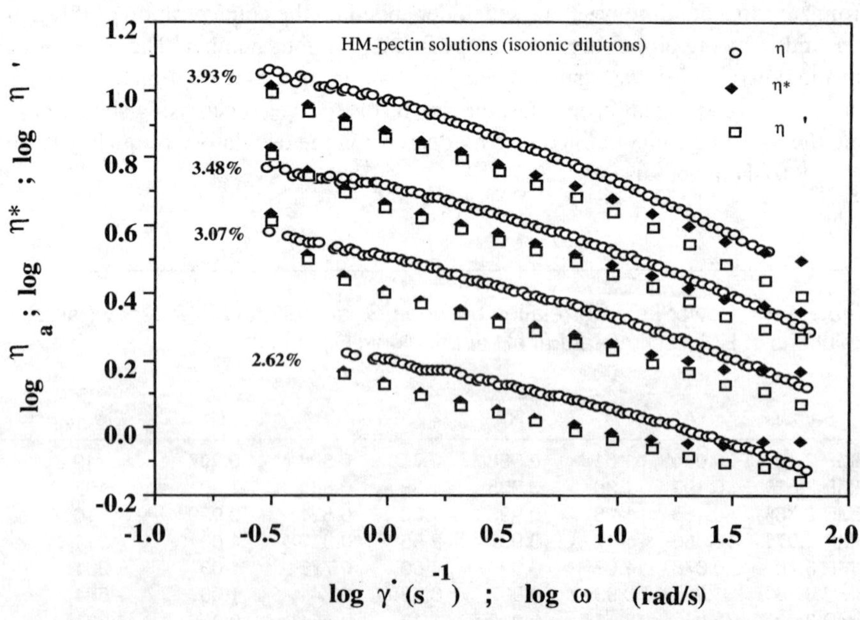

Figure 4–9 Cox-Merz Plot for Rheological Data on High-Methoxyl Pectin Dispersions. Lopes da Silva, 1993.

of the deviations from the Cox-Merz rule are attributed to structure decay due to the effect of strain deformation applied to a system being low in oscillatory shear and sufficiently high in steady shear to break down intermolecular associations. In these cases, the magnitudes of η^* (ω) are higher than those of η_a ($\dot{\gamma}$) (Lopes da Silva et al., 1993). The observed departure from the Cox-Merz rule may be attributed to aggregation phenomena of the HM pectin in solution (Davis et al., 1980; Paoletti et al., 1986; Sawayama et al., 1988). Consequently, the HM pectin dispersions were not true solutions but two-phase systems with pectin micro-aggregates dispersed into the solvent.

In addition, for dispersed polymer systems at small strain amplitudes, Matsumoto et al. (1975) observed a pronounced departure from the Cox-Merz rule, with η_a much smaller than η^* at the same rate of shear and angular frequency. They also observed that η^* decreased with increase in the strain amplitude such that for high strain amplitudes, η^* was smaller than η_a.

Low-Methoxyl Pectin Dispersions

The LM pectin solutions followed quite satisfactorily the correlation of Cox and Merz (Lopes da Silva et al., 1993) (Figure 4–10). The different behavior relative

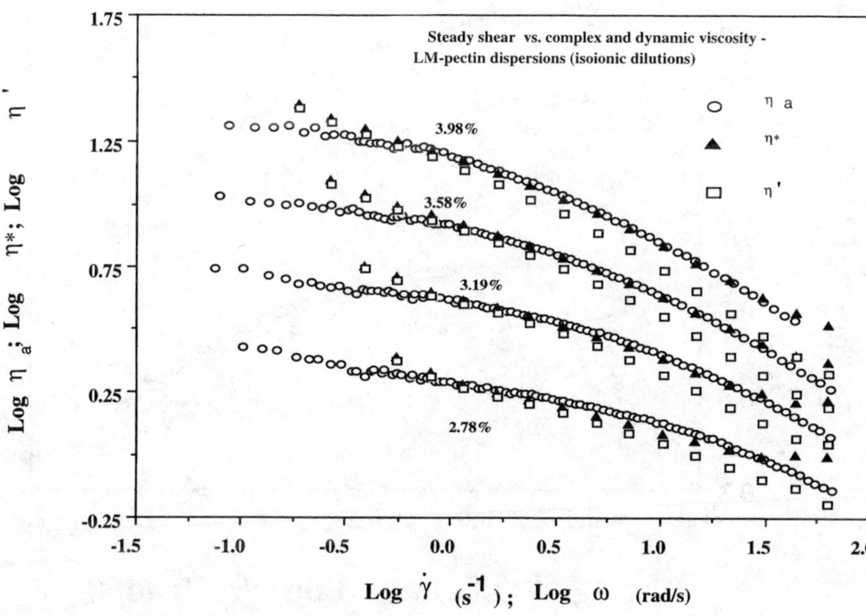

Figure 4–10 Cox-Merz Plot for Rheological Data on Low-Methoxyl Pectin Dispersions. Lopes da Silva, 1993.

to the HM pectin solutions can be attributed to less extension of intermolecular association, as a result of electrostatic repulsions due to the presence of ionized free carboxyl groups (higher charge density).

The viscosity values in oscillatory shear at low frequency, especially for the LM pectin solutions for which the oscillatory tests were extended to lower frequency magnitude, may be suggestive of the existence of a yield stress, which can be substantiated by the results obtained in our earlier study in steady shear flow (Lopes da Silva et al., 1992). On the other hand, the effects of friction in our controlled stress rheometer (Krieger, 1990) also cannot be ruled out.

Locust Bean Gum Dispersions

For locust bean (LB) gum aqueous solutions, at low shear rate and frequency, η^*, η', and η_a have similar magnitudes but η^* and η' are slightly smaller than η_a; at higher shear rate and frequency, η^* and η' diverged while η^* and η_a converged, with η^* becoming higher than η_a (Lopes da Silva et al., 1993) (Figure 4–11). Sim-

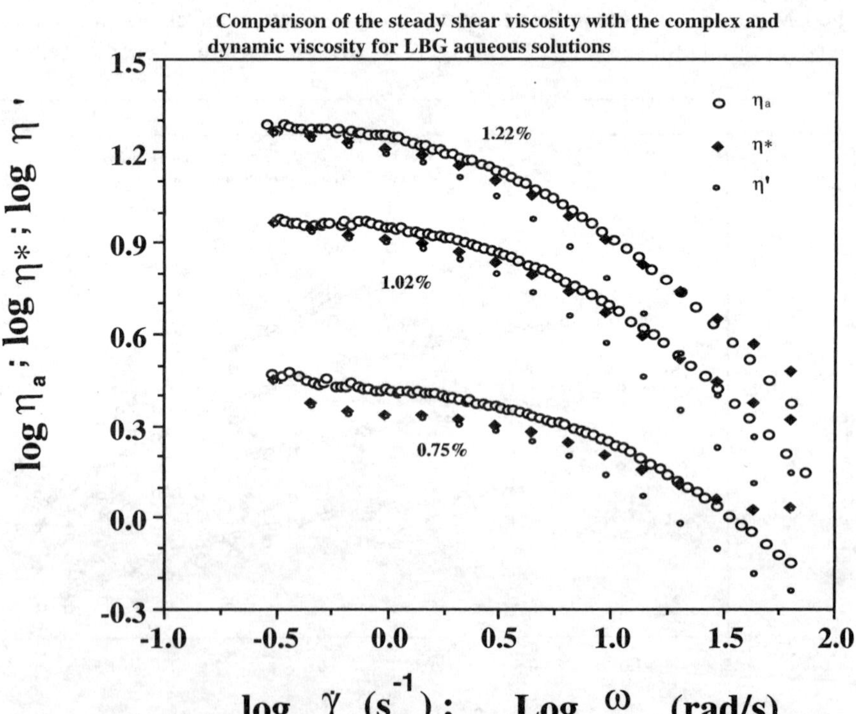

Figure 4–11 Cox-Merz Plot for Rheological Data on Locust Bean Gum Dispersions. Lopes da Silva, 1993.

ilar behavior at high shear rate and frequency was previously reported for guar gum solutions (Mills and Kokini, 1984). Departures from the Cox-Merz rule were observed for galactomannan samples at low shear rates and frequencies (Morris et al., 1981), which were attributed to the presence of high density of entanglements ("hyper-entanglements") resultant from very specific polymer-polymer interactions. In the LB gum solutions, the departure seems to increase with decreasing concentration. Mesquite seed gum dispersions (Yoo et al., 1994) showed reasonable agreement with the Cox-Merz rule unlike the departures found for galactomannan samples at low shear rates and frequencies.

Constitutive Equations Based on Molecular Structure

Constitutive equations relate stress tensor to various kinematic tensors and can be used to solve flow and other engineering problems. The development of constitutive equations to describe the state of stress continues to be an active area of research. The equations have been developed based on either molecular structure or continuum mechanics where the molecular structure is not considered explicitly and the response of a material is independent of the coordinate system (principle of material indifference). In the former, the polymer molecules are represented by mechanical models and a probability distribution of the molecules, and relationships between macroscopic quantities of interest are derived. Three models have found extensive use in rheology: the bead-spring model for dilute polymer solutions and the transient net work and the reptation models for concentrated polymer solutions and polymer melts.

Dilute Polymer Solution Molecular Theory

In a dilute polymer solution, there are no entanglements of the individual molecules and the molecules can be considered to be essentially isolated from each other, so that treatment of an individual molecule is sufficient. Much work was done to understand the rheological behavior of dilute polymer solutions using the concept that a polymer molecule can be idealized as dumbbells, that is, two beads connected by a spring. The hydrodynamic, Brownian, and intramolecular forces acting on the beads are considered when the dumbbells are suspended in a Newtonian fluid. Because experimental data cannot be taken at concentrations considered to be sufficiently low for the absence of intermolecular interactions, data obtained at low polymer concentrations (e.g., $c < 10^{-2}$ g mL^{-1}) must be extrapolated to infinite dilution (Ferry, 1980).

The appropriate viscoelastic functions are the dynamic rheological properties extrapolated to infinite dilution and are called the intrinsic dynamic rheological properties (storage modulus G' and the loss modulus G'', and the dynamic viscosities η' and η''):

$$[G'] = \lim_{c \to 0} (G'/c) \tag{4.20}$$

$$[G''] = \lim_{c \to 0} [(G'' - \omega\eta_s)/c] \tag{4.21}$$

$$[\eta'] = \lim (\eta' - \eta_s)/\eta_s c \tag{4.22}$$

$$[\eta''] = \lim_{c \to 0} \eta''/\eta_s c \tag{4.23}$$

The results of theories of several molecular models (e.g., rigid and flexible molecules) are predicted to be proportional to RT/M, where M is the molecular weight that is often assumed to be uniform. The theoretical predictions are conveniently expressed in terms of the reduced dimensionless intrinsic moduli:

$$[G']_R = [G']M/RT \tag{4.24}$$

$$[G'']_R = [G'']M/RT \tag{4.25}$$

Two theories based on the bead-spring model have found extensive use in the polymer literature. In Rouse's theory, the hydrodynamic interactions between the motions of the submolecule junctions is ignored and is based on the change in free energy associated with the entropy decrease for nonrandom configurations and the tendency of a system to diffuse toward a random state (Ferry, 1980). In Zimm's theory, inertial forces are neglected, and the force on each bead is calculated as the sum of (1) hydrodynamic drag force by the solvent, (2) the force associated with the Brownian motion, and (3) the Hookean springlike forces exerted by two neighboring submolecules (Ferry, 1980). Rouse's theory predicts

$$[G']_R = [G'']_R = \left(\frac{\pi}{2\sqrt{2}}\right) (\omega\tau_1)^{1/2} \tag{4.26}$$

The predictions of Zimm's theory are

$$[G']_R = 1.05\ (\omega\tau_1)^{2/3} \tag{4.27}$$

$$[G'']_R = 1.82\ (\omega\tau_1)^{2/3} \tag{4.28}$$

Figure 4–12 contains the predictions of the Rouse theory on the left and of the Zimm theory on the right. As is to be expected, the Zimm theory, which takes the hydrodynamic interactions into account, predicts experimental data well.

Concentrated Polymer Dispersions

For a long time, based on rheological data, some kind of intermolecular coupling was suspected. Pioneering studies by Tobolsky and others led to the concept of an entangled network structure with two sets of relaxation or retardation times. The

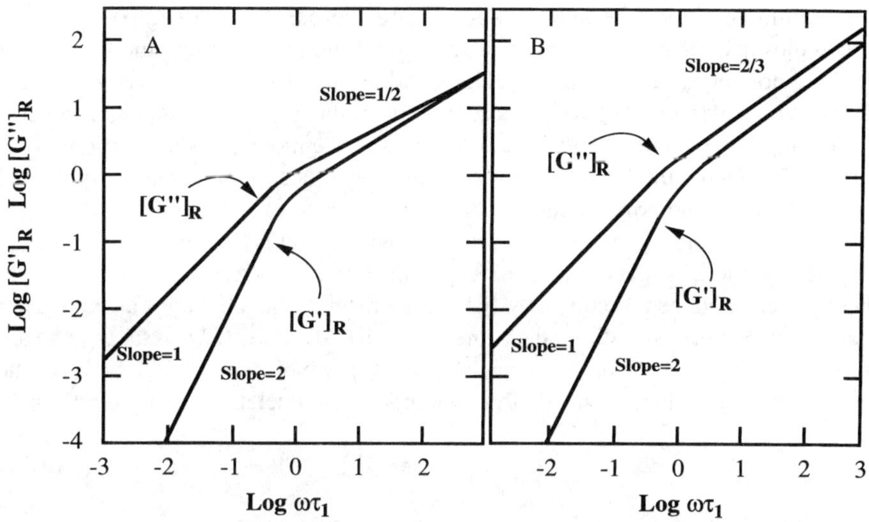

Figure 4–12 Predictions of Reduced Storage and Loss Moduli by Models of Rouse (left) and Zimm (right). *Source:* J.D. Ferry, *Viscoelastic Properties of Polymers*. © 1980, John Wiley and Sons, Inc. Reprinted with permission of John Wiley and Sons, Inc.

transient net work model is an adaptation of the network theories of rubber elasticity. In concentrated polymer solutions and polymer melts, the network junctions are temporary and not permanent as in chemically cross-linked rubber, so existing junctions can be destroyed to form new junctions. Earlier, it was mentioned that a critical concentration, c^*, exists, below which a polymer dispersion exhibits dilute solution behavior and above which it exhibits concentrated solution behavior. In terms of molecular weight dependence, a low-molecular weight region exists where η_0 increases almost in direct proportion to molecular weight M and a high–molecular weight region exists where η_0 increases with the 3.4 power of the molecular weight. Further studies led to the conclusion that polymer chain entanglements are topological and not dissociating junctions (Ferry, 1980). The transient net work model can predict many of the linear viscoelastic phenomena and shear-thinning behavior; the rates of creation and loss of segments can be considered to be functions of shear rate.

The concept of polymer entanglements represents intermolecular interaction different from that of coil overlap-type interaction. However, it is difficult to define the exact topological character of entanglements. The entanglements concept was aimed at understanding the important nonlinear rheological properties, such as the shear rate dependence of viscosity. However, viscoelastic properties could not be defined quantitatively as is possible with the reptation model. Because an entanglement should be able to slide along the contours of the participating molecules and change its position with thermal motions, entanglement points or loci are not very useful for better understanding of rheological behavior. Instead, one can consider a model with topological restraints such as a tube placed to surround a polymer molecule and that conforms to the molecule's contortions (Figure 4–13). Such a model was introduced by Edwards and mathematically treated by de Gennes, who pointed out that the easiest escape for sufficiently long chains is achieved by sliding along the contorted contour, that is, by reptation.

The tube model avoids dealing with specification of an average molecular weight between entanglement points (M_e), but the diameter of the tube (d) is an equivalent parameter. Further, the ratio of the contour length to diameter (L/d) is taken to be approximately the magnitude of the entanglements per molecule (M/M_e). Some of the important relationships predicted from the reptation model include (Tirrell, 1994)

$$\lambda_{\text{rep}} \sim M^3 \tag{4.29}$$

where, λ_{rep} is the longest relaxation time, that is, time for the chain to diffuse one chain length or renew its configuration. The self-diffusion coefficient

$$D_s \sim M^{-2} \tag{4.30}$$

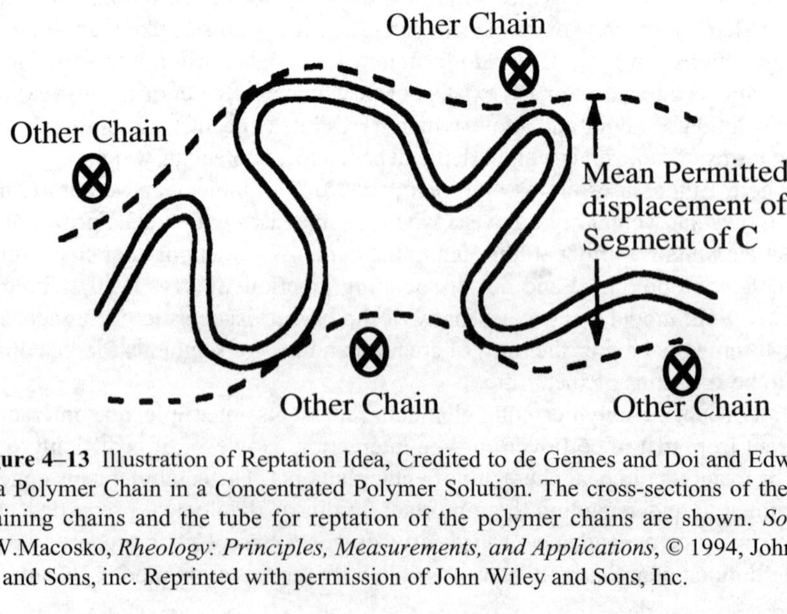

Figure 4–13 Illustration of Reptation Idea, Credited to de Gennes and Doi and Edwards, of a Polymer Chain in a Concentrated Polymer Solution. The cross-sections of the constraining chains and the tube for reptation of the polymer chains are shown. *Source*: C.W.Macosko, *Rheology: Principles, Measurements, and Applications*, © 1994, John Wiley and Sons, inc. Reprinted with permission of John Wiley and Sons, Inc.

and viscosity

$$\eta \sim M^3 \tag{4.31}$$

The prediction of the above equation is the closest of any molecular theory to experimentally observed 3.4 power. A constitutive equation based on the reptation model can be written as

$$\bar{\bar{\tau}} = \int_{\infty}^{t} \{M_1[t - t', I_1(t,t'), I_2(t,t')]\, \bar{\bar{C}}_t^{-1} + M_1[t - t', I_1(t,t'), I_2(t,t')]\, (\bar{\bar{C}}_t^{-1})^2\}\, dt \tag{4.32}$$

where $\bar{\bar{\tau}}$ is the extra stress tensor and I_1 and I_2 are the invariants of the Finger deformation tensor $\bar{\bar{C}}_t^{-1}$ that incorporates both stretch and rotation in the deformation gradient. The memory functions M_1 and M_2 can be written as products of functions of the past time $(t - t')$ and of deformation only. In continuum mechanics, the above equation is known as the Kaye-BKZ equation (De Kee and Wissburn, 1998). More discussion on constitutive equations can be found in the texts of Bird et al. (1977), Dealy and Wissburn (1990), and Carreau et al. (1997).

RHEOLOGY OF HEATED STARCH DISPERSIONS

The composition and characteristics of starches have been well covered elsewhere (Whistler and Daniel, 1985; Hoseney, 1998), and they will be reviewed only briefly here. Starch is the major energy reserve of seeds, roots, and tubers and is produced in small individual packets called granules (Whistler and Daniel, 1985; Chamberlain, 1996). The characteristics of starch granules vary in an individual plant depending on the tissue environment. For example, granules in corn's central floury endosperm are round, yet those in the outer, proteinacious, horny endosperm are angular (Whistler and Daniel, 1985). As discussed under the rheological behavior of dispersions in Chapter 2, the rheological behavior of starch dispersions also is influenced very strongly by the volume fraction and state of the granules in the dispersion.

Starch granules have a nucleation point, the hilium, around which the granule is formed. A Maltese cross pattern, characteristics of birefringency, with its center at the hilium can be seen when the granule is viewed with polarized light. Granules contain molecules of amylose and/or amylopectin that seem to be uniformly arranged (Whistler and Daniel, 1985). All starch molecules are natural polymers of anhydroglucose units and are present in either a linear or branched configuration. Linkages between the glucose molecules occur through enzymatic condensation in the plant at carbons 1 and 4 (forming a linear polymer) or carbons 1 and 6 (forming a branched polymer) (Langan, 1986).

Amylose, the linear starch polymer, typically has molecular weights ranging from 10^5 to 10^6 and 500 to 5000 glucose residues per molecule. It has been determined that some α-(1→6)-branching exists in the amylose molecule on the order of 2–8 branch points per molecule, with side chains ranging in length from 4 to $>$ 100 glucose units (Galliard and Bowler, 1987).

Native starches contain approximately 25% amylose, yet there are some mutant varieties of corn that contain 85% amylose (high-amylose corn) or no amylose (waxy maize) (Whistler and Daniel, 1985). Amylose leaches from starch granules during the gelatinization stage and affects both the viscosity of the continuous phase and the rate of retrogradation on cooling.

Retrogradation is the process whereby linear regions of the polysaccharide align and reduce the solubility of the molecule (Galliard and Bowler, 1987). Amylopectin characteristically has a molecular weight greater than 10^8 with 5% of its structure containing α-(1→6)-branch points (Galliard and Bowler, 1987). It is generally agreed that the amylopectin chains are assembled in a cluster structure, yet a consensus has not been found for the fine structure, nature of branching, and proportions of side-chains in amylopectin (Galliard and Bowler, 1987).

In food products, starch can be either a mere ingredient used in low concentrations to thicken the product texture or the main component responsible for the texture. The former kind of products usually contain water in excess of gelatinization requirements, while in the latter, such as baked products, water is available in a limited amount. In rheological studies on dispersions of low starch concentrations, because of settling of the ungelatinized granules, considerable caution should be exercised to ensure that different samples taken from a dispersion contain the same concentration of starch granules. Likewise, granule settling will occur in viscometer geometries that are vertically oriented, such as concentric cylinder and vertical capillary/tube.

Starch granules show birefringence in the form of a Maltese cross when viewed under polarized light. The amylopectin molecules in the starch granule are partially crystalline. At room temperature, the polymers in the amorphous phase of the starch granule are glassy under dry conditions. At about 22% moisture and at 25°C, the polymer will go through a glass transition (Hoseney, 1998). Irreversible changes, often referred to as gelatinization, occur when a starch-water mixture is heated above a critical temperature, which depends on the source of starch. These changes include loss of birefringence and, more importantly, crystallinity that allows the granules to swell under excess water conditions.

A gelatinized starch dispersion may be regarded as a composite material consisting of swollen granules and granular fragments dispersed in a continuous biopolymer matrix (Evans and Haisman, 1979; Eliasson, 1986; Morris, 1990; Noel et al., 1993). Therefore, the properties of the dispersed phase, the continuous phase, and interactions between the components are three important considerations for insights into the rheology of gelatinized starch suspensions.

Rheological changes in heated starch dispersions (STDs) have been studied under isothermal and nonisothermal heating conditions. Under isothermal conditions, the apparent first-order reaction rate and the Arrhenius models described the rate of starch gelatinization and the influence of temperature on the rate, respectively (Kubota et al., 1979; Lund, 1984; Okechukwu et al., 1991). However, because of the important role of temperature history, changes in η_a during gelatinization cannot be described accurately by the models obtained under isothermal conditions. Temperature and heating rates are important variables during the sol-gel transition period; in addition, the shear rate ($\dot{\gamma}$) or dynamic frequency (ω) has a significant effect on η_a or complex viscosity (η^*), respectively. A good appreciation of the transition phase is vital to a full understanding of the rheological properties of gelatinized starch dispersions.

VISCOUS AND VISCOELASTIC PROPERTIES DURING HEATING OF STARCH DISPERSIONS

In many food processes, foods containing starches are heated continuously over a range of temperatures, so viscosity versus temperature or viscosity versus time profiles are of interest. Results on an 8% (w/w) unmodified corn STD obtained by differential scanning colorimetry (DSC) are compared with dynamic rheological

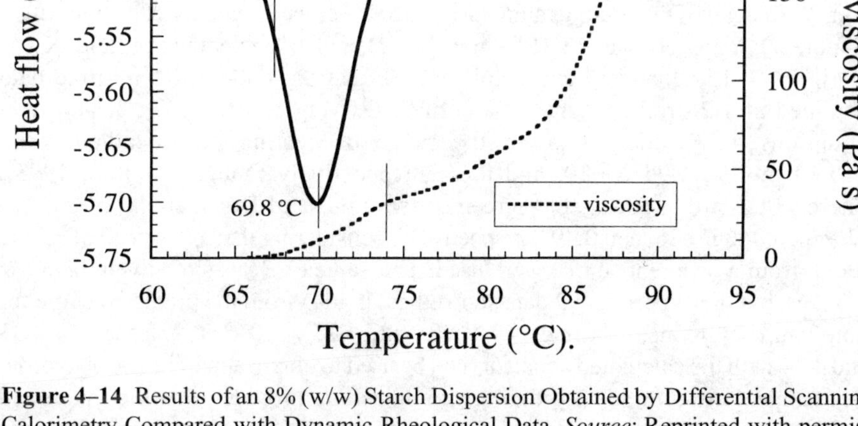

Figure 4–14 Results of an 8% (w/w) Starch Dispersion Obtained by Differential Scanning Calorimetry Compared with Dynamic Rheological Data. *Source*: Reprinted with permission from W.H. Yang, A.K. Datta, and M.A. Rao, Behavior of Starch Gelatinization in Simulation of Heat Transfer, *Engineering and Food at ICEF 7/Part 2*, pp. K1–K5, International Congress on Engineering and Food.

data (Figure 4–14). The initial temperature of gelatinization from both methods was about the same, ~67°C, but the differences in peak and conclusion temperatures from the two methods were considerable. The DSC peak temperature, due to the melting of starch crystallites cooperatively in excess water, coincided with that of first rise in viscosity where the rate of development of viscosity was the greatest. Because of the continued swelling of the starch granules, the viscosity increased until 90°C. After the peak viscosity, further increase in temperature resulted in a decrease in viscosity (Yang et al., 1997).

Model of Dolan and Colleagues

Dolan et al. (1989) developed a model (Equations 4.33 and 4.34) to describe apparent viscosity as a function of time during starch gelatinization under non-isothermal conditions, and additional discussion on its application to starch dispersions can be found in Dolan and Steffe (1990). The model contains an exponential function of the temperature-time history and the Arrhenius equation to describe the gelatinization reaction. The special form of the model for constant shear rate and starch concentration is as follows:

$$\eta_{\text{dim}} = \frac{\eta_a - \eta_{ug}}{\eta_\infty - \eta_{ug}} = [1 - \exp(-k\Psi)]^\alpha \tag{4.33}$$

$$\Psi = \int T(t)\exp\left(\frac{-\Delta E_g}{RT(t)}\right) dt \tag{4.34}$$

where η_a is apparent viscosity (Pa s); η_{dim}, dimensionless apparent viscosity; η_{ug}, ungelatinized apparent viscosity (Pa s); η_∞, highest magnitude of η_a during gelatinization; Ψ, integral of time-temperature history; ΔE_g, activation energy of gelatinization (J mole^{-1}); R, gas constant (8.314 J mol^{-1} K^{-1}); and T, temperature of fluid (K).

Figure 4–15 illustrates applicability of Equation 4.33 to the η^*-time data obtained at 1.26 rad s^{-1} on a 8% corn STD. Magnitudes of E_g, k, and α in Equations 4.33 and 4.34, over the range of heating rates studied, were 240 kJ mol^{-1}, 1.95 E+30, and 0.668, respectively (Yang and Rao, 1998). These values are in reasonable agreement with those of Dolan et al. (1989): 210 kJ/mol, 0.846 E+26, and 0.494, respectively, considering that η^* instead of η_a and starch from a different source were used. The value of R^2 was relatively high (R^2 = 0.96), but the experimental data showed deviations from the predicted curve for time-temperature history values 5.0×10^{-31} to 1.25×10^{-30} K s. Equations 4.33 and 4.34 with the calculated constants can be used to obtain apparent viscosity from any temperature-time history during gelatinization, provided that the initial and maximum viscosities are known; however, because of granule sedimentation, the former is difficult to determine experimentally. Better fit may be obtained by expressing ΔE_g as a function of temperature and the Arrhenius relationship to be temperature-time history–dependent (Dolan et al., 1989; Dolan and Steffe, 1990).

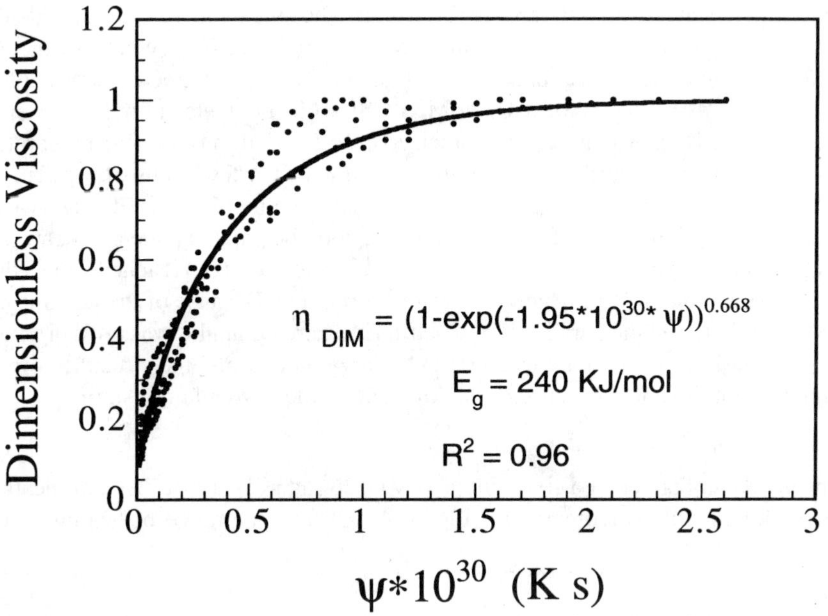

Figure 4–15 Applicability of the Model of Dolan et al. (1989) to an 8% Corn Starch Dispersion η^* Data. *Source*: Reprinted with permission from W.H. Yang, *Rheological Behavior and Heat Transfer to a Canned Starch Dispersion: Computer Simulation and Experiment*, PhD Thesis, © 1997, Cornell University.

Model for η^*-Temperature Data

An η^*-temperature master curve was derived for corn starch dispersions (Yang and Rao, 1998) based on separate characteristics: (1) η_a versus temperature profiles of dispersions of a specific starch (e.g., corn, bean) are usually similar over narrow ranges of starch concentrations (Launay et al., 1986; Champenois et al., 1998), so it may be possible to develop simple relationships for estimating viscosity over specific ranges of concentration. (2) Because η^* data are obtained at low strains with minimal alteration of the STD structure, they provide unique opportunities for studying applicable models. (3) Empirically obtained frequency shift factor (Ferry, 1980) has been used successfully in time-temperature superposition studies on food polymer dispersions (Lopes da Silva et al., 1994), and the applicability of similar, if not identical, scaling of frequency was explored for STDs.

Two critical factors for obtaining reliable η^* data were (Yang and Rao, 1998) (1) control of the sample volume that was loaded carefully on the bottom rheometer plate and removal of excess dispersion from the edge of the top plate, and (2) minimizing water evaporation from the sample when the temperature was high and avoiding displacement of paraffin oil by tiny air bubbles from the edge of the top plate. Dynamic rheological data were obtained using 4-cm diameter parallel plate (gap 500 μm) geometry of a Carri-Med CSL-100 rheometer (TA Instruments, New Castle, DE). A plate-cone geometry was also tried, but the narrow gap (52 μm at the center) contributed to capillary suction of the few drops of paraffin oil that were placed on the exposed edge of the plate to minimize loss of water vapor during heating into the corn STD. In contrast, with the parallel plate geometry, the drops of paraffin oil placed soon after the storage modulus (G') reached values that were measurable did not penetrate into the corn STD. Because of the narrow gap (500 μm) between the parallel plates, settling of starch granules was not a problem as it might have been in a concentric cylinder geometry and the structure formation was not disrupted because of the low magnitude of oscillatory strain.

Influence of Heating Rates

The η^* data obtained at a fixed frequency (1.26 rad s^{-1}) and at different heating rates plotted against temperature (Figure 4–16) over the range of heating rates,

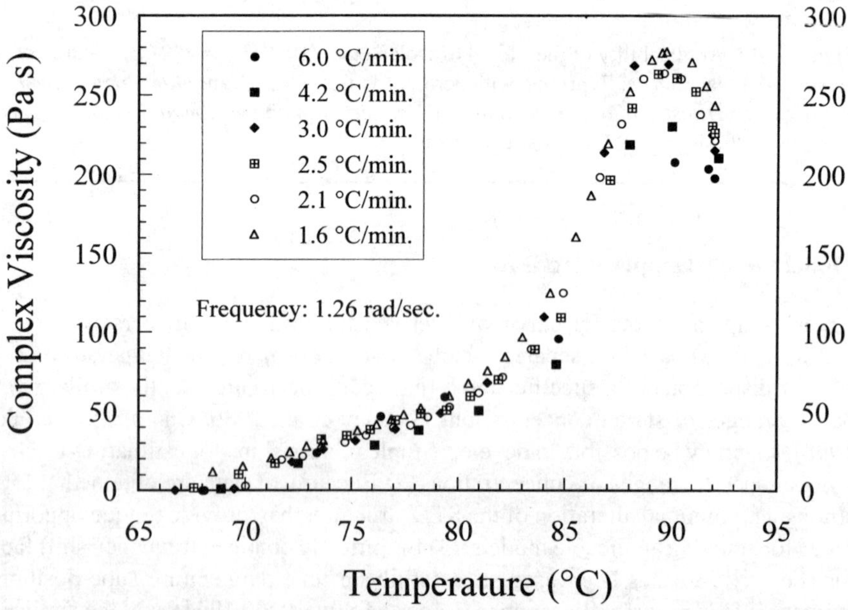

Figure 4–16 The η^* Data of an 8% Corn Starch Dispersion at a Fixed Frequency (1.26 rad s^{-1}) and at Different Heating Rates Plotted against Temperature. *Source*: Reprinted with permission from W.H. Yang, *Rheological Behavior and Heat Transfer to a Canned Starch Dispersion: Computer Simulation and Experiment*, PhD Thesis, © 1997, Cornell University.

1.6°C min^{-1} to 6.0°C min^{-1}, collapsed to a single curve. The maximum difference in the magnitudes of complex viscosity between the highest (6.0°C min^{-1}) and the lowest (1.6°C min^{-1}) heating rates was about 25%.

The slight decrease in η^* that occurred after the first increase (Figure 4–16) was found to be concomitant with end of the DSC endothermic peak due to melting of crystallites (Eliasson, 1986). Because of continued swelling of the starch granules, the viscosity increased until about 90°C, leading to the formation of a transient network of granules touching each other and sometimes resulting in granule disruption (Blanshard, 1979) and amylose leaching (Eliasson, 1986). Similar viscosity-temperature behavior has been noted for other cereal starches (Launay et al., 1986; Doublier, 1987).

Heating at a Fixed or Increasing Temperature

After the peak viscosity was reached, increasing the temperature further resulted in decrease in viscosity (Eliasson, 1986; Doublier, 1987), which can be attributed to rupture of the granules. Two distinct types of curves were obtained after the

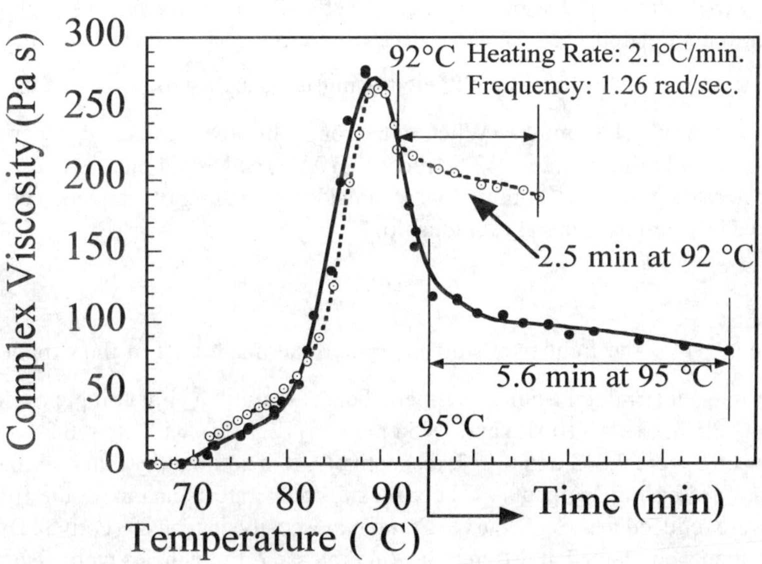

Figure 4–17 Two Distinct Curves Obtained after Peak Viscosity Was Reached (8% corn starch dispersion). (1) The temperature was maintained at 92°C for about 2.5 minutes or at 95°C for about 5.5 minutes, or (2) the temperature was increased up to 95°C continuously after the peak viscosity temperature. *Source*: Reprinted with permission from W.H. Yang, *Rheological Behavior and Heat Transfer to a Canned Starch Dispersion: Computer Simulation and Experiment*, PhD Thesis, © 1997, Cornell University.

peak viscosity was reached (Figure 4–17) depending on whether (1) the temperature was maintained at 92°C for about 2.5 minutes or at 95°C for about 5.5 minutes, or (2) the temperature was increased up to 95°C continuously after the peak viscosity temperature. The former was characterized by a slower rate of decrease in η^* after the temperature was held constant and the latter by a rapid rate of decrease in η^* up to the final temperature (95°C), suggesting that the network structure formed during gelatinization was drastically weakened by increase in the temperature.

It is theoretically possible that at very high heating rates, starch granule swelling would be limited by moisture diffusion. However, a more practical limitation would be to obtain reliable rheological data at a high heating rate because the temperature range over which viscosity of the corn STD increases would be covered in a short time period.

Influence of Frequency

At a fixed heating rate of 2.1°C min^{-1} and over the range of frequencies employed from 1.26 to 47.12 rad s^{-1}, the profiles of η^* at a specific frequency versus temperature were similar in shape, so that by choosing an arbitrary reference frequency (ω_r), all the η^*-temperature curves at the different frequencies could be reduced to a single curve. A master curve of reduced complex viscosity $\eta_R^* = \eta^* \left(\frac{\omega}{\omega_r}\right)$ versus temperature was obtained using $\omega_r = 6.28$ rad s^{-1} (1.0 Hz) at the heating rate of 2.1°C min^{-1}. When data from a limited number of experiments conducted at high ω values—62.83 rad s^{-1} (10 Hz) and 78.54 rad s^{-1} (12.5 Hz)—at the heating rate of 2.1°C min^{-1} were considered, scaling of frequency could be achieved by a more general relationship:

$$\eta_R^* = \eta^* \left(\frac{\omega}{\omega_r}\right)^{\beta} \tag{4.35}$$

where $\left(\frac{\omega}{\omega_r}\right)^{\beta}$ is the frequency shift factor and the magnitude of the exponent, β, needs to be determined from experimental data at ω and ω_r. For complex viscosity data at 62.83 rad s^{-1} (10 Hz) and 78.54 rad s^{-1} (12.5 Hz), with $\omega_r = 6.28$ rad s^{-1}, values of β were 0.913 and 0.922, respectively. It is noteworthy that both increasing and decreasing segments of the viscosity-temperature data at all the frequencies were reduced to a single curve. Complex viscosity data on 8% corn STDs from experiments conducted at different heating rates and frequencies were also superposed using 6.28 rad s^{-1} (1.0 Hz) as ω_r (Figure 4–18). Figure 4–18 is the master curve of reduced complex viscosity η_R^* data obtained on the 8% corn STD at different heating rates and frequencies, and it represents data at several shear rates as a function of temperature. The above modeling procedure was found to be applicable to the η^* data obtained on 6% corn STDs at a heating rate of 3.0°C min^{-1} and at dynamic frequencies 3.14, 12.57, and 47.12 rad s^{-1}. The superposi-

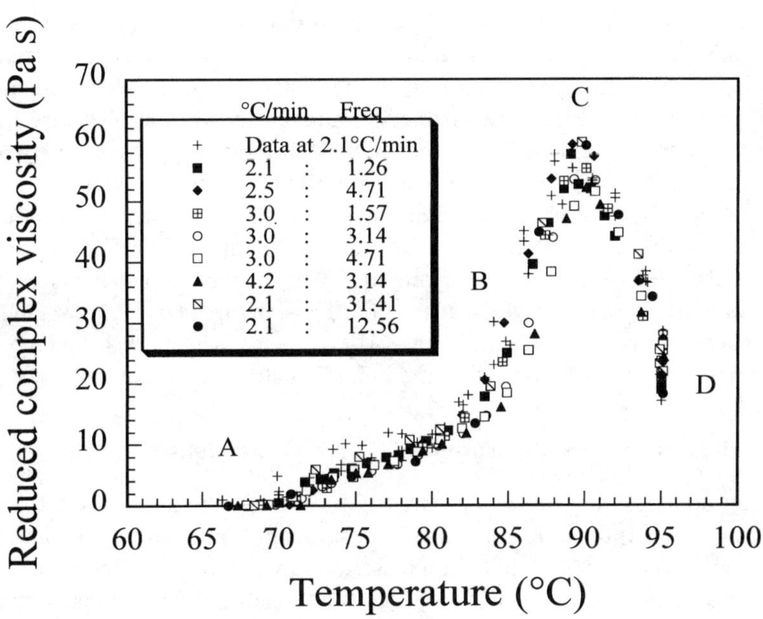

Figure 4–18 Master Curve of Reduced Complex Viscosity $\eta_R^* = \eta^* (\omega/\omega_r)$ Data Obtained on 8% Corn Starch Dispersion at Several Heating Rates and Frequencies. The letters A–D refer to stages in granule growth. *Source*: Reprinted with permission from W.H. Yang, *Rheological Behavior and Heat Transfer to a Canned Starch Dispersion: Computer Simulation and Experiment*, PhD Thesis, © 1997, Cornell University.

tion of the η^*-temperature data suggests that the molecular mechanisms involved were similar at the different heating rates and oscillatory frequencies used and depended mainly on the temperature history of the corn STD. It also suggests that the studied corn STDs exhibited thermorheologically simple behavior (Plazek, 1996).

The same modeling procedure was applied to 3.5% corn STDs and the results used to numerically simulate heat transfer to STDs in a can (Yang, 1997). From recent studies (Yang and Rao, 1998; Liao et al., 1999), the general shape of viscosity versus temperature for native, as opposed to cross-linked, starch dispersions is as shown in Figure 4–18. It is important to note that the values of viscosity of the gelatinized dispersions at the higher temperatures are substantially higher than those of the ungelatinized dispersions.

State of Granules during Heating

The important role of suspended solids in the viscosity of a dispersion was discussed in Chapter 2. The shape of the curve in Figure 4–18 reflects the changes in the volume fraction of starch granules, which in turn is due to the changes in the size of the granules, as a dispersion is heated. Initially, at low temperatures, the granules are in the raw state and the volume fraction is low. Because the granules swell due to water absorption as they are heated (Figure 4–18, segment A-B-C), the volume fraction increases and reaches a maximum value (Figure 4–18, C). With further heating, the granules rupture and disintegrate, resulting in a gradual decrease in the volume fraction of the granules in the dispersion (Figure 4–18, segment C-D). The rupture of granules also results in the release of amylose, which contributes to the viscosity of the continuous phase of the starch dispersion. Thus, in Figure 4–18, the segment C-D is not a mirror image of the segment A-B-C. The leached amylose and the granule remnants contribute to a viscosity at point D that is higher than that at point A.

Rheological Properties of Gelatinized Starch Dispersions

When an aqueous suspension of starch is heated isothermally above the gelatinization onset temperature, the granules absorb large amounts of water and swell considerably to impart substantial increase to the viscosity of the suspension. As granule swelling progresses, the suspension is transformed into a paste that exhibits rheological properties very different from the initial suspension. The swelling of granules is also accompanied by leaching or solubilization of granular constituents, mainly amylose, into the continuous suspending matrix. Granules swell to a maximum value at elevated temperatures, followed by granular disruption and exudation of granule contents into the suspension matrix.

Rheology of Starch Dispersions with Intact Granules

Starch gelatinization up to peak viscosity studied by various techniques, such as viscometry and particle size measurement by laser diffraction, is essentially a first-order kinetic process whose extent is determined principally by pasting temperature and time (Kubota et al., 1979; Lund, 1984; Okechukwu et al., 1991; Okechukwu and Rao, 1995).

Mild thixotropy has been observed in gelatinized wheat starch dispersions characterized at low temperatures such as 25°C but not at 60°C and above (Doublier, 1981; Harrod, 1989; Bagley and Christianson, 1982). The time-dependent rheological behavior shown by starch is attributed to structure formation by amylose leached into the suspending matrix on cooling to low gelling temperatures (Ellis et al., 1989).

Gelatinized starch suspensions often exhibit yield stresses at low shear rates (Evans and Haisman, 1979; Kubota et al., 1979; Doublier, 1981; Christianson and Bagley, 1984; Colas, 1986; Harrod, 1989; Giboreau et al., 1994). At high shear rates, the shear responses follow the power law relations, prompting the use of the

Herschel-Bulkey and the power law models for the analysis and description of the rheological behavior of starch suspensions. Yield stresses in gelatinized corn starch suspensions at sediment volume fraction of about 0.6 were shown to increase with granule swelling (Christianson and Bagley, 1984).

At low starch concentration, most authors have used the simple power law model without a yield stress for describing shear responses over a wide range of shear. The flow behavior index *n* and the consistency coefficient *K* are useful parameters for describing the flow behavior of starch suspensions. Gelatinized starch pastes preheated to temperatures of about 90°C are generally reported to be shear-thinning (pseudoplastic) fluids with values of *n* considerably less than 1.0 (Evans and Haisman, 1979; Doublier, 1981; Colas, 1986; Ellis et al., 1989; Noel et al., 1993). This value was also reported by Colas (1986) to decrease linearly with increasing starch concentration in cross-linked corn starches. This observation appears consistent with those of Evans and Haisman (1979), who showed that in suspensions of corn, potato, modified corn, and tapioca starches, as well as wheat flour, the index decreased rapidly from unity and approached a constant value at high concentration.

Shear Thickening in Starch Dispersions

Several studies on starch gelatinization have indicated that some starches, notably wheat, corn, waxy maize, and cross-linked waxy maize may show shear-thickening (dilatant) behavior with values of the power law flow behavior index, *n,* greater than 1.0 during the early stages of gelatinization (Bagley and Christianson, 1982; Dail and Steffe, 1990a, 1990b; Okechukwu and Rao, 1995). Shear-thickening flow behavior was reported in wheat starch dispersions heated at 60°C for 15 min (Bagley and Christianson, 1982) using a 2.61 mm gap concentric cylinder viscometer measured at 23°C. Dintzis and Bagley (1995) reported shear-thickening behavior in Amioca (98% amylopectin) (7.7%) dispersions in cans heated at 140°C and the rheological behavior examined at 50°C in a 2.61 mm gap concentric cylinder viscometer. It was also reported in cross-linked waxy maize starch (National 465) (1.82% and 2.72%) at 121–143°C (Dail and Steffe, 1990a, 1990b) using a tube (1.27 cm diameter) flow viscometer.

Typical granule size distribution (GSID) data of heated corn starch dispersions (Okechukwu and Rao, 1995) are shown in Figure 4–19. Similar data were presented for other starch dispersions: cowpea (Okechukwu and Rao, 1996), cross-linked waxy maize (Chamberlain, 1996), and tapioca (Tattiyakul, 1997). These data seem to suggest a relationship between *n* and the change in polydispersity of granules during the sol-gel transition process. Specifically, the flow behavior index appears to decrease linearly with increase in the standard deviation (SD) of heated corn starch granules (Okechukwu and Rao, 1995). With an R^2 of 0.98, Equation 4.36, which is independent of pasting temperature, provides the basis for

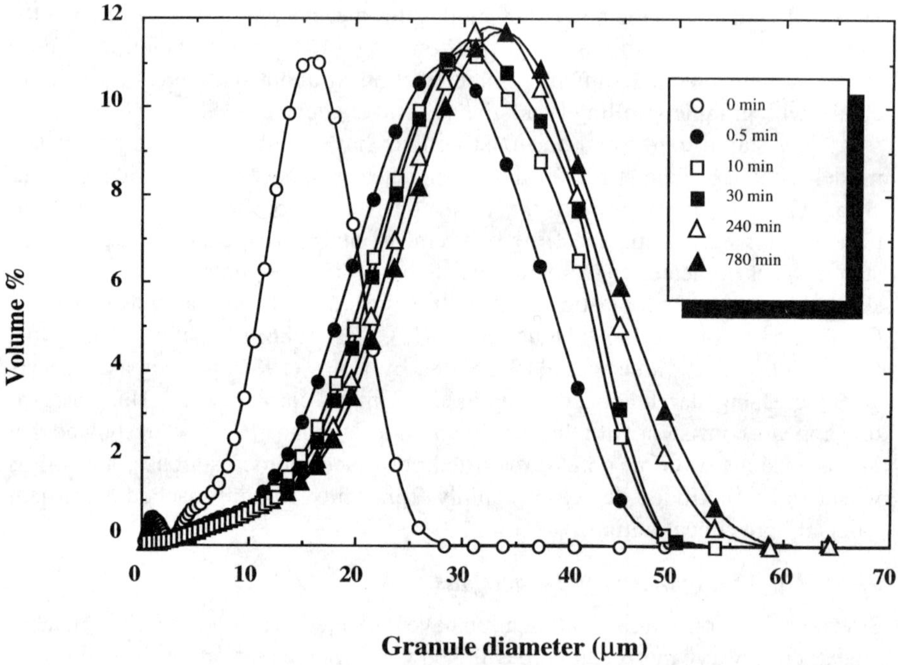

Figure 4–19 Typical Granule Size Distribution Data of Heated Corn Starch Dispersions. Significant increase in granule size resulted due to heating. *Source:* Reprinted with permission from P.E. Okechukwu and M.A. Rao, Kinetics of Cornstarch Granule Swelling in Excess Water in Gums and Stabilizers, *The Food Industry 8*, G.O. Phillips, P.A. Williams, and D.J. Wedlock, eds., pp. 49–57, © 1996, Oxford University Press.

resolving the changes in the value of n observed at various stages of starch gelatinization.

$$n = 2.17 - 0.0968\ \mathrm{SD} \tag{4.36}$$

The above equation suggests a maximum n value of about 2.2 for corn starch. The minimum values of n, usually within the pseudoplastic range, would depend in part on the particle size distribution of the gelatinizing granules, which increases as gelatinization progresses. By arbitrarily choosing gelatinization to be complete at 95°C, it was shown that a range of values of n can be obtained depending on the extent of granule swelling of corn starch (Okechukwu and Rao, 1995).

Apparently, the different values of n found in the literature for suspensions of the same starch paste under various temperature-time profiles reflect various extents of starch granule swelling. Early investigations on the rheology of raw starch suspensions reveal that dispersions of corn starch in water and aqueous mixtures of ethylene glycol and glycerin exhibited dilatant behavior with values of *n* in the range 1.2–2.5 (Griskey and Green, 1971). It would therefore appear that gelatinization may be described as a transition process that transforms the rheological behavior of starch suspensions from dilatant to pseudoplastic.

Hoffman's study (1972) on polyvinyl chloride spheres in dioctyl phthalate dispersions pointed out that shear-thickening flow behavior is a reproducible phenomenon. Hoffman suggested that "the discontinuous viscosity behavior which occurs at solids concentrations above 0.50 is caused by a flow instability in which surfaces of spheres, packed in two dimensional hexagonal packing at low shear rates, breakup into less ordered arrays of spheres. In any given viscometric flow the ordered surfaces of spheres are parallel to surfaces of constant shear and one axis of the packing always points in the direction of flow" (1972, p. 167). However, computer simulation studies on dispersions of rigid particles suggest formation of clusters of particles that leads to initiation of shear-thickening behavior (Bossis and Brady, 1989). As shear rate is increased, the loss of long-range orientational order that results in discontinuous viscosity behavior is also referred to as "shear melting" (Chow and Zukoski, 1995b). Particle size distribution; volume fraction; shape; particle-particle interaction; continuous phase viscosity; and type, rate, and time of deformation are important factors in dilatancy (Barnes, 1989; Boersma et al., 1991, 1992; Laun et al., 1991).

Role of Particles in Shear-Thickening Nonfood Dispersions

In shear-thickening behavior of nonfood colloidal dispersions, mostly TiO_2 particles, the role of particle size distribution has been examined only recently (D'Haene et al., 1993). These studies on rigid particles with dia. 1-600 nm and relatively high viscosity suspending fluids (e.g., dioctyl phthalate) reveal several important features of shear-thickening (Barnes, 1989; Boersma et al., 1991, 1992; Laun et al., 1991; Chow and Zukoski, 1995a, 1995b).

In synthetic dispersions, the smaller diameter particles and higher viscosity suspending fluids, such as dioctyl phthalate, result in very low particle Reynolds numbers, defined as (Russel et al., 1989)

$$\mathrm{Re} = \frac{D^2 \dot{\gamma} \rho}{\eta} \tag{4.37}$$

To illustrate the order of magnitudes, at a fixed $\dot{\gamma} = 100\ \mathrm{s}^{-1}$, the particle Reynolds number for typical 500 nm particles in dioctyl phthalate ($\rho = 980\ \mathrm{kg\ m}^{-3}$, $\eta = 0.054$ Pa s) is 4.54×10^{-7}, while that for a typical 30 μm starch granule in water ($\rho = 1000\ \mathrm{kg\ m}^{-3}$, $\eta = 0.001$ Pa s) at the same $\dot{\gamma}$ is 9.0×10^{-2}, so the flow regimes are substantially different.

From electrostatic interaction between two particles and the hydrodynamic interaction between two particles approaching each other along their line of centers, Boersma et al. (1991) developed an equation for $\dot{\gamma}_c$ of high volume fraction dispersions, that is, dispersions with small interparticle distances, *h:*

$$\dot{\gamma}_c = \frac{2\pi\varepsilon_0\varepsilon_1\Psi_0}{6\pi\eta_s a^2}\frac{\kappa h}{2} \tag{4.38}$$

where ε_0 is permitivity of vacuum; ε_1 is relative dielectric constant; Ψ_o is surface potential of sphere, here equal to ξ-potential of the particles in the medium; η_s is the medium viscosity; a is particle radius; κ is the reciprocal Debye double-layer thickness; and h is distance between two particles. This equation, restricted to small values of h, is valid to $h/a < 0.05$, where a is the particle radius, within $\pm 25\%$ error. A simpler estimate of the shear rate at which shear-thickening appears was derived using a dimensionless number (N_d) from the characteristic magnitudes of the interparticle and shear forces. Shear-thickening will occur at values of $N_d > \sim 1.0$:

$$\dot{\gamma}_c = \frac{2\pi\varepsilon_0\varepsilon_1\Psi_0}{6\pi\eta_s a^2} \tag{4.39}$$

Boersma et al. (1991) suggested that experimental and calculated values of $\dot{\gamma}_c$ be compared at a volume fraction ϕ_c given by

$$\phi_c = \left(\frac{a}{a + 1/\kappa}\right)^3 \cdot 0.605 \tag{4.40}$$

However, the model overpredicted the shear rate for thickening and also failed to capture the volume fraction dependence of the critical shear rate (Chow and Zukoski, 1995a, 1995b).

Chow and Zukoski (1995a, 1995b) observed that although all suspensions that were ordered at rest melted with increase in shear rate, thickening was observed only at a volume fraction greater than a critical value ϕ_c, which occurred near 0.5 for particles with diameters 250 nm. Shear melting occurred for all samples with $\phi > \phi_{melt}$ (= 0.38), while shear-thickening occurred only for $\phi > \phi_c$ (~ 0.5). Near ϕ_c, the thickening was weakened and reversible, and as ϕ approached 0.6, the thickening became "nearly violent." For shear-thickening samples, starting at a critical stress the shear rate remained at a nearly constant value over a range of applied stresses that was limited by fracture of the sample (D'Haene et al., 1993). This suggests that controlled stress rheometers are better suited to study the transition to shear-thickening (D'Haene et al., 1993).

For oscillatory strains less than approximately 0.03–0.04, the suspensions showed linear viscoelastic response with the plateau modulus, G_0, being independent of frequency (Chow and Zukoski, 1995b). Because G_0 increases exponentially with volume fraction, by investigating properties over a wide volume fraction range, a correlation can be established indicating that stresses marking the onset of microstructural or rheological transitions were independent of ϕ if scaled on the suspension's elastic modulus G_0. By scaling stress on G_0, the constitutive response of the latex suspension was reduced to a single volume fraction–independent curve when plotted as a function of the Deborah number:

$$\mathrm{De} = (\eta_s\, \dot{\gamma}/G_0) \tag{4.41}$$

where η_s is the continuous phase viscosity and $\dot{\gamma}$ is the shear rate. Chow and Zukoski (1995b) suggested that this correlation results because η_s/G_0 represented the natural relaxation time of the suspension.

Size Distribution of Intact Starch Granules

Granule swelling during starch gelatinization progresses in a manner that increases the granule size distribution when measured by the standard deviation of granules in suspension (Okechukwu and Rao, 1995). This change in the polydispersity of the starch granules introduces important rheological consequences on the suspension behavior. Over a wide range of swelling of cowpea starch granules heated at 67–80°C, n decreased from 1.3 to 0.56 with increase in ($\mathrm{SD}/\mathrm{SD}_0$) of unruptured granules in the starch dispersion, where SD_0 is the standard deviation of the unheated granules. Over a granule SD of 15–32 μm, a power law equation correlated the two dimensionless variables n and ($\mathrm{SD}/\mathrm{SD}_0$):

$$n = 2.36\left[\frac{\mathrm{SD}}{\mathrm{SD}_0}\right]^{-1.4} \qquad R^2 = 0.95 \tag{4.42}$$

Therefore, in addition to the concentration of starch that influences the magnitude of n (Colas, 1986; Evans and Haisman, 1979), increase in polydispersity of the gelatinized granules appears to be an important factor in decreasing the severity of shear-thickening and increasing pseudoplasticity of gelatinized starch dispersions.

Antithixotropic Behavior of Cross-Linked Starch Dispersions

The antithixotropic nature of the thermally processed for 8, 15, and 30 min at 120°C cross-linked waxy maize (CWM) starch dispersions was observed clearly in shear cycles: 5-minute ascending and 5-minute descending. The response curves of

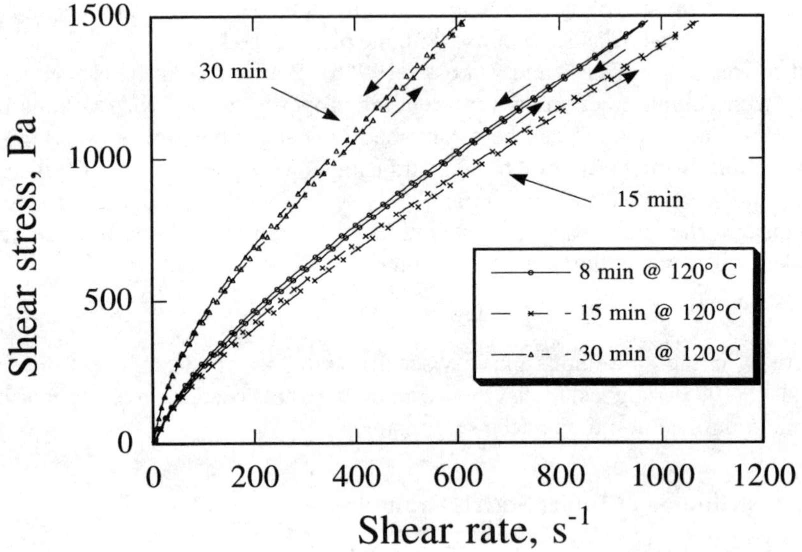

Figure 4–20 The Flow Curves of Cross-Linked Waxy Maize Samples Heated 8, 15, and 30 Minutes Became More Viscous during Each Shear Cycle. That is, they exhibited time-dependent shear-thickening (antithixotropic) behavior. *Source:* Reprinted with permission from E.K. Chamberlain, *Characterization of Heated and Thermally Processed Cross-Linked Waxy Maize Starch Utilizing Particle Size Analysis, Microscopy, and Rheology,* MS Thesis. © 1996, Cornell University.

the sample (Figure 4–20) indicate that each sample became more viscous with shearing; that is, it exhibited time-dependent shear-thickening (antithixotropic) behavior. Further, the antithixotropic behavior was irreversible in that after the shearing was stopped, the structure did not return to the structure prior to shearing.

While it is possible that the hysteresis loops (Figure 4–20) were affected by sample preparation procedure, such as the stirring prior to rheological testing, there were distinct differences between the sheared and unsheared samples, thus suggesting shear-induced structure. CWM starch dispersions have visible, possibly shear-induced, structure in the form of clusters. The CWM starch granule heated at 120°C for 30 min (Figure 4–21) had dents and crevices throughout the surface and appeared to have pores on the surface. It is possible that the granule crevices served as linkage sites to aid in attachment to other granules. The membranes also have many microscopic finger-like protrusions on the surface. After shearing for three cycles, the thermally processed samples did not appear to have sustained any surface damage, reflecting the strength and resilience of the granules due to cross-linking. However, the morphology of the gelatinized granule

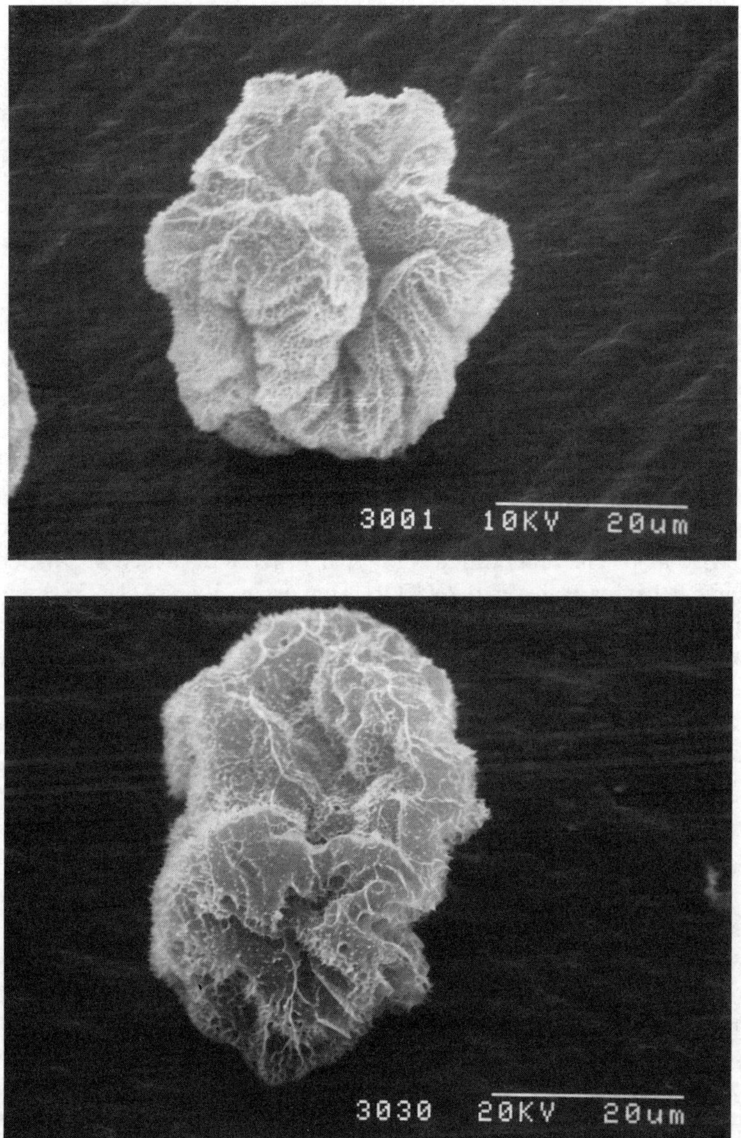

A

B

Figure 4–21 Cross-Linked Waxy Maize Granules Heated at 120°C for 30 Minutes Unsheared (A) and Sheared on a Rheometer (B) Had Dents and Crevices throughout Their Surfaces and Appeared To Have Pores on the Surfaces. *Source:* Reprinted with permission from E.K. Chamberlain, *Characterization of Heated and Thermally Processed Cross-Linked Waxy Maize Starch Utilizing Particle Size Analysis, Microscopy, and Rheology*, MS Thesis, © 1996, Cornell University.

population has been modified by shearing with concomitant time-dependent shear-thickening.

Aqueous dispersions of cross-linked waxy maize heated at 120°C for 8 min, 15 min, and 30 min all showed a shear-thinning behavior (Chamberlain, 1996; Chamberlain et al., 1998). In addition, an antithixotropic loop was found in the shear rate versus shear stress curves for the three samples with the stresses of the descending shear curve being higher than those of the descending shear curve for each sample. The loops for 8 min and 15 min were similar, but the stress values of the 30 min loop were significantly higher than those of the 8 and 15 min loops, indicating that the latter was more viscous than the other two dispersions.

Time-dependent shear-thickening behavior was also seen with autoclaved 2.0% dispersions of unmodified waxy maize starch in 0.2*N* potassium hydroxide (KOH) (Dintzis and Bagley, 1995). Two hysteresis loops were seen in the first shear cycle: One loop, at shear rates $< 200\ s^{-1}$, was anticlockwise, indicating time-dependent shear-thickening, and the other loop, at shear rates $> 250\ s^{-1}$, was clockwise, indicating thixotropic behavior. A second shear cycle resulted in only one clockwise loop. The unmodified waxy maize starch dispersed in DMSO showed the same behavior as that in 0.2*N* KOH. Dintzis and Bagley (1995) showed that a 2% gently dispersed waxy maize starch dispersion exhibited both thixotropy (at shear rates $> 60\ s^{-1}$) and antithixotropy (at shear rates $< 40\ s^{-1}$). With additional stirring, the 2% waxy maize starch dispersion had an anticlockwise loop, indicative of antithixotropic behavior. Phase microscopy revealed that the sheared waxy maize starch preparations had a mottled appearance, indicative of refractive index inhomogeneities.

Starch Granule Mass Fraction and Viscosity

The mass fraction of granules plays an important role in the magnitudes of rheological properties of starch dispersions (Evans and Haisman, 1979). Because of their deformable nature, it is difficult to determine volume fraction of starch dispersions accurately, and it is preferable to work with starch granule mass (SGM) fraction, (cQ), where c is dry starch concentration, w/w, and Q is the mass of hydrated starch granules per unit weight of dry starch. To determine the mass of the hydrated granules, a 20 mL sample of evenly dispersed starch suspension is centrifuged for 15 minutes at $1000 \times g$ (Eliasson, 1986; Leach et al., 1959), and the top clear supernatant is removed, leaving behind the swollen sediment. In addition, the liquid in the interstices should be drained carefully; otherwise, high values of cQ will result. SGM fraction takes into consideration both solids concentration and volume of starch granules after water absorption and plays an important role in consistency and yield stress (Bagley and Christianson, 1982; Okechukwu and Rao, 1995) as well as the viscoelastic properties (G', G'') of starch dispersions, and has been referred to as "notional volume fraction" (Evans and Lips, 1992).

The mean diameter of the granules and the size distribution of the granules influence the magnitude of consistency index and the flow behavior (shear-thinning or shear-thickening), respectively (Okechukwu and Rao, 1995). Increase in the volume fraction of solids in a starch dispersion can be achieved by either increasing the concentration of the solids or through volume expansion of the solids. Linear relationships were found between cQ and $(D_t/D_0)^3$ (D_t is the diameter of the heated starch granule and D_0 is that of the unheated granule) over a wide range of the latter for heated STDs of corn (Okechukwu and Rao, 1995), cowpea (Okechukwu and Rao, 1996), and cross-linked waxy maize starches (Chamberlain, 1996) (Table 4–6). For the tapioca STDs (Rao and Tattiyakul, 1999), two linear relationships provided a better fit of the data: one for values of $(D_t/D_0)^3$ from about 15 to 40 and the other from 40 to about 50 (Figure 4–22). In determination of Q, centrifuging and siphoning off the top layer of supernatant fluid are not precise procedures, which contributed to the spread of the cQ data points in Figure 4–22. For tapioca starch dispersions, a large increase in cQ at high values of $(D_t/D_0)^3$ is noteworthy. Because tapioca starch had a higher value of slope, especially at high $(D_t/D_0)^3$ values, the starch granule diameter ratio contributed to a higher increase of cQ of tapioca STDs.

The ratio of mean granule diameters (D_t/D_0) can be used to correlate consistency index data of starch dispersions. At low values of cQ, the granules were sufficiently apart and did not affect dispersion viscosity substantially. As cQ increased above a threshold value, granule-granule interaction increased, accompanied by

Table 4–6 Relationships between $(D_t/D_0)^3$ and Volume Fraction of Heated 2.6% Starch Dispersions

Starch	*Intercept*	*Slope*	R^2	*Reference*
Unmodified corn	0.16	0.011	0.86	Okechukwu and Rao (1995)
Cowpea	0.09	0.013	0.84	Okechukwu and Rao (1996)
Cross-linked waxy maize	0.23	0.0015	0.98	Chamberlain et al. (1996)
Tapioca ($15<(D_t/D_0)^3<40$)	~0*	0.012	NA	Rao and Tattiyakul (1999)
Tapioca ($40<(D_t/D_0)^3<50$)	~0*	0.044	NA	Rao and Tattiyakul (1999)

* Physically meaningless values −0.07 and −1.4 were obtained.

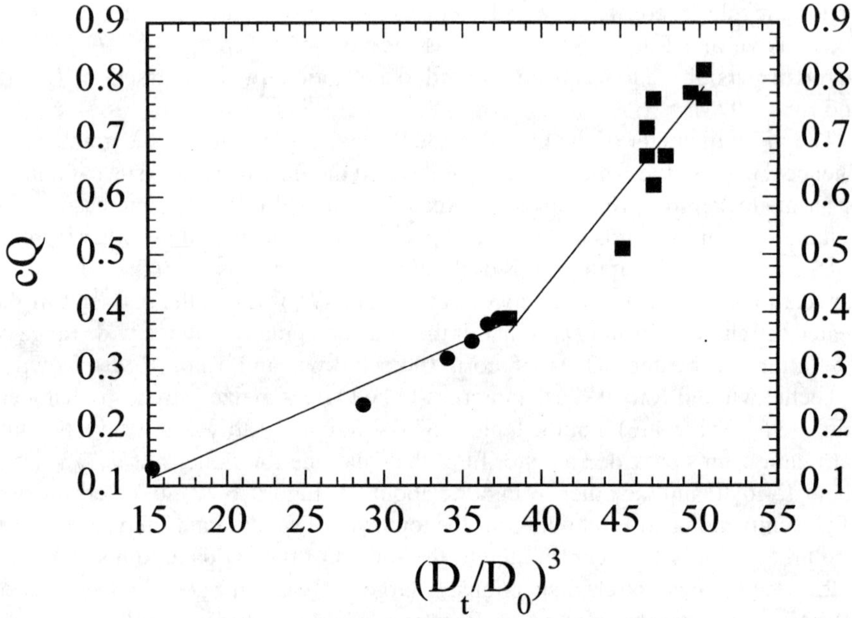

Figure 4–22 For Tapioca Starch Dispersions, Two Linear Equations or a Single Exponential Equation (not shown) Provided a Better Fit of the Data between cQ and $(D_t/D_0)^3$. *Source*: Reprinted from *Carbohydrate Polymers*, M.A. Rao and J. Tattiyakul. Granule Size and Rheological Behavior of Heated Tapioca Starch Dispersions, © 1999, with permission from Elsevier Science.

large increase in dispersion viscosity. Figure 4–23 shows the relationship between the granule diameter ratio (D_t/D_0) and the consistency index (K) of gelatinized 2.6% tapioca STDs, which can be expressed by the equation

$$K = K_0 \exp\left\{\varepsilon\left(\frac{D_t}{D_0}\right)\right\} \tag{4.43}$$

where K_0 and ε are empirical constants. K increased gradually at the beginning and rapidly as the granule diameter ratio approached its maximum value. At $(D_t/D_0) = 3.4$, values of K started to increase rapidly. The estimated values of K_0, ε, and R^2 for various 2.6% STDs are given in Table 4–7. The higher value of ε for tapioca starch indicates that, among these four starches, the starch granule ratio had the highest influence on the consistency index (K) of the tapioca starch dispersions. During the later stages of gelatinization involving the bursting of granules, a progressive reduction in suspension viscosity takes place and Equation 4.43 does not describe the suspension viscosity data.

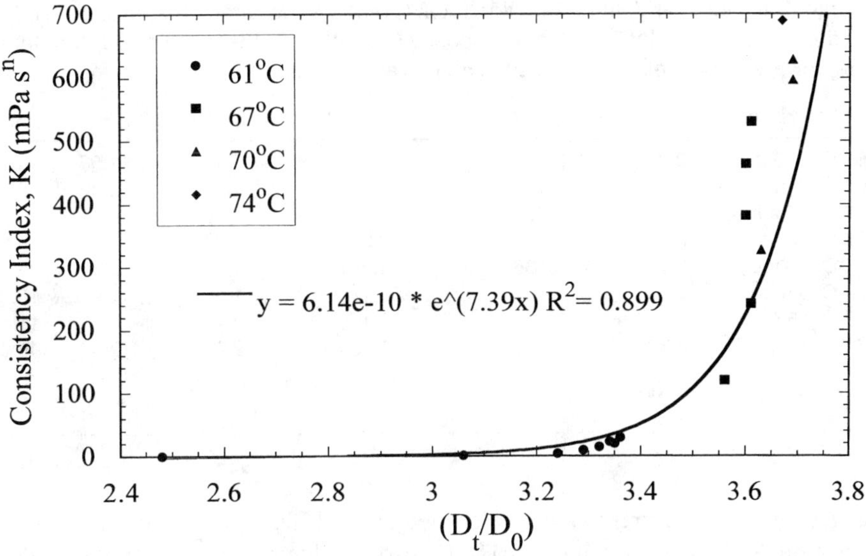

Figure 4–23 The Relationship between the Granule Diameter Ratio (D_t/D_0) and the Consistency Index (K) of Gelatinized 2.6% Tapioca Starch Dispersions. *Source*: Reprinted from *Carbohydrate Polymers*, M.A. Rao and J. Tattiyakul. Granule Size and Rheological Behavior of Heated Tapioca Starch Dispersions, © 1999, with permission from Elsevier Science.

Mass Fraction versus Relative Viscosity

The relative viscosity of the gelatinized 2.6% tapioca STDs was calculated using Equation 4.44:

$$\eta_r = \frac{\eta_\infty}{\eta_s} \tag{4.44}$$

where η_r is the relative viscosity, η_∞ is the Casson viscosity (mPa s) (Table 4–8), and η_s is the viscosity (mPa s) of the supernatant determined at 20°C; Table 4–8

Table 4–7 Relationships between (D_t/D_0) and the Power Law Consistency Index of Heated Starch Dispersions:

$$K = K_0 \exp\left\{\varepsilon\left(\frac{D_t}{D_0}\right)\right\}$$

Starch	K_0 *(mPa s^n)*	ε	R^2	*Reference*
Unmodified corn	2.07×10^{-5}	3.69	0.99	Okechukwu and Rao (1995)
Cowpea	2.40×10^{-5}	4.84	0.96	Okechukwu and Rao (1996b)
Cross-linked waxy maize	2.13×10^{-7}	3.36	0.92	Chamberlain et al. (1998)
Tapioca	6.14×10^{-10}	7.40	0.90	Rao and Tattiyakul (1999)

Table 4–8 Notional Volume Fraction (*cQ*), Casson Yield Stress (σ_{0c}, Pa, $m = 0.25$ and $m = 0.50$) and Infinite Shear Viscosity (η_∞, mPa s), Viscosity of Supernatant (η_s, mPa s), and Relative Viscosity (η_r) of Tapioca Starch Dispersions

*Dispersion**	cQ	σ_{0c}, m = 0.25	η_∞, m = 0.25	σ_{0c}, m = 0.50	η_s	η_r
61°C, 240 min	0.39	0.002	7.2	0.11	1.53	4.7
67°C, 5 min	0.62	0.120	12.1	1.13	2.02	6.0
67°C, 15 min	0.67	0.071	27.0	1.48	2.48	10.9
67°C, 30 min	0.72	0.181	13.9	1.64	2.60	5.3
67°C, 60 min	0.77	0.188	14.3	1.78	2.76	5.2
70°C, 5 min	0.77	0.078	30.1	1.96	1.97	15.3
70°C, 15 min	0.82	0.098	32.1	2.10	2.82	11.4
74°C, 1 min	0.78	2.77	139.45	2.25	2.85	48.9

* Heating temperature and time.

also contains yield stress values calculated using different exponents in the Casson model. The η_r values of tapioca STDs strongly depended on their volume fraction, and there was considerable scatter in the values as a result of uncertainties in estimating η_∞ and cQ values (Figure 2–8). The line in Figure 2–8 represents values predicted by the model (Equation 4.45) of Quemada et al. (1985) with the structural parameter $k = 1.80$:

$$\eta_r = \left(1 - \frac{1}{2}k\,(cQ)\right)^{-2} \qquad (4.45)$$

Quemada et al. (1985) reported values of k in the range 2.50 to 3.82 for dispersions of rigid solids, and 1.70 to 1.85 for red blood suspensions. Because heated starch granules are deformable, the value $k = 1.80$ of red blood cells was selected for the data on tapioca STDs. Given that the Krieger-Dougherty model predictions were substantially different than experimental η_r values of STDs (Ellis et al., 1989; Noel et al., 1993; Tattiyakul, 1997), the reasonable applicability of Equation 4.45 (Quemada et al., 1985) to tapioca STDs is noteworthy. However, the model of Quemada et al. (1985) is a phenomenological model, while that of Krieger-Dougherty is based on intrinsic viscosity of a single sphere.

Effect of Starch Concentration

Power law relations of the form shown in Equation 4.46 have been used to describe the influence of starch concentration on the viscosity of gelatinized wheat starch suspensions over the shear rate range 20 and 200 s^{-1}.

$$K = K_c c^b \qquad (4.46)$$

where K is in Pa s^n, and starch concentration was defined in mg/mL (Doublier, 1981). For gelatinized starch characterized at 70°C, $K_c = 8.74 \times 10^{-7}$ and $b = 3.88$ for starch concentration, c, in the range 2.5–8%. Evans and Haisman (1979) reported that the apparent viscosity at any shear rate $\eta_{\dot{\gamma}}$ raised to power 0.4 was linearly correlated with concentration for potato, tapioca, and corn starches within the shear rates 10^1 to 10^3 s^{-1}:

$$(\eta_{\dot{\gamma}})^{0.4} = K^{11}(c - c_o) \qquad (4.47)$$

where c_o is the concentration at which the viscosity extrapolated to zero and K^{11} is a proportionality constant. Thus at various shear rates it is possible to construct linear plots relating apparent viscosity with starch concentration. Plots of apparent viscosity as a function of starch concentration were later shown by Bagley and Christianson (1982) to be greatly affected by the pasting temperature and time. However, they obtained a temperature-independent master plot that correlated the viscosity over cQ ratio (η/cQ) with the mass fraction of swollen starch granules in gelatinized wheat starch suspension, cQ. A similar master plot was also obtained by plotting the ratio K/cQ against volume fraction of swollen corn starch granules, since at low starch concentrations volume fraction approximates mass fraction (Okechukwu and Rao, 1995); for a 2.6% corn starch suspension the plot could be represented by Equation 4.48:

$$\frac{K}{cQ} = 0.48 \exp(16.94cQ)\text{: } R^2 = 0.85 \qquad (4.48)$$

Effect of Temperature on Flow Behavior

The viscosity of fully gelatinized starch suspensions generally decreases with increase in temperature. This rheological response of preheated starches has been shown to follow an Arrhenius dependency of the consistency coefficient on temperature, as shown in Equation 4.49.

$$K = K_{TO} \exp(E_a/RT) \qquad (4.49)$$

where E_a is the activation energy and K_{TO} an empirical constant. For a wheat starch dispersion heated to 96°C, values of $E_a = 5.13$ kcal/mole and $K_{TO} = 1.51 \times 10^{-4}$ Pa were reported (Doublier, 1981).

Dynamic Rheological Behavior of Starch Dispersions

Viscoelastic properties of starch dispersions are of interest to better understand their structure (Biliaderis, 1992). In addition, studies on synthetic dispersions (Boersma et al., 1992; Chow and Zukoski, 1995b) showed that they may play a role in flow behavior.

Cross-Linked Waxy Maize Starch Dispersions

Figure 4–24 shows the dynamic properties, storage modulus (G', Pa) and loss modulus (G'', Pa), versus frequency (rad s^{-1}) of gelatinized cross-linked waxy maize (CLWM) starch dispersions of 3%, 4%, and 5% solids. In all dispersions, G' was much higher than G''; that is, the dispersions exhibited gel-like behavior over the studied range of frequencies. Both G' and G'' increased with concentration and showed a weak dependence on frequency at low values. This dependency was more pronounced in dispersions of higher concentration.

The Cox-Merz rule (Equation 4.50) was not too applicable to gelatinized CLWM starch dispersions in the range of concentrations studied, in that the complex viscosity (η^*) was higher than the flow viscosity over the range of frequencies and shear rates studied (Figure 4–25).

$$\eta^*(\omega) = \eta(\dot{\gamma})\big|_{\omega=\dot{\gamma}} \tag{4.50}$$

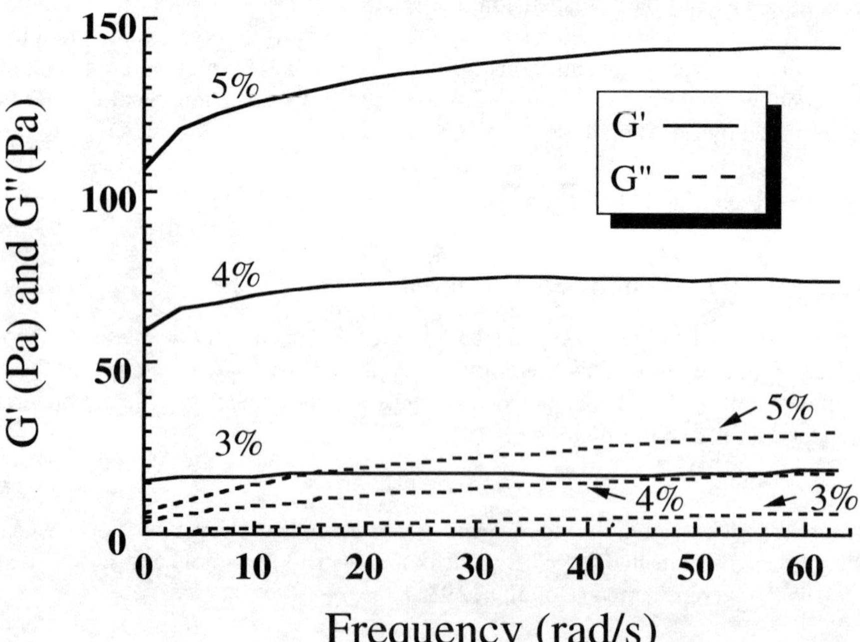

Figure 4–24 The Dynamic Properties, Storage Modulus (G', Pa) and Loss Modulus (G'', Pa), versus Frequency (rad s^{-1}) of Gelatinized Cross-Linked Waxy Maize Starch Dispersions of 3%, 4%, and 5% Solids. *Source*: Reprinted with permission from P.M.S. da Silva, J.C. Oliveira, and M.A. Rao, The Effect of Granule Size Distribution on the Rheological Behavior of Heated Modified and Unmodified Maize Starch Dispersions, *Journal of Texture Studies*, Vol. 28, pp.123–138, © 1997, Food and Nutrition Press.

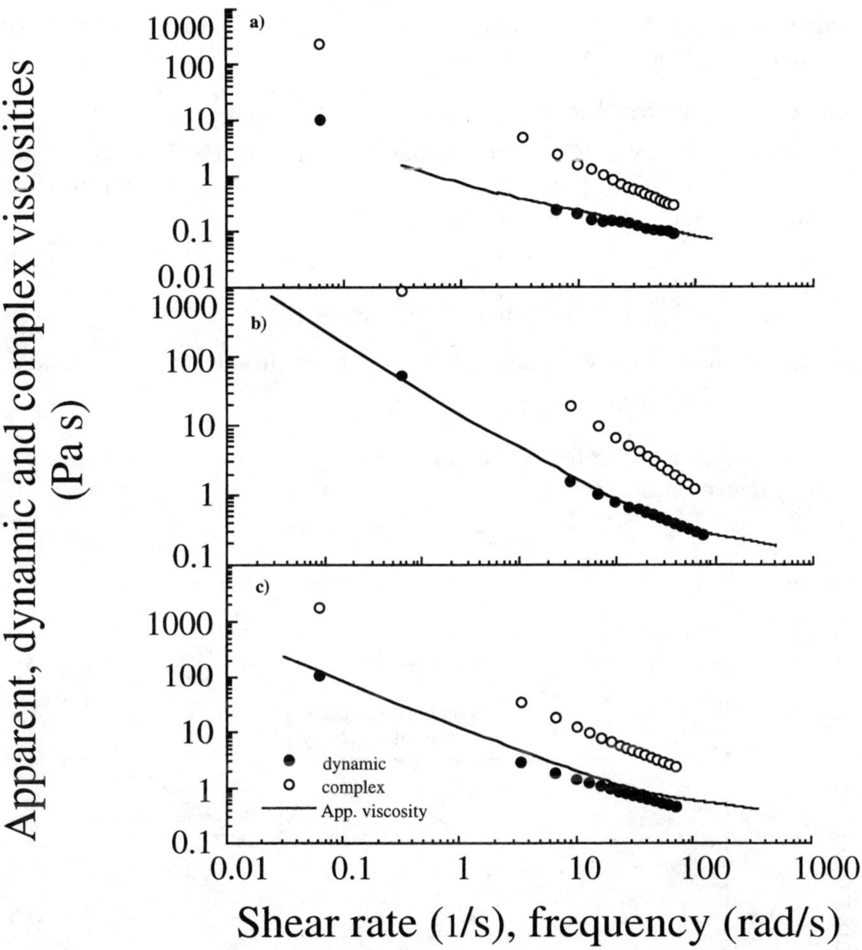

Figure 4–25 The Cox-Merz Rule Was Not Applicable to Gelatinized Cross-Linked Waxy Maize Starch Dispersions in the Range of Concentrations Studied in That the Complex Viscosity (η^*) Was Higher Than the Flow Viscosity over the Range of Frequencies and Shear Rates Studied. *Source*: Reprinted with permission from P.M.S. da Silva, J.C. Oliveira, and M.A. Rao, The Effect of Granule Size Distribution on the Rheological Behavior of Heated Modified and Unmodified Maize Starch Dispersions, *Journal of Texture Studies*, Vol. 28, pp.123–138, © 1997, Food and Nutrition Press.

where $\eta^* = (G^*/\omega)$, ω is the frequency of oscillation (rad s^{-1}), and $G^* = \sqrt{(G')^2 + (G'')^2}$. Deviation from Cox-Merz rule also reflects the gel-like nature of the gelatinized dispersions (Ross-Murphy, 1984). The complex viscosity can be further expressed in terms of real and imaginary parts, $\eta^* = \eta' - i\eta''$; it is interesting to note that the dynamic viscosity (η') is in better agreement with the apparent viscosity than η^* (Figure 4–25).

Four Percent Tapioca Starch Dispersions

Log-log plots of η^* versus dynamic frequency (ω, rad s^{-1}) and η_a versus shear rate ($\dot{\gamma}$) of a 4% tapioca starch dispersion heated at 70°C for 30 min, were described by Equations 4.51 and 4.52, respectively.

$$\eta^* = 18.32\omega^{-0.74} \qquad R^2 = 1.00 \tag{4.51}$$

$$\eta_a = 8.36\dot{\gamma}^{-0.73} \qquad R^2 = 1.00 \tag{4.52}$$

A modified Cox-Merz relationship was found to be applicable between η^* and η_a:

$$\eta^*(\omega) = 2.24[\eta_a(\dot{\gamma})]^{1.04}\big|_{\omega=\dot{\gamma}} \qquad R^2 = 1.00 \tag{4.53}$$

This relationship is useful for calculation of apparent viscosity data from complex viscosity data or vice versa.

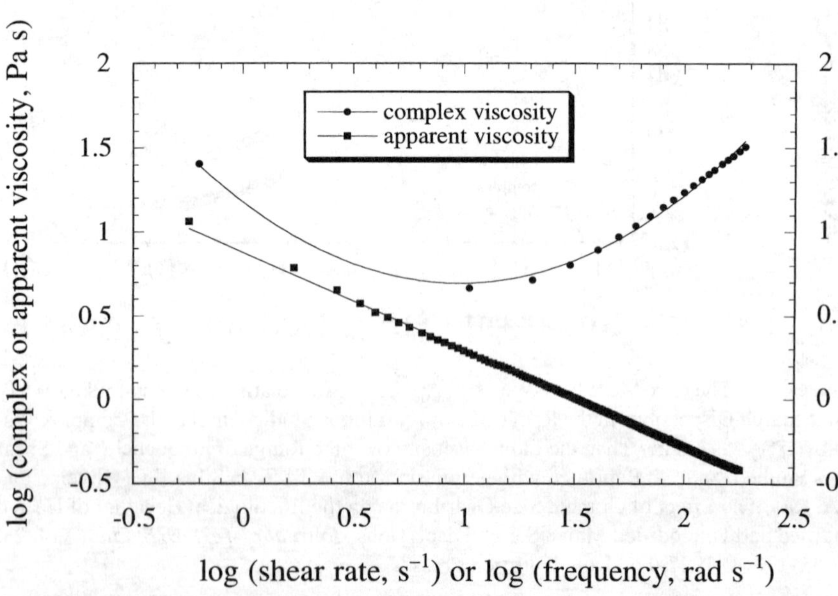

Figure 4–26 For 4% Tapioca Starch Dispersion Heated at 70°C for 5 Minutes, Values of η_a Decreased with $\dot{\gamma}$, but Those of η^* Increased with ω after an Initial Decrease. *Source*: Reprinted from *Carbohydrate Polymers*, M.A. Rao and J. Tattiyakul. Granule Size and Rheological Behavior of Heated Tapioca Starch Dispersions, © 1999, with permission from Elsevier Science.

For the 4% tapioca starch dispersion heated at 70°C for 5 min, values of η_a decreased with $\dot{\gamma}$, but those of η^* increased with ω after an initial decrease (Figure 4–26). For the data shown in Figure 4–26, the critical frequency (ω_c) at which the transition occurred was about 100 rad s^{-1}. In comparison to $\dot{\gamma}_c$ values for shear-thinning to shear-thickening flow behavior of 2.6% tapioca and cowpea starch (Okechukwu and Rao, 1996) dispersions, the value of $\omega_c = 100$ rad s^{-1} is lower. It is possible that the η_a versus $\dot{\gamma}$ data of the 4% dispersion heated at 70°C for 5 min had a $\dot{\gamma}_c$ at a very high shear rate. However, because the objective was to examine applicability of the Cox-Merz relationship, very high values of $\dot{\gamma}$ were not used.

Because $\eta^* = (G^*/\omega)$ and $G^* = \sqrt{(G')^2 + (G'')^2}$, further examination of G' and G'' of the two dispersions (Figure 4–27) revealed that whereas both G' and G'' of the 70°C/30 min dispersion and G'' of the 70°C/5 min dispersion increased gradually in a convex up manner with ω, G' of the 70°C/5 min dispersion increased dramatically with ω in a convex down manner. The above described dynamic rheological behavior of both dispersions was reproduced in several replicate rheological measurements, so it appears that the shear-thickening behavior of η^* of 70°C/5 min dispersion was the result of dramatic increase in elastic behavior of the dispersion at the small strain ($\gamma = 3\%$) value used in the experiment. At the shear rates ($d\gamma/dt$) used to obtain η_a data on the same 70°C/5 min 4% dispersion, shear-thickening was not observed, probably because shear-thinning to shear-thickening transition occurred at higher shear rates than used in the experiment; Okechukwu and Rao (1996) observed that such transition occurred in cowpea starch dispersions at about 340 s^{-1}. Given that heated starch dispersions were found to exhibit shear-thinning and shear-thickening (Bagley and Christianson, 1982; Okechukwu and Rao, 1995, 1996), as well as time-dependent shear-thickening (Dintzis and Bagley, 1995; Chamberlain, 1996; da Silva et al., 1997) flow behavior, perhaps it should not be surprising that shear-thickening should be seen in η^* versus ω data.

Effect of Particle Characteristics on Modulus

Evans and Lips (1992) extended the work of Buscall et al. (1982a, 1982b) to relate the shear modulus, G, of a dispersion of spherical particles to various characteristics of the particles and the continuous medium:

$$G = \frac{\phi_{cp} n G_p}{5\pi(1-\mu_p)}\left[\phi_{red}^{1/3}(1-\phi_{red}^{-1/3})^{1/2} - \frac{8}{3}\phi_{red}^{2/3}(1-\phi_{red}^{-1/3})^{3/2}\right] \quad (4.54)$$

where G_p is shear modulus of material of spherical particle, μ_p is Poisson's ratio, ϕ_{cp} is the volume fraction for close-packing, and ϕ_{red} is the reduced phase volume fraction (ϕ/ϕ_{cp}) that is given as $\phi_{red} = (a_0/h)^3$; a_0 is undeformed particle radius, and h is distance to nearest particle. Evans and Lips suggested that for fully swelled starch granules, phase volume fraction $\phi = cQ$. Agreement between experimental data on starch dispersions and theory was poor, but the theoretical trends were followed by Sephadex spheres used in chromatography columns

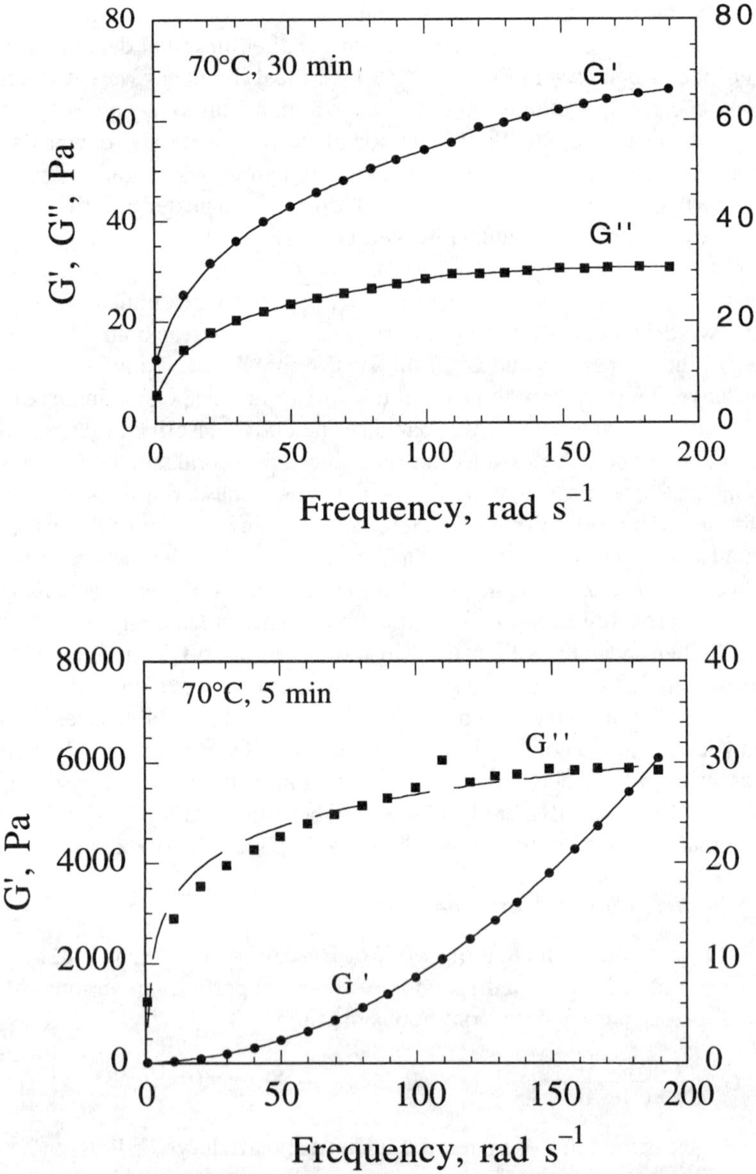

Figure 4–27 *G′* and *G″* of Two Tapioca Dispersions Revealed That Whereas Both *G′* and *G″* of 70°C/30 min Dispersion and *G″* of the 70°C/5 min Dispersion Increased Gradually in a Convex Up Manner with *ω*, *G′* of the 70°C/5 min Dispersion Increased Dramatically with *ω* in a Convex Down Manner. *Source*: Reprinted from *Carbohydrate Polymers*, M.A. Rao and J. Tattiyakul. Granule Size and Rheological Behavior of Heated Tapioca Starch Dispersions, © 1999, with permission from Elsevier Science.

(Evans and Lips, 1992, 1993). The difference between theory and experiment for heated starch dispersions was because the granules were not truly spherical and rigid ("macroscopic elastic spheres") and they were polydisperse in nature. Also, it was not possible to determine the representative viscoelastic behavior of all the granules from the stress-strain data on a few granules using the cell poker technique (McConnaughey and Petersen, 1980), which still provides mainly qualitative results (Zahalak et al., 1990). Plots of SGM fraction versus normalized, based on values of G' at $cQ = 1$, G' data on starches resulted in three groups—chemically modified, root and tuber, and cereal (Evans and Lips, 1992)—that were attributed to "structural origins" of the starches.

Effect of Sugar, Salt, and Lipids on Starch Gelatinization

Sucrose and other sugars increase the gelatinization temperature of starches; the increase depends on the water activity of the sugar solution and the sugar: The gelatinization temperature of starch increased with decrease in water activity of the sugar solution, and at the same water activity, sugars with longer chain lengths (sucrose, maltose, and maltotriose) increased the gelatinization temperature more than sugars with shorter chain lengths (fructose and glucose). At very high sugar content, the starch does not gelatinize at all and the product collapses during baking, as with cookies and fudge brownies (Hoseney, 1998).

In limited water ($< \approx 30\%$), the gelatinization temperature increases as the moisture content decreases. In a sugar solution, the sugar displaces part of the water and, in effect, decreases the moisture content; therefore, the gelatinization temperature increases in sugar solutions. Hoseney (1998) pointed out that the increase in the onset gelatinization temperature is also due to lowering of the water activity.

Chaotropic salts, such as calcium chloride, decrease the starch gelatinization temperature to the extent that gelatinization may take place at room temperature. In contrast, nonchaotropic salts raise the gelatinization temperature. While fats and emulsifiers have minimal effect on the starch gelatinization temperature, they affect the rate of starch granule swelling after gelatinization (Hoseney, 1998).

Rheological Behavior of Starch-Protein Dispersions

In many foods, both starch and protein can be encountered, so understanding interactions between them is important. The selectivity in interaction between proteins and starches is best seen in results of dynamic rheological studies. The results depend upon the molecular structure of protein and the starch: The amylose/amylopectin ratio, the composition of protein and starch, as well as the phase transition temperatures are important factors influencing protein-starch interaction. Because proteins and starches are thermodynamically different polymers, their presence together may lead to phase separation, inversion, or mutual interaction with significant consequences on texture (Morris, 1990).

In an aqueous medium, a given protein and polysaccharide can be compatible or incompatible. If they are thermodynamically incompatible, a two-phase system is obtained in which the two molecules are mostly in separate phases. However, if they are thermodynamically compatible, two types of mixtures can be formed: (1) a homogeneous and stable system in which the two molecules coexist in a single phase, or (2) a two-phase system where the molecules interact and both are in essentially concentrated phases. Further, there is a coacervation due to an electrostatic association between the two polymers. Anionic polysaccharide and protein are incompatible at pH values above the protein's isoelectric point (point of minimum solubility, pH $\approx$ 5.1) and completely compatible below it due to the net opposite charges they carry. Factors affecting protein-polysaccharide compatibility and the characteristics of their complexes include the molecular characteristics of the two molecules (e.g., molecular weight, net charge, and chain flexibility), the pH, ionic strength, temperature, the protein/polysaccharide ratio, rate of acidification, and rate of shear during acidification (Tolstoguzov, 1985, 1991). The rheological behavior of specific starch/protein systems is discussed next.

Gluten/Starch

In food products, starch can be either a mere ingredient used in low concentrations to thicken the product texture or the main component responsible for the texture. The former kind of products usually contain water in excess of gelatinization requirements, while in the latter, such as baked products, water is available in a limited amount. In these products, made from wheat flour, gluten is considered to be the main component responsible for viscoelastic properties. Nevertheless, experiments on dough made with only starch and gluten revealed that starch does not act only as an inert filler but interacts with gluten and is active in determining the viscoelastic behavior of both the dough and the cooked product (Faubion and Hoseney, 1990; Petrofsky and Hoseney, 1995).

Addition of gluten to 6.5–6.6% (w/w) dispersions of different starches at a 1% level calculated on dry starch basis increased the G' of wheat and rye starches and lowered that of maize starch but did not affect the modulus of barley, triticale, and potato starches (Lindahl and Eliasson, 1986). G' was also found to increase with increasing amounts of gluten (1–4%) added to the wheat starch. Studies on starches with different amylose/amylopectin ratios suggest that gelatinization of amylopectin and not amylose favors a synergistic interaction with proteins, leading to an increase in dispersion viscosity (Chedid and Kokini, 1992; Madeka and Kokini, 1992).

Addition of gluten strongly changed the viscoelastic properties of starch pastes and gels. In pure starch dispersions, the rheological unit was a starch granule that determined the viscoelastic behavior of the whole network due to its ability to be deformed or packed (Champenois et al., 1998). When gluten was present, the rheological units were no longer the starch granules but the "cells" formed by gluten fibrils and filled by starch granules. In the case of flour, the internal lipids obviously play an important role, so it cannot be considered rheologically to be a simple blend of starch and gluten. The range of gluten/starch ratios studied (0 to 0.20) included the natural ratio in flour of 12/88 (Chedid and Kokini, 1992). When gluten was added to starch, the water content was calculated based on the weight of both starch and gluten. Therefore, compared to a pure starch dispersion of the same concentration, there was less starch in a gluten/starch dispersion, but the water/starch ratio was almost constant. The gluten delayed or partly reduced the contact between granules.

The most probable effect of adding gluten was to hinder granule-granule interactions because, at the temperatures corresponding to the early stage of swelling, the amylose had not yet leached out of the granules. It is hypothesized that the

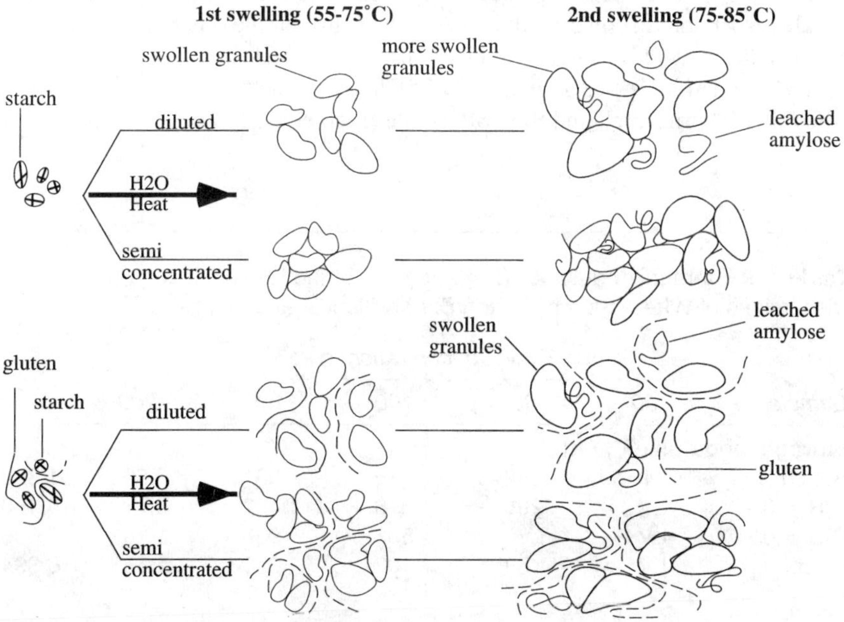

Figure 4–28 Schematic of a Network of Wheat Starch Granules and Gluten. *Source*: Y. Champenois, M.A. Rao, and L.P. Walker, Influence of Gluten on the Viscoelastic Properties of Starch Pastes and Gels, *Journal of Science, Food and Agriculture*, Vol. 78, pp. 119–126, © 1998. Permission is granted on behalf of SCI who hold the copyright to this material.

gluten fibrils became embedded between the starch granules and consequently hindered the formation of the granule network (Figure 4–28). This structure may be described as a phase-separated network of starch granules and gluten (Morris, 1986). The gluten fibrils hinder the formation of the starch network that normally occurs either by granule-granule interactions (early stage) or by amylose chain entanglements (later stage) (Figure 4–28). The presence of these fibers weakened the starch network, so gluten probably plays the role of fault zones in the network structure. It is thus possible to imagine that the presence of both gluten and an emulsifier or a lipid could result in a synergistic effect among them. Indeed, values of G' and G'' for flour were much lower than those for gluten/starch mixtures even though G/S was equal to 0.20 (Table 4–9).

Soy Protein/Corn Starch

The shape of the G' versus ω curves for all corn starch (CS)–soy protein (SP) isolate dispersions was similar, where G' exhibited a small dependence on frequency (Figure 4–29). The 10% corn starch dispersion had the highest and the 10% SP isolate dispersion the lowest value of G'. The values of the other dispersions were between these two limits. Gel compositions in order of decreasing G' values were 10% CS $>$ 1SI:9CS $\approx$ 3SI:7CS $\approx$ 5SI:5CS $>$ 7SI:3CS $>$ 9SI:1CS $>$ 10% SI, where CS is corn starch and SI is SP isolate (Liao et al., 1996).

Table 4–9 Storage Modulus G' (Pa) and tan δ at the End of Heating (EH) of Dispersions of Wheat Starch, Gluten/Starch Blends, and Flour

	Starch/Flour Concentration, g kg^{-1}					
Sample	*60*	*80*	*100*	*150*	*200*	*300*
Storage modulus (G′, Pa)						
Starch	140	342	779	2,531	7,565	31,898
G/S = 0.12	LV	295	616	1,422	3,713	19,150
G/S = 0.20	LV	LV	533	986	1,101	6,606
Flour	LV	68	102	320	1,567	5,894
tan δ						
Starch	0.36	0.19	0.21	0.19	0.09	0.04
G/S = 0.12	>1	0.23	0.20	0.15	0.10	0.06
G/S = 0.20	>1	>1	0.30	0.18	0.22	0.13
Flour	>1	0.44	0.46	0.40	0.22	0.21

Note: G/S is gluten to starch ratio; LV = Low value (< 10 Pa).

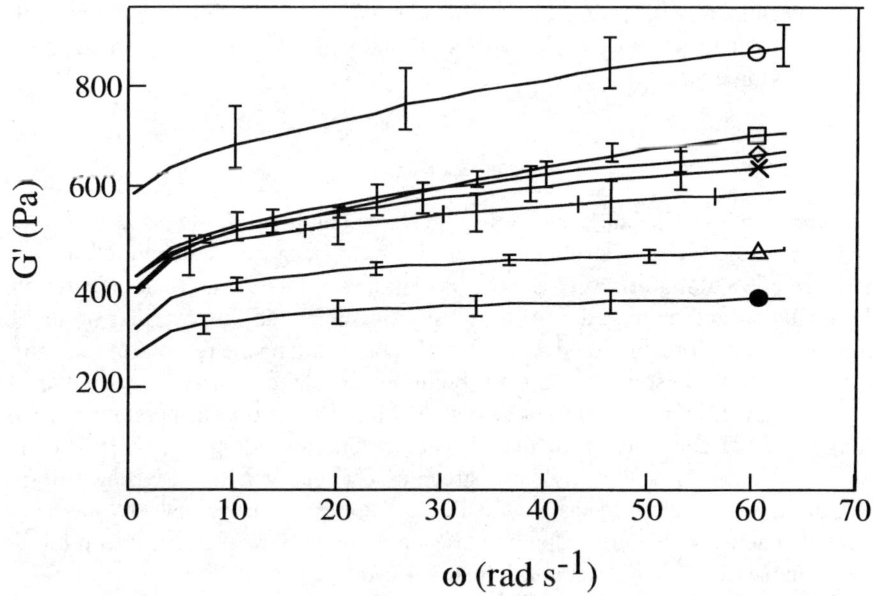

Figure 4–29 The Shape of the G' versus ω Curves for All Corn Starch–Soy Protein Isolate Dispersions Was Similar, Where G' Exhibited a Small Dependence on Frequency.

Table 4–10 Magnitudes of Intercepts (K', K'') and Slopes (n', n'') from Linear Regression of Log ω vs Log G' and Log G'' of Different Ratios of Corn Starch (CS)–Soybean Protein Isolate (SPI) mixtures

	G′ (Pa)		*G″ (Pa)*	
Mixture	K′	n′	K″	n″
10% Cornstarch (CS)	562.3	0.10	29.5	0.24
10% SPI	281.8	0.07	39.8	0.002
9% CS + 1% SPI	398.1	0.12	29.5	0.23
7% CS + 3% SPI	426.6	0.11	44.7	0.19
5% CS + 5% SPI	416.9	0.10	51.3	0.14
3% CS + 7% SPI	398.1	0.09	53.7	0.08
1% CS + 9% SPI	338.8	0.08	47.9	0.013

The corn starch–SP isolate dispersions displayed true gel-like behavior. Table 4–10 contains magnitudes of the slopes (n' and n'') and intercepts (K' and K'') in the following equations:

$$G' = K' (\omega)^{n'} \tag{4.55}$$

$$G'' = K'' (\omega)^{n''} \tag{4.56}$$

In the starch–SP isolate dispersions, the starch continuous network was formed first as a result of its lower transition temperature. Because the denaturation temperature of 7S globulin was close to the gelatinization temperature of corn starch, the diffusion and aggregation of amylose molecules and the swelling of starch granules were a little hindered and reduced. This could be the reason for the value of G' of starch–SP isolate dispersions being lower than that of 10% starch dispersion. However, comparison with G' values of 6–10% starch dispersions reveals the substantial contribution of the SP isolate. The unfolding protein molecules continued to penetrate the amylose aggregates, especially in the early stages of gel formation when the starch paste was still weak and two continuous networks were formed. The two structures supplement each other, and the properties of proteins add to those of the starch network already present.

Figure 4–30 is a plot of the experimental values of G', at a frequency of 30 rad s^{-1}, of dispersions of various SP isolate/corn starch ratios and of corn starch with 6–10% solids. It also has a straight line that interpolates G' values between those of corn starch and SP isolate. When the proportion of starch was less than 5%, the measured values were higher than the interpolated values. While above 5% starch content, the interpolated values were higher than the measured.

The asymmetry in the G' versus corn starch content (Figure 4–30) is typical of several protein-polysaccharide systems in which phase separation and inversion have occurred (Tolstoguzov, 1985; Ross-Murphy, 1984). However, the magnitudes of G' of the individual components with 10% solids differ only by a factor of two and are lower than in other such systems (Tolstoguzov, 1985). More severe asymmetry was observed in dispersions containing corn starch on the one hand, and 11S and 7S globulins on the other (Chen et al., 1996). At starch concentrations less than 5%, the continuous phase made up predominantly by SP isolate is weaker, and the effect of adding the stronger (higher modulus) starch granules is to increase G', the so-called isostress (Ross-Murphy, 1984; Morris, 1986) condition. As the starch concentration was increased further, phase inversion occurred, in that the stronger starch dispersion was the continuous phase and the SP isolate dispersion became the dispersed phase (Figure 4–30). Additional insight into the interaction of corn starch–SP isolate was found in creep-compliance data (Liao et al., 1996).

Cowpea Protein/Cowpea Starch

The G' development in 10% solids gels from blends of cowpea protein and cowpea starch over a 10-hour aging period at 20°C is shown in Figure 4–31

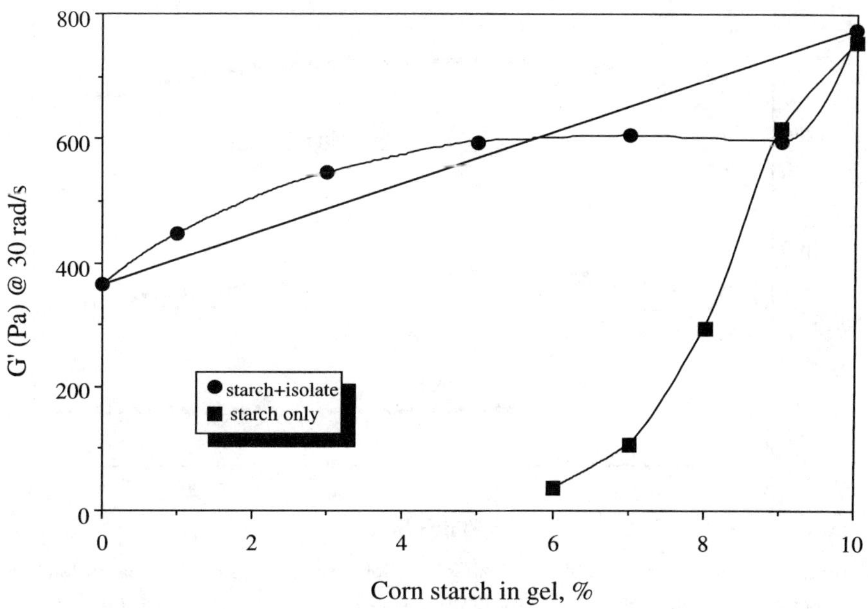

Figure 4–30 Plot of the Experimental Values of G', at a Frequency of 30 rad s^{-1}, of Dispersions of Various Soy Protein Isolate/Corn Starch Ratios and of Corn Starch with 6–10% Solids. Straight line interpolates G' values between those of corn starch and soy protein isolate. *Source:* Reprinted with permission from H-J Liao et al., Rheological and Calorimetric Properties of Heated Corn Starch-Soybean Protein Isolate Dispersions, *Journal of Texture Studies*, Vol. 27, pp. 403–418, © 1996, Food and Nutrition Press.

(Okechukwu and Rao, 1997); G' increased from an initial value, G'_0, to a maximum plateau value, G'_{max}, that increased with the proportion of starch in the mixture. A slight decrease in G' was observed during the later part of aging, probably due to weakening of starch granules. The rise in G' up to the maximum value followed an apparent first-order rate process, Equation 4.57.

$$\frac{dX}{dt} = -kX \tag{4.57}$$

where $X = \dfrac{G'_{max} - G(t)}{G'_{max} - G'_0}$, k is a rate constant, and t is aging time. Estimated values of the apparent rate constants at 20°C were 0.016, 0.022, and 0.023 min^{-1} for the blends at protein/starch ratios (R) of 5/5, 3/7, and 1/9, with R^2 values 0.959, 0.997, 0.995, respectively. The variation of G'_{max} (Pa) with mass fraction (x_S) of starch in the blend was described by the following equation:

$$G'_{max} = 7.69(x_s)^{3.35};\ R^2 = 0.998 \tag{4.58}$$

The nearly equal values of the exponents for the variation of G' with starch con-

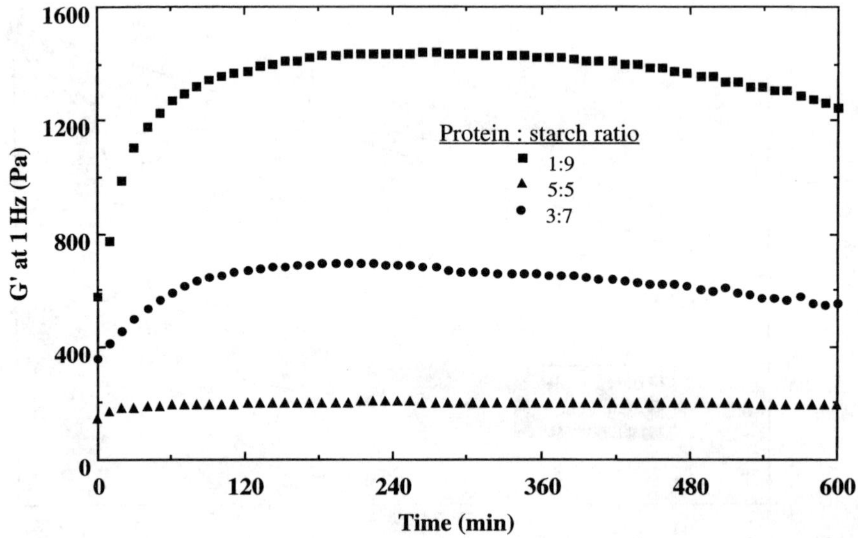

Figure 4–31 Development of G' in 10% Solids Gels from Blends of Cowpea Protein and Cowpea Starch over a 10-Hour Aging Period at 20°C. *Source*: Reprinted from F*ood Hydrocolloids*, P.E. Okechukwu and M.A. Rao, Vol. 11, Calorimetric and Rheological Behavior of Cowpea Protein Plus Starch Cowpea and Corn Gels, pp. 339–345, © 1997, with permission from Elsevier Science.

centration in cowpea starch/protein mixtures (within the range of $R < 1/9$) and in cowpea starch gels emphasizes the important contribution of starch to the structure of the former. G' development in the protein-starch gels followed an essentially apparent first-order process and did not show either the dominant two-stage growth pattern seen in starch gels or the almost linear pattern observed in the cowpea protein gels.

At pH values close to 7.0, the cowpea protein solubilized and formed the continuous phase in which raw starch granules were dispersed. On heating, the granules absorbed water, and as they swelled they exuded some amylose into the continuous matrix prior to the protein denaturation. The higher transition temperature of the protein compared to that of starch created the environment for the leaching of amylose into the solution of the protein (Muhrbeck and Eliasson, 1991). The coexistence of the protein and the amylose in the continuous phase may be responsible for the distinct kinetic pattern in the observed development of G' in the cowpea protein-starch blends. For such a system, the protein/starch ratio (R) would reflect the protein/amylose ratio in the continuous matrix and may be considered to be an important parameter for assessing G' of the mixture gels.

Figure 4–32 shows the variation of G' at 1 Hz of protein/starch gels (10% solids)

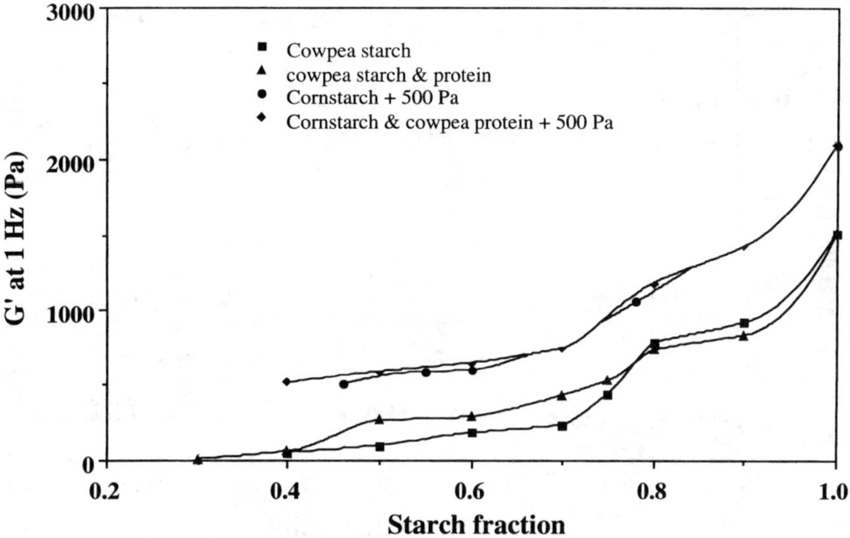

Figure 4–32 Variation of G' at 1 Hz of Protein/Starch Gels (10% Solids) from Frequency Sweeps on Gels after 1 Hour of Aging with Starch Fraction (x_s). *Source*: Reprinted from *Food Hydrocolloids*, P.W. Okechukwu and M.A. Rao, Vol. 11, Calorimetric and Rheological Behavior of Cowpea Protein Plus Starch Cowpea and Corn Gels, pp. 339–345, © 1997, with permission from Elsevier Science.

---

from frequency sweeps on gels after 1 hr of aging with starch fraction (x_S). Up to $x_S < 0.8$ ($R < 2/8$), G' of the mixed gels was higher than that of cowpea starch gel at the same concentration. For $x_S > 0.8$, G' of the starch gel was higher than G' of the mixed gel. Figure 4–32 also contains G' at 1 Hz from frequency sweeps on gels of cowpea protein and corn starch after one hour of aging at 25°C. Addition of the cowpea protein to corn starch at high levels did not affect the magnitude of G' of gel and at low levels resulted in a slight decrease. Similar decrease in G' was observed after the addition of 1% gluten to corn starch (Lindahl and Eliasson, 1986). Recognizing that G' of 10% solids cowpea protein gels was too low to be measured, the shape of the modulus-starch fraction curves of the mixed gels containing either corn starch or cowpea starch suggests phase separation. The higher G' values of the protein/cowpea starch gels at low levels of starch appear to be due to favorable kinetics during the early stages of G' development (Figure 4–33). For aging periods less than 250 minutes in 5% solids gels, G' of the mixed gel (5% solids and 5 parts protein and 5 parts starch) was more than that of the starch gel of equivalent solids content. At a higher starch and solids level (9%), G' of the mixed gel (9% solids and 1 part protein and 9 parts starch) was lower throughout the aging period.

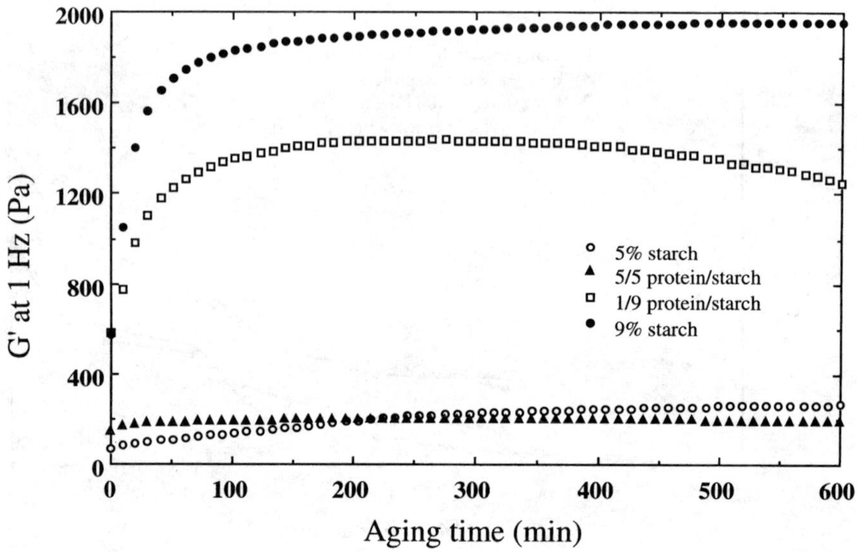

Figure 4–33 The Higher G' Values of the Protein/Cowpea Starch Gels at Low Levels of Starch Appear To Be Due to Favorable Kinetics during the Early Stages of G' Development. For aging periods less than 250 minutes in 5% solids gels, G' of the mixed gel (5% solids and 5 parts protein and 5 parts starch) was more than that of the starch gel of equivalent solids content. At a higher starch and solids level (9%), G' of the mixed gel (9% solids and 1 part protein and 9 parts starch) was lower throughout the aging period. *Source*: Reprinted from *Food Hydrocolloids*, P.E. Okechukwu and M.A. Rao, Vol. 11, Calorimetric and Rheological Behavior of Cowpea Protein Plus Starch Cowpea and Corn Gels, pp. 339–345, © 1997, with permission from Elsevier Science.

Other Protein/Starch Systems

Differences in the rheological behavior of gels from mixtures of fish protein and starch were attributed to the variation in the gelatinization temperatures of the starches in relation to the transition temperature of the protein. In such binary systems, the continuous gel network would be formed by the component with the lower transition temperature (Wu et al., 1985). Based on studies on starches and bovine serum albumin, Muhrbeck and Eliasson (1991) showed that when a starch network is formed before the gel formation of the protein, the two structures supplement each other without any specific interaction, to form a complex system with two continuous networks. When the protein network was formed first, the diffusion and aggregation of the amylose molecules and the swelling of the granules were hindered; the starch cannot form a continuous network but will instead act as a filler reinforcing the protein network.

REFERENCES

Axelos, M.A.V., J.F. Thibault, and J. Lefebvre. 1989. Structure of citrus pectins and viscometric study of their solution properties. *Int J Biol Macromol* 11: 186–191.

Bagley, E.B., and D.D. Christianson. 1982. Swelling capacity of starch and its relationship to suspension viscosity: Effect of cooking time, temperature and concentration. *J Texture Studies* 13: 115–126.

Barnes, H.A. 1989. Shear thickening "dilatancy" in suspensions of non aggregating solid particles dispersed in Newtonian liquids. *J Rheol* 33: 329–366.

Biliaderis, C.G. 1992. Characterization of starch networks by small strain dynamic rheometry. In *Developments in carbohydrate chemistry,* eds. R.J. Alexander and H.F. Zobel, 87–135. St. Paul, MN: American Association of Cereal Chemists.

Bird, R.B., R.C. Armstrong, and O. Hassager. 1977. *Dynamics of polymeric liquids—Fluid mechanics.* New York: John Wiley & Sons.

Blanshard, J.M.V. 1979. Physicochemical aspects of starch gelatinization. In *Polysaccharides in food,* eds. J.M.V. Blanshard and J.R. Mitchell, 139–152. London: Butterworth.

Blanshard, J.M.V. 1987. Starch granule structure and function: A physicochemical approach. In *Starch: Properties and potential,* ed. T. Galliard, 16–54. New York: John Wiley & Sons.

Boersma, W.H., P.J.M. Baets, J. Laven, and H.N. Stein. 1991. Time-dependent behavior and wall slip in concentrated shear thickening dispersions. *J Rheol* 35: 1093–1120.

Boersma, W.H., J. Laven, and H.N. Stein. 1992. Viscoelastic properties of concentrated shear-thickening dispersions. *J Colloid Interface Sci* 149: 10–22.

Bossis, G., and J.F. Brady. 1989. The rheology of Brownian suspensions. *J Chem Phys* 91: 1866–1879.

Buscall, R., J.W. Goodwin, M.W. Hawkins, and R.H. Ottewell. 1982a. Viscoelastic properties of concentrated lattices I. Methods of examination. *J Chem Soc Faraday Trans* 78: 2873–2887.

Buscall, R., J.W. Goodwin, M.W. Hawkins, and R.H. Ottewell. 1982b. Viscoelastic properties of concentrated lattices II. Theoretical analysis. *J Chem Soc Faraday Trans* 78: 2889–2899.

Carreau, P.J., D. De Kee, and R.P. Chhabra. 1997. *Rheology of polymeric systems: Principles and applications.* Brooklyn, NY: Hanser Publishers.

Chamberlain, E.K. 1996. Characterization of heated and thermally processed cross-linked waxy maize starch utilizing particle size analysis, microscopy and rheology. M.S. thesis, Cornell University, Ithaca, NY.

Chamberlain, E.K., M.A. Rao, and C. Cohen. 1998. Shear thinning and antithixotropic behavior of a heated cross-linked waxy maize starch dispersion. *Int J Food Properties.* In press.

Champenois, Y., M.A. Rao, and L.P. Walker. 1998. Influence of gluten on the viscoelastic properties of starch pastes and gels. *J Sci Food Agric* 78: 119–126.

Chedi, L.L., and J.L. Kokini. 1992. Influence of protein addition on rheological properties of amylose- and amylopectin-based starches in excess water. *Cereal Chem* 69: 551–555.

Chen, C-J., P.E. Okechukwu, S. Damodaran, and M.A. Rao. 1996. Rheological properties of heated corn starch + soybean 7S and 11S globulin dispersions. *J Texture Studies* 27: 419–432.

Chou, T.D., and J.L. Kokini. 1987. Rheological properties and conformation of tomato paste pectins, citrus and apple pectins. *J Food Sci* 52: 1658–1664.

Chow, M.K., and C.F. Zukoski. 1995a. Gap size and shear history dependencies in shear thickening of a suspension ordered at rest. *J Rheol* 39: 15–32.

Chow, M.K., and C.F. Zukoski. 1995b. Nonequilibrium behavior of dense suspensions of uniform particles: Volume fraction and size dependence of rheology and microstructure. *J Rheol* 39: 33–59.

Christianson, D.D., and E.B. Bagley. 1984. Yield stresses in dispersions of swollen deformable cornstarch granules. *Cereal Chem* 61: 500–503.

Colas, B. 1986. Flow behavior of crosslinked cornstarches. *Lebensmittel Wissenschaft Technol* 19: 308–311.

Cox, W.P., and E.H. Merz. 1958. Correlation of dynamic and steady flow viscosities. *J Polymer Sci* 28: 619.

Dail, R.V., and J.F. Steffe. 1990a. Dilatancy in starch solutions under low acid aseptic processing conditions. *J Food Sci* 55: 1764–1765.

Dail, R.V., and J.F. Steffe. 1990b. Rheological characterization of crosslinked waxy maize starch solutions under low acid aseptic processing conditions using tube viscometry techniques. *J Food Sci* 55: 1660–1665.

Da Silva, P.M.S., J.C. Oliveira, and M.A. Rao. 1997. The effect of granule size distribution on the rheological behavior of heated modified and unmodified maize starch dispersions. *J Texture Studies* 28: 123–138.

Davidson, R.L. 1980. *Handbook of water-soluble gums and resins.* New York: McGraw-Hill.

Davis, M.A.F., M.J. Gidley, E.R. Morris, D.A. Powell, and D.A. Rees. 1980. Intermolecular association in pectin solutions. *Int J Biol Macromol* 2: 330–332.

De Kee, D., and K.F. Wissburn. 1998. Polymer rheology. *Physics Today 51,* no. 6: 24–29.

Dealy, J.M., and K.F. Wissburn. 1990. *Melt rheology and its role in plastics processing: Theory and applications.* New York: Van Nostrand Reinhold.

D'Haene, P., J. Mewis, and G.G. Fuller. 1993. Scattering dichroism measurements of flow-induced structure of a shear thickening suspension. *J Colloid Interface Sci* 156: 350–358.

Dintzis, F.R., and E.B. Bagley. 1995. Shear-thickening and transient flow effects in starch solutions. *J Appl Polymer Sci* 56: 637–640.

Dolan, K.D., and J.F. Steffe. 1990. Modeling rheological behavior of gelatinizing starch solutions using mixer viscometry data. *J Texture Studies* 21: 265–294.

Dolan, K.D., J.F. Steffe, and R.G. Morgan. 1989. Back extrusion and simulation of viscosity development during starch gelatinization. *J Food Process Eng* 11: 79–101.

Doublier, J.L. 1981. Rheological studies on starch. Flow behavior of wheat starch pastes. *Starch/Stärke* 33: 415–420.

Doublier, J.L. 1987. A rheological comparison of wheat, maize, fava bean and smooth pea starches. *J Cereal Sci* 5: 247–262.

Elbirli, B., and M.T. Shaw. 1978. Time constants from shear viscosity data. *J Rheol* 22: 561–570.

Eliasson, A.C. 1986. Viscoelastic behavior during the gelatinization of starch: 1. Comparison of wheat, maize, potato and waxy barley starches. *J Texture Studies* 17: 253–265.

Ellis, H.S., S.G. Ring, and M.A. Whittam. 1989. A comparison of the viscous behavior of wheat and maize starch pastes. *J Cereal Sci* 10: 33–44.

Evans, I.D., and D.R. Haisman. 1979. Rheology of gelatinized starch suspensions. *J Texture Studies* 10: 347–370.

Evans, I.D., and A. Lips. 1992. Viscoelasticity of gelatinized starch dispersions. *J Texture Studies* 23: 69–86.

Evans, I.D., and A. Lips. 1993. Influence of soluble polymers on the elasticity of concentrated dispersions of deformable food microgel particles. In *Food colloids and polymers: Stability and mechanical properties,* eds. E. Dickinson and P. Walstra, 214–223. Cambridge, England: The Royal Society of Chemistry.

Faubion, J.M., and R.C. Hoseney. 1990. The viscoelastic properties of wheat flour doughs. In *Dough rheology and baked product texture,* eds. H. Faridi and J.M. Faubion, 29–66. New York: Van Nostrand Reinhold.

Ferry, J.D. 1980. *Viscoelastic properties of polymers.* New York: John Wiley & Sons.

Galliard, T., and P. Bowler. 1987. Morphology and composition of starch. In *Starch: Properties and potential, critical reports on applied chemistry,* Vol. 13, ed. T. Galliard, 54–78. New York: John Wiley & Sons.

Giboreau, A., G. Cuvelier, and B. Launay. 1994. Rheological behavior of three biopolymer/water systems with emphasis on yield stress and viscoelastic properties. *J Texture Studies* 25: 119–137.

Glicksman, M. 1969. *Gum technology in the food industry.* New York: Academic Press.

Graessley, W.W. 1967. Viscosity of entangling polydisperse polymers. *J Chem Phys* 47: 1942–1953.

Graessley, W.W. 1974. The entanglement concept in polymer rheology. *Adv Polymer Sci* 16: 1–179.

Graessley, W.W. 1980. Polymer chain dimensions and the dependence of viscoelastic properties on concentration, molecular weight and solvent power. *Polymer* 21: 258–315.

Griskey, R.G., and R.G. Green. 1971. Flow of dilatant shear-thickening fluids. *Am Inst Chem Engrs J* 17: 725–728.

Harris, E.K. Jr. 1970. Viscometric properties of polymer solutions and blends as functions of concentration and molecular weight. Ph.D. diss., University of Wisconsin, Madison.

Harrod, M. 1989. Modelling of flow properties of starch pastes prepared by different procedures. *J Food Process Eng* 11: 257–275.

Hoffman, R.L. 1972. Discontinuous and dilatant viscosity behavior in concentrated suspensions. I. Observation of a flow instability. *Trans Soc Rheol* 16: 155–173.

Hoseney, R.C. 1998. Gelatinization phenomena of starch. In *Phase/state transitions in foods: Chemical; structural, and rheological changes,* eds. M.A. Rao and R.W. Hartel, 95–110. New York: Marcel Dekker.

Kaletunc-Gencer, G., and M. Peleg. 1986. Rheological characteristics of selected food gum mixtures in solution. *J Texture Studies* 17: 61–70.

Krieger, I.M. 1985. Rheology of polymer colloids. In *Polymer colloids,* eds. R. Buscall, T. Corner, and J.F. Stageman, 219–246. New York: Elsevier Applied Science.

Krieger, I.M. 1990. 1989 Bingham medal address: The role of instrument inertia in controlled-stress rheometers. *J Rheology* 34: 471–483.

Kubota, K., Y. Hosakawa, K. Suziki, and H. Hosaka. 1979. Studies on the gelatinization rate of rice and potato starches. *J Food Sci* 44: 1394–1397.

Kulicke, W.M., and R.S. Porter. 1980. Relation between steady shear flow and dynamic rheology. *Rheologica Acta* 19: 601–605.

Langan, R.E. 1986. Food industry. In *Modified starches: Properties and uses,* ed. O.B. Wurzburg, 199–212. Boca Raton, FL: CRC Press.

Lapasin, R., S. Pricl, and P. Tracanelli. 1991. Rheology of hydroxyethyl guar gum derivatives. *Carbohydrate Polymers* 14: 411–427.

Laun, H.M, R. Bung, and F. Schmidt. 1991. Rheology of extremely shear thickening polymer dispersions passively viscosity switching fluids. *J Rheol* 35: 999–1034.

Launay, B., J.L. Doublier, and G. Cuvelier. 1986. Flow properties of aqueous solutions and dispersions of polysaccharides. In *Functional properties of food macromolecules,* eds. J.R. Mitchell and D.A. Ledward, 1–78. London: Elsevier Applied Science.

Leach, H.W., L.D. McGowen, and T.J. Schoch. 1959. Structure of starch granule. I. Swelling and solubility patterns of various starches. *Cereal Chem* 36: 534–544.

Liao, H-J., P.E. Okechukwu, S. Damodaran, and M.A. Rao. 1996. Rheological and calorimetric properties of heated corn starch-soybean protein isolate dispersions. *J Texture Studies* 27: 403–418.

Liao, H-J., J. Tattiyakul, and M.A. Rao. 1999. Superposition of complex viscosity curves during gelatinization of starch dispersion and dough. *J Food Process Eng* (In press).

Lindahl, L., and A.C. Eliasson. 1986. Effects of wheat proteins on the viscoelastic properties of starch gels. *J Sci Food Agric* 37: 1125–1132.

Lopes da Silva, J.A.L. 1994. Rheological characterization of pectin and pectin-galactomannan dispersions and gel. Ph.D. diss., Escola Superior de Biotecnologia, Porto, Portugal.

Lopes da Silva, J.A.L., and M.A. Rao. 1992. Viscoelastic properties of food gum dispersions. In *Viscoelastic properties of foods,* eds. M.A. Rao and J.F. Steffe, 285–316. London: Elsevier Applied Science.

Lopes da Silva, J.A.L., M.P. Gonçalves, and M.A. Rao. 1992. Rheological properties of high-methoxyl pectin and locust bean gum solutions in steady shear. *J Food Sci* 57: 443–448.

Lopes da Silva, J.A.L., M.P. Gonçalves, and M.A. Rao. 1993. Viscoelastic behavior of mixtures of locust bean gum and pectin dispersions. *J Food Eng* 18: 211–228.

Lopes da Silva, J.A.L., M.P. Gonçalves, and M.A. Rao. 1994. Influence of temperature on dynamic and steady shear rheology of pectin dispersions. *Carbohydrate Polymers* 23: 77–87.

Lund, D. 1984. Influence of time, temperature, moisture, ingredients and processing conditions on starch gelatinization. *Crit Rev Food Sci Nutr* 20: 249–273.

Madeka, H., and J.L. Kokini. 1992. Effect of addition of zien and gliadin on the rheological properties of amylopectin starch with low-to-intermediate moisture. *Cereal Chem* 69: 489–494.

Matsumoto, T., C. Hitomi, and S. Onogi. 1975. Rheological properties of disperse systems of spherical particles in polystyrene solution at long time scales. *Trans Soc Rheol* 19: 541–545.

McConnaughey, W.B., and N.O. Petersen. 1980. Cell poker: An apparatus for stress-strain measurements on living cells. *Rev Sci Instrum* 51: 575–580.

Miller, S.A., and C.A. Mann. 1944. Agitation of two-phase systems of immiscible liquids. *Trans Am Inst Chem Engrs* 40: 709.

Mills, P.L., and J.L. Kokini. 1984. Comparison of steady shear and dynamic viscoelastic properties of guar and karaya gums. *J Food Sci* 49: 1–4, 9.

Morris, E.R. 1981. Rheology of hydrocolloids. In *Gums and stabilisers for the food industry 2,* eds. G.O. Philips, D.J. Wedlock, and P.A. Williams, 57. Oxford, England: Pergamon Press.

Morris, E.R., A.N. Cutler, S.B. Ross-Murphy, and D.A. Rees. 1981. Concentration and shear rate dependence of viscosity in random coil polysaccharide solutions. *Carbohydrate Polymers* 1: 5–21.

Morris, E.R., and B. Ross-Murphy. 1981. Chain flexibility of polysaccharides and glicoproteins from viscosity measurements. In *Techniques in carbohydrate metabolism,* B310 ed. D.H. Northcote, p. 1–46. Amsterdam: Elsevier.

Morris, V.J. 1986. Multicomponent gels. In *Gums and stabilisers for the food industry 3,* eds. G.O. Philips, D.J. Wedlock, and P.A. Williams, 87–99. London: Elsevier Applied Science.

Morris, V.J. 1990. Starch gelation and retrogradation. *Trends in Food Science and Technology,* July, 2–6.

Muhrbeck, P., and A.C. Eliasson. 1991. Rheological properties of protein/starch mixed gels. *J Texture Studies* 22: 317–332.

Noel, T.R., S.G. Ring, and M.A. Whittam. 1993. Physical properties of starch products: Structure and function. In *Food colloids and polymers: Stability and mechanical properties*, eds. E. Dickinson and P. Wolstra, 126–137. Cambridge, England: Royal Society of Chemistry.

Okechukwu, P.E., and M.A. Rao. 1995. Influence of granule size on viscosity of cornstarch suspension. *J Texture Studies* 26: 501–516.

Okechukwu, P.E., and M.A. Rao. 1996. Role of granule size and size distribution in the viscosity of cowpea starch dispersions heated in excess water. *J Texture Studies* 27: 159–173.

Okechukwu, P.E., and M.A. Rao. 1997. Calorimetric and rheological behavior of cowpea protein plus starch cowpea and corn gels. *Food Hydrocolloids* 11: 339–345.

Okechukwu, P.E., M.A. Rao, P.O. Ngoddy, and K.H. McWatters. 1991. Flow behavior and gelatinization of cowpea flour and starch dispersions. *J Food Sci* 56: 1311–1315.

Paoletti, S., A. Cesaro, F. Delben, and A. Ciana. 1986. Ionic effects on the conformation, equilibrium, properties, and rheology of pectate in aqueous solution and gels. In *Chemistry and function of pectins*, eds. M.L. Fishman and J.J. Jen, 73–87. ACS Symposium Series. Washington, DC: American Chemical Society.

Petrofsky, K.E., and R.C. Hoseney. 1995. Rheological properties of dough made with starch and gluten from several cereal sources. *Cereal Chem 72,* no. 1: 53–58.

Plazek, D.J. 1996. 1995 Bingham medal address: Oh, thermorheological simplicity, wherefore art thou? *J Rheology* 40: 987–1014.

Plutchok, G.J., and J.L. Kokini. 1986. Predicting steady and oscillatory shear rheological properties of CMC and guar gum blends from concentration and molecular weight data. *J Food Sci* 51: 1284–1288.

Quemada, D., P. Fland, and P.H. Jezequel. 1985. Rheological properties and flow of concentrated disperse media. *Chem Eng Comm* 32: 61–83.

Rao, M.A., and J. Tattiyakul. 1999. Granule size and rheological behavior of heater tapioca starch dispersions. *Carbohydrate Polymers* 38: 123–132.

Robinson, G., S.B. Ross-Murphy, and E.R. Morris. 1982. Viscosity-molecular weight relationships, intrinsic chain flexibility and dynamic solution properties of guar galactomannan. *Carbohydrate Res* 107: 17–32.

Rochefort, W.E., and S. Middleman. 1987. Rheology of xanthan gum: Salt, temperature and strain effects in oscillatory and steady shear experiments. *J Rheol* 31: 337–369.

Rodriguez, F. 1989. Principles of polymer systems. 3d ed. New York: Hemisphere Publishing.

Ross-Murphy, S.B. 1984. Rheological methods. In *Biophysical methods in food research*, ed. H.W.S. Chan, 138–199. London: Blackwell Scientific.

Russel, W.B., D.A. Saville, and W.R. Schowalter. 1989. *Colloidal dispersions.* Cambridge, England: Cambridge University Press.

Sawayama, S., A. Kawabata, H. Nakahara, and T. Kamata. 1988. A light scattering study on the effects of pH on pectin aggregation in aqueous solution. *Food Hydrocolloids* 2: 31–37.

Tam, K.C., and C. Tiu. 1993. Improved correlation for shear-dependent viscosity of polyelectrolyte solutions. *J Non-Newtonian Fluid Mechanics* 46: 275–288.

Tattiyakul, J. 1997. Studies on granule growth kinetics and characteristics of tapioca starch dispersion during gelatinization using particle size analysis and rheological methods. M.S. thesis, Cornell University, Ithaca, NY.

Tirrell, M. 1994. Rheology of polymeric liquids. In *Rheology: Principles, measurements, and applications*, ed. C.W. Macosko, 475–514. New York: VCH Publishers.

Tolstoguzov, V.B. 1985. Functional properties of protein-polysaccharide mixtures. In *Functional properties of food macromolecules*, eds. J. Mitchell and D.A. Ledward, 385–415. London: Elsevier Applied Science.

Tolstoguzov, V.B. 1991. Functional properties of food proteins and role of protein-polysaccharide interaction—Review. *Food Hydrocolloids* 4: 429–468.

Whistler, R.L., and J.R. Daniel. 1985. Carbohydrates. In *Food chemistry*, ed. O.R. Fennema, 69–138. New York: Marcel Dekker.

Whitcomb, P.J., and C.W. Macosko. 1978. Rheology of xanthan gum. *J Rheol* 22: 493–505.

Wu, M.C., D.D. Hamann, and T.C. Lanier. 1985. Rheological and calorimetric investigations of starch-fish protein systems during thermal processing. *J Texture Studies* 16: 53–74.

Yang, W.H. 1997. Rheological behavior and heat transfer to a canned starch dispersion: Computer simulation and experiment. Ph.D. diss., Cornell University, Ithaca, NY.

Yang, W.H., A.K. Datta, and M.A. Rao. 1997. Rheological and calorimetric behavior of starch gelatinization in simulation of heat transfer. In *Engineering and food at ICEF 7/Part 2*, ed. R. Jowitt, K1–K5. London: Sheffield Academic Press.

Yang, W.H., and M.A. Rao. 1998. Complex viscosity-temperature master curve of cornstarch dispersion during gelatinization. *J Food Process Eng* 21: 191–207.

Yoo, B., A.A. Figueiredo, and M.A. Rao. 1994. Rheological properties of mesquite seed gum in steady and dynamic shear. *Lebensmittel Wissenschaft und Technologie* 27: 151–157.

Zahalak, G.I., W.B. McConnaughey, and E.L. Elson. 1990. Determination of cellular mechanical properties by cell poking, with an application to leukocytes. *J Biomechanical Eng* 112: 283–294.

CHAPTER 5

Rheological Behavior of Processed Fluid and Semisolid Foods

M. Anandha Rao

In this chapter, the rheological properties of processed fluid and semisolid foods will be discussed. Where data are available, the role of the composition of the foods on their rheological behavior will be emphasized.

FRUIT JUICES AND PUREES: ROLE OF SOLUBLE AND INSOLUBLE SOLIDS

Fruit juices and plant food (PF) dispersions are important items of commerce. Single-strength and concentrated clear fruit juices, that is, juices from which insoluble solids and dissolved polymers (e.g., pectins) have been removed, exhibit Newtonian flow behavior or close to it, and the sugar content plays a major role in the magnitude of the viscosity and the effect of temperature on viscosity (Saravacos, 1970; Rao et al., 1984; Ibarz et al., 1987, 1992a, 1992b). Magnitudes of rheological parameters of fruit juices and purees are summarized in Table 5–A2 (see Appendix 5–A), of tomato pastes in Table 5–A3, and the effect of temperature on apparent viscosity and consistency index are summarized in Table 5–A4. It should be noted that the viscosity (or apparent viscosity, consistency index) of concentrated fruit juices decreases with increase in temperature and the magnitude of the activation energy of clear fruit juices increases markedly with increase in sugar content.

Because depectinized and filtered apple and grape juices are Newtonian fluids, equations derived by Bayindirli (1992, 1993) can be used to estimate their viscosities as a function of concentration and temperature. The equation for viscosity of both apple and grape juices is of the form

$$\frac{\eta}{\eta_{wa}} = \exp\left\{\frac{A\,(^{\circ}\mathrm{Brix})}{[100 - B(^{\circ}\mathrm{Brix})]}\right\} \tag{5.1}$$

where η is the viscosity (mPa s) of depectinized and filtered apple juice, η_{wa} is the viscosity of water at the same temperature (mPa s), and A and B are constants.

For apple juice in the concentration range 14–39 °Brix and temperature 293–353K (20–80°C), the constants A and B are

$$A = -0.24 + (917.92/T) \qquad B = 2.03 - 0.00267\,T \tag{5.2}$$

For grape juice in the concentration range 19–35 °Brix and temperature 293–353K (20–80°C), the constants A and B are

$$A = -3.79 + (1821.45/T) \qquad B = 0.86 + 0.000441\,T \tag{5.3}$$

In both Equations 5.2 and 5.3, the temperature, T, is in K. For the purpose of illustration, from Equations 5.2 and 5.3, the viscosity of 30 °Brix apple juice at 20°C was estimated to be 4.0 mPa s and that of grape juice 2.8 mPa s, respectively.

Rao et al. (1984) studied the role of concentration and temperature on the viscosity of concentrated depectinized and filtered apple and grape juices. At a constant temperature, the effect of concentration could be well described by an exponential relationship. For example, at 20°C, the effect of concentration (~41 to 68 °Brix) on the viscosity of apple concentrates was described by the Equation 5.4 with $R^2 = 0.947$:

$$\eta = 1.725 \times 10^{-5} \exp(0.136 \times {}^\circ\text{Brix}) \tag{5.4}$$

For concentrated grape juice samples (~41 to 68 °Brix) at 20°C, the effect of concentration was described by Equation 5.5 with $R^2 = 0.966$:

$$\eta = 2.840 \times 10^{-5} \exp(0.137 \times {}^\circ\text{Brix}) \tag{5.5}$$

The effect of temperature and concentration on the viscosity of concentrated apple juice can be combined to obtain a single approximate equation that can be used for estimating viscosity as a function of both temperature and °Brix:

$$\eta_a = A \exp(E_a/RT) \exp(\beta\,{}^\circ\text{Brix}) \tag{5.6}$$

For apple juice, the constants A, $\overline{E}_a$, and β were 3.153×10^{-14} Pa s, 45.0 kJ mol^{-1}, and 0.167 °Brix^{-1}, respectively, and the $R^2 = 0.94$. For the viscosity of concentrated grape juice concentrates, the constants A, $\overline{E}_a$, and β were 6.824×10^{-14} Pa s, 46.6 kJ mol^{-1}, and 0.151 °Brix^{-1}, respectively, and the $R^2 = 0.97$. The above equation should be considered to provide only approximate values of viscosity because $\overline{E}_a$ is an average activation energy for the entire range of concentrations studied. In fact, the E_a of concentrated apple and grape juices are exponentially dependent on the °Brix values (Figure 5–1). Figure 5–1 illustrates the important role of sugar content on the activation energy of concentrated apple and grape juices. The respective equations describing the influence of °Brix on the E_a (kJ mol^{-1}) of the apple and grape juice are

$$E_a = 5.236 \exp[0.0366({}^\circ\text{Brix})];\ R^2 = 0.94 \tag{5.7}$$

$$E_a = 8.587 \exp[0.0294({}^\circ\text{Brix})];\ R^2 = 0.97 \tag{5.8}$$

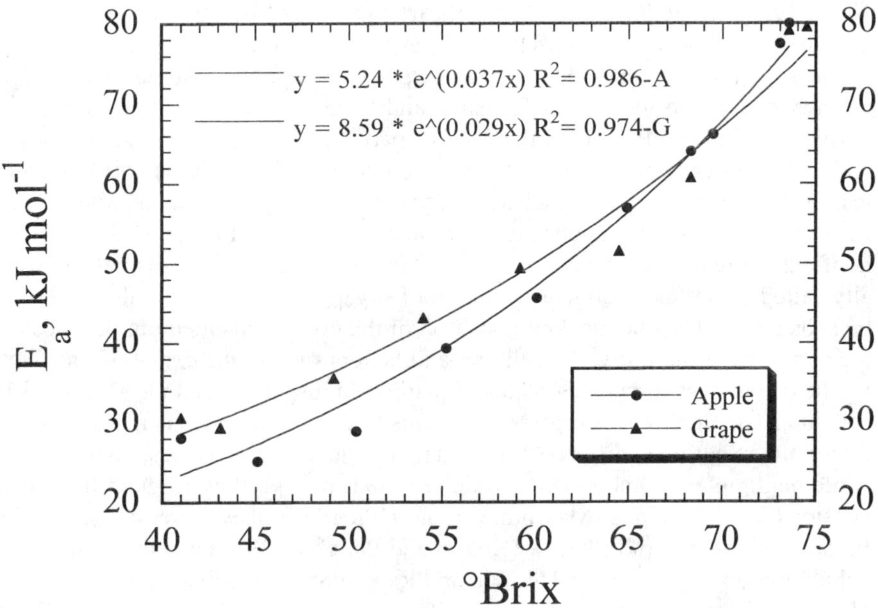

Figure 5–1 The Flow Activation Energy (E_a) of Concentrated Apple and Grape Juices Are Exponentially Dependent on Their °Brix Values.

Thus a more accurate estimate of the viscosities of concentrated apple and grape juices can be obtained by incorporating the value of E_a calculated using either Equation 5.7 or 5.8 at a specific concentration in Equation 5.6. Although two separate equations were derived for the apple and grape juice concentrates, one could easily justify a single equation that would be reasonably applicable to both concentrates.

Role of Plant Food Insoluble Solids

Several observations useful to concentrated purees can be found in studies on the rheological behavior of nonfood dispersions that have been reviewed earlier in Chapter 2 and elsewhere (Jinescu, 1974; Jeffrey and Acrivos, 1976; Metzner, 1985); other useful studies include those on the rheology of rigid fiber suspensions by Ganani and Powell (1985) and on slurries of irregular particles (Wildemuth and Williams, 1984). Also, there has been what may be called a kinetic or structural approach to rheology of suspensions (Michaels and Bolger, 1962; Hunter and

Nicol, 1968) where the basic flow units are assumed to be small clusters, or aggregates, that at low shear rates give the suspension a finite yield stress.

The solid particles in plant food (PF) dispersions are not of simple shapes (e.g., spheres, rods), and they are deformable and have multimodal size distributions (Tanglertpaibul and Rao, 1987a). Also, the particles are hydrated and are in physical and chemical equilibrium with the continuous medium, so they differ significantly from artificial fibers such as of glass or of synthetic polymers. The continuous phases of PF dispersions also have features that are different from those of nonfood suspensions. The continuous medium of a typical food dispersion, usually called serum, is an aqueous solution of sugars, organic acids, salts, and pectic substances. The chemical composition of the continuous medium depends on the particular commodity, the cultivar, and factors such as the extent of ripening.

The differences in characteristics of plant food dispersions can lead to significant interparticle forces compared to viscous forces, and Brownian motion can be important, resulting in different flow and rheological characteristics. For example, while negligible wall slip effects were reported for pipe flow of glass fiber suspensions in silicone oils (Maschmeyer and Hill, 1977), they were of concern for PF dispersions in capillary flow (Kokini and Plutchok, 1987) and in a concentric cylinder geometry (Higgs, 1974; Qiu and Rao, 1988). Food dispersions are complex materials whose characteristics with respect to the nature of the insoluble solids as well as those of the fluid media often are determined a priori to experimentation. Nonfood suspensions, such as glass fibers in mineral oils and in polymers, can be custom-made with one's choice of continuous and dispersed media having the desired characteristics. In contrast, for preparing a set of engineered PF dispersion samples, the source for the continuous and dispersed phases is another PF dispersion.

Nevertheless, to understand the role of various components of PF dispersions, systematic studies on the role of the amount and size of insoluble solids can be conducted by careful preparation of the test samples. Figure 5–2 illustrates the processing steps used to vary pulp particle size distribution in tomato concentrates (Tanglertpaibul and Rao, 1987a). Similar processing technique was used in the preparation of apple sauce samples by Rao et al. (1986) and Qiu and Rao (1988). The amount and the size distribution of the insoluble solids in the PF dispersions depend to some extent on the size of the screen employed in the finishing operation during their manufacture (Tanglertpaibul and Rao, 1987a) and to a limited extent on the speed of the finisher (Rao et al., 1986).

In the rheological behavior of concentrated pulpy fruit juices and purees (e.g., orange and grapefruit juices), both the pulp and sugar contents were found to be the key components (Mizrahi and Berk, 1972; Durán and Costell, 1982; Vitali and Rao, 1984a, 1984b). In tomato concentrates also, where sugar content is not as high as in concentrated fruit juices (e.g., orange juice), both pulp and sugars are key components (Harper and El-Sahrigi, 1965; Rao et al., 1981; Fito et al., 1983).

Fresh Tomatoes, Sorted, Washed

↓

Crushing

↓

Hot Break, 98°C, 4 min, 5 s

↓

Finishing (Screens opening 0.51 mm to 1.14 mm)

↓

Evaporate Juice to Different Solids Content

Figure 5–2 Illustration of the Processing Steps in Tomato Concentrates, Using Finisher Screens with Different Diameter Openings To Vary Particle Size and Content. Tanglertpaibul and Rao, 1987a.

The viscosity (or apparent viscosity) of a suspension can be related to the viscosity of the continuous medium (i.e., in terms of the relative viscosity, η_r)

$$\eta_r = \frac{\text{Viscosity of suspension}}{\text{Viscosity of continuous medium}} \tag{5.9}$$

and the volume fraction of solids, ϕ (Jinescu, 1974; Metzner, 1985; Ganani and Powell, 1985). Because of the compressible nature of PF dispersions, the direct determination of the magnitude of ϕ is difficult as it depends on the centrifugal force employed in the separation of the phases. In samples of tomato concentrates, volume fraction of solids was determined from volumetric measurements to be about 0.15 to 0.45 (Tanglertpaibul and Rao, 1987b). In the case of concentrated orange juice (COJ), for the purpose of quality control, the amount of dispersed solid matter (pulp content) in a 12 °Brix sample is determined at 360 *g* so that the pulp content in different samples is determined at a standard centrifugal force and at a standard concentration that sets, to a reasonable extent, the continuous medium's viscosity (Praschan, 1981). The pulp content of COJ, as determined by the standard method, ranges from about 2% to 16%. Because of the highly viscous nature of apple sauce and tomato concentrates as well as the large amount of pulp content, the technique used for COJ may not be satisfactory for these products. Therefore, ratios such as pulp:total sample weight (Takada and Nelson, 1983) and pulp:serum (Tanglertpaibul and Rao, 1987b) have been employed.

Size Distribution of Fruit Juice Solids

Size distribution data on solid particles of PF dispersions are sparse. Carter and Buslig (1977) determined size distribution of relatively small pulp particles in

COJ with a Coulter counter. Wet sieving technique was employed on apricot puree, apple sauce, and tomato concentrates (Trifiro et al., 1987; Rao et al., 1986; Tanglertpaibul and Rao, 1987a) where volume fractions of solids retained on standard sieves were used to calculate the approximate average particle sizes. It is clear that the solid particles of PF dispersions have multimodal size distributions. Also, changes in particle sizes induced by homogenization of apricot puree affected consistency of the puree (Durán and Costell, 1985), while changes induced by using finisher screens of different sizes affected the rheological properties of a tomato puree (Tanglertpaibul and Rao, 1987a). Because the volume fraction of some of the tomato concentrates was in the range 0.15–0.45 (Tanglertpaibul and Rao, 1987a), these results are consistent with the general observation (Metzner, 1985) that distribution of particle sizes affects dispersion viscosities when volumetric loading of solids is above 20%.

Because relatively low-cost powerful desktop computers and instruments have become available, size distribution of tomato pulp particles can be examined based on laser diffraction (den Ouden and van Vliet, 1993). Although most tomato pulp particles were smaller than 900 μm, the upper particle size limit of the LS130 instrument, a few particles were larger (den Ouden and van Vliet, 1993; Yoo and Rao, 1996). Particle size distributions of 21 °Brix concentrates (Figure 5–3, top) illustrate the general shape of size distribution versus particle diameter profiles. The distributions reveal differences, although minor, between the concentrates from 0.69 mm and 0.84 mm finisher screens. Interestingly, serum obtained from a commercial concentrate after centrifugation at 100,000 $\times$ g for one hour also contained relatively large particles, >100–200 μm as measured by laser diffraction (Figure 5–3, bottom). The presence of the few large particles could be due to the finishing operation in which the tomato pulp is forced through finisher screen holes, resulting in elongated shreds of cell wall and skin of the fruit.

While the size distribution of reasonably well-defined smaller (max. dia ~100 μm) gelatinized corn starch granules could be determined reliably with the Coulter LS130 and its role in rheological behavior established (Okechukwu and Rao, 1995), similar efforts for tomato concentrates require further work. Also, what role these few large particles may play in the rheological behavior of the concentrates is not very obvious, so determination of their exact sizes (e.g., by using an instrument with a wider range of particle sizes) may provide only marginal understanding of the size distribution but not of the rheology of the concentrates. Because of the relatively wide range of size distribution of the tomato pulp particles (Figure 5–3), it is unlikely that shear-thickening flow behavior will be encountered with tomato concentrates as in gelatinized starch dispersions that can exhibit relatively narrow particle size distributions (Okechukwu and Rao, 1995).

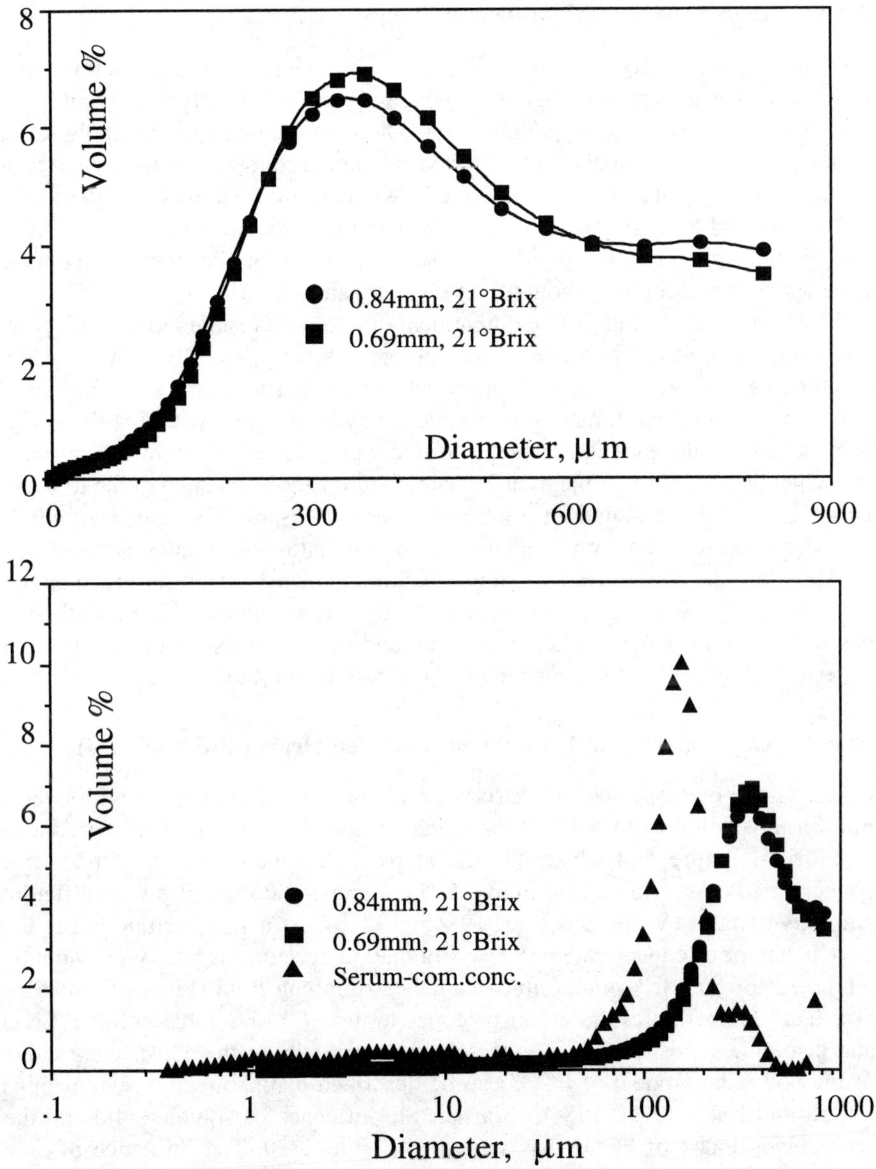

Figure 5–3 Particle Size Distributions of 21 °Brix Tomato Concentrates (top) Derived from Using 0.69 mm and 0.84 mm Finisher Screens, and in Serum Obtained from a Commercial Concentrate after Centrifugation at 100,000 × *g* for One Hour (bottom). *Source:* Reprinted with permission from B. Yoo and M.A. Rao, Creep and Dynamic Rheological Behavior of Tomato Concentrates: Effect of Concentration and Finisher Screen Size, *Journal of Texture Studies*, Vol. 27, pp. 451–459, © 1996, Food and Nutrition Press.

Serum Viscosity of Plant Food Dispersions

The viscosities of the continuous media of PF dispersions depend on their composition, that is, sugar and pectin content, which in turn may be affected by cultivars, fruit maturity, and processing conditions. The continuous media of apple sauce with sugar concentration of about 16 °Brix and tomato concentrates with sugar concentrations in the range 6–25 °Brix were Newtonian fluids (Rao et al., 1986; Tanglertpaibul and Rao, 1987b). At high concentrations, they can be expected to be mildly shear-thinning fluids as was the case with the serum of concentrated orange juice with concentration of about 65 °Brix (Vitali and Rao, 1984a, 1984b). Caradec and Nelson (1985) found that heat treatment of tomatoes resulted in loss of serum viscosity. Correlations in terms of the relative viscosity, η_{rel}, for relatively low magnitudes and narrow ranges of sugar concentrations, such as the continuous media of apple sauce and tomato concentrates, may be possible with just the pectin content as the independent variable. Information regarding serum intrinsic viscosity, which is a measure of the hydrodynamic volume and is related to the molecular weight of the biopolymer in solution, will be very useful. Magnitudes of intrinsic viscosity can be determined either from extrapolation techniques, such as with the Huggins equation or from the slopes of straight lines of quantities based on relative viscosity. Magnitudes of intrinsic viscosity varied slightly with the method of determination, the extent of heat treatment, and to a lesser extent on the shearing experienced by the food suspension (Tanglertpaibul and Rao, 1987c).

Rheological Properties of Frozen Concentrated Orange Juice (FCOJ)

Concentrated orange juice (65 °Brix) is a major item of commerce with over a million tons being exported by Brazil and another million tons being produced in Florida. Figure 5–4 illustrates the steps in the manufacture of 65 °Brix concentrated orange juice, and Figure 5–5 the preparation samples with different amounts of pulp (Vitali and Rao, 1984a, 1984b). At a fixed temperature, the flow behavior of concentrated 65 °Brix orange juice serum is nearly Newtonian, while that of 65 °Brix orange juice is a shear-thinning fluid (Figure 5–6). Also at a fixed temperature, the effect of concentration (c) of soluble solids (°Brix) and insoluble solids (pulp) on either apparent viscosity or the consistency index of the power law model of FCOJ can be described by exponential relationships (Vitali and Rao, 1984a, 1984b). Earlier, the influence of soluble solids on the consistency index of FCOJ was shown in Figure 2–10. The influence of pulp content on the consistency index is shown in Figure 5–7, where ln K is plotted versus percentage pulp content. Equations 5.10 and 5.11 are applicable to the consistency index (K) of the power law model (in Chapter 2, Equations 2.35 and 2.36, respectively). In the case of FCOJ, insoluble solids are expressed in terms of pulp content determined on a 12 °Brix sample by centrifugation for 10 min at $360 \times g$.

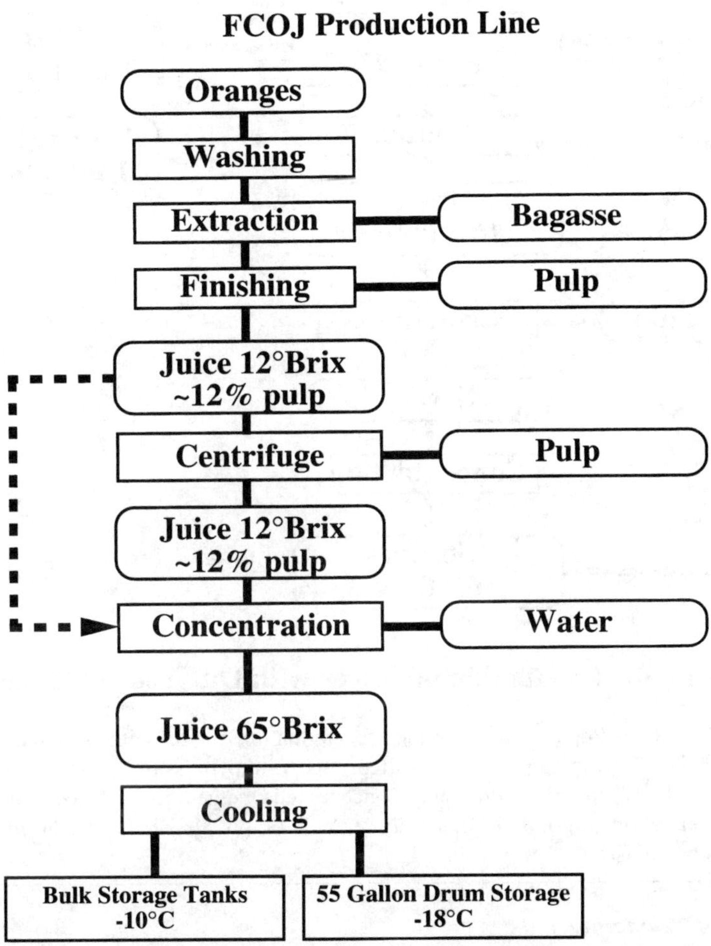

Figure 5–4 Unit Operations in the Manufacture of 65 °Brix Concentrated Orange Juice Samples with Different Amounts of Pulp. *Source*: Reprinted with permission from A.A. Vitali and M.A. Rao, Flow Properties of Low-pulp Concentrated Orange Juice: Effect of Temperature and Concentration, *Journal of Food Science*, Vol. 49, pp. 882–888, © 1984, Institute of Food Technologists.

$$K = K^{c} \exp (B_{K}^{c} \text{ °Brix}) \qquad (5.10)$$

$$K = K^{P} \exp (B_{K}^{P} \text{ Pulp}) \qquad (5.11)$$

where K^c, K^P, B_K^c, and B_K^P are constants. As pointed out in Chapter 2, to study the effect of composition variables, one can also work with apparent viscosity instead of the power law consistency index.

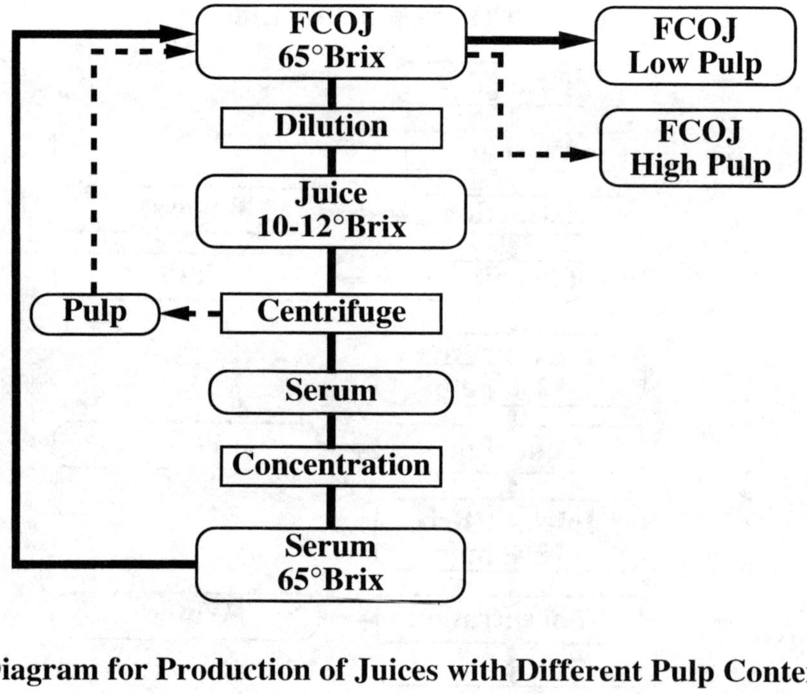

Figure 5–5 Steps (top to bottom) in the Preparation of Samples with Different Amounts of Pulp from 65 °Brix Sample. *Source*: Reprinted with permission from A.A. Vitali and M.A. Rao, Flow Properties of Low-pulp Concentrated Orange Juice: Effect of Temperature and Concentration, *Journal of Food Science*, Vol. 49, pp. 882–888, © 1984, Institute of Food Technologists.

Effect of Temperature

The flow behavior of FCOJ is strongly influenced by temperature, as seen in Figure 5–8 where data at several temperatures from Vitali and Rao (1984a, 1984b) are shown. A wide range of temperatures are encountered during FCOJ processing and storage, so the effect of temperature on rheological properties needs to be documented. The effect of temperature on either apparent viscosity or the consistency index, K, of the power law model (Equation 5.12) of FCOJ can be described by the Arrhenius relationship. The equation for the consistency index is

$$K = K_{\infty} \exp (E_{aK}/RT) \tag{5.12}$$

where K_{∞} is the frequency factor, E_{aK} is the activation energy (J mol^{-1}), R is the gas constant, and T is temperature (K). A plot of ln K (ordinate) versus (T^{-1}) (abscissa) results in a straight line, and E_{aK} = (slope $\times$ R), and K_{∞} is exponential of

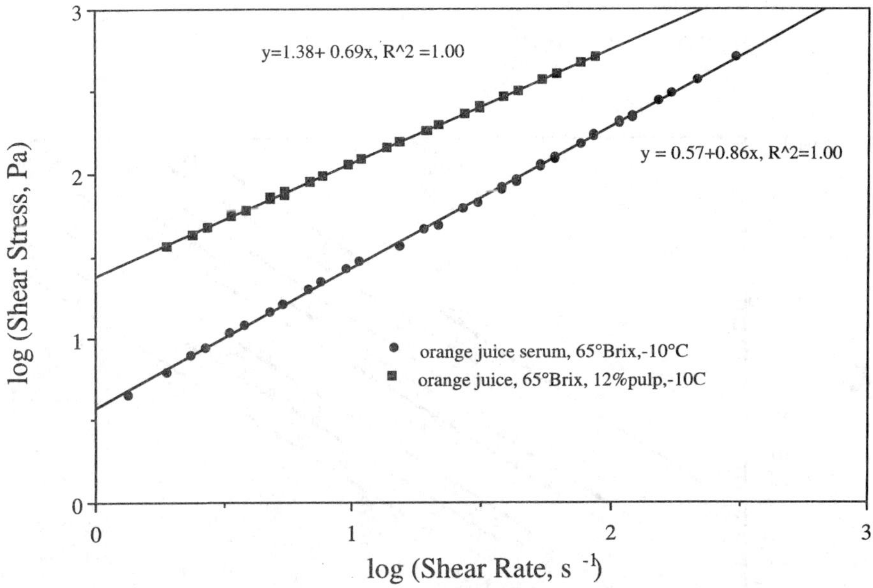

Figure 5–6 The Flow Behavior of Concentrated 65 °Brix Orange Juice Serum Is Nearly Newtonian (slope = 0.86), While That of 65 °Brix Orange Juice Is a Mildly Shear-Thinning Fluid (slope = 0.69). Data from Vitali and Rao, 1984a, 1984b.

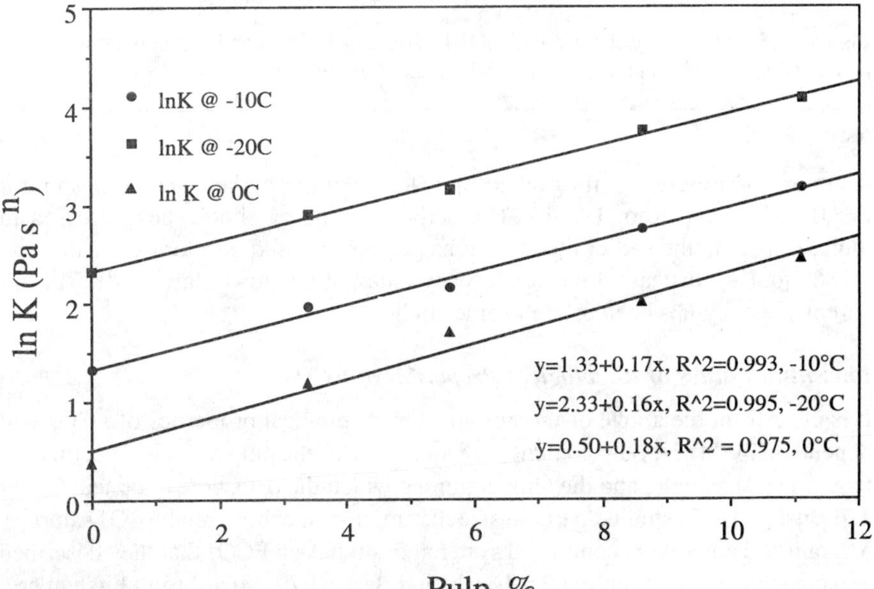

Figure 5–7 Influence of Pulp Content on the Consistency Index (*K*), where ln *K* Plotted versus Percentage Pulp Content. Vitali and Rao, 1984ab.

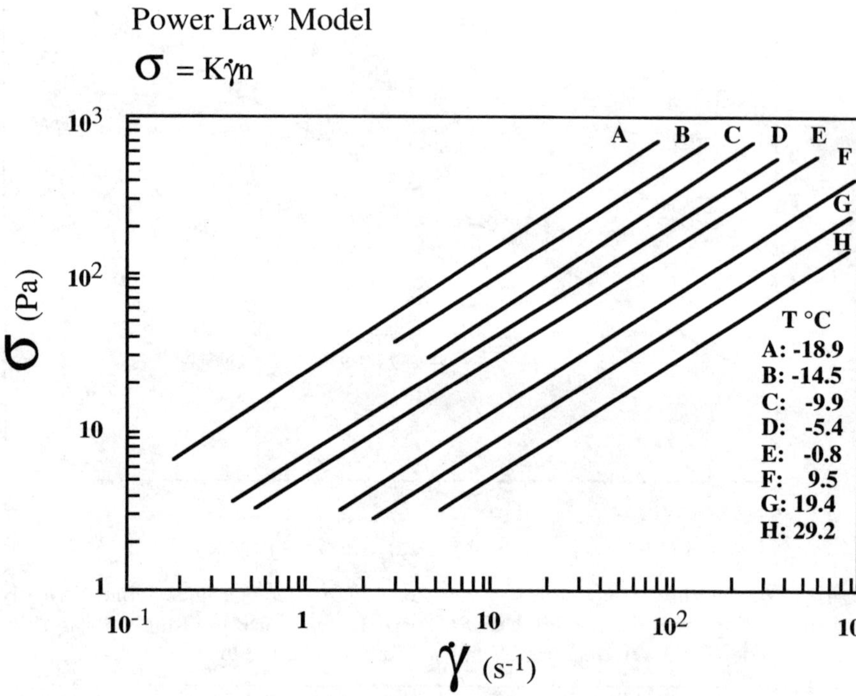

Figure 5–8 The Flow Behavior of FCOJ Is Strongly Influenced by Temperature. Lines represent data at several temperatures. Vitali and Rao, 1984ab.

the intercept. Figure 5–9 illustrates applicability of the Arrhenius relationship for FCOJ (Vitali and Rao, 1984b). The activation energy should be expressed in joules (J), but in the earlier literature it has been expressed in calories (1 calorie = 4.1868 joules). Instead of the consistency index of the power law model, the apparent viscosity has been used in some studies.

Literature Values of Rheological Properties of FCOJ

It is clear from the above discussion that the rheological properties of FCOJ will depend on the °Brix, pulp content, size and shape of the pulp particles, pectin content of FCOJ sample, and the temperature at which the data were obtained. Carter and Buslig (1977) studied particle size distribution in commercial FCOJ samples. Mizrahi and coworkers conducted systematic studies on FCOJ that they described as a physicochemical approach (Mizrahi and Berk, 1970; Mizrahi and Firstenberg, 1975; Mizrahi, 1979). They used a modified Casson equation to describe the flow behavior of FCOJ.

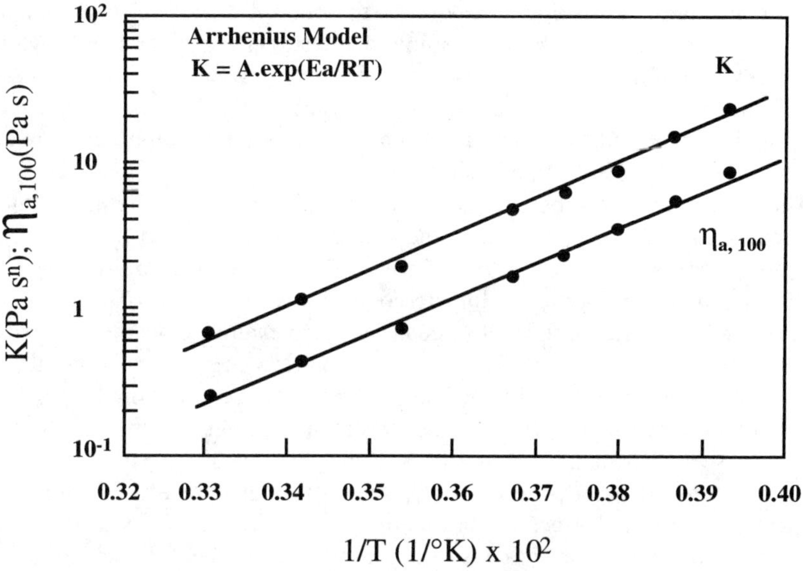

Figure 5–9 Arrhenius Plots for the Effect of Temperature on Apparent Viscosity at 100 s^{-1} (η_a, 100) and Consistency Index of Pera FCOJ with 5.7% Pulp Content. *Source:* Reprinted with permission from A.A. Vitali and M.A. Rao, Flow Properties of Low-Pulp Concentrated Orange Juice: Serum Viscosity and Effect of Pulp Content, *Journal of Food Science*, Vol. 49, pp. 876–880, © 1984, Institute of Food Technologists.

Because the power law model (Equation 2.3) is used in determination of pumping and mixing power requirements, literature values of the power law parameters of several FCOJ samples reported by Crandall et al. (1982) and Vitali and Rao (1984a, 1984b) are given in Table 5–A2. The different values in Table 5–A2 reflect the influence of fruit varieties, °Brix, pulp content, and temperature. In particular, the strong influence of temperature on the consistency index of the power law model should be noted. The constants in Equations 5.10 through 5.12 that describe the effect of temperature, °Brix, and pulp content on the consistency index of the power law model presented by Vitali and Rao (1984a, 1984b) are summarized in Table 5–A2. These data should be useful, albeit with caution, in estimation of ballpark values of the power law parameters of FCOJ from temperature, °Brix, and pulp content data.

Rheological Properties of Tomato Concentrates

Numerous studies have been conducted on tomato concentrates using quality control instruments, such as the Bostwick consistometer. Unfortunately, for a specific set of test samples, values of Bostwick consistency decrease with increase in solids concentration, making it impossible to obtain data at high solids loadings. Attempts at correlating Bostwick readings and basic viscosity data on Newtonian fluids in terms of theoretical analysis based on gravity wave flow were successful (McCarthy and Seymour, 1994a); however, a similar attempt with non-Newtonian fluids (McCarthy and Seymour, 1994b) and an attempt at empirical correlation met with limited success (Rao and Bourne, 1977).

The power law parameters (consistency index and flow behavior index) of tomato concentrates reported in the literature can be found in Table 5–A2. Tanglertpaibul and Rao (1987b) examined the contribution of soluble solids to the rheological properties of tomato concentrates (Figure 5–10) with concentrates made from a single finisher screen. They also examined the influence of different finisher screen sizes (0.51 to 1.14 mm) on flow properties. From the shear rate versus shear stress data obtained on 20% solids concentrates made from juices that were produced using different finisher screen openings shown in Figure 5–11, it can be seen that using a 0.69 mm screen resulted in concentrates with the highest viscosity; use of the 0.69 mm screen probably resulted in juice with highest pulp content. Because the composition of tomatoes and other fruits depends on the particular varieties being employed and their maturity, the effect

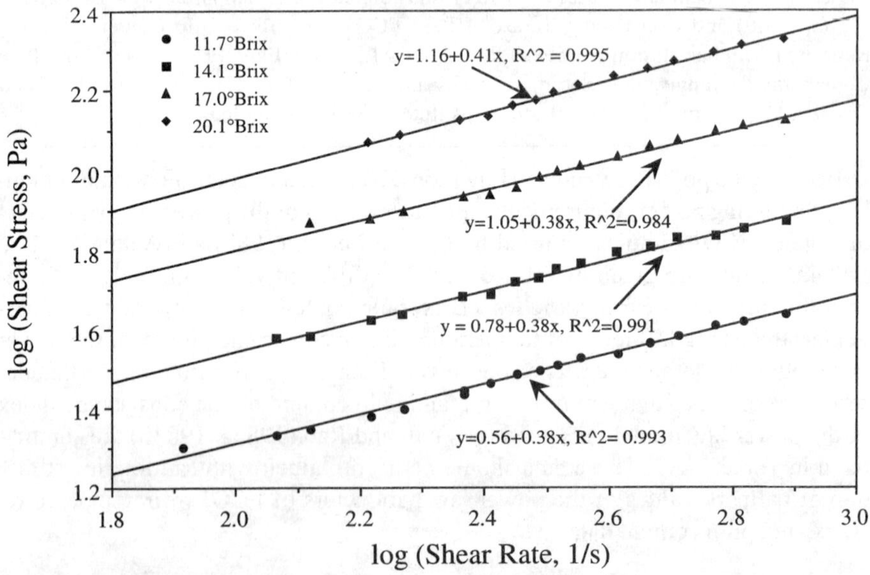

Figure 5–10 Influence of Soluble Solids on Shear Rate versus Shear Stress Data of Tomato Concentrates Using the Same Finisher Screen. *Source:* Reprinted with permission from T. Tanglertpaibul and M.A. Rao, Rheological Properties of Tomato Concentrates as Affected by Particle Size and Methods of Concentration, *Journal of Food Science*, Vol. 52, pp. 141–145, © 1987, Institute of Food Technologists.

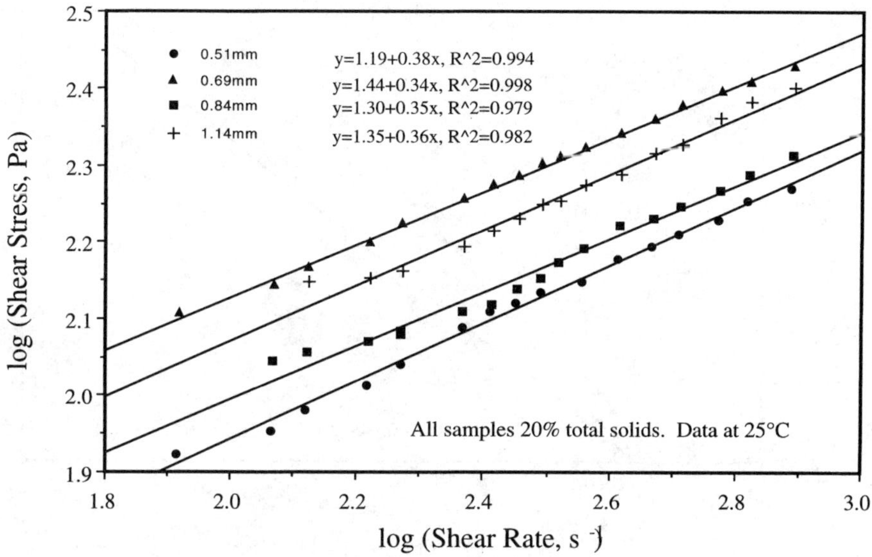

Figure 5–11 Shear Rate versus Shear Stress Data Obtained on 20% Solids Concentrates Made from Juices That Were Produced Using Different Finisher Screen Openings. Using a 0.69 mm screen resulted in concentrates with the highest viscosity.

---

of finisher screen size on the pulp content and size distribution should be further explored.

The role of pulp content on flow behavior was examined by adding tomato pulp to concentrates with different soluble solids: 5.6, 10.0, 15.0, and 20.0 °Brix. A plot of the pulp content versus the apparent viscosity at 100 s^{-1} (Figure 5–12) is shown where the linear relationship is valid only over the narrow range of pulp contents. Because the influence of pulp content was approximated and the apparent viscosity at 100 s^{-1}, a meaningless negative intercept was obtained. More importantly, the data in Figure 5–11 reflect the important role of solids loading in rheological behavior.

Rao et al. (1981) studied the influence of temperature and total solids content (c) on the apparent viscosity at 100 s^{-1} ($\eta_{a,100}$) of tomato concentrates prepared from the varieties grown at the New York State Agricultural Experiment Station: Nova, New Yorker, #475, #934 hot break process, and #934 cold break process. The results of Nova hot break process concentrates were expressed by a model slightly different than that used to examine concentrated apple and grape juices.

$$\eta_{a,100} = \alpha \exp(E_a/RT)(c^{\beta}) \tag{5.13}$$

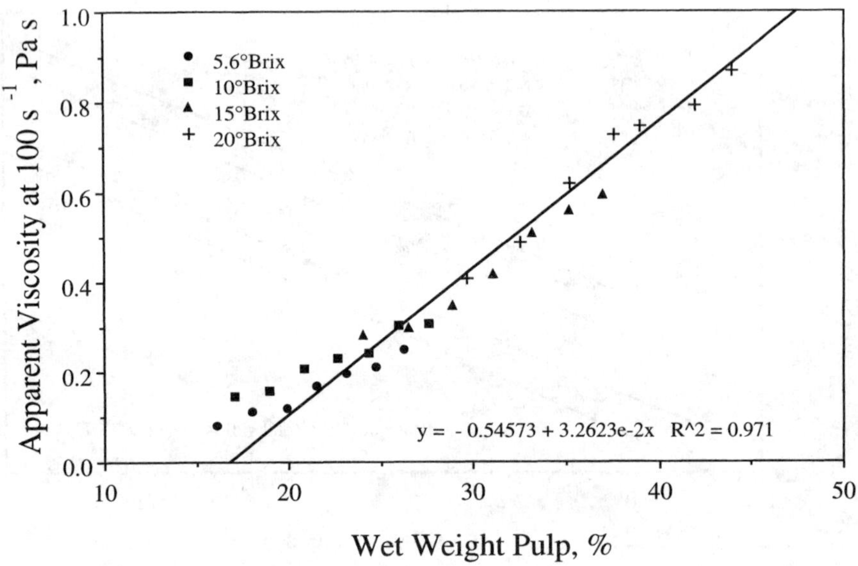

Figure 5–12 Plot of Pulp Content versus Apparent Viscosity at 100 s^{-1} of Tomato Concentrates Illustrates the Important Role of Pulp Content. The meaningless negative intercept indicates that the linear relationship cannot be extended to low pulp contents.

The results for Nova concentrates were $\alpha = 1.3 \times 10^{-5}$, $E_a = 9.21$ kJ mol^{-1}, and $\beta = 2.6$. It is very encouraging that these values were close to those reported by Harper and El-Sahrigi (1965) for the tomatoes of variety VF6: $\alpha = 1.05 \times 10^{-5}$, $E_a = 9.63$ kJ mol^{-1}, and $\beta = 2.0$. Magnitudes of β for the concentrates made from New Yorker hot break process, #475 hot break process, #934 hot break process, and #934 cold break process were 2.5, 2.4, 2.5, and 2.4, respectively, while that of Nova tomato concentrates was slightly higher, ~2.6 (Figures 5–13 and 5–14). Thus a useful general result is that the viscosity of tomato concentrates can be scaled by the factor (total solids)$^{2.5}$. For example, at a fixed temperature, one can predict the viscosity η_2 of a tomato concentrate having the concentration c_2 from the magnitude η_1 of a concentrate of the same variety having the concentration c_1 from the relationship:

$$\eta_2 = \eta_1 (c_2/c_1)^{2.5} \tag{5.14}$$

Rheological Properties of Tomato Pastes

Rao and Cooley (1992) examined the flow properties of tomato pastes and their serums, as well as the dynamic rheological properties of the pastes. From the

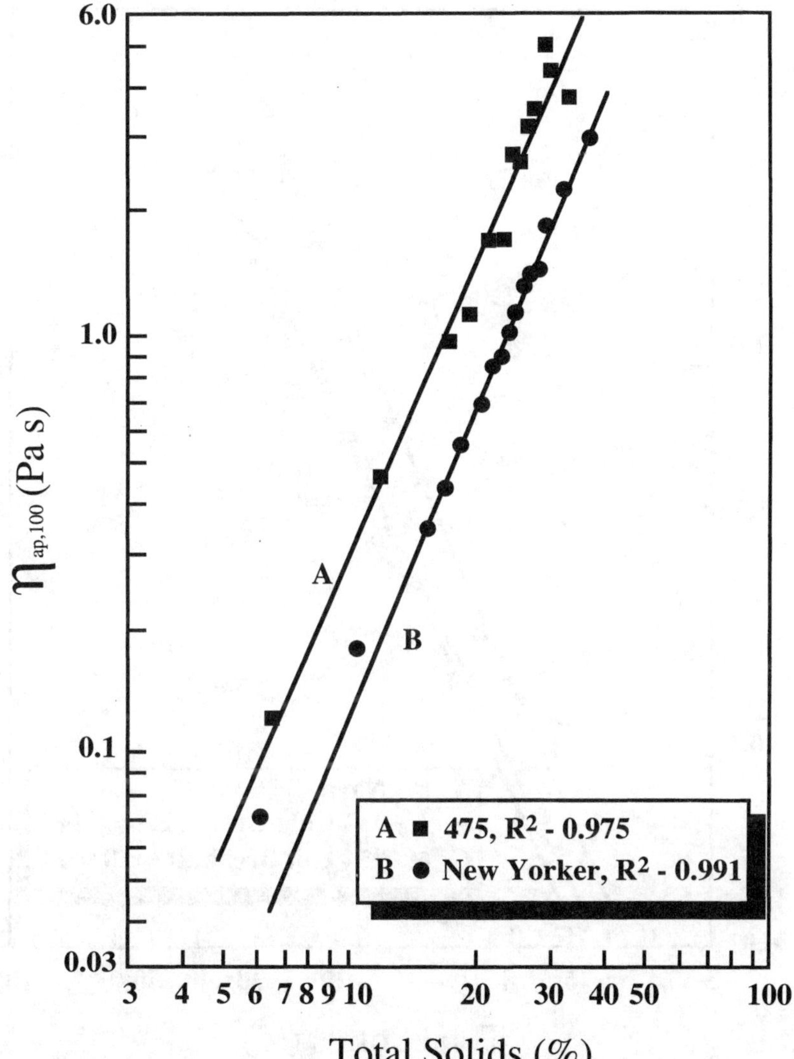

Figure 5–13 Log-Log Plot of the Apparent Viscosity at 100 s^{-1} versus Total Solids of Tomato Concentrates from #475 and New Yorker Tomatoes Using Hot-Break Juice (slopes of lines = 2.4 and 2.5, respectively). *Source:* Reprinted with permission from T. Tanglertpaibul and M.A. Rao, Rheological Properties of Tomato Concentrates as Affected by Particle Size and Methods of Concentration, *Journal of Food Science*, Vol. 52, pp. 141–145, © 1987, Institute of Food Technologists.

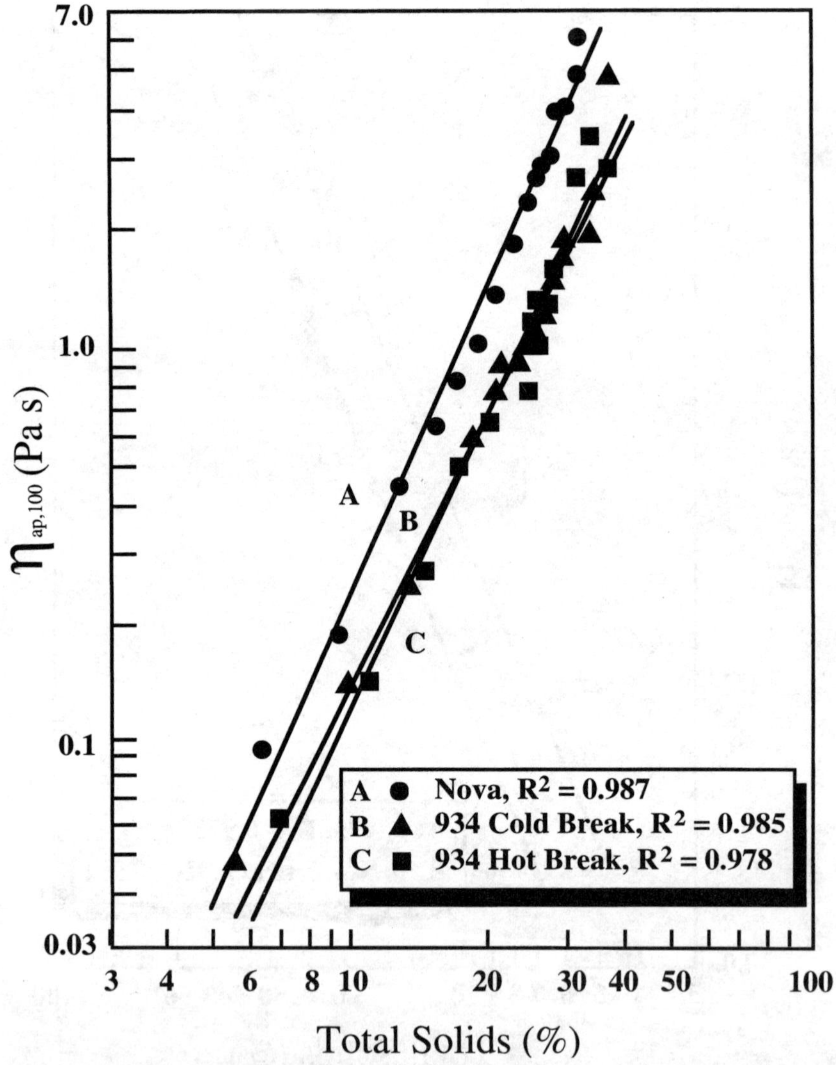

Figure 5–14 Log-Log Plot of Apparent Viscosity at 100 s^{-1} versus Total Solids of Tomato Concentrates from Hot-Break Nova Tomato Concentrates and #934 Hot-Break and Cold-Break Process (slope of lines = 2.7, 2.4, and 2.5, respectively). *Source:* Reprinted with permission from M.A. Rao, M.C. Bourne, and H.J. Cooley, Flow Properties of Tomato Concentrates, *Journal of Texture Studies*, Vol. 12, pp. 521–538, © 1981, Food and Nutrition Press.

steady shear rheological data, obtained with a concentric cylinder viscometer, given in Table 5–A3 it can be seen that the tomato paste (TP) samples had high magnitudes of yield stress. Further, they were highly shear-thinning in nature: The range of magnitudes of the flow behavior index (n) was 0.13–0.40; the average value was 0.28 with $\sigma = 0.062$. Magnitudes of Casson yield stresses were in the range 78–212 Pa. The tomato paste serum samples were low viscosity Newtonian fluids; their viscosities ranged from 4.3–140 mPa s.

Dynamic Shear Data

Plots of log ω versus log G' and log G'', such as that shown in Figure 5–15, showed that G' was higher than G'' at all values of ω employed; in addition, the decrease in η^* confirmed the shear-thinning nature of the TP samples. Further, linear regression of ln ω versus ln G' and n G'' data showed that the resulting straight lines had small magnitudes of slopes (0.075 to 0.210 for G' and 0.110 to 0.218 for G'').

$$G' = K' (\omega)^{n'} \tag{5.15}$$

$$G'' = K'' (\omega)^{n''} \tag{5.16}$$

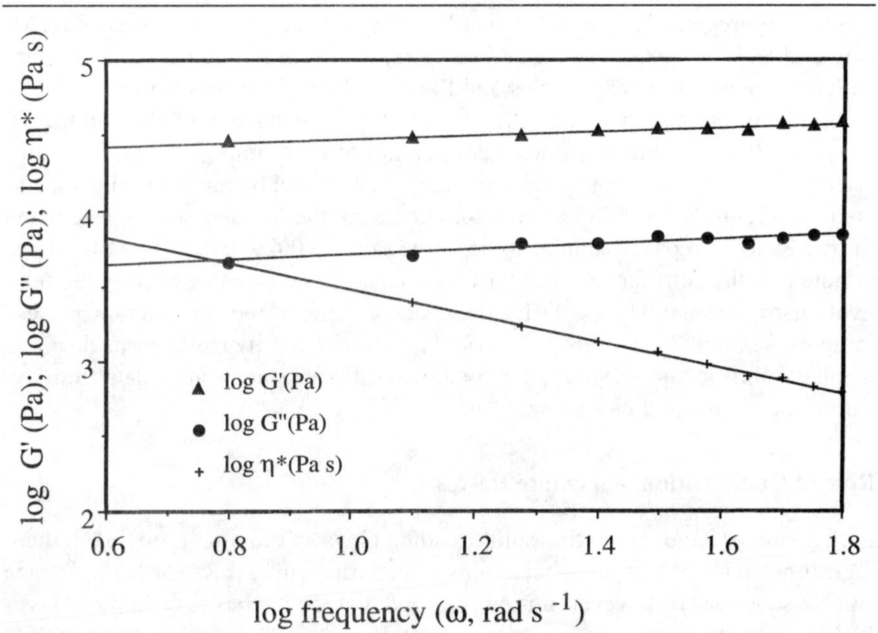

Figure 5–15 Plots of Log ω versus Log G', Log G'', and Log η^* of Tomato Pastes. *Source:* Reprinted with permission from M.A. Rao, and H.J. Cooley, Flow Properties of Tomato Concentrates, *Journal of Texture Studies*, Vol. 23, pp. 415–425, © 1981, Food and Nutrition Press.

From a structural point of view, it is known that for true gels, log ω versus log G' or log G'' plots have nearly zero slopes (Ross-Murphy, 1984), while for weak gels and highly concentrated solutions such plots have positive slopes and G' is higher than G'' over large ranges of ω. In addition, as discussed in detail below, magnitudes of η^* were greater than η_a for all magnitudes of shear rates and oscillatory frequencies. This means that TP samples behaved more like true gels than weak gels.

Applicability of Cox-Merz Rule

With automated rheometers it is relatively easier to obtain dynamic shear data than steady shear data. For this reason, the interrelationship between ω and η^* on one hand and $\dot{\gamma}$ and η_a on the other is of interest. The Cox-Merz rule, that is, equal magnitudes of η_a and η^* at equal values of $\dot{\gamma}$ and ω, respectively (Equation 5.17), was obeyed by several synthetic and biopolymer dispersions (Lopes da Silva et al., 1993).

$$\eta^* (\omega) = \eta\dot{\gamma}\big|_{\omega=\dot{\gamma}} \tag{5.17}$$

However, it is not obeyed by biopolymer dispersions with either hyperentanglements or aggregates (Lopes da Silva et al., 1993) and complex food systems (Bistany and Kokini, 1983), such as TP samples, that have strong network structure with interspersed tomato particles and fibers. Indeed, there was considerable deviation between the log $\dot{\gamma}$ versus log η_a and log ω versus log η^* data on the TP samples. However, the two lines were parallel to each other. Further, by multiplying ω by a shift factor (α), the two sets of data could be made to follow a single line (Figure 5–16). Shift factors for correcting the frequency were calculated for the 25 tomato paste samples to be in the range 0.0029–0.029. It is of interest to note that the shift factors for 18 of the tomato paste samples were in the relatively narrow range 0.0041–0.012; the average value of the shift factor for these samples was 0.0074 with $\sigma = 0.0019$. Therefore, from dynamic shear data and tabulated shift factors, it would be possible to estimate steady shear data from dynamic shear data, and vice versa.

Role of Composition of Tomato Pastes

Earlier studies have shown that solids loading plays an important role in the rheological behavior of tomato concentrates (Tanglertpaibul and Rao, 1987b; Takada and Nelson, 1983). However, it is emphasized that tomato pastes cannot be classified as distinct suspensions in which the solids are suspended in continuous fluid media. The difference in the structures of tomato concentrates and pastes is very well indicated by the magnitudes of centrifugal forces required to separate the solids (dispersed phase) from the serum (continuous phase). For the tomato con-

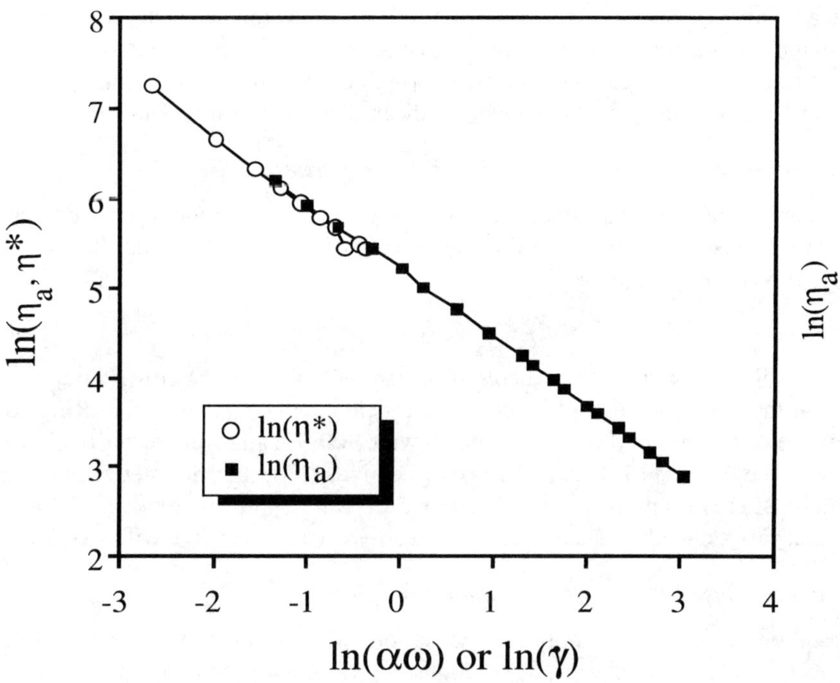

Figure 5–16 Modified Cox-Merz Plot for Tomato Concentrates. *Source:* Reprinted with permission from M.A. Rao and H.J. Cooley, Rheology of Tomato Pastes in Steady and Dynamic Shear, *Journal of Texture Studies*, Vol. 23, pp. 415–425, © 1992, Food and Nutrition Press.

---

centrates, magnitudes of centrifugal force <30,000 × *g* were sufficient to separate the solids and the serum (Takada and Nelson, 1983; Tanglertpaibul and Rao, 1987b). In contrast, a centrifugal force of 100,000 × *g* was necessary to separate the two phases of TP samples. Therefore, the classical relative viscosity-volume fraction relationships used for correlating data on suspensions would not be appropriate for TP samples. Indeed, attempts at correlating relative viscosity, based on apparent viscosity of tomato pastes at 10 s^{-1}, with volume fraction of solids alone were not successful in that very low values of R^2 were obtained. It is likely that the shape of the particles should be taken into consideration (Yoo and Rao, 1994).

The rheological properties of the complex network structure of tomato pastes may be assumed to be made up of two contributions: (1) one network structure contributed by the solids phase, in proportion to ϕ_s, and (2) another network structure contributed by the liquid (continuous phase), in proportion to $\phi_l = 1 - \phi_s$. However, the effective continuous phase is not the low-viscosity serum itself but a highly viscous liquid that is an integral part of the tomato paste. Also implicit is that the solids fraction plays a major role in the structure of the TP samples; that is, it can be considered to be the structuring component. These assumptions are also in line with the gel-like behavior indicated by the dynamic shear data.

Structure-Based Models for Steady Shear Properties

For correlating the consistency index (K) of the power law model and the Casson yield stress (σ_{0c}), again the lower bound form of Takayanagi model (Ross-Murphy, 1984) was satisfactory:

$$(1/\text{SSP}) = [(\phi_s)/A] + [(1 - \phi_s)/B] \tag{5.18}$$

where SSP is a steady shear rheological parameter, either the consistency index (K) of the power law model or the Casson yield stress (σ_{0c}), and A and B are constants. For K, magnitudes of A, B, and R^2 were 900, 72, and 1.00, respectively, and for σ_{0c} they were 135, 112, and 1.00, respectively. For the consistency index, 21 (out of 25) values predicted by Equation 5.18 were within ± 20% of the experimental results, while for the Casson yield stress, 19 values were within ± 25%.

Structure-Based Models for Dynamic Properties

Based on the above assumptions, correlations were developed first for the intercepts (K', K'') and slopes (n', n'') of log ω–log G' and log ω–log G'' data. The correlation form that was successful was again the lower bound form of Takayanagi model (Ross-Murphy, 1984):

$$(1/\text{DRP}) = [(\phi_S)/A] + [(1 - \phi_S)/B] \tag{5.19}$$

where, DRP is a dynamic rheological property, and A and B are constants. Magnitudes of the constants and R^2 were determined by nonlinear regression (GENSTAT, Numerical Algorithms Group, Oxford, England). For K', values of A, B, and R^2 were 9540, 4807, and 1.00, respectively, and for K'' they were 1552, 6197, and 1.00, respectively. The mean value of n' was 0.177 and that of n'' was 0.152. Agreement between predicted K' and K'' values was not as good as for the steady shear properties: 17 predicted K' and 19 K'' values were within ± 30%.

Because of the wide variability in tomato pastes due to different cultivars and processing methods, the constants in the structure-based correlations would not be universally applicable. However, the correlations can be used to obtain order of magnitude estimates of the rheological properties of tomato pastes without extensive and expensive experimentation. The various constants are valid within the ranges of the variables from which they were derived. More importantly, it is hoped that studies will be conducted to make use of the concepts presented here with respect to the gel-like nature of tomato pastes, shift factors for Cox-Merz rule, and structure-based (ϕ_s and $\phi_1 = 1 - \phi_s$) relationships for correlating the rheological parameters.

Creep-Compliance Data on Tomato Concentrates

Creep compliance, $J(t)$, data were obtained on tomato concentrates made from Pito #696 hot-break juice that was produced using two finisher screens having holes of diameter 0.69 mm (0.027 in.) and 0.84 mm (0.033 in.) (Yoo and Rao, 1996); the finisher was operated at 1000 rpm. $J(t)$ at any time was described by a six-element model made up of an instantaneous compliance, two Kelvin-Voigt bodies, and Newtonian compliance (Sherman, 1970):

$$J(t) = J_0 + J_1(1 - e^{-t/\tau_1}) + J_2\,(1 - e^{-t/\tau_2}) + t/\eta \tag{5.20}$$

where $J_0 = 1/G_0$ is the instantaneous elastic compliance; $J_1 = 1/G_1$ and $J_2 = 1/G_2$ are the retarded elastic compliances; τ_1 and τ_2 are retardation times; η is viscosity; and G_0, G_1, and G_2 are moduli (Sherman, 1970; Rao, 1992).

The auto mode of analysis of the experimental $J(t) - t$ data provided magnitudes of model parameters that followed the experimental data better than those obtained in the manual mode of data analysis. The latter were very sensitive to estimates of instantaneous modulus, and in some instances, there was considerable difference between experimental data and model predictions (not shown here). Therefore, the results of auto mode data analysis are presented in Table 5–A5; the standard deviations of the parameters are also in Table 5–A5 in parentheses. The standard deviation of J_1 was higher than standard deviations of the other parameters, reflecting greater uncertainty in its estimation.

The effect of the two finisher screen sizes on the $J(t) - t$ data was examined at 21 °Brix and the concentrate from the 0.84 mm screen was more viscoelastic than that from the 0.69 mm screen. Because the particle size distributions of the two concentrates were similar, it appears that the higher pulp content, 38.6% in the former as opposed to 32.5% in the latter, is the primary reason for the difference in the magnitudes of the rheological parameters.

As expected, magnitudes of instantaneous elastic modulus, $G_0 = (1/J_0)$, and of storage modulus G' at 1 rad s^{-1} (Yoo and Rao, 1996) were nearly equal. Also, as noted by Giboreau et al. (1994) for a gel-like modified starch paste, the two parameters were of the same order of magnitude. Further, as can be inferred from data in Table 5–A5, the other moduli from creep-compliance data ($G_1 = 1/J_1$, $G_2 = 1/J_2$) and the Newtonian viscosity increased with increase in °Brix.

Model for Apparent Viscosity of Plant Food Dispersions

Because no general theories exist even for concentrated nonfood suspensions of well-defined spherical particles (Jeffrey and Acrivos, 1976; Metzner, 1985), approaches to studying the influence of the viscosity of the continuous medium (serum) and the pulp content of PF dispersions, just as for nonfood suspensions, have been empirical. In PF dispersions, the two media can be separated by centrifugation and their characteristics studied separately (Mizrahi and Berk, 1970).

One model that was proposed for relating the apparent viscosity of food suspensions is (Rao, 1987)

$$\eta_{a,\,PFD} = \eta_{a,\,serum} + A\,(\text{pulp})^{B} \tag{5.21}$$

where $\eta_{a,\,PFD}$ is the apparent viscosity of suspension, $\eta_{a,\,serum}$ is the apparent viscosity of serum, and pulp refers to the pulp content. The model is simple and predicts the viscosity of the serum when the pulp content is zero. The coefficient A indicates the contribution of a unit amount of pulp content, and the coefficient B is the slope of a plot of log pulp versus log $\eta_{specific}$ of the dispersion and reflects the relative intensity of the effect of pulp content in different food suspensions or the effect of different processes and/or cultivars that could influence the pulp content. However, innovative and broadly applicable models that describe the role of pulp content are needed for PF dispersions.

Yield Stress of Apple Sauce

Magnitudes of the yield stresses from the vane method (Table 5–A6) were much higher than those from the flow models (not shown); in a few instances, the former were more than twice the latter (Qiu and Rao, 1988). In an attempt to understand the difference between the two types of yield stresses, the model presented by Michaels and Bolger (1962) may be useful. Because the underlying assumptions of the model have been reviewed by others (Metz et al., 1979; Durán and Costell, 1982), only the pertinent aspects of the model will be covered here. In the model, the stress necessary to produce deformation at a constant rate is divided into three parts:

$$\sigma = \sigma_n + \sigma_{cr} + \sigma_{vi} \tag{5.22}$$

where, σ_n is the stress required to break the aggregate network, σ_{cr} is the stress required to break the structural bonds, and σ_{vi} is the stress required to overcome the purely viscous drag. At zero shear rate the network yield stress σ_n equals σ_0, the yield stress obtained by extrapolation of the shear rate–shear stress data to zero shear rate. Another yield stress, the Bingham yield stress (σ_B) can be obtained by extrapolation of only the linear portion of the shear rate versus shear stress diagram to zero shear rate (Michaels and Bolger, 1962). The Bingham yield stress is part of the total stress at high shear rates. One contribution of the model of Michaels and Bolger (1962) is that σ_0 is proportional to the volume fraction of solids raised to the power of three and that σ_B is proportional to the square of the volume fraction of solids. Also, it is known that the magnitude of σ_B is higher than that of σ_0. In addition to the two yield stresses, the viscosity at infinite shear rate (η_∞) also has been shown to be important.

For apple sauce (Qiu and Rao, 1988), magnitudes of σ_B were determined from plots of shear rate–shear stress obtained as the intercept as shown in Figure 5–17. In comparison to the magnitudes of the yield stresses obtained by extrapolation of shear rate–shear stress data to zero shear rates, the magnitudes of σ_B were much higher. Further, while they are close, they are still higher than the magnitudes of yield stresses determined by the vane method (Table 5–A6). Nevertheless, considering the uncertainties involved in obtaining σ_B and the range of values that can be obtained with the vane method, it appears that the vane method gives magnitudes that are very close to those of σ_B.

One result of the work of Michaels and Bolger (1962) is the prediction that σ_0 is proportional to the volume fraction of solids raised to the power of three and that σ_B is proportional to the square of the volume fraction of solids. They confirmed their predictions with data on dispersions that had magnitudes of σ_B less than 0.9 Pa. The apple sauce samples of Qiu and Rao (1988) were highly concentrated dispersions with some magnitudes of σ_B of about 80 Pa. In addition, the pulp content determined by centrifugation at 360g had a relatively narrow range of magnitudes.

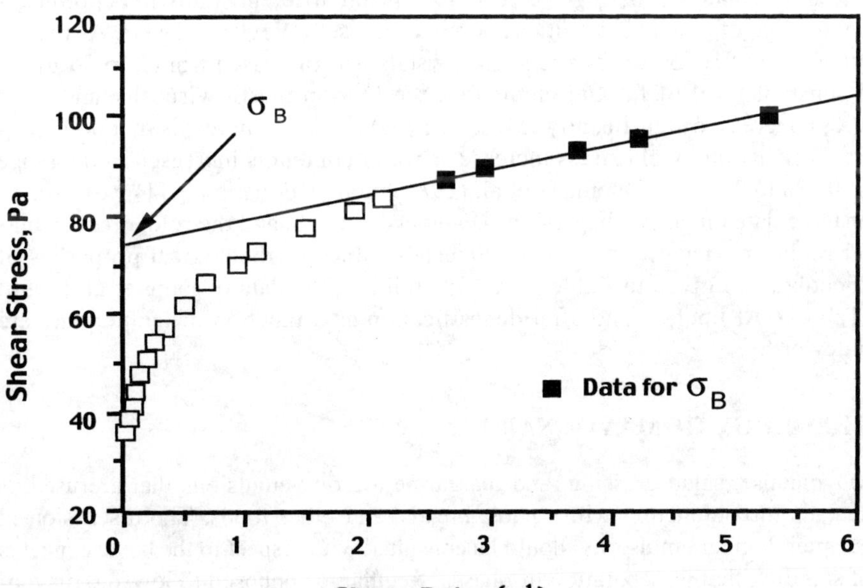

Figure 5–17 Magnitudes of Bingham Yield Stress (σ_B) Determined from Plots of Shear Rate–Shear Stress Data Obtained for Apple Sauce. *Source:* Reprinted with permission from M.A. Rao and H.J. Cooley, Rheology of Tomato Pastes in Steady and Dynamic Shear, *Journal of Texture Studies*, Vol. 23, pp. 415–425, © 1992, Food and Nutrition Press.

For this reason, verification of the Michaels-Bolger model was shown with data on the samples in which the amount of pulp was adjusted.

RHEOLOGICAL PROPERTIES OF CHOCOLATE

The basic ingredients of chocolate are crystalline sugar, cocoa solids, a liquid phase, and an emulsifier, lecithin (Hartel, 1998). Most chocolate coatings contain less than 1% water. In melted chocolate, the solid particles are dispersed throughout the lipid carrier, and the rheological properties depend on the interactions between the solid particles and the continuous lipid phase. The lecithin coats the solid particles thus affecting the Casson yield stress and plastic viscosity. The polar group of lecithin interacts at the solid surfaces, particularly the sucrose crystals, where most of the water is located, and provides a bridge between the hydrophilic surface and the lipid phase. Thus, lecithin allows the movement of solids under shear (Hartel, 1998). Increasing the amount of lecithin results in an initial decrease in yield stress but a significant increase subsequently (Chevally, 1975). Addition of small amounts of cocoa butter reduces chocolate viscosity (Chevally, 1975), and different emulsifiers produce a wide range of plastic viscosities and yield stresses (Weyland, 1994). As the average size of cocoa solids and sugar crystals was increased from 2 to 50 μm at a constant level of fat and emulsifier, the Casson plastic viscosity and yield stress decreased significantly (Musser, 1973). Also of interest is that both plastic viscosity and yield stress increase as water content is increased in the range of 0.1% to 2%. The Quemada et al. (1985) model (Equation 2.11) was used to analyze data on cocoa dispersions (Fang et al., 1996) and the role of cocoa butter replacers (Fang et al., 1997). Selected values of rheological properties of chocolate are given in Table 5–A7. In addition, the data of Fang et al. (1996) (Table 5–A8) before and after degasification as a function of temperature are noteworthy.

RHEOLOGY OF MAYONNAISE

Mayonnaise, salad dressing, and margarine are o/w emulsions that are used as spreads in conjunction with salads, breads, and other foods. The discussion in Chapter 2 under emulsions should be consulted with respect to the basic concepts of stability, that is, separation of phases. Regular, as opposed to low-fat, mayonnaise has a dispersed phase (oil) concentration of about 70–80%. Often the emulsifying agent in these foods is egg yolk. The roles of oil, water, egg yolk, thickening agents (usually gums), salt, and sugar have been studied by food technologists and summarized in recent publications (Harrison and Cunningham, 1985, 1986).

The physical changes in the dispersed droplets occur through the processes of creaming, flocculation, and coalescence. Creaming denotes the movement of oil droplets under the action of gravity; flocculation denotes the clustering of droplets, and coalescence means the spontaneous joining of small droplets into larger ones. Unless phase separation occurs, creaming and flocculation in o/w emulsions may not lead to detectable changes in the appearance and rheological behavior. However, coalescence of droplets, which is associated with the free fat from o/w emulsions, is generally unacceptable at all levels (Dickinson and Stainsby, 1982, 1987).

Small droplet size is created by high-pressure homogenization, and the premature coalescence of the droplets is minimized by adsorption of an emulsifier at the interface. Generally egg yolk is used as the emulsifier, that is, to lower the o/w interfacial tension. Gums such as CMC, propylene glycol, guar gum, and xanthan gum are used as emulsion stabilizers in mayonnaise and salad dressings. Therefore, the time required to hydrate gums is an important factor in the stability of emulsions.

Rheological properties of mayonnaise have been studied using different rheological techniques: steady shear rate–shear stress, time-dependent shear rate–shear stress, stress growth and decay at a constant shear rate, dynamic viscoelastic behavior, and creep-compliance viscoelastic behavior. More studies have been devoted to the study of rheological properties of mayonnaise than of salad dressings, probably because the former is a more stable emulsion and exhibits complex viscous and viscoelastic rheological behavior.

Because of the existence of yield stress as well as time-dependent rheological behavior of mayonnaise, it would seem reasonable to expect that traditional relationships between steady shear properties on one hand and small amplitude dynamic properties on the other that were found for polymeric liquids will not hold for mayonnaise. Bistany and Kokini (1983) showed that the Cox-Merz rule and other relationships at low shear rates and frequencies did not hold for mayonnaise and other semisolid foods. Further, one must also be careful when using dynamic tests on emulsions, such as mayonnaise, in view of the potential for aiding coalescence of the oil droplets and consequently altering the structure of the sample.

Stress Overshoot Data of Mayonnaise

Several studies were conducted on the stress overshoot and/or decay at a constant shear rate. Kokini and Dickie (1981) obtained stress growth and decay data on mayonnaise and other foods with a Rheometrics Mechanical Spectrometer at 0.1, 1.0, 10.0, and 100 s^{-1}. As expected from studies on polymers, shear stresses for mayonnaise and other food materials displayed increasing degrees of overshoot with increasing shear rates. The Bird-Leider empirical equation was used to model the transient shear stresses.

$$\sigma = -K\dot{\gamma}^n \left[1 + (b\dot{\gamma}t - 1) \exp(-t/an\lambda)\right] \tag{5.23}$$

where $\dot{\gamma}$ is the shear rate, t is time, λ is time constant, a and b are constants, and K and n are power law constants. At very small times for a constant shear rate, the model reduces to

$$\sigma = \text{constant } (b\,\dot{\gamma}t) \tag{5.24}$$

Equation 5.23 simulates the elastic response or sudden overshoot when a constant shear rate is applied. The peak shear stress is reached at a time t_{max} given by

$$t_{max} = an\lambda + \frac{1}{b\dot{\gamma}} \tag{5.25}$$

The magnitude of σ_{max} can be calculated by substituting the above expression for t_{max} in Equation 5.23. After σ_{max} is reached, the term $\exp -(1/an\lambda)$ simulates the exponential decay; the product $an\lambda$ as a whole controls the exponential decay.

Kokini and Dickie (1981) found that the Bird-Leider equation provided moderately good predictions for peak shear stresses (σ_{max}) and the corresponding times (t_{max}), but the prediction of shear stress decay was poor. They suggested that a series of relaxation times, as opposed to a single exponential term, would be needed for mayonnaise and other materials.

Campanella and Peleg (1987) presented stress growth and decay data on mayonnaise at shear rates of 1.8, 5.4, 9.9, and 14.4 s^{-1} with a Haake Rotovisco RV3 concentric cylinder viscometer system. They modeled the data by a three-constant model that was a modification of Larson's (1985) model that was successfully employed for polyethylene melts with a wide distribution of molecular weights. The model employed by Campanella and Peleg was

$$\frac{\sigma(t)}{\sigma_\infty} = \frac{\Gamma(m_1)\sin m_1\pi}{\pi(m_2\dot{\gamma})^{m_1-1}} \left(\frac{t^{1-m_1}\, e^{-m_2\dot{\gamma}t}}{m_1} + \int_0^t s^{-m_1}\, e^{-m_2\dot{\gamma}s}\, ds \right) \tag{5.26}$$

where σ_∞ is the steady state stress, Γ is the gamma function, m_1 and m_2 are constants, $\dot{\gamma}$ is the shear rate, $s = t - t'$, and the deformation history of the experiment is described by

$$\gamma = \left| \begin{array}{ll} \dot{\gamma}t & -\infty < t' < 0 \\ \dot{\gamma}(t - t') & 0 < t' < t \end{array} \right. \tag{5.27}$$

The constant m_1 has a value $0 < m_1 < 1$ and it is representative of the relaxation modulus, when it is expressed as

$$G(t - t') = C(t - t')^{-m_1} \tag{5.28}$$

The parameter m_2 is obtained from the time at which the maximum shear stress is reached, t_{max}:

$$m_2 = \frac{1}{\dot{\gamma} t_{max}} \tag{5.29}$$

The above relationship implies that t_{max} is inversely proportional to the imposed shear rate, and this was found to be true for the experimental data. From experimental data, the values of $m_1 = 0.3$ and $m_2 = 0.07$ were determined. Even though the damping constant (m_2) was assumed to be shear rate dependent, a single representative value of m_2 was found to be satisfactory because m_2 was a weak function of shear rate.

Figoni and Shoemaker (1983) also found that a single first-order function did not adequately describe the stress decay data of a mayonnaise sample. They suggested that two independent first-order processes, each with a different rate constant but acting simultaneously, are major contributors to the stress decay data on mayonnaise. The first process was attributed to the reversible flocculation-deflocculation process of oil droplets that occurs during the steady shear process, while the second term describes the changes occurring in the continuous phase. The possibility that the two first-order processes were applicable to the flocculation and deflocculation of aggregates of two size groups of oil droplets was also suggested.

Creep-Compliance Behavior of Mayonnaise

Creep-compliance behavior of mayonnaise and other emulsions has been investigated by Sherman and coworkers. The chief advantage of the creep-compliance test is that data can be obtained with minimal disturbance to the test samples and under linear-viscoelastic regime. Nearly all, if not all, of the previous studies were conducted with commercial (supermarket) samples whose compositions were not determined. Therefore, the role of composition on the rheological parameters could not be determined. In contrast, Kiosseoglou and Sherman (1983b) employed the formulations shown in Table 5–A9 to study the effect of egg white on the rheological properties of mayonnaise. In addition, the effects of pH, salt, and sugar on rheological properties were studied.

The creep-compliance data on the mayonnaises were modeled by means of the six-parameter model:

$$J(t) = J_0 + J_1\,[1 - \exp(-t/\tau_1)] + J_2\,[1 - \exp(-t/\tau_2)] + \frac{t}{\eta_N} \tag{5.30}$$

The creep-compliance parameters can be used to calculate the moduli $E_i = 1/J_i$ and viscosities $\eta_i = E_i \tau_i$. In the following, the magnitudes of the viscoelastic parameters, that is, the moduli and the viscosities, will be discussed.

The viscoelastic parameters reached maximum values within one hour after preparation of the emulsions, and then they decreased continuously as the emulsions were aged. The E_0 values (Pa) of freshly prepared mayonnaise increased as the pH decreased: pH 6.2, E_0 750; pH 4.5, E_0 885; pH 3.9, E_0 1220; and pH 3.3, E_0 1351. There was a substantial decrease in the viscoelasticity parameters as the emulsions were aged at room temperature at all pHs except pH 6.2.

At a constant pH 3.9, salt increased the initial viscoelasticity, the effect being greater at the 0.25% (w/w) level than at the 1.25% level. However, the rate of decrease in E_0 was lower at 1.25% NaCl than at 0.25%. Sugar had much smaller effect than salt on the viscoelasticity of mayonnaise. Replacing egg yolk with egg white, while keeping the total solids content constant, increased E_0. When the egg yolk content was kept constant and additional egg white was incorporated, all the viscoelastic parameters increased substantially. With an egg yolk/egg white ratio of 0.5, the viscoelasticity parameters decreased least when the mayonnaises were aged.

The addition of salt to the egg yolk solution during the preparation of the solutions disrupts the granules and provides additional surface active material for adsorption, such as α- and β-lipovitellins and phosvitin. In this regard, Chang et al. (1972) observed with an electron microscope a continuous speckled layer around oil drops and also a fibrous membrane on the surface of drops after they were washed with water. The electron-dense particles in the speckled layer were interpreted to be egg yolk lipoproteins and the fibrous material to be livetin-phosvitin complex.

The low viscoelasticity of mayonnaise in the presence of sugar was considered to be due to the shielding effect exerted by sugar molecules on the protein groups that were involved in interaction and network formation between the oil drops. The higher viscoelasticity of mayonnaise containing egg white was thought to be due to strengthening of the interlinked network structure by interaction with the egg yolk components of the interfacial membrane around the oil drops.

RHEOLOGY OF SALAD DRESSINGS

Rheological properties of salad dressings also were studied using different techniques: steady shear rate–shear stress, stress growth and decay at a constant shear rate, dynamic viscoelastic behavior, and creep-compliance viscoelastic behavior.

Tanaka and Fukuda (1976) studied the steady shear rate–shear stress behavior, thixotropic behavior, and dynamic viscoelastic behavior of salad dressing prepared with different stabilizing agents (0.6%, w/w): guar gum, tragacanth gum, pectin, starch paste, tamarind seed extracts, and xanthan gum. The power law flow behavior index ranged from 0.20 when xanthan gum was used to 0.68 when corn starch was added. A thixotropy index (TI) was calculated as the sum of the difference between the magnitudes of the shear stress in increasing (S_{ui}) and decreasing (S_{di}) of shear rates:

$$\mathrm{TI} = \sum_{i=1}^{4} (S_{ui} - S_{di}) \tag{5.31}$$

where the magnitudes of S_{ui} and S_{di} were read off at four unspecified shear rates. Magnitudes of TI, of apparent viscosity at a shear rate of 0.1 s^{-1}, of dynamic elasticity (G'), and of dynamic viscosity (η') were highest for salad dressings with xanthan gum and lowest for dressings with corn starch.

Elliott and Ganz (1977) noted that steady shear and dynamic rheological properties of commercial salad dressings covered a wide range of magnitudes. Because the products were being sold successfully on the commercial market, they suggested that a strict control of rheological properties is not necessary. A modified Bingham body, shown in Figure 3–34, was used to account for the observed response to dynamic tests of the samples that exhibited yield stresses. Stress growth and decay at a constant shear rate was analyzed as the excess work (W) of structure breakdown, discussed in more detail in Chapter 3:

$$W = \dot{\gamma} \int_{A}^{B} (\sigma - \sigma_{\infty}) dt \tag{5.32}$$

where $\dot{\gamma}$ is the constant shear rate, σ is the time-dependent stress, σ_{∞} is the equilibrium shear stress value, and A and B are the times at which the stress curve overshot the equilibrium stress value and the time at which the equilibrium value was reached, respectively. The slopes of log steady shear apparent viscosity versus log apparent viscosity of the salad dressings were between -0.72 and -0.86.

Paredes et al. (1988) found commercial bottled salad dressings to be more viscous than dressings prepared from dry mixes. The Arrhenius activation energy of flow for the commercial dressings ranged between 10.5 and 20.5 kJ/mole. Thixotropic behavior at a constant shear rate was described better by the Weltman equation (Equation 5.33) than the equation of Hahn et al. (Equation 5.34).

$$\sigma = A_1 - B_1 \log t \tag{5.33}$$

$$\log(\sigma - \sigma_e) = A_2 - B_2 t \tag{5.34}$$

where σ is the shear stress, t is time in seconds, σ_e is the equilibrium shear stress, and A_1, A_2, B_1, and B_2 are constants.

Commercial bottled salad dressings showed higher viscoelasticity than dressings prepared from dry mixes, and bottled reduced calorie dressings exhibited intermediate viscoelasticity (Paredes et al., 1989). A four-parameter model described the creep-compliance behavior of the commercial salad dressings and the effect of storage on a prepared salad dressing containing xanthan gum:

$$J(t) = J_0 + J_1 [1 - \exp(-t/\tau_1)] + t/\eta_N \tag{5.35}$$

The viscoelastic parameters of the model salad dressing (Table 5–A10) decreased with storage time until 21 days and then increased from the 28th to the 42nd day. Specifically, the dramatic changes in the magnitudes of the viscoelastic parameters during the first two weeks are noteworthy. Kiosseoglou and Sherman (1983b) also reported decrease in viscoelasticity of o/w emulsions during storage up to the maximum storage time of 20 days. The observed increase in viscoelasticity after 21 days by Paredes et al. (1989) was most likely related to the hydration of xanthan gum in the salad dressings and that an eventual decrease and stabilization in the viscoelasticity of the salad dressings would be inevitable. The study of Paredes et al. (1989) points out, among other things, that the role of hydrocolloids as salad dressing ingredients must be studied (Andon, 1987).

MELTING CHARACTERISTICS OF MARGARINES

Melting characteristics of margarines and table spreads are important for flavor release and consumer acceptance. Oscillatory measurements as a function of temperature, and also drop points, were used to quantify rheological changes accompanying melting, and differential scanning calorimetry (DSC) measurements were used to quantify melting of fat crystals in these products (Borwankar et al., 1992). For low-fat spreads, rheological changes characteristic of melting could occur at a temperature higher than when fat melting occurs, and the fat melting is largely unaffected by emulsification. The rheology of low-fat spreads was governed by emulsion characteristics such as the proportion of the aqueous phase and the size of the water droplets. Drop point measurements provide a simple method to determine melting in the rheological sense. Perceived meltability of these products, in general, was due to a combined perception of cooling sensation accompanying the melting of fat crystals and the sensation of flow accompanying the rheological transitions. The cooling sensation was significant only for the case of butter and high-fat margarines; in these cases, the fat melting and rheological transitions both occurred at about body temperature. In reduced fat products, the cooling sensation was not significant, and perceived meltability correlated with the viscoelastic and drop point data.

REFERENCES

Andon, S.A. 1987. Interaction of salad dressing ingredients and hydrocolloids. *Food Eng* 59: 47–48.

Bayindirli, L. 1992. Mathematical analysis of variation of density and viscosity of apple juice with temperature and concentration. *J Food Process Preservation* 16: 23–28.

Bayindirli, L. 1993. Density and viscosity of grape juice as a function of concentration and temperature. *J Food Process Preservation* 17: 147–151.

Bistany, K.L., and J.L. Kokini. 1983. Dynamic viscoelastic properties of foods in texture control. *J Rheol* 27: 605–620.

Borwankar, R.P., L.A. Frye, A.E. Blaurock, and F.J. Sasevich, 1992. Rheological characterization of melting of margarines and tablespreads. *J Food Eng* 16: 55–74.

Campanella, O.H., and M. Peleg. 1987. Determination of yield stress of semi-liquid foods from squeezing flow data. *J Food Sci* 52: 214–215, 217.

Caradec, P.L., and P.E. Nelson. 1985. Effect of temperature on the serum viscosity of tomato juice. *J Food Sci* 50: 1497–1498.

Carter, R.D., and B.S. Buslig. 1977. Viscosity and particle size distribution in commercial Florida frozen concentrated orange juice. *Proc Fla State Hort Soc* 90: 130–132.

Chang, C.M., W.D. Powrie, and O. Fennema. 1972. Electron microscopy of mayonnaise. *Can Inst Food Sci Technol J* 5: 134–137.

Chevally, J. 1975. Rheology of chocolate. *J Texture Studies* 6: 177–196.

Crandall, P.G., C.S. Chen, and R.D. Carter. 1982. Models for predicting viscosity of orange juice concentrate. *Food Technol 36,* no. 5: 245–252.

den Ouden, F.W.C., and T. van Vliet. 1993. Determination of particle size distribution in tomato concentrate. In *Food colloids and polymers: Stability and mechanical properties,* eds. E. Dickinson and P. Walstra, 285–288. Cambridge, England: The Royal Society of Chemistry.

Dickinson, E., and G. Stainsby. 1982. *Colloids in food.* New York: Applied Science Publishers.

Dickinson, E., and G. Stainsby. 1987. Progress in the formulation of food emulsions and foams. *Food Technol* 41: 74, 76–81, 116.

Durán, L., and E. Costell. 1982. Rheology of apricot puree: Characterization of flow. *J Texture Studies* 13: 43–58.

Durán, L., and E. Costell. 1985. Influence of homogenization on the rheological behaviour of apricot puree. *Acta Alimentaria* 14: 201–210.

Elliott, J.H., and A.J. Ganz. 1977. Salad dressings—Preliminary rheological characterization. *J Texture Studies* 8: 359–377.

Fang, T.N., C. Tiu, X. Wu, and S. Dong, 1996. Rheological behaviour of cocoa dispersions. *J Texture Studies* 26: 203–215.

Fang, T., H. Zhang, T.T. Hsieh, and C. Tiu. 1997. Rheological behavior of cocoa dispersions with cocoa butter replacers. *J Texture Studies* 27: 11–26.

Figoni, P.I., and C.F. Shoemaker. 1983. Characterization of time dependent flow properties of mayonnaise under steady shear. *J Texture Studies* 14: 431–442.

Fito, P.J., G. Clemente, and F.J. Sanz. 1983. Rheological behaviour of tomato concentrates hot break and cold break. *J Food Eng* 2: 51–62.

Ganani, E., and R.E. Powell. 1985. Suspensions of rodlike particles: Literature review and data correlations. *J Composite Mater* 19: 194–215.

Giboreau, A., G. Cuvelier, and B. Launay. 1994. Rheological behavior of three biopolymer/water systems, with emphasis on yield stress and viscoelastic properties. *J Texture Studies* 25: 199–137.

Harper, J.C., and A.F. El-Sahrigi. 1965. Viscometric behavior of tomato concentrates. *J Food Sci* 30: 470–476.

Harrison, L.J., and F.E. Cunningham. 1985. Factors influencing the quality of mayonnaise: A review. *J Food Quality* 8: 1–20.

Harrison, L.J., and F.E. Cunningham. 1986. Influence of frozen storage time on properties of salted yolk and its functionality in mayonnaise. *J Food Quality* 9: 167–174.

Hartel, R.W. 1998. Phase transitions in chocolate and coatings. In *Phase/state transitions in foods: Chemical, structural, and rheological changes,* eds. M.A. Rao and R.W. Hartel, 217–251. New York: Marcel Dekker.

Higgs, S.J. 1974. An investigation into the flow behaviour of complex non-Newtonian foodstuffs. *J Phys D: Appl Phys* 7: 1184–1191.

Hunter, R.J., and Nicol, S.K. 1968. The dependence of plastic flow behavior of clay suspensions on surface properties. *J Colloid Interface Sci* 28: 250–259.

Ibarz, A., M. Vicente, and J. Graell. 1987. Rheological behavior of apple juice and pear juice and their concentrates. *J Food Eng* 6: 257–267.

Ibarz, A., C. Gonzalez, S. Esphugas, and M. Vicente. 1992a. Rheology of clarified juices. I: Peach juices. *J Food Eng* 15: 49–61.

Ibarz, A., J. Pagan, and R. Miguelsanz. 1992b. Rheology of clarified fruit juices. II: Blackcurrant juices. *J Food Eng* 15: 63–73.

Jeffrey, D.J., and A. Acrivos. 1976. The rheological properties of suspensions of rigid particles. *Am Inst Chem Engineers J* 22: 417–432.

Jinescu, V.V. 1974. The rheology of suspensions. *Int Chem Eng* 14: 397–420.

Kiosseoglou, V.D., and P. Sherman. 1983. Influence of egg yolk lipoproteins on the rheology and stability of o/w emulsions and mayonnaise. 1. Viscoelasticity of groundnut oil-in-water emulsions and mayonnaise. *J Texture Studies* 14: 397–417.

Kokini, J.L., and A. Dickie. 1981. An attempt to identify and model transient viscoelastic flow in foods. *J Texture Studies* 12: 539–557.

Kokini, J.L., and G.J. Plutchok. 1987. Viscoelastic properties of semisolid foods and their biopolymeric components. *Food Technol 41,* no. 3: 89–95.

Larson, R.G. 1985. Constitutive relationships for polymeric materials with power-law distributions of relaxation times. *Rheol Acta* 24: 327–334.

Lopes da Silva, J.A.L., M.P. Goncalves, and M.A. Rao. 1993. Viscoelastic behavior of mixtures of locust bean gum and pectin dispersions. *J Food Eng* 18: 211–228.

Maschmeyer, R.O., and C.T. Hill. 1977. Rheology of concentrated suspensions of fibers in tube flow. *Trans Soc Rheol* 21: 183–194.

McCarthy, K.L., and J.D. Seymour. 1994a. A fundamental approach for the relationship between the Bostwick measurement and Newtonian fluid viscosity. *J Texture Studies* 24: 1–10.

McCarthy, K.L., and J.D. Seymour, 1994b. Gravity current analysis of the Bostwick consistometer for power law foods. *J Texture Studies* 25: 207–220.

Metz, B., N.W.F. Kossen, and J.C. van Suijdam. 1979. The rheology of mold suspensions. In *Advances in biochemical engineering,* eds. T.K. Ghose, A. Fiechter, and N. Blakebrough, Vol. 11, 103–156. New York: Springer-Verlag.

Metzner, A.B. 1985. Rheology of suspensions in polymeric liquids. *J Rheol* 29: 739–775.

Michaels, A.S., and J.C. Bolger. 1962. The plastic behavior of flocculated kaolin suspensions. *Ind Eng Chem Fundamentals* 1: 153–162.

Mizrahi, S. 1979. A review of the physicochemical approach to the analysis of the structural viscosity of fluid food products. *J Texture Studies* 10: 67–82.

Mizrahi S., and Z. Berk. 1970. Flow behavior of concentrated orange juice. *J Texture Studies* 1: 342–355.

Mizrahi, S., and Z. Berk. 1972. Flow behaviour of concentrated orange juice: Mathematical treatment. *J Texture Studies* 3: 69–79.

Mizrahi, S., and R. Firstenberg. 1975. Effect of orange juice composition on flow behaviour of six-fold concentrate. *J Texture Studies* 6: 523–532.

Musser, J.C. 1973. Gloss on chocolate and confectionery coatings. In *Proceedings of the 27th Pennsylvania Manufacturing Confectioners' Association Production Conference,* 46–50. Lancaster, PA.

Okechukwu, P.E., and M.A. Rao. 1995. Influence of granule size on viscosity of corn starch suspension. *J Texture Studies* 26: 501–516.

Paredes, M.D.C., M.A. Rao, and M.C. Bourne. 1988. Rheological characterization of salad dressings. 1. Steady shear, thixotropy and effect of temperature. *J Texture Studies* 19: 247–258.

Paredes, M.D.C., M.A. Rao, and M.C. Bourne. 1989. Rheological characterization of salad dressings. 2. Effect of storage. *J Texture Studies* 20: 235–250.

Praschan, V.C. 1981. *Quality control manual for citrus processing plants.* Safety Harbor, Florida: Intercit.

Qiu, C-G., and M.A. Rao. 1988. Role of pulp content and particle size in yield stress of apple sauce. *J Food Sci* 53: 1165–1170.

Quemada, D., P. Fland, and P.H. Jezequel. 1985. Rheological properties and flow of concentrated diperse media. *Chem Eng Comm* 32: 61–83.

Rao, M.A. 1987. Predicting the flow properties of food suspensions of plant origin: Mathematical models help clarify the relationship between composition and rheological behavior. *Food Technol 41,* no. 3: 85–88.

Rao, M.A. 1992. Measurement of viscoelastic properties of fluid and semisolid foods. In *Viscoelastic properties of foods,* eds. M.A. Rao and J.F. Steffe, 207–232. London: Elsevier Applied Science Publishers.

Rao, M.A., and M.C. Bourne. 1977. Analysis of the plastometer and correlation of Bostwick consistometer data. *J Food Sci* 42: 261–264.

Rao, M.A., M.C. Bourne, and H.J. Cooley. 1981. Flow properties of tomato concentrates. *J Texture Studies* 12: 521–538.

Rao, M.A., and H.J. Cooley. 1984. Determination of effective shear rates of complex geometries. *J Texture Studies* 15: 327–335.

Rao, M.A., and H.J. Cooley. 1992. Rheology of tomato pastes in steady and dynamic shear. *J Texture Studies* 23: 415–425.

Rao, M.A., H.J. Cooley, and A.A. Vitali. 1984. Flow properties of concentrated fruit juices at low temperatures. *Food Technol 38,* no. 3: 113–119.

Rao, M.A., H.J. Cooley, J.N. Nogueira, and M.R. McLellan. 1986. Rheology of apple sauce: Effect of apple cultivar, firmness, and processing parameters. *J Food Sci* 51: 176–179.

Rao, M.A. and O. Palomino, L.N. 1974. Flow properties of tropical fruit purees. *J Food Sci* 39: 160–161.

Ross-Murphy, S.B. 1984. Rheological methods. In *Biophysical methods in food research,* ed. H.W.S. Chan, 138–199. London: Blackwell Scientific Publications.

Saravacos, G.D. 1970. Effect of temperature on viscosity of fruit juices and purees. *J Food Sci* 35: 122–125.

Sherman, P. 1970. *Industrial rheology.* New York: Academic Press.

Takada, N., and P.E. Nelson. 1983. A new consistency method for tomato products: The precipitate weight ratio. *J Food Sci* 48: 1460–1462.

Tanaka, M., and H. Fukuda. 1976. Studies on the texture of salad dressings containing xanthan gum. *Can Inst Food Sci Technol J* 9: 130–134.

Tanglertpaibul, T., and M.A. Rao. 1987a. Rheological properties of tomato concentrates as affected by particle size and methods of concentration. *J Food Sci* 52: 141–145.

Tanglertpaibul, T., and M.A. Rao. 1987b. Flow properties of tomato concentrates: Effect of serum viscosity and pulp content. *J Food Sci* 52: 318–321.

Tanglertpaibul, T., and M.A. Rao. 1987c. Intrinsic viscosity of tomato serum as affected by methods of determination and methods of processing concentrates. *J Food Sci* 52: 1642–1645, 1688.

Trifiro, A., G. Saccani, S. Gherardi, and D. Bigliardi. 1987. Effect of content and sizes of suspended particles on the rheological behavior of apricot purees. *Industria Conserve* 62: 97–104.

Vitali, A.A., and M.A. Rao. 1984a. Flow properties of low-pulp concentrated orange juice: Serum viscosity and effect of pulp content. *J Food Sci* 49: 876–880.

Vitali, A.A., and M.A. Rao. 1984b. Flow properties of low-pulp concentrated orange juice: Effect of temperature and concentration. *J Food Sci* 49: 882–888.

Weyland, M. 1994. Functional effects of emulsifiers in chocolate. *Manuf Confect* 745: 111–117.

Wildemuth, C.R., and Williams, M.C. 1984. Viscosity of suspensions modeled with a shear-dependent maximum packing fraction. *Rheol Acta* 23: 627–635.

Yoo, B., and M.A. Rao. 1994. Effect of uninodal particle size and pulp content on rheological properties of tomato puree. *J Texture Studies* 25: 421–436.

Yoo, B., and M.A. Rao. 1996. Creep and dynamic rheological behavior of tomato concentrates: Effect of concentration and finisher screen size. *J Texture Studies* 27: 451–459.

Appendix 5–A

Literature Values of Rheological Properties of Foods

Paul E. Okechukwu and M. Anandha Rao

Literature values of rheological properties of fluid foods are useful in design and handling applications. Holdsworth (1971, 1993) presented compilations of the literature values of rheological properties of fluid foods. Magnitudes of the flow properties of several foods from the literature that were selected after examination of the measurement technique used are listed in several tables: (1) viscosity of water in Table 5–A1; (2) rheological properties of fruit juices and purees in Table 5–A2; (3) flow properties of tomato pastes (Rao and Cooley, 1992) in Table 5–A3; (4) the effect of temperature on apparent viscosity and consistency index of fruit juices and purees in Table 5–A4; (5) creep-compliance parameters of tomato concentrates (Yoo and Rao, 1996) in Table 5–A5; (6) values of yield stress of prepared apple sauce samples using the vane method and extrapolated Bingham yield stress in Table 5–A6 (Qiu and Rao, 1988); (7) flow properties of chocolate in Table 5–A7; (8) flow properties of chocolate before and after degasification in Table 5–A8 (Fang et al., 1996); (9) formulations of three mayonnaises in Table 5–A9; (10) effect of storage on the viscoelastic properties of a model salad dressing (Paredes et al., 1989) in Table 5–A10; (11) flow properties of mustard, mayonnaise, fats, and oils in Table 5–A11; (12) the effect of temperature on flow properties of oils as reported by Lang et al. (1992) in Table 5–A12, in which an unusual model, that is, $\ln \eta = A + [B/(T + C)]$, was used to describe the effect of temperature; (13) flow properties of egg products in Table 5–A13; (14) flow properties of meat batters in Table 5–A14; (15) flow properties of dairy products in Table 5–A15; (16) flow properties of syrups and honeys in Table 5–A16; and (17) viscosity of wines in Table 5–A17. Some of the tables listed above were not referred to in Chapter 5.

Table 5–A1 Viscosity of Water at Different Temperatures. Perry, 1950.

Temperature (°C)	*Viscosity (mPa s)*	*Temperature (°C)*	*Viscosity (mPa s)*
0	1.7921	38	0.6814
1	1.7313	39	0.6685
2	1.6728	40	0.6560
3	1.6191	41	0.6439
4	1.5674	42	0.6321
5	1.5188	43	0.6207
6	1.4728	44	0.6097
7	1.4284	45	0.5988
8	1.3860	46	0.5883
9	1.3462	47	0.5782
10	1.3077	48	0.5683
11	1.2713	49	0.5588
12	1.2363	50	0.5494
13	1.2028	51	0.5404
14	1.1709	52	0.5315
15	1.1404	53	0.5229
16	1.1111	54	0.5146
17	1.0828	55	0.5064
18	1.0559	56	0.4985
19	1.0299	57	0.4907
20	1.0050	58	0.4832
20.2	1.0000	59	0.4759
21	0.9810	60	0.4688
22	0.9579	61	0.4618
23	0.9358	62	0.4550
24	0.9142	63	0.4483
25	0.8937	64	0.4418
26	0.8737	65	0.4355
27	0.8545	66	0.4293
28	0.8360	67	0.4233
29	0.8180	68	0.4174
30	0.8007	69	0.4117
31	0.7840	70	0.4061
32	0.7679	71	0.4006
33	0.7523	72	0.3952
34	0.7371	73	0.3900
35	0.7225	74	0.3849
36	0.7085	75	0.3799
37	0.6947	76	0.3750

continues

Table 5–A1 *continued*

Temperature (°C)	*Viscosity (mPa s)*	*Temperature (°C)*	*Viscosity (mPa s)*
77	0.3702	89	0.3202
78	0.3655	90	0.3165
79	0.3610	91	0.3130
80	0.3565	92	0.3095
81	0.3521	93	0.3060
82	0.3478	94	0.3027
83	0.3436	95	0.2994
84	0.3395	96	0.2962
85	0.3355	97	0.2930
86	0.3315	98	0.2899
87	0.3276	99	0.2868
88	0.3239	100	0.2838

Table 5–A2 Rheological Properties of Fruit and Vegetable Products

Product	*Conc. (% solids)*	*Method*	*Temp (°C)*	*Rate Shear (s^{-1})*	*K* (Pa s^n)*	n	*Yield Stress (Pa)*	*Reference*
Tomato juice (pH 4.3)	5.80	Conc cylinder	32.2	500–800	0.22	0.59		Harper and El-Sahrigi, 1965
			48.9		0.27	0.54		
			65.6		0.37	0.47		
	12.80		32.3		2.10	0.43		
			48.9		1.18	0.43		
			65.6		2.28	0.34		
			82.2		2.12	0.35		
	16.00		32.3		3.16	0.45		
			48.9		2.27	0.45		
			65.6		3.18	0.40		
			82.2		3.27	0.38		
	25.00		32.3		12.9	0.41		
			48.9		10.5	0.42		
			65.6		8.0	0.43		
			82.2		6.1	0.44		
Tomato juice	30.00	Conc cylinder	32.3	500–800	18.7	0.40		Harper and Leberman, 1962
			48.9		15.1	0.42		
			65.6		11.7	0.43		
			82.2		7.9	0.44		
Tomato conc.	35.74	Parallel plate carimed	25.0	4–576	297.0	0.38		Alviar et al., 1990
	24.16				65.7	0.27		
	17.86				26.1	0.26		
	11.78				6.0	0.33		

continues

Table 5–A2 *continued*

	35.74	Back extrusion	25.0	4–576	249.0	0.29		
	24.16				58.5	0.21		
	17.86				24.3	0.23		
	11.78				5.82	0.27		
Tomato paste	29.7 °Brix	Haake mixer	32.0		208.0	0.27	206	Rao et al., 1993
	29.7 °Brix		39.0		179.0	0.31	180	
	23.8 °Brix		33.0		48.0	0.47	40	
	23.8 °Brix		39.0		34.0	0.52	29	
	16.3 °Brix		25.0		24.0	0.23	38	
Tomato sauce	7.2 °Brix	Conc cylinder	20.0	0–10	4.0	0.40		McCarthy and Seymour, 1994
	8.9 °Brix	Haake RV20	20.0	0–10	7.3	0.33		
Pear puree	18.30	Conc cylinder	32.3	0.1–2600	2.3	0.49	3.5	
			48.9		1.9	0.48		
			65.6		1.6	0.48		
			82.2		1.5	0.48		
	26.13		32.3		6.2	0.45	8.1	
			48.9		5.0	0.45		
			65.6		4.2	0.46		
			82.2		3.6	0.46		
	31.00		32.3		10.9	0.45	19.1	
			48.9		8.80	0.45		
			65.6		6.8	0.46		
			82.2		5.6	0.46		
	37.16		32.3		17.0	0.46		
			48.9		13.5	0.46		

continues

Table 5–A2 *continued*

Product	*Conc. (% solids)*	*Method*	*Temp (°C)*	*Rate Shear (s^{-1})*	*K* (Pa s^n)*	n	*Yield Stress (Pa)*	*Reference*
	45.75		32.3		35.5	0.48	33.9	
			48.9		26.0	0.48		
			65.6		20.0	0.48		
			82.2		16.0	0.48		
Pear puree	14.60	Tube visc	27.0	100–2000	5.3	0.38		Saravacos, 1968
	16.00		30.0		5.6	0.35		Harper, 1960
	24.30	Conc cylinder	26.6	1–1000	5.8	0.39		
	33.40				38.5	0.38		
	37.60				49.7	0.38		
	37.50				64.8	0.38		
	47.60				120.0	0.33		
	51.30				205.0	0.34		
	49.30				170.0	0.34		
	48.80				150.0	0.34		
	15.20				4.25	0.35		
Apple sauce	11.00	Brookfield	30.0	5–50	11.6	0.34		Saravacos, 1970
	11.00	RTV	82.0	5–50	9.0	0.34	11.6	Saravacos and Moyer, 1967
		Tube visc	27.0	100–2000	12.7	0.28		Saravacos, 1968
Apple sauce		Haake mixer	32.0		200.0	0.42	240	Rao et al., 1993
Depectin.	75 °Brix		20–70			1.00		Saravacos, 1970
Apple juice	50 °Brix		20–70			1.00		
	30 °Brix		20–70			1.00		
	15 °Brix		20–70			1.00		

continues

continued

Cloudy apple juice	65.5 °Brix		20–70			0.65	
	50 °Brix		20–70			0.85	
	40 °Brix		20–70			1.00	
	30 °Brix		20–70			1.00	
	15 °Brix		20–70			1.00	
Apple juice	69.8 °Brix	Haake RV2/ Deer Rheom.	25.0		0.24	1.00	Rao et al., 1993
	65.3 °Brix		23.0		0.09	1.00	
	69.8 °Brix		43.0		0.03	1.00	
	51.1 °Brix		25.0		0.02	1.00	
Apricot puree	16	Brookfield RTV	30.0	5–50	6.80	0.30	Saravacos and Moyer, 1967
	13.8	Tube visc.	27.0		7.2	0.41	Saravacos, 1970
Apricot puree	17.7	Conc cylinder	26.7	1–1000	5.4	0.29	Harper, 1960
	44.3				56.0	0.39	
	51.4				108.0	0.36	
	55.2				152.0	0.34	
	59.3				300.0	0.32	
	41.4				540.0	0.35	
	23				11.2	0.35	
Peach puree	40.1				58.5	0.35	
	49.8				85.5	0.34	
	49.6				152.0	0.34	
	58.4				440.0	0.34	
	17				13.8	0.35	
	21.9				21.1	0.55	
	26				13.4	0.40	
	29.6				18.0	0.40	
	37.5				44.0	0.38	
	10.9				9.40	0.44	

continues

Table 5–A2 *continued*

Product	*Conc. (% solids)*	*Method*	*Temp (°C)*	*Rate Shear (s^{-1})*	*K* (Pa s^n)*	n	*Yield Stress (Pa)*	*Reference*
Peach puree	11.7	Brookfield	30.0	5–50	7.20	0.28		Saravacos, 1970
Plum	14	RTV tube	30.0	5–50	2.2	0.34		Saravacos, 1970
Plum	14	visc.	82.2	5–50	2.0	0.34		
Concord conc.	64		20–70			0.90		Saravacos, 1970
	50					1.00		
	30					1.00		
	15					1.00		
Passion fruit	15.6 °Brix	Tube visc.	21.6		0.07	0.74		Rao, Otoya,
	20.7 °Brix		21.4		0.29	0.53		Palomino, and
	25.3 °Brix		20.0		0.80	0.52		Bernhardt, 1974
	30.6 °Brix		23.2		1.62	0.49		
	33.4 °Brix		20.0		2.62	0.45		
Papaya	7.3 °Brix		26.0		0.91	0.53		Rao, 1977
Mango	9.3 °Brix		24.2		2.06	0.33		
Guava	10.3 °Brix		23.4		3.90	0.49		
Guava puree	7.2 °Brix	Tube visc.	24.0	100–10,000	0.26	0.68		Brekke and Myers,
	7.4 °Brix		24.0		0.28	0.68		1978
Guava conc.	12.3 °Brix		24.0		1.60	0.55		
	15.9 °Brix		24.0		5.00	0.47		
	22.7 °Brix		24.0		41.00	0.32		

continues

continued

Depectin. guava puree	8.8 °Brix		24.0		<0.1	1.65		
	8.9 °Brix		24.0		<0.1	1.78		
Depectin. guava conc.	15.1 °Brix		24.0		0.30	0.71		
	24.3 °Brix		24.0		1.60	0.62		
	31.4 °Brix		24.0		2.30	0.59		
Conc orange	42.5 °Brix	Haake RV12	25.0	0–500	0.41	0.59		Crandall et al., 1982
juice	15.0 °Brix	conc		0–500	0.60	0.60	0.0	
Hamlin		cylinder		0–500	0.92	0.68		
early			−10.0	0–500	1.43	0.71		
Hamlin late	41.4 °Brix		25.0	0–500	0.19	0.735		
			15.0	0–500	0.81	0.56		
			0.0	0–500	0.18	0.62		
			−10.0	0–500	1.39	0.71		
Pineapple early	40.3 °Brix		25.0	0–500	0.26	0.64		
			15.0	0–500	0.59	0.59		
			0.0	0–500	0.89	0.68		
			−10.0	0–500	1.22	0.71		
Pineapple late	41.8 °Brix		25.0	0–500	0.86	0.53		
			15.0	0–500	1.34	0.54		
			0.0	0–500	1.86	0.64		
			−10.0	0–500	3.64	0.63		

continues

Table 5–A2 *continued*

Product	*Conc. (% solids)*	*Method*	*Temp (°C)*	*Rate Shear (s^{-1})*	*K* (Pa s^n)*	n	*Yield Stress (Pa)*	*Reference*
Valencia early	43.1 °Brix		25.0	0–500	0.51	0.54		
			15.0	0–500	0.67	0.61		
			0.0	0–500	1.40	0.62		
			−10.0	0–500	2.72	0.61		
Valencia late	41.9 °Brix		25.0	0–500	0.84	0.54		
			15.0	0–500	1.18	0.57		
			0.0	0–500	1.86	0.64		
			−10.0	0–500	4.14	0.63		
Naval	65.1 °Brix		−18.5		29.20	0.71		Rozema and Beverloo, 1974
			−14.1		14.60	0.76		
			−9.3		10.80	0.74		
			−5.0		7.90	0.72		
			−0.7		5.90	0.71		
			10.1		2.70	0.73		
			19.9		1.60	0.72		
			29.5		0.9	0.74		
Orange juice Pera orange, 3.4% pulp	64.9 °Brix	Haake RV2 conc cylinder	−18.0		18.3	0.80		Vitali and Rao, 1984a
			−14.0		10.6	0.81		
			−10.0		7.1	0.79		
			−5.0		4.9	0.78		
			0.0		3.2	0.79		
			10.0		1.6	0.79		
			20.0		0.7	0.83		
			30.0		0.4	0.82		

continues

Table 5–A2 *continued*

Pera orange, 5.7% pulp	65 °Brix	−18.0		24.5	0.76
		−14.0		15.6	0.76
		−10.0		8.8	0.79
		−5.0		6.5	0.77
		0.0		4.5	0.76
		10.0		2.1	0.78
		20.0		1.3	0.77
		30.0		0.7	0.80
Pera orange, 8.6% pulp	65.3 °Brix	−18.0		42.8	0.70
		−14.0		25.3	0.73
		−10.0		15.8	0.74
		−5.0		11.1	0.73
		0.0		7.3	0.74
		10.0		3.9	0.73
		20.0		1.9	0.76
		30.0		1.4	0.73
Pera orange, 11.1% pulp	64.8 °Brix	−18.0		58.9	0.67
		−14.0		34.6	0.70
		−10.0		23.9	0.69
		−5.0		16.0	0.69
		0.0		11.5	0.69
		10.0		5.6	0.70
		20.0		2.9	0.72
		30.0		2.1	0.70

continues

Table 5–A2 *continued*

Product	*Conc. (% solids)*	*Method*	*Temp (°C)*	*Rate Shear (s^{-1})*	K* *(Pa s^n)*	n	*Yield Stress (Pa)*	*Reference*
Valencia	65.3 °Brix		−18.3		109.9	0.55		
orange			−14.2		59.7	0.61		
21.2% pulp			−9.7		40.6	0.60		
			−5.1		24.5	0.63		
			−0.4		18.5	0.62		
			10.2		8.3	0.65		
			19.7		6.1	0.61		
			29.6		2.5	0.68		
Mango pulp								
3.49% db	16 °Brix	Conc cylinder	30	20–250	5.1	0.97		Manohar et al.,
pectin		$r_1/r_2 = 0.94$	40		4.4	0.98		1991
(pH 3.56)		Rheotest2	50		3.8	0.99		
			60		3.3	0.99		
			70		2.8	0.99		
	20 °Brix		30	20–250	7.9	0.28		
			40		6.8	0.28		
			50		5.4	0.27		
			60		4.5	0.28		
			70		3.6	0.27		
	25 °Brix		30	20–250	9.8	0.29		
			40		8.2	0.29		
			50		7.5	0.28		
			60		6.4	0.28		
			70		5.6	0.28		

continues

Table 5–A2 *continued*

	30 °Brix	30	20–250	19.3	0.28
		40		14.7	0.28
		50		12.3	0.28
		60		10.3	0.28
		70		8.5	0.28
3.38% db pectin	16 °Brix	30	20–250	2.9	0.29
		40		2.5	0.29
		50		2.3	0.29
		60		1.9	0.29
		70		1.7	0.28
3.29% db pectin	16 °Brix	30	20–250	2.7	0.29
		40		2.4	0.28
		50		2.1	0.29
		60		1.8	0.28
		70		1.5	0.29
3.23% db pectin	16 °Brix	30	20–250	1.7	0.28
		40		1.4	0.28
		50		1.1	0.29
		60		0.9	0.29
		70		0.8	0.29
		30	20–250	2.4	0.29
		40		2.0	0.29
		50		1.6	0.29
		60		1.4	0.29
		70		1.2	0.29

continues

Table 5–A2 *continued*

Product	*Conc. (% solids)*	*Method*	*Temp (°C)*	*Rate Shear (s^{-1})*	K* *(Pa s^n)*	n	*Yield Stress (Pa)*	*Reference*
			30	20–250	3.0	0.29		
			40		2.6	0.29		
			50		2.2	0.29		
			60		1.9	0.29		
			70		1.7	0.29		
			30	20–250	5.3	0.28		
			40		4.4	0.29		
			50		3.7	0.28		
			60		3.1	0.29		
			70		2.5	0.29		
Mango pulp juice	16 °Brix	Conc cylinder Rheotest2	25		10.0	0.32		Rao et al., 1985
Mango pulp juice	17.0 °Brix	Rotational RVT Brookfield	25		27.8	0.33		Gunjal and Waghmare, 1987
Mango pulp juice	16.0 °Brix	Rotational LVF Brookfield	25		12.30	0.280		Garcia et al., 1974
Tamarind juice	7 °Brix	Conc cylinder Rheotest2	35.0	10–400	0.002	0.99		Manohar et al., 1991
		*R*1/*R*2 = 0.98	45.0		0.001	1.0		
			55.0		0.001	1.00		
	9 °Brix		25.0		0.006	1.00		
			45.0		0.004	1.00		
			55.0		0.003	1.00		

continues

Table 5–A2 *continued*

	23 °Brix	25.0		0.087	0.76
		35.0		0.062	0.75
		45.0		0.050	0.73
		55.0		0.039	0.75
	28 °Brix	25.0		0.35	0.63
		35.0		0.26	0.62
		45.0		0.19	0.63
		55.0		0.16	0.63
		65.0		0.14	0.63
	36 °Brix	25.0		0.46	0.63
		35.0		0.35	0.61
		45.0		0.26	0.63
		55.0		0.23	0.62
		65.0		0.21	0.61
	44 °Brix	30.0		1.23	0.63
		40.0		0.83	0.63
		50.0		0.63	0.64
		60.0		0.53	0.61
		65.0		1.53	0.63
		70.0		0.43	0.63
	50 °Brix	35.0		3.40	0.62
		45.0		2.65	0.63
		55.0		1.92	0.63
		65.0		1.53	0.63
		70.0		1.38	0.63
	62 °Brix	27.0		9.07	0.63
		35.0		6.28	0.63
		45.0		4.46	0.63
		55.0		3.32	0.62
		70.0		2.23	0.63

continues

Table 5–A2 *continued*

Product	*Conc. (% solids)*	*Method*	*Temp (°C)*	*Rate Shear* (s^{-1})	K* ($Pa\ s^n$)	n	*Yield Stress (Pa)*	*Reference*
Pineapple juice	7 °Brix	Conc cylinder	30.0	100–600	0.03	1.00		Varshney and Kumbhar, 1978
	22 °Brix	narrow gap	30.0		0.06	0.91		
	36 °Brix	viscometer	30.0		0.13	0.81		
	7 °Brix	Conc cylinder	45.0		0.02	1.03		
	22 °Brix	narrow gap	45.0		0.05	0.90		
	36 °Brix	viscometer	45.0		0.12	0.81		
Orange juice	13 °Brix	Conc cylinder	30.0	100–600	0.06	0.86		
	22 °Brix	narrow gap	30.0		0.12	0.79		
	33 °Brix	viscometer	30.0		0.14	0.78		
	13 °Brix	Conc cylinder	45.0		0.02	1.00		
	22 °Brix	narrow gap	45.0		0.08	0.79		
	33 °Brix	viscometer	45.0		0.12	0.98		
Pulp free orange juice % pectin (galacturonic acid)		Brookfield LVT conc cylinder						Mizrahi, 1979
0.047	60 °Brix			1–1142	0.016	0.55		
0.019	65 °Brix				0.003	0.91		
0.034	60 °Brix				0.005	0.73		
Raspberry juice	15 °Brix	Conc cylinder	5.0	3–2000	0.029	0.90		Ibarz and Pagan, 1987
		Haake RV12	10.0		0.022	0.92		

continues

Table 5–A2 *continued*

	NV	15.0		0.017	0.93
		20.0		0.014	0.94
		25.0		0.010	0.96
		30.0		0.009	0.96
		35.0		0.006	0.99
		45.0		0.004	1.01
		55.0		0.003	1.00
		60.0		0.002	1.02
20 °Brix	Conc cylinder	5.0	3–2000	0.058	0.90
	Haake RV12	10.0		0.045	0.92
	NV	15.0		0.032	0.94
		20.0		0.030	0.94
		25.0		0.023	0.96
		30.0		0.019	0.96
		35.0		0.014	0.99
		45.0		0.011	0.99
		55.0		0.010	1.01
		60.0		0.008	1.00
25 °Brix	Conc cylinder	5.0	3–2000	0.344	0.76
	Haake RV12	10.0		0.169	0.86
	NV	15.0		0.111	0.91
		20.0		0.072	0.94
		25.0		0.064	0.94
		30.0		0.050	0.95
		35.0		0.033	1.00
		45.0		0.024	1.00
		55.0		0.021	1.00
		60.0		0.019	1.01
30 °Brix	Conc cylinder	5.0	3–2000	0.739	0.74
	Haake RV12	10.0		0.437	0.82

continues

Table 5–A2 *continued*

Product	*Conc. (% solids)*	*Method*	*Temp (°C)*	*Rate Shear (s^{-1})*	K* *(Pa s^n)*	n	*Yield Stress (Pa)*	*Reference*
		NV	15.0		0.365	0.81		
			20.0		0.202	0.90		
			25.0		0.142	0.91		
			30.0		0.124	0.91		
			35.0		0.101	0.92		
			45.0		0.067	0.95		
			55.0		0.049	0.95		
			60.0		0.040	0.96		
	35 °Brix	Conc cylinder	5.0	3–2000	1.86	0.69		
		Haake RV12	10.0		1.30	0.72		
		NV	15.0		0.81	0.80		
			20.0		0.67	0.79		
			25.0		0.56	0.79		
			30.0		0.34	0.86		
			35.0		0.28	0.87		
			45.0		0.17	0.91		
			55.0		0.12	0.92		
			60.0		0.08	0.95		
	41 °Brix	Conc cylinder	5.0	3–2000	6.25	0.59		
		Haake RV12	10.0		4.30	0.64		
		NV	15.0		3.36	0.66		
			20.0		2.57	0.68		
			25.0		1.61	0.73		
			30.0		1.34	0.75		
			35.0		0.95	0.79		
			45.0		0.46	0.88		
			55.0		0.35	0.87		
			60.0		0.30	0.88		

continues

Table 5–A2 *continued*

Apple juice	71 °Brix	Haake RV12	5.0	1.26	1.00	Ibarz et al., 1987
		conc cylinder	15.0	0.26	1.00	
			25.0	0.22	1.00	
			35.0	0.13	1.00	
			40.0	0.088	1.00	
			45.0	0.050	1.00	
			60.0	0.027	1.00	
Apple juice	35 °Brix		25.0	0.001	1.00	
	45 °Brix		25.0	0.008	1.00	
	50 °Brix		25.0	0.018	1.00	
	60 °Brix		25.0	0.066	1.00	
	71 °Brix		25.0	0.217	1.00	
Pear juice	70 °Brix		5.0	1.46	1.00	Ibarz et al., 1987
			10.0	0.876	1.00	
			15.0	0.527	1.00	
			25.0	0.233	1.00	
			35.0	0.110	1.00	
			45.0	0.061	1.00	
			60.0	0.031	1.00	
Pear juice	30 °Brix		25.0	0.003	1.00	
	40 °Brix		25.0	0.005	1.00	
	45 °Brix		25.0	0.008	1.00	
	50 °Brix		25.0	0.013	1.00	
	55 °Brix		25.0	0.019	1.00	
	60 °Brix		25.0	0.041	1.00	
	65 °Brix		25.0	0.074	1.00	
	70 °Brix		25.0	0.233	1.00	

continues

Table 5–A2 *continued*

Product	*Conc. (% solids)*	*Method*	*Temp (°C)*	*Rate Shear (s^{-1})*	*K* (Pa s^n)*	n	*Yield Stress (Pa)*	*Reference*
Black currant juice clarified	35 °Brix	Haake RV12 conc cylinder	5.0		0.0096	1.00		Ibarz et al., 1992b
			10.0		0.0077	1.00		
			15.0		0.0064	1.00		
			20.0		0.0056	1.00		
			25.0		0.0049	1.00		
			30.0		0.0044	1.00		
			35.0		0.0038	1.00		
			40.0		0.0035	1.00		
			45.0		0.0031	1.00		
			50.0		0.0027	1.00		
			55.0		0.0026	1.00		
			60.0		0.0024	1.00		
	40 °Brix		5.0		0.0138	1.00		
			10.0		0.0105	1.00		
			15.0		0.0090	1.00		
			20.0		0.0075	1.00		
			25.0		0.0065	1.00		
			30.0		0.0057	1.00		
			35.0		0.0050	1.00		
			40.0		0.0046	1.00		
			45.0		0.0041	1.00		
			50.0		0.0038	1.00		
			55.0		0.0033	1.00		
			60.0		0.0031	1.00		

continues

Table 5–A2 *continued*

45 °Brix	5.0	0.0234	1.00
	10.0	0.0174	1.00
	15.0	0.0136	1.00
	20.0	0.0112	1.00
	25.0	0.0095	1.00
	30.0	0.0079	1.00
	35.0	0.0069	1.00
	40.0	0.0061	1.00
	45.0	0.0055	1.00
	50.0	0.0050	1.00
	55.0	0.0047	1.00
	60.0	0.0042	1.00
50 °Brix	5.0	0.0371	1.00
	10.0	0.0284	1.00
	15.0	0.0215	1.00
	20.0	0.0175	1.00
	25.0	0.0143	1.00
	30.0	0.0116	1.00
	35.0	0.0100	1.00
	40.0	0.0084	1.00
	45.0	0.0077	1.00
	50.0	0.0067	1.00
	55.0	0.0061	1.00
	60.0	0.0056	1.00
55 °Brix	5.0	0.0744	1.00
	10.0	0.0549	1.00
	15.0	0.0400	1.00
	20.0	0.0325	1.00
	25.0	0.0245	1.00

continues

Table 5–A2 *continued*

Product	*Conc. (% solids)*	*Method*	*Temp (°C)*	*Rate Shear (s^{-1})*	K* *(Pa s^n)*	n	*Yield Stress (Pa)*	*Reference*
			30.0		0.0195	1.00		
			35.0		0.0161	1.00		
			40.0		0.0135	1.00		
			45.0		0.0130	1.00		
			50.0		0.0113	1.00		
			55.0		0.0099	1.00		
			60.0		0.0091	1.00		
	60 °Brix		5.0		0.1589	1.00		
			10.0		0.1093	1.00		
			15.0		0.0802	1.00		
			20.0		0.0558	1.00		
			25.0		0.0449	1.00		
			30.0		0.034	1.00		
			35.0		0.027	1.00		
			40.0		0.022	1.00		
			45.0		0.020	1.00		
			50.0		0.018	1.00		
			55.0		0.015	1.00		
			60.0		0.013	1.00		
	64.5 °Brix		5.0		0.500	1.00		
			10.0		0.298	1.00		
			15.0		0.208	1.00		
			20.0		0.138	1.00		
			25.0		0.101	1.00		
			30.0		0.073	1.00		
			35.0		0.055	1.00		
			40.0		0.044	1.00		

continues

Table 5–A2 *continued*

			45.0		0.035	1.00	
			50.0		0.029	1.00	
			55.0		0.024	1.00	
			60.0		0.020	1.00	
Clarified peach juice	40 (°Brix)	Conc cylinder Haake Rotovisco	5.0	50–700	0.016	1.00	Ibarz et al., 1992a
	45				0.025	1.00	
	50				0.044	1.00	
	55				0.085	1.00	
	60				0.181	1.00	
	65				0.526	1.00	
	69				2.087	1.00	
	40		10.0		0.013	1.00	
	45				0.019	1.00	
	50				0.033	1.00	
	55				0.063	1.00	
	60				0.123	1.00	
	65				0.404	1.00	
	69				1.288	1.00	
	40		15.0		0.011	1.00	
	45				0.016	1.00	
	50				0.027	1.00	
	55				0.048	1.00	
	60				0.092	1.00	
	65				0.249	1.00	
	69				0.743	1.00	
	40		20.0		0.009	1.00	
	45				0.013	1.00	
	50				0.021	1.00	

continues

Table 5–A2 *continued*

Product	*Conc. (% solids)*	*Method*	*Temp (°C)*	*Rate Shear (s^{-1})*	K* *(Pa s^n)*	n	*Yield Stress (Pa)*	*Reference*
	55				0.037	1.00		
	60				0.062	1.00		
	65				0.175	1.00		
	69				0.485	1.00		
	40		25.0		0.008	1.00		
	45				0.011	1.00		
	50				0.017	1.00		
	55				0.029	1.00		
	60				0.049	1.00		
	65				0.126	1.00		
	69				0.324	1.00		
	40		35.0		0.006	1.00		
	45				0.008	1.00		
	50				0.012	1.00		
	55				0.020	1.00		
	60				0.033	1.00		
	65				0.070	1.00		
	69				0.155	1.00		
	40		45.0		0.005	1.00		
	45				0.006	1.00		
	50				0.009	1.00		
	55				0.013	1.00		
	60				0.021	1.00		
	65				0.041	1.00		
	69				0.084	1.00		

continues

Table 5–A2 *continued*

	40		60.0	0.004	1.00	
	45			0.005	1.00	
	50			0.006	1.00	
	55			0.009	1.00	
	60			0.013	1.00	
	65			0.023	1.00	
	69			0.042	1.00	
Tomato conc	10 °Brix	Contraves	15.0	6.98	0.37	Fito et al., 1983
(hot-		Rheomat 15	20.0	7.00	0.36	
break			30.0	0.240	0.36	
samples),			40.0	6.03	0.37	
0.857%			50.0	4.84	0.38	
pectin			60.0	4.75	0.37	
(as galac-			70.0	4.49	0.35	
turonic acid	14 °Brix		15.0	14.6	0.38	
anhydride)			20.0	13.9	0.37	
			30.0	12.6	0.34	
			40.0	12.3	0.34	
			50.0	11.7	0.32	
			60.0	12.2	0.32	
			70.0	12.2	0.24	
	15 °Brix		15.0	19.7	0.33	
			20.0	21.1	0.33	
			30.0	18.9	0.33	
			40.0	17.0	0.33	
			50.0	14.2	0.34	
			60.0	12.0	0.33	
			70.0	11.3	0.32	

continues

Table 5–A2 *continued*

Product	*Conc. (% solids)*	*Method*	*Temp (°C)*	*Rate Shear (s^{-1})*	*K* (Pa s^n)*	*n*	*Yield Stress (Pa)*	*Reference*
	20 °Brix		15.0		40.0	0.31		
			20.0		34.4	0.31		
			30.0		33.6	0.32		
			40.0		28.8	0.33		
			50.0		25.2	0.33		
			60.0		23.2	0.29		
			70.0		17.2	0.29		
	25 °Brix		15.0		76.9	0.28		
			20.0		61.5	0.29		
			30.0		76.7	0.28		
			40.0		60.3	0.27		
			50.0		57.7	0.26		
			60.0		47.2	0.31		
			70.0		39.4	0.32		
	28 °Brix		15.0		65.1	0.40		
			20.0		81.3	0.33		
			30.0		10.5	0.25		
			40.0		86.8	0.27		
			50.0		67.7	0.28		
			60.0		87.6	0.22		
			70.0		43.6	0.26		
Tomato conc	10 °Brix	Contraves	15.0		0.45	0.43		
Cold break		Rheomat 15	20.0		0.38	0.43		
0.857%			30.0		0.56	0.38		
pectin (as			40.0		0.27	0.44		
galacturonic			50.0		0.40	0.41		
acid			60.0		0.26	0.42		
anhydride)			70.0		0.12	0.45		

continues

Table 5–A2 *continued*

14 °Brix	15.0	4.1	0.35
	20.0	3.2	0.34
	30.0	3.0	0.38
	40.0	2.8	0.33
	50.0	2.3	0.39
	60.0	2.3	0.36
	70.0	2.0	0.35
15 °Brix	15.0	6.8	0.34
	20.0	7.4	0.36
	30.0	6.6	0.38
	40.0	5.8	0.39
	50.0	5.1	0.38
	60.0	4.3	0.36
	70.0	4.1	0.38
20 °Brix	15.0	8.6	0.35
	20.0	6.9	0.39
	30.0	7.1	0.36
	40.0	6.4	0.40
	50.0	5.4	0.37
	60.0	5.5	0.37
	70.0	4.9	0.36
25 °Brix	15.0	17.2	0.34
	20.0	15.4	0.35
	30.0	14.0	0.35
	40.0	12.3	0.36
	50.0	11.2	0.35
	60.0	8.9	0.38
	70.0	9.4	0.34

continues

Table 5–A2 *continued*

Product	*Conc. (% solids)*	*Method*	*Temp (°C)*	*Rate Shear (s^{-1})*	*K* (Pa s^n)*	*n*	*Yield Stress (Pa)*	*Reference*
	28 °Brix		15.0		29.3	0.34		
			20.0		27.5	0.35		
			30.0		25.9	0.34		
			40.0		23.0	0.33		
			50.0		19.4	0.35		
			60.0		17.9	0.33		
			70.0		15.1	0.35		
Apple sauce (Mott's)	18.2 °Brix	Haake RV3	25.0		34.0	0.42	34	Barbosa-Cánovas and Peleg, 1983
Apple sauce (Stop and Shop)	18.1				6.4	0.43	30	
Baby food banana (Beech-Nut)	15		25.0		5.4	0.62	21	
Baby food banana (Gerber)	15.1		25.0		28.0	0.59	28	
Baby food peach (Beech-Nut)	10.8		25.0		0.600	0.83	14	
Baby food peach (Gerber)	16		25.0		1.40	0.60	13	
Mustard (Golden's)			25.0		5.50	0.52	35	

continues

Table 5–A2 *continued*

Mustard (Stop & Shop)			25.0		3.40	0.56	20	
Tomato conc.	% TS-25.74	Parallel plates	25.0		297.0	0.38		Alviar et al., 1990
	24.16	Carri-Med	25.0		65.7	0.27		
	17.86		25.0		26.1	0.26		
	11.78		25.0		6.00	0.33		
	% TS	Back						Alviar et al., 1990
	25.74	extrusion	25.0		249.0	0.29		
	24.16		25.0		58.5	0.21		
	17.86		25.0		24.3	0.23		
	11.78		25.0		5.8	0.27		
Tomato ketchup		Bohlin VOR conc cylinder	25.0	50–2000	6.1	0.40		Autio, 1991
		Haake RV3 conc cylinder	25.0	50–2000	5.6	0.41		
Conc. orange juice		Haake RV12 conc cylinder						
Hamlin early	42.5 °Brix		25.0	0–500	0.08	0.86	1.36	Crandall et al., 1982
			15.0	0–500	0.13	0.86	1.88	
			0.0	0–500	0.17	0.97	3.72	
			−10.0	0–500	0.35	0.95	5.64	
Hamlin late	41.4 °Brix		25.0	0–500	0.031	1.06	0.94	
			15.0	0–500	0.229	0.77	2.07	
			0.0	0–500	0.370	0.83	3.72	
			−10.0	0–500	0.339	0.96	5.64	

continues

Table 5–A2 *continued*

Product	*Conc. (% solids)*	*Method*	*Temp (°C)*	*Rate Shear (s^{-1})*	K* *(Pa s^n)*	n	*Yield Stress (Pa)*	*Reference*
Pineapple	40.3 °Brix		25	0–500	0.06	0.91	0.90	
early			15	0–500	0.15	0.83	1.7	
			0	0–500	0.26	0.90	3.1	
			−10	0–500	0.08	1.19	7.5	
Pineapple	41.8 °Brix		25	0–500	0.10	0.91	2.8	
late			15	0–500	0.18	0.88	4.3	
			0	0–500	0.46	0.88	6.2	
			−10	0–500	0.47	0.99	15.0	
Valencia	43.1 °Brix		25	0–500	0.12	0.84	1.5	
early			15	0–500	0.19	0.82	1.9	
			0	0–500	0.36	0.86	4.3	
			−10	0–500	0.89	0.81	7.5	
Valencia	41.9 °Brix		25	0–500	0.15	0.84	24.4	
late			15	0–500	0.26	0.82	3.4	
			0	0–500	0.48	0.88	6.2	
			−10	0–500	0.77	0.92	15.0	
Clarified	20.0	Conc cylinder	30.0	50–750	0.001			Lopez et al., 1989
banana	30.0	Rheotest2	30.0		0.003			
juice	39.6		30.0		0.006			
	49.0		30.0		0.010			
Tamarind	2	Conc cylinder	24.0	3–1300	0.00045	0.87	0.003	Bhattacharya et al.,
kernel flour,	4	Rheotest2	24.0		0.00056	0.84	0.004	1991
pH = 7,	6		24.0		0.00167	0.82	0.015	
suspension	8		24.0		0.00488	0.78	0.019	
	10		24.0		0.00681	0.75	0.021	

* Some values were converted from cp or mPa s units.

Table 5–A3 Solids Fraction and Steady Shear Rheological Properties of Tomato Pastes and Serum. Rao and Cooley, 1992.

Solids Fraction* %	*Flow Behavior Index,* n (–)	*Consistency Index,* K (Pa s^n)	*Yield Stress* σ_{0c} (Pa)	*Casson Viscosity,* η_{Ca} (Pa s)	*Serum Viscosity* (mPa s)
0.81	0.25	252.4	176.9	7.8	43.3
0.78	0.18	206.4	196.3	1.1	96.5
0.73	0.29	222.6	133.5	10.2	62.7
0.72	0.30	188.8	109.3	9.6	36.5
0.54	0.22	164.5	157.9	1.4	4.3
0.62	0.24	168.7	123.0	3.8	—
0.71	0.27	220.5	138.1	8.8	56.4
0.74	0.29	240.2	129.3	15.0	29.0
0.71	0.29	231.2	130.6	12.8	25.4
0.72	0.29	207.6	125.5	9.5	53.8
0.82	0.30	245.1	149.3	13.0	140.3
0.77	0.35	234.2	127.6	17.3	76.5
0.73	0.34	245.7	113.9	22.2	61.8
0.63	0.32	205.3	99.5	16.3	5.5
0.69	0.32	204.7	105.0	14.2	53.6
0.72	0.40	324.5	77.9	85.6	76.0
0.66	0.31	238.4	144.2	13.1	125.8
0.65	0.29	200.5	140.5	7.0	59.0
0.69	0.28	197.3	120.0	8.3	48.2
0.66	0.25	179.2	131.2	3.8	5.0
0.66	0.33	273.1	112.4	32.5	—
0.71	0.33	174.0	121.0	7.1	48.2
0.53	0.16	139.0	170.2	0.1	29.1
0.50	0.13	177.1	211.6	0.1	5.7
0.53	0.22	77.8	92.7	0.2	32.3

* Solids content of pastes; serum was obtained by removing insoluble solids by centrifugation.

Table 5–A4 Effect of Temperature on Rheological Properties of Fruit and Vegetable Products

Effect of temperature model based on viscosity, $\eta_a = \eta_\infty \exp(E_{av}/RT)$
Effect of temperature model based on consistency index, $K = K_\infty \exp(E_{ak}/RT)$
where T is temperature (°K), η_∞ and K_∞ are collision factors or values at infinite temperature, R is gas constant, and E_{av} and E_{ak} are activation energies based in viscosity and consistency index, respectively.

Sample	η_∞ *(Pa s)*	E_{av} *(kJ mol^{-1})*	*Shear Rate (s^{-1})*	K_∞ *Pa s^n*	E_{ak} *(kJ mol^{-1})*	*Temp. Range (°C)*	*Reference*
Pera Orange Juice, 65 °Brix							
0% Pulp	1.299E-10	51.5	100	7.672E−11	54.0	30–(−18)	Vitali and Rao, 1984b
3.4%	7.755E-10	48.2		9.881E−10	49.8		
5.7%	4.400E-09	43.1		3.130E−08	43.1		
8.6%	9.526E-09	44.0		1.742E−08	45.2		
11.1%	2.202E-08	44.0		4.195E−08	44.4		
Pera Orange Juice, 5.7% Pulp							
65 °Brix	4.410E-09	44.8	100	3.13E−08	43.1	30–(−18)	
62 °Brix	2.670E-08	39.4		1.39E−07	40.6		
58 °Brix	9.340E-08	34.8		1.31E−07	36.8		
56 °Brix	1.280E-07	33.5		8.95E−08	37.3		
52 °Brix	1.640E-07	31.4		6.62E−08	35.6		
Pera Orange Juice, 4.6% Pulp							
65 °Brix	3.45E-09	45.6	100	1.86E−08	10.6	30–(−18)	
62 °Brix	2.48E-08	39.8		3.77E−08	40.2		
58 °Brix	9.56E-08	35.6		2.30E−07	36.0		
56 °Brix	1.29E-07	33.9		1.09E−07	36.8		
52 °Brix	1.83E-07	31.8		1.66E−07	34.8		

continues

Concentrated Apple Juice from McIntosh Apples						
45.1 °Brix	3.394E−07	25.1				
50.4 °Brix	1.182E−07	28.9				
55.2 °Brix	2.703E−09	39.4				
60.1 °Brix	3.935E−10	45.6				
64.9 °Brix	7.917E−12	56.9				
68.3 °Brix	1.156E−12	64.1				
Concentrated Grape Juice from Concord Grapes						
43.1 °Brix	8.147E−08	29.3				
49.2	1.074E−08	35.6				
54	9.169E−10	43.1				
59.2	1.243E−10	49.4				
64.5	1.340E−12	51.5				
68.3	6.086E−12	60.7				
Mango Juice						
16 °Brix, 3.49% pectin			0.0145	15.0	30–70	Manohar et al., 1991
20 °Brix, 3.49% pectin			0.01677	15.5	30–70	
25 °Brix, 3.49% pectin			0.03275	14.6	30–70	
30 °Brix, 3.49% pectin			0.0339	15.8	30–70	
16 °Brix, 3.38% pectin			0.00894	14.9	30–70	
16 °Brix, 3.29% pectin			0.00855	14.7	30–70	
16 °Brix, 3.23% pectin			0.004	15.2	30–70	
21 °Brix, 3.23% pectin			0.00614	15.1	30–70	
25 °Brix, 3.23% pectin			0.00799	15.1	30–70	
28 °Brix, 3.23% pectin			0.0122	15.3	30–70	

continues

Table 5–A4 *continued*

Sample	η_∞ *(Pa s)*	E_{av} *(kJ mol^{-1})*	*Shear Rate (s^{-1})*	K_∞ *Pa s^n*	E_{ak} *(kJ mol^{-1})*	*Temp. Range (°C)*	*Reference*
Depectinized Apple Juice							Saravacos, 1970
75 °Brix		59.5					
50 °Brix		35.2					
30 °Brix		26.4					
15 °Brix		22.2					
Cloudy Apple Juice							Saravacos, 1970
65.5 °Brix		38.1					
50 °Brix		25.5					
40 °Brix		24.3					
30 °Brix		21.4					
15 °Brix		14.7					
Peach Puree, 11.7 °Brix		7.1					Saravacos, 1970
Concord Grape Juice							Saravacos, 1970
64 °Brix		46.9					
50 °Brix		28.9					
30 °Brix		26.0					
15 °Brix		22.2					
Passion Fruit Juice—Concentrated							Vitali et al., 1974
15.6 °Brix		18.8					
20.7 °Brix		17.2					
25.3 °Brix		16.7					
30.6 °Brix		15.9					
33.4 °Brix		13.4					

continues

Table 5–A4 *continued*

Mango Pulp Juice, 16 °Brix	16.0		Rao et al., 1985
Mango Pulp Juice, 17 °Brix	11.0		Gunjal and Waghmare, 1987
Combined Concentration and Temperature Models & Data			
Tamarind Juice, 7–62 °Brix	18.7	30–70	Manohar et al., 1991
$K_o = AC^b$; $A = 1.091$ E-11, $b = 4.815$, C is concentration in °Brix			
$K = K_o \exp(E_{ak}/RT)$; $E_{ak} = 18.7$ kJ mol^{-1}			
Raspberry Juice, 15–41 °Brix	37.9	5–60	Ibarz and Pagan, 1987
$K = K_\infty \exp(E_{ak}/RT)$; $E_{ak} = 37.9$ kJ mol^{-1}			
$K_\infty = AC^b$; $A = 2.076$E-15 and $b = 5.018$			
Concentrated Apple Juice, 71 °Brix			Ibarz et al., 1987
$\eta_\infty = 3.385$E-10 $E_{ar} = 50.2$			
$K = AC^b$, $A = 9.304$E-14, $b = 4.892$			
$K = K_1 \exp(BC)$, $K_1 = 7.9$E-5, $B = 0.106$			
Clarified Banana Juice	25.1	50–750	Lopez et al., 1989

Table 5–A5 Creep-Compliance Parameters of Tomato Concentrates. Yoo and Rao, 1996.

Sample (°Brix)	J_0 *(Pa^{-1})*	J_1 *(Pa^{-1})*	τ_1 *(s)*	J_2 *(Pa^{-1})*	τ_2 *(s)*	η *(Pa s)*
Concentrates from 0.69 mm (0.027 in.) Finisher Screen						
21.0	1.93×10^{-4}	1.37×10^{-4}	3.96	2.05×10^{-4}	24.1	1.11×10^{6}
	(5.73×10^{-5})	(1.28×10^{-4})	(1.08)	(8.99×10^{-5})	(4.07)	(5.34×10^{5})
22.7	2.04×10^{-4}	1.64×10^{-4}	2.84	1.90×10^{-4}	26.6	8.83×10^{5}
	(8.03×10^{-6})	(3.67×10^{-5})	(0.95)	(2.23×10^{-5})	(2.20)	(6.42×10^{4})
25.9	9.58×10^{-5}	4.51×10^{-5}	3.23	1.14×10^{-4}	23.2	1.76×10^{6}
	(1.26×10^{-5})	(1.42×10^{-5})	(1.10)	(3.38×10^{-5})	(4.24)	(3.67×10^{5})
29.8	8.42×10^{-5}	4.38×10^{-5}	2.83	9.56×10^{-5}	20.6	1.88×10^{6}
	(2.43×10^{-6})	(1.11×10^{-5})	(1.10)	(2.92×10^{-5})	(2.56)	(2.91×10^{5})
32.2	7.77×10^{-5}	4.60×10^{-5}	3.13	1.06×10^{-4}	21.0	1.96×10^{6}
	(7.72×10^{-6})	(1.46×10^{-5})	(0.47)	(3.31×10^{-5})	(1.67)	(2.21×10^{5})
Concentrates from 0.84 mm (0.033 in.) Finisher Screen						
21.0	1.68×10^{-4}	7.22×10^{-5}	3.17	1.87×10^{-4}	20.6	9.66×10^{5}
	(3.15×10^{-5})	(1.97×10^{-5})	(0.78)	(4.31×10^{-5})	(3.94)	(2.99×10^{5})
24.4	1.25×10^{-4}	5.27×10^{-5}	2.61	1.12×10^{-4}	19.4	1.40×10^{6}
	(2.48×10^{-5})	(1.04×10^{-5})	(0.91)	(3.00×10^{-5})	(1.12)	(2.34×10^{5})
30.4	7.51×10^{-5}	4.54×10^{-5}	3.68	7.04×10^{-5}	22.44	2.84×10^{6}
	(7.29×10^{-6})	(1.16×10^{-5})	(0.73)	(2.09×10^{-5})	(3.33)	(3.58×10^{5})

Numbers in parentheses are standard deviations of parameters.

Table 5–A6 Magnitudes of Yield Stress of Apple Sauce Determined Using the Vane Method and Extrapolated (Bingham) Values. Qiu and Rao, 1988.

	Vane† and RPM				
*Sample**	*I1, 1.0 rpm*	*I1, 0.4 rpm*	*I2, 1.0 rpm*	*I2, 0.4 rpm*	*Bingham Yield Value*
RGA25	63.9	55.8	66.1	58.0	70.6
RGA27	58.9	53.1	55.9	51.8	66.2
RGA29	87.1	76.0	84.1	74.2	81.6
RGA45	83.7	73.7	80.9	76.9	81.0
RGA47	86.0	72.1	87.6	81.4	82.2
RGA49	55.8	52.0	54.5	51.3	73.1
RGB25	56.6	46.5	56.1	47.2	70.8
RGB27	57.4	48.1	54.0	45.9	69.2
RGB29	67.0	57.0	68.8	62.0	71.6
RGB45	51.5	45.8	52.6	47.5	62.6
RGB47	58.9	49.3	54.0	48.0	69.4
RGB49	76.0	65.1	72.3	64.2	74.4

* Sample code: RG stands for Rhode Island Greening cultivar; A indicates firmness of 37.4 N, and B for 74.3 N; the first number indicates the finisher screen opening size: 2 for 1.6 mm screen, 4 for 3.2 mm screen; the last number indicates the finisher speed (500, 700, or 900 rpm).

† I1 is an eight-blade vane described by Rao and Cooley (1984); I2 is a six-blade vane described in Qiu and Rao (1988).

Table 5–A7 Casson Model Parameters of Commercial Chocolate Samples. Chevally, 1991.

Brand Name	*Temp. (°C)*	*Shear Rate (s^{-1})*	*Plastic Viscosity, η_∞ (Pa s)*	*Yield Stress (Pa)*
ArniCoop	40	5–60	2.5	10.9
		60–5	2.5	9.1
Cailler (Nestle)		5–60	3.7	16.5
Cadbury Dairy		5–60	4.4	33.7
Milk GB		60–5	4.8	20.5
Galaxy GB		5–60	4.7	78.0
		60–5	4.9	49.2
Lindt F		5–60	1.8	10.3
		60–5	1.9	8.9
Nestle F		5–60	2.6	19.8
		60–5	2.6	15.9
Poulain F		5–60	2.7	19.4
Suchard Milka		5–60	1.9	12.2
		60–5	1.9	9.4
Yorkie GB Dark chocolate		5–60	4.2	10.0
Bournville		5–60	3.9	42.9
Black Magic		5–60	2.4	10.1
Cote d'Or F		5–60	4.0	23.8
Cremant		5–60	2.70	25.8
Terry Plain GB		5–60	1.90	20.4
		60–5	1.90	19.0
Nestle White		5–60	4.70	14.2
Rowntree White		5–60	6.20	9.6
Suchard White		5–60	2.40	38.5

Note: The rheological method is Haake RV20.

Table 5–A8 Casson Model Parameters of Cocoa Mass as a Function of Temperature. Fang et al., 1996.

	Predegasified Sample		*Degasified Sample*	
Temp	σ_{OC}	η_∞	σ_{OC}	η_∞
40	19.0	4.55	2.45	4.64
60	19.6	1.77	2.36	1.94
80	14.8	1.13	2.36	0.84
93	7.31	1.12	2.12	0.573
110			1.35	0.331
130			0.97	0.204

Table 5–A9 Formulations of Three Mayonnaises. Kiosseoglu and Sherman, 1983b.

	% wt/wt		
Ingredient	*1*	*2*	*3*
Egg yolk	6.00	3.99	6.00
Egg white	—	8.28	12.00
Water	12.00	5.73	—
Acetic acid	0.20	0.20	0.20
Sugar	1.55	1.55	1.55
Salt	0.25	0.25	0.25
Oil	80.00	80.00	80.00

Table 5–A10 Viscoelastic Parameters of Model Salad Dressing During Storage at 2.8°C. Paredes et al., 1989.

Storage Time (Days)	E_0 *(dyne cm^{-2} × 10^{-3})*	E_1 *(dyne cm^{-2} × 10^{-3})*	η_1 *(poise × 10^{-3})*	η_N *(poise × 10^{-3})*
1	4.17	1.08	7.5	74.4
3	1.64	1.92	16.7	114.6
7	1.49	2.38	31.3	51.6
14	0.72	0.89	52.8	154.2
21	0.65	0.43	132.8	492.4
28	0.63	1.00	180.0	493.8
42	4.17	1.45	15.0	303.0

Table 5–A11 Rheological Properties of Mustard, Mayonnaise, Oils, and Fats

Item	*Temp. (°C)*	*Shear Rate (s^{-1})*	*Consistency* Index (Pa s^n)*	*Flow Behavior Index*	*Reference*
Mustard	25	30–1300	18.50	0.39	Kokini, 1992
	25	30–1300	19.10	0.39	
	25	40–1100	27.00	0.34	
	25	40–1100	33.00	0.28	
Castor	10		2.42	1.0	Kokini, 1992
	30		0.451	1.0	
	40		0.231	1.0	
	100		0.017	1.0	
Corn	38		0.317	1.0	Kokini, 1992
	25		0.057	1.0	

continues

Table 5–A11 *continued*

Item	*Temp. (°C)*	*Shear Rate (s^{-1})*	*Consistency* Index (Pa s^n)*	*Flow Behavior Index*	*Reference*
Cottonseed	20		0.070	1.0	Kokini, 1992
	38		0.039	1.0	
Linseed	50		0.018	1.0	Kokini, 1992
	90		0.007	1.0	
Olive	10		0.138	1.0	Kokini, 1992
	40		0.036	1.0	
	70		0.012	1.0	
Peanut	25		0.066	1.0	
	38		0.025	1.0	
	21.1	0.32–64	0.065	1.0	
	37.8	0.32–64	0.039	1.0	
	54.4	0.32–64	0.027	1.0	
Rapeseed	0		2.530	1.0	
	20		0.163	1.0	
	30		0.096	1.0	
Safflower	38		0.029	1.0	
	25		0.092	1.0	
Sesame	38		0.032	1.0	
Soybean	30		0.041	1.0	
	50		0.021	1.0	
	90		0.008	1.0	
Sunflower	38		0.031	1.0	
Mayonnaise	25	30–1300	6.40	0.55	
	25	30–1300	6.60	0.54	
	25	40–1100	4.20	0.60	
	25	40–1100	4.70	0.59	
Cream (10% fat)	40		0.00148	1.0	Kokini, 1992
	60		0.00107	1.0	
	80		0.00083	1.0	
Cream (20% fat)	40		0.00238	1.0	
	60		0.00171	1.0	
	80		0.00129	1.0	
Cream (30% fat)	40		0.00395	1.0	
	60		0.00289	1.0	
	80		0.0022	1.0	
Cream (40% fat)	40		0.0069	1.0	
	60		0.0051	1.0	
	80		0.00395	1.0	

continues

Table 5–A11 *continued*

Item	*Temp. (°C)*	*Shear Rate (s^{-1})*	*Consistency* Index (Pa s^n)*	*Flow Behavior Index*	*Reference*
Canola	4		0.1627	1.0	Lang et al., 1992
	10		0.1172	1.0	
	25		0.057	1.0	
	40		0.0334	1.0	
Refined, bleached	4		0.1568	1.0	
Winterized	10		0.1197	1.0	
	25		0.0589	1.0	
	40		0.0333	1.0	
Crambe	23.9		0.0867	1.0	Noureddini et al., 1992
	37.8		0.0487	1.0	
	48.9		0.0330	1.0	
	60.0		0.0234	1.0	
	82.2		0.0132	1.0	
	100		0.00906	1.0	
	110		0.00755	1.0	
Rapeseed	23.9		0.0788	1.0	
	37.8		0.0449	1.0	
	48.9		0.0303	1.0	
	60		0.0214	1.0	
	82.2		0.0121	1.0	
	100		0.00841	1.0	
	110		0.00715	1.0	
Corn	23.9		0.0523	1.0	
	37.8		0.0308	1.0	
	48.9		0.0227	1.0	
	60		0.0157	1.0	
	82.2		0.00928	1.0	
	100		0.00657	1.0	
	110		0.0056	1.0	
Soybean	23.9		0.0543	1.0	
	37.8		0.0318	1.0	
	48.9		0.0233	1.0	
	60		0.00161	1.0	
	82.2		0.0951	1.0	
	100		0.00673	1.0	
	110		0.00571	1.0	

continues

Table 5–A11 *continued*

Item	*Temp. (°C)*	*Shear Rate (s^{-1})*	*Consistency* Index (Pa s^n)*	*Flow Behavior Index*	*Reference*
Milkweed	23.9		0.0563	1.0	
	37.8		0.0327	1.0	
	48.9		0.0238	1.0	
	60.0		0.0164	1.0	
	82.2		0.0096	1.0	
	100		0.00679	1.0	
	110		0.00571	1.0	
Coconut	37.8		0.028	1.0	
	48.9		0.0198	1.0	
	60		0.0133	1.0	
	82.2		0.00755	1.0	
	100		0.00522	1.0	
	110		0.00436	1.0	
Lesquerella	23.9		0.275	1.0	
	37.8		0.121	1.0	
	48.9		0.0698	1.0	
	60		0.0435	1.0	
	82.2		0.0204	1.0	
	100		0.0127	1.0	
	110		0.0102	1.0	
Nonanoic	23.9		0.00725	1.0	
	37.8		0.00497	1.0	
	48.9		0.00392	1.0	
	60		0.00303	1.0	
	71.1		0.00248	1.0	
	82.2		0.00202	1.0	
	98.9		0.00156	1.0	
	110		0.00137	1.0	
Capric	37.8		0.00633	1.0	
	48.9		0.00496	1.0	
	60		0.00372	1.0	
	71.1		0.00302	1.0	
	82.2		0.00242	1.0	
	98.9		0.00184	1.0	
	110		0.0016	1.0	
Lauric	48.9		0.00747	1.0	
	60		0.00542	1.0	
	71.1		0.00428	1.0	
	82.2		0.00338	1.0	
	98.9		0.0025	1.0	
	110		0.0021	1.0	

continues

Table 5–A11 *continued*

Item	*Temp. (°C)*	*Shear Rate (s^{-1})*	*Consistency* Index (Pa s^n)*	*Flow Behavior Index*	*Reference*
Myristic	60		0.00748	1.0	
	71.1		0.00583	1.0	
	82.2		0.00450	1.0	
	98.9		0.00326	1.0	
	110		0.00278	1.0	
Palmitic	71.1		0.00764	1.0	
	82.2		0.00580	1.0	
	98.9		0.00413	1.0	
	110		0.00347	1.0	
Stearic	82.2		0.00731	1.0	
	98.9		0.00508	1.0	
	110		0.00424	1.0	
	121.1		0.00350	1.0	
Oleic	23.9		0.02900	1.0	
	37.8		0.01770	1.0	
	48.9		0.01320	1.0	
	60		0.00932	1.0	
	71.1		0.00727	1.0	
	82.2		0.00561	1.0	
	98.9		0.00403	1.0	
	110		0.00341	1.0	
Erucic	37.8		0.03230	1.0	
	48.9		0.02240	1.0	
	60		0.01590	1.0	
	71.1		0.01190	1.0	
	82.2		0.00890	1.0	
	98.9		0.00618	1.0	
	110		0.00516	1.0	
Palm oil	35.0		0.0963	1.0	Timms, 1985
	40.0		0.0404	1.0	
	45.0		0.0333	1.0	
	50.0		0.0270	1.0	
Palm olein	20.0		0.0853	1.0	
	25.0		0.0692	1.0	
	30.0		0.0563	1.0	
	35.0		0.0434	1.0	
	40.0		0.0386	1.0	
	45.0		0.0326	1.0	
	50.0		0.0265	1.0	

continues

Table 5–A11 *continued*

Item	*Temp. (°C)*	*Shear Rate (s^{-1})*	*Consistency* Index (Pa s^n)*	*Flow Behavior Index*	*Reference*
Palm kernel	30.0		0.0440	1.0	
	35.0		0.0356	1.0	
	40.0		0.0300	1.0	
	45.0		0.0254	1.0	
	50.0		0.0210	1.0	
Coconut	25.0		0.0481	1.0	
	30.0		0.0398	1.0	
	35.0		0.0329	1.0	
	40.0		0.0274	1.0	
	45.0		0.0233	1.0	
	50.0		0.0190	1.0	
Soybean	20.0		0.0636	1.0	
	25.0		0.0516	1.0	
	30.0		0.0431	1.0	
	35.0		0.0368	1.0	
	40.0		0.0317	1.0	
	45.0		0.0268	1.0	
	50.0		0.0225	1.0	
Peanut oil	30.0		0.0345	1.0	Krishna, 1993
Peanut +6.8% free fatty acids	30.0		0.0360	1.0	
Peanut +3% rice bran wax	30.0		0.0626	1.0	
Peanut +1% oryzanol	30.0		0.0374	1.0	
Peanut +1% glyceride mono-stearate	30.0		0.0375	1.0	
Peanut +2% soy lecithin	30.0		0.0414	1.0	

continues

Table 5–A11 *continued*

Item	*Temp. (°C)*	*Shear Rate (s^{-1})*	*Consistency* Index (Pa s^n)*	*Flow Behavior Index*	*Reference*
Rice bran oil triglyceride	30.0		0.0366	1.0	
Comm. rice bran oil	30.0		0.0370	1.0	
Cocoa butter using Ubbelohde-type glass capillary viscometer					Sterling and Wuhrmann, 1960
	30.0		0.0722	1.0	
	28.0		0.0864	1.0	
	27.0		0.0986	1.0	
	26.0		0.1114	1.0	
	25.0		0.1241	1.0	
	24.0		0.1380	1.0	

* Some values were converted from cp or mPa s units.

Table 5–A12 Effect of Temperature on Rheological Properties of Oils and Fats. Lang et al., 1992.

Oil	*Temp Range*	*Temperature Model*	*Model Constants*
Canola (refined)	4–100°C	$\vartheta = \exp(C_o + C_1 T + C_2 T^2)$; ϑ = kinematic viscosity, $[(mm)^2\ s^{-1}]$	$C_0 = 0.0519$, $C_1 = 0.0002$
Crambe	23.9–110°C	$\ln \eta = A + \frac{B}{T + C}$; η = viscosity (mPa s), T(K)	$A = -2.0277E+00$, $B = 9.2659E+02$, $C = 1.5427E+02$
Rapeseed	23.9–110°C	Same	$A = -2.44621E+01$, $B = 6.21021E+03$, $C = 2.6671E-02$
Corn	23.9–110°C	Same	$A = -1.95604E+01$, $B = 5.22544E+03$, $C = 1.9943E-02$
Soybean	23.9–110°C	Same	$A = -2.00596E+01$, $B = 5.32374E+03$, $C = 2.0634E-02$
Milkweed	23.9–110°C	Same	$A = -2.03845E+01$, $B = 5.40303E+03$, $C = 2.0948E-02$
Coconut	23.9–110°C	Same	$A = 1.0945E+01$, $B = -2.25154E+03$, $C = 9.1778E+05$
Lesquerella	23.9–110°C	Same	$A = 6.6123E+00$, $B = -6.28877E+03$, $C = 1.7800E+06$
Nonanoic	23.9–110°C	Same	$A = -1.28399E+01$, $B = 3.44593E+03$, $C = 1.0839E-02$
Capric	37.8–110°C	Same	$A = -2.6280E+00$, $B = 7.1762E+02$, $C = -1.5129E+02$
Lauric	48.9–110°C	Same	$A = -2.5030E+00$, $B = 7.0888E+02$, $C = -1.6475E+02$
Myristic	60.0–110°C	Same	$A = 6.1590E-01$, $B = -1.94571E+03$, $C = 8.0399E+05$
Palmitic	71.1–110°C	Same	$A = -3.43507E+01$, $B = 7.86679E+03$, $C = 3.9312E-02$
Stearic	82.2–110°C	Same	$A = 2.6780E+00$, $B = 3.43749E+03$, $C = 1.1344E+06$
Oleic	23.9–110°C	Same	$A = -1.64441E+01$, $B = 4.55414E+03$, $C = 1.5075E-02$
Erucic	37.8–110°C	Same	$A = -1.82905E+01$, $B = 5.10332E+03$, $C = 1.7229E-2$

Note: The rheological method is Cannon-Fenske.

Table 5–A13 Rheological Properties of Egg Products and Activation Energies of Flow. Rao, 1997.

Name of Product	*Conc. (% Solids)*	*pH*	*Method*	*Temp (°C)*	K *(Pa s^n)*	n	σ_o *(Pa)*	E_a *(kJ mol^{-1})*
Whole egg	25.0	7.5	Capillary visc	30	0.0064	1.0		24.7
Stabilized egg white	12.0	7.0	Capillary visc	30	0.0027	1.0		16.2
Plain yolk	45.0	6.8	Capillary visc	30	0.0920	1.0		26.8
Sugared yolk (10% sucrose)	50.0	6.3	Capillary visc	30	0.0560	1.0		27.5
Salted yolk (10% salt)	50.0	6.0	Capillary visc	30	0.0400	1.0		34.9

Table 5–A14 Rheological Properties of Meats

Product	*Conc. (% solids)*	*Method*	*Temp (°C)*	*Shear Rate (s^{-1})*	K *(Pa s^n)*	n	*Yield Stress (Pa)*	*Reference*
Minced beef	26.8	Haake MV1	20	4.7–1694	4.6	0.33	4.4	Bianchi et al., 1985
21.7% protein								
4.7% fat								
73.2% water								
Minced beef/soy protein mixtures								
9:1	26.8				2.1	0.49	4.4	
7:3	26.8				9.2	0.30	7.3	
5:5	26.8				12.0	0.31	18	

Table 5–A15 Rheological Properties of Dairy Products

Product	*Conc. (% Solids)*	*Method*	*Temp (°C)*	*Shear Rate (s^{-1})*	*K* (Pa s^n)*	n	*Yield Stress (Pa)*	*Reference*
Milk			20		0.0020	1.00		Kokini, 1992
			30		0.0015	1.00		
			40		0.0011	1.00		
			50		0.0010	1.00		
			60		0.0008	1.00		
			70		0.0007	1.00		
			80		0.0006	1.00		
			0		0.0034	1.00		
			5		0.0031	1.00		
			10		0.0026	1.00		
			15		0.0023	1.00		
			20		0.0020	1.00		
			25		0.0017	1.00		
			30		0.0015	1.00		
			35		0.0013	1.00		
			40		0.0012	1.00		
Skim milk	Normal	Unknown	25		0.0014	1.00		Hayes, 1987
Whole milk	12.3							Wayne and Shoemaker, 1988
Whole milk	Normal	Unknown	20		0.0212	1.00		
Whole milk	12.3	Rotational	25		0.0018	1.00		
0.05482%	11.9	Rheometer	25		0.0018	1.00	0.021	
Butter fat milk	12.1	Weissenberg	25		0.0018	1.00	0.0211	

continues

Table 5–A15 *continued*

Product	*Conc. (% Solids)*	*Method*	*Temp (°C)*	*Shear Rate (s^{-1})*	*K* (Pa s^n)*	n	*Yield Stress (Pa)*	*Reference*
Skim milk	9.7	1° cone and	25		0.0016	1.00	0.0465	
	15.5	plate	25		0.0024	1.00	0.0747	
	16.3		25		0.0026	1.00	0.0357	
	17.7		25		0.0028	1.00	0.0268	
	18.9		25		0.0031	1.00	0.0341	
	20.2		25		0.0034	1.00	0.0371	
Stirred yoghurts								
0.05% Fat	12.5	Rheomat 30	10	15–300	8.7000	.250	2	Benezech and
	11.6	& Carrimed	10	15–300	13.4000	.160	2	Maingonnat,
	12.0	CS 100	10	15–300	8.9000	.250	2	1993
1.6% Fat	16.0	(cone and	10	15–300	14.3000	.195	2	
		plate)						
Yoplait	Unknown	Haake	10	0–500	1.4300	.531	6.94	Ramaswamy
			15	0–500	1.2100	.532	7.16	and Basak,
			20	0–500	0.8600	.568	6.34	1991
			25	0–500	0.6100	.596	5.45	
Delisle	Unknown		10	0–500	3.6200	.418	13.7	
Whey protein solution								
	10	Bohlin VOR						
	20	(cone and	22	0–230	2.7700	1.00		
	30	plate)			12.2	.960		
					62.1	.950		

continues

Whole Milk Retentates							
6.29% protein, 6.88% fat, 1.03% ash, 6.08% lactose							Roscane and Smith, 1993
	20.27	Haake Rotovisco	5	1–1170	0.01084	0.97	
6.44% protein, 7.06% fat, 1.03% ash, 5.96 lactose							
	20.49		5		0.01109	0.96	
6.49% protein, 7.03% fat, 1.06% ash, 6.45% lactose							
	21		5		0.01862	0.90	
10.19% protein, 10.12% fat, 1.36% ash, 6.46% lactose							
	28.13		5		0.04074	0.95	
10.70% protein, 10.60% fat, 1.42% ash, 6.90% lactose							
	29.62		5		0.04841	0.93	
9.83% protein, 10.33% fat, 1.06% ash, 6.45% lactose							
	28.82		5		0.0389	0.94	
10.19% protein, 10.12% fat, 1.36% ash, 6.46% lactose							
	28.13		50		0.00385	1.03	
10.70% protein, 10.60% fat, 1.42% ash, 6.90% lactose							
	29.62		50		0.00359	1.04	
9.83% protein, 10.33% fat, 1.06% ash, 6.45% lactose							
	28.82		50		0.004416	1.01	
10.19% protein, 10.12% fat, 1.36% ash, 6.46% lactose							
	28.13		75		0.00385	1.18	
10.70% protein, 10.60% fat, 1.42% ash, 6.90% lactose							
	29.62		75		0.00359	1.090	
9.83% protein, 10.33% fat, 1.06% ash, 6.45% lactose							
	28.82		75		0.00441	1.130	
11.63% protein, 14.4% fat, 1.67% ash, 8.28% lactose							
	35.99		5		0.6607	0.810	
11.07% protein, 13.02% fat, 1.60% ash, 8.50% lactose							
	34.19		5		0.3715	0.810	

continues

Product	Conc. (% Solids)	Method	Temp (°C)	Shear Rate (s^{-1})	K* (Pa s^n)	n	Yield Stress (Pa)	Reference
10.48% protein, 13.10% fat, 1.61% ash, 9.62% lactose	34.61		5		0.3758	0.810		
11.63% protein, 14.4% fat, 1.67% ash, 8.28% lactose	35.99		50		0.0254	0.900		
11.07% protein, 13.02% fat, 1.60% ash, 8.50% lactose	34.19		50		0.01428	0.950		
10.48% protein, 13.10% fat, 1.61% ash, 9.62% lactose	34.61		50		0.008128	1.040		
11.63% protein, 14.4% fat, 1.67% ash, 8.28% lactose	35.99		75		0.1778	0.690		
11.07% protein, 13.02% fat, 1.60% ash, 8.50% lactose	34.19		75		0.1884	0.650		
10.48% protein, 13.10% fat, 1.61% ash, 9.62% lactose	34.61		75		0.6918	0.540		
Dulce de leche milk, household sample								
	71, pH = 6.1	Haake	25	1.35–28.35	28.38	0.6	45.46	Rovedo et al., 1991
	71	Conc cyl.	40		15.69	0.61	18.28	
	71	MV11	55		5.7	0.65	9.65	
Confectionery sample								
	71, pH = 6.1	Haake	25		78.3	0.43	4.57	
	71	Conc cyl.	40		36.7	0.47	4.85	
	71	MV11	55		15.76	0.47	0.0	

Effect of Temperature on Concentrated Freeze Skim Milk Viscosity Data Obtained over the Shear Rate Range 0.1–14 s^{-1} and Temperature Range 20° to −3.4°C — Chang and Hartel, 1997

Total Solids (%)	η_∞ (Pa s)	E_a (kJ mol^{-1})
9.56	5.62E−8	25.1
15	1.92E−9	28.9
20	5.93E−8	32.9
25	2.83E−8	30.7
30	4.68E−10	42.0
35	1.03E−11	52.8

* Some values were converted from cp or mPa s units.

Table 5–A16 Rheological Properties of Foods Containing Sugars

Product	*Conc. (% Solids)*	*Method*	*Temp. (°C)*	*Shear Rate (s^{-1})*	*K* (Pa s^n)*	*n*	*Reference*
Corn syrup	48.4	Conc cyl.	26.6	1–1000	0.0053	1.0	Harper, 1960
	54.5				0.0106	1.0	
	63.6				0.0338	1.0	
	72.5				0.1600	1.0	
	29.6				0.0205	1.0	
	36.4				0.0228	1.0	
	40.8				0.0290	1.0	
	36.9				0.0245	1.0	
	41.5				0.0228	1.0	
	23.0				0.0122	1.0	
	25.0				0.0146	1.0	
	29.0				0.0151	1.0	
	31.0				0.0149	1.0	
	30.0				0.0148	1.0	
	36.2				0.0186	1.0	
	39.4				0.0218	1.0	
	42.0				0.0240	1.0	
Corn syrup	77.4 °Brix	Conc cyl.	20	1–10	5.46	1.0	McCarthy and Seymour, 1994
Honey	18.2		8.9		52.8	1.0	Munro, 1943
			9.8		46.8	1.0	
			10.9		40.3	1.0	
			12.1		34.08	1.0	
			12.8		30.00	1.0	
			13.4		27.12	1.0	
			14.2		23.76	1.0	
			14.8		20.88	1.0	
			15.8		17.04	1.0	
			16.8		14.4	1.0	
			17.6		12.24	1.0	
			18.3		10.8	1.0	
			19.1		9.6	1.0	
			20.1		8.4	1.0	
			22.0		6.72	1.0	
			23.3		5.76	1.0	
			25.0		4.8	1.0	
			28.6		3.24	1.0	
			30.8		2.4	1.0	
			34.5		1.87	1.0	

continues

Table 5–A16 *continued*

Product	*Conc. (% Solids)*	*Method*	*Temp. (°C)*	*Shear Rate (s^{-1})*	K* *($Pa\ s^n$)*	n	*Reference*
			36.8		1.51	1.0	
			40.0		1.20	1.0	
			41.6		0.98	1.0	
			42.8		0.86	1.0	
			43.3		0.82	1.0	
			44.5		0.80	1.0	
			45.7		0.75	1.0	
			47.1		0.68	1.0	
			49.7		0.58	1.0	
			51.8		0.52	1.0	
			56.5		0.37	1.0	
			60.1		0.30	1.0	
			63.9		0.24	1.0	
			67.4		0.19	1.0	
			71.6		0.15	1.0	
			75.9		0.11	1.0	
			78.5		0.10	1.0	
			80.0		0.09	1.0	
Buckwheat	18.6		3.9		70.56	1.0	
			5.2		61.20	1.0	
			7.0		52.32	1.0	
			8.1		46.56	1.0	
			9.3		40.80	1.0	
			10.3		36.96	1.0	
			11.1		33.12	1.0	
			12.2		27.60	1.0	
			12.7		25.20	1.0	
			13.8		20.16	1.0	
			15		16.20	1.0	
			16		12.82	1.0	
			17.2		10.32	1.0	
			18.1		8.88	1.0	
			18.5		8.16	1.0	
			19.2		7.44	1.0	
			20.5		6.24	1.0	
			21.9		5.28	1.0	
			22.9		4.68	1.0	
			24.8		3.86	1.0	
			26.5		3.26	1.0	
			28.1		2.83	1.0	
			31		2.16	1.0	
			36.6		1.44	1.0	

continues

Table 5–A16 *continued*

Product	*Conc. (% Solids)*	*Method*	*Temp. (°C)*	*Shear Rate (s^{-1})*	*K* (Pa s^n)*	n	*Reference*
			46.9		0.72	1.0	
			48.7		0.61	1.0	
			54		0.44	1.0	
			58		0.34	1.0	
			65.6		0.23	1.0	
			66.9		0.20	1.0	
			69.4		0.17	1.0	
			72.2		0.15	1.0	
			74.1		0.14	1.0	
			76.2		0.14	1.0	
			77.8		0.14	1.0	
			78.9		0.15	1.0	
			79.6		0.15	1.0	
Goldenrod	19.4		6.6		33.6	1.0	
			7.7		30.0	1.0	
			8.8		26.18	1.0	
			9.5		23.52	1.0	
			10.6		19.92	1.0	
			10.9		15.60	1.0	
			12.7		12.96	1.0	
			13.5		11.28	1.0	
			14		10.32	1.0	
			14.4		9.36	1.0	
			15.7		7.92	1.0	
			16.1		7.44	1.0	
			16.9		6.48	1.0	
			18.0		6.0	1.0	
			19.9		5.04	1.0	
			21.6		3.96	1.0	
			22.9		3.38	1.0	
			24.3		2.93	1.0	
			26.7		2.35	1.0	
			29.8		1.82	1.0	
			32.4		1.51	1.0	
			34.8		1.20	1.0	
			37.0		0.93	1.0	
			39.0		0.82	1.0	
			40.5		0.75	1.0	
			43.4		0.60	1.0	
			46.7		0.49	1.0	

continues

Table 5–A16 *continued*

Product	*Conc. (% Solids)*	*Method*	*Temp. (°C)*	*Shear Rate (s^{-1})*	K* *(Pa s^n)*	n	*Reference*
			48.7		0.41	1.0	
			52.4		0.32	1.0	
			55.8		0.26	1.0	
			59.8		0.21	1.0	
			64.0		0.17	1.0	
			68.0		0.14	1.0	
			71.4		0.12	1.0	
			74.9		0.10	1.0	
			78.2		0.09	1.0	
			80.4		0.07	1.0	
Sweetclover	17		12.0		43.20	1.0	
			13.0		37.44	1.0	
			14.8		30.24	1.0	
			15.7		25.80	1.0	
			16.8		21.12	1.0	
			17.3		18.96	1.0	
			18.6		15.60	1.0	
			19.6		13.68	1.0	
			21.1		11.28	1.0	
			23.0		8.88	1.0	
			24.7		7.20	1.0	
			26.5		6.00	1.0	
			28.2		4.80	1.0	
			30.0		4.08	1.0	
			31.8		3.36	1.0	
			34.6		2.52	1.0	
			35.8		2.16	1.0	
			37.6		1.97	1.0	
			38.7		1.87	1.0	
			40.4		1.56	1.0	
			43.1		1.25	1.0	
			45.9		1.01	1.0	
			48.3		0.86	1.0	
			50.4		0.77	1.0	
			54.5		0.63	1.0	
			58.4		0.52	1.0	
			62.0		0.44	1.0	
			65.9		0.34	1.0	
			68.2		0.28	1.0	
			70.6		0.25	1.0	
			73.4		0.19	1.0	

continues

Table 5–A16 *continued*

Product	*Conc. (% Solids)*	*Method*	*Temp. (°C)*	*Shear Rate (s^{-1})*	K* *(Pa s^n)*	n	*Reference*
			75.6		0.17	1.0	
			77.9		0.14	1.0	
			80.1		0.12	1.0	
Sage	18.6		11.7		72.96	1.0	
			12.0		66.72	1.0	
			12.2		63.12	1.0	
			12.5		59.52	1.0	
			12.8		56.64	1.0	
			12.9		54.24	1.0	
			13.5		49.44	1.0	
			13.9		44.88	1.0	
			14.2		42.00	1.0	
			14.6		39.36	1.0	
			15.1		35.76	1.0	
			15.8		31.56	1.0	
			16.2		29.52	1.0	
			16.8		27.12	1.0	
			17.3		25.20	1.0	
			17.9		23.52	1.0	
			18.8		21.60	1.0	
			20.2		18.48	1.0	
			21.6		15.36	1.0	
			23.4		12.24	1.0	
			25.9		8.88	1.0	
			27.8		6.96	1.0	
			30.7		5.52	1.0	
			32.1		4.56	1.0	
			34.0		3.84	1.0	
			36.2		3.12	1.0	
			38.8		2.40	1.0	
			40.9		1.92	1.0	
			42.0		1.64	1.0	
			44.1		1.48	1.0	
			45.7		1.30	1.0	
			48.0		1.15	1.0	
			50.7		0.93	1.0	
			53.4		0.76	1.0	
			56.0		0.62	1.0	
			60.1		0.48	1.0	
			63.6		0.39	1.0	

continues

Table 5–A16 *continued*

Product	*Conc. (% Solids)*	*Method*	*Temp. (°C)*	*Shear Rate* (s^{-1})	K* (*Pa* s^n)	n	*Reference*
			67.4		0.30	1.0	
			72.0		0.23	1.0	
			73.6		0.21	1.0	
			75.5		0.19	1.0	
			76.8		0.17	1.0	
			77.6		0.16	1.0	
			78.7		0.15	1.0	
			80.5		0.14	1.0	
Sucrose solution	20	Unknown	0.0		0.003818	1.0	Perry, 1950
			5.0		0.003166	1.0	
			10.0		0.002662	1.0	
			15.0		0.002275	1.0	
			20.0		0.001967	1.0	
			25.0		0.001710	1.0	
			30.0		0.001510	1.0	
			35.0		0.001336	1.0	
			40.0		0.001197	1.0	
			45.0		0.001074	1.0	
			50.0		0.000974	1.0	
			55.0		0.000887	1.0	
			60.0		0.000811	1.0	
			65.0		0.000745	1.0	
			70.0		0.000688	1.0	
			75.0		0.000637	1.0	
			80.0		0.000892	1.0	
			85.0		0.000552	1.0	
	40		0.0		0.014820	1.0	
			5.0		0.011600	1.0	
			10.0		0.009830	1.0	
			15.0		0.007496	1.0	
			20.0		0.006223	1.0	
			25.0		0.005206	1.0	
			30.0		0.004398	1.0	
			35.0		0.003776	1.0	
			40.0		0.003261	1.0	
			45.0		0.002858	1.0	
			50.0		0.002506	1.0	
			55.0		0.002277	1.0	
			60.0		0.001989	1.0	
			65.0		0.001785	1.0	

continues

Table 5–A16 *continued*

Product	*Conc. (% Solids)*	*Method*	*Temp. (°C)*	*Shear Rate (s^{-1})*	*K* (Pa s^n)*	n	*Reference*
			70.0		0.001614	1.0	
			75.0		0.001467	1.0	
			80.0		0.001339	1.0	
			85.0		0.001226	1.0	
			90.0		0.001127	1.0	
			95.0		0.001041	1.0	
	60		10.0		0.113900	1.0	
			15.0		0.074900	1.0	
			20.0		0.056700	1.0	
			25.0		0.044020	1.0	
			30.0		0.034010	1.0	
			35.0		0.026620	1.0	
			40.0		0.021300	1.0	
			45.0		0.017240	1.0	
			50.0		0.014060	1.0	
			55.0		0.011710	1.0	
			60.0		0.009870	1.0	
			65.0		0.008370	1.0	
			70.0		0.007180	1.0	
			75.0		0.006220	1.0	
			80.0		0.005420	1.0	
			85.0		0.004750	1.0	
			90.0		0.004170	1.0	
			95.0		0.003730	1.0	
Molasses			21.0		6.600	1.0	Hayes, 1987
			37.8		1.872	1.0	
			49.0		0.920	1.0	
			66.0		0.374	1.0	

* Some values were converted from cp or mPa s units.

Table 5–A17 Rheological Properties of Alcohols and Wines and Effect of Temperature (Activation Energy, E_a)

Name of Product	Sugar (%)	Alcohol (%)	Method	Temp. (°C)	Shear Rate (s^{-1})	K* (mPa s^n)	n	Reference
Ethanol				20		1.2	1.0	Hayes, 1987
density = 760 kg/m^3								
Methyl alcohol				0		0.81	1.0	Hayes, 1987
density = 810 kg/m^3				20		0.59	1.0	
Wines Model Solutions								
Model 1, % m/v glycerin = 0								
	0.0	12.0	Rheomat RM-15	25		1.36	1.0	Rao et al., 1977
E_a = 20.2 kJ mol^{-1}								
Model 2, % m/v glycerin = 1.3								
	1.0	12.0				1.32	1.0	
E_a = 23.1 kJ mol^{-1}								
Model 3, % m/v glycerin = 1.3								
	5.0	12.0				1.30	1.0	
E_a = 25.0 kJ mol^{-1}								
Model 4, % m/v glycerin = 1.3								
	10.0	12.0				2.06	1.0	
E_a = 26.7 kJ mol^{-1}								
White Wines								
Rhine, % m/v glycerin = 0.78								
	2.7	12.6				1.55	1.0	
E_a = 22.0 kJ mol^{-1}								
Aurora Blanc, % m/v glycerin = 1.06								
	1.2	11.5				1.41	1.0	
E_a = 23.3 kJ mol^{-1}								
Elvira, % m/v glycerin = 0.76								
	0.33	12.6				1.44	1.0	

continues

E_a = 23.8 kJ mol^{-1}				
Emerald Riesling, % m/v glycerin = 0.87				
	0.64	10.9	1.46	1.0
E_a = 24.7 kJ mol^{-1}				
Chablis, % m/v glycerin = 0.71				
	0.21	11.1	1.43	1.0
E_a = 18.8 kJ mol^{-1}				
Red Wines				
Ives, % m/v glycerin = 0.76				
	0.4	11.20	1.37	1.0
E_a = 23.7 kJ mol^{-1}				
Baco Noir, % m/v glycerin = 0.90				
	0.05	12.40	1.52	1.0
E_a = 21.5 kJ mol^{-1}				
Rubicon, % m/v glycerin = 0.69				
	3.35	12.61	1.48	1.0
E_a = 22.3 kJ mol^{-1}				
Hybrid chancellor, % m/v glycerin = 0.72				
	0.05	12.20	1.45	1.0
E_a = 19.4 kJ mol^{-1}				
Ruby Cabernet, % m/v glycerin = 0.64				
	0.83	11.30	1.48	1.0
E_a = 19.9 kJ mol^{-1}				
Burgundy, % m/v glycerin = 1.32				
	1.53	12.77	1.43	1.0
E_a = 19.1 kJ mol^{-1}				
Rosé Wines				
Catawba, % m/v glycerin = 0.85				
	6.7	12.2	1.67	1.0
E_a = 24.7 kJ mol^{-1}				

continues

Table 5–A17 *continued*

Name of Product	*Sugar (%)*	*Alcohol (%)*	*Method*	*Temp. (°C)*	*Shear Rate (s^{-1})*	K* *($mPa\ s^n$)*	n	*Reference*
Fortified Wines								
Cream sherry, % m/v glycerin = 0.67								
	13.4	16.6				2.45	1.0	
E_a = 23.5 kJ mol^{-1}								
Cream sherry, % m/v glycerin = 0.37								
	11.63	17.1				1.95	1.0	
E_a = 28.2 kJ mol^{-1}								
Beer, density = 1000 kg/m^3				0.0		1.30	1.0	Hayes, 1987
Chardonnay wine, density = 1010				14		3.90	1.0	Lopez et al., 1989
E_a = 14.2 kJ mol^{-1}				16		3.74	1.0	
				20		3.45	1.0	
				22		3.32	1.0	
Chardonnay wine, density = 1000				14		3.82	1.0	
				16		3.67	1.0	
				20		3.38	1.0	
				22		3.25	1.0	
Chardonnay wine, density = 995				14		3.78	1.0	
				16		3.63	1.0	
				20		3.35	1.0	
				22		3.22	1.0	

* Some values were converted from cp or mPa s units.

REFERENCES

Alviar, M., B. Salome, and D.S. Reid. 1990. Determination of rheological behavior of tomato concentrates using back extrusion. *J Food Sci* 55: 554–555.

Autio, K. 1991. Measurement of flow curves for model liquids and real food systems with two commercial viscometers. *J Food Eng* 13: 57–66.

Barbosa-Cánovas, G.V., and M. Peleg. 1983. Flow parameters of selected commercial semi-liquid products. *J Texture Studies* 14: 213–234.

Benezech, T., and J.F. Maingonnat. 1993. Flow properties of stirred yoghurt: Structured parameter approach in describing time dependency. *J Texture Studies* 24: 455–473.

Bhattacharya, S., S. Bal, R.K. Mukherjee, and S. Battacharya. 1991. Rheological behaviour of Tamarind *Tamarindus indica* kernel suspension. *J Food Eng* 13: 151–158.

Bhattacharya, S., K.K. Bhatt, and Raghuveer. 1992. Rheology of Bengal gram Cicer arietinum flour suspensions. *J Food Eng* 17: 83–96.

Bianchi, M.A., A.M.R. Pilosof, and G.B. Bartholomai. 1985. Rheological behaviour of comminuted meat systems containing soy protein isolates. *J Texture Studies* 16: 193–206.

Brekke, J.E., and A.L. Myers. 1978. Viscometric behavior of guava purees and concentrates. *J Food Sci* 43: 272–273.

Crandall, P.G., C.S. Chen, and R.D. Carter. 1982. Models for predicting viscosity of orange juice concentrate. *Food Technol 36,* no. 5: 245–252.

Fang, T.N., C. Tiu, X. Wu, and S. Dong, 1996. Rheological behaviour of cocoa dispersions. *J Texture Studies* 26: 203–215.

Fito, P.J., G. Clemente, and F.J. Sanz. 1983. Rheological behaviour of tomato concentrates hot break and cold break. *J Food Eng* 2: 51–62.

Garcia, R., J. Rivera, and C. Rolz, 1974. Rheological properties of tropical products and their enzymic classification. *Proc IV Int Cong Food Sci Techol* 2: 18–26.

Gibroeau, A., G. Cuvelier, and B. Launay. 1994. Rheological behavior of three biopolymer/water systems, with emphasis on yield stress and viscoelastic properties. *J Texture Studies* 25: 199–137.

Gunjal, B.B., and N.J. Waghmare. 1987. Flow characteristics of pulp, juice and nectar of "Baneshan" and "Neelum" mangoes. *J Food Sci Technol India* 24: 20–23.

Harper, J.C. 1960. Viscometric behavior in relation to evaporation of fruit purees. *Food Technol* 14: 557–560.

Harper, J.C., and A.F. El-Sahrigi. 1965. Viscometric behavior of tomato concentrates. *J Food Sci* 30: 470–476.

Harper, J.C., and K.W. Leberman. 1962. Rheological behavior of pear purees. *Proc 1st Int Cong Food Sci Technol* 1: 719–728.

Hayes, G.D. 1987. *Food engineering data handbook*. New York: John Wiley & Sons.

Holdsworth, S.D. 1971. Applicability of rheological models to the interpretation of flow and processing behavior of fluid food products. *J Texture Studies* 2: 393–418.

Holdsworth, S.D. 1993. Rheological models used for the prediction of the flow properties of food products: A literature review. *Trans Institution Chem Engineers* 71, Part C: 139–179.

Ibarz, A., and J. Pagan. 1987. Rheology of raspberry juices. *J Food Eng* 6: 269–289.

Ibarz, A., M. Vicente, and J. Graell. 1987. Rheological behavior of apple juice and pear juice and their concentrates. *J Food Eng* 6: 257–267.

Ibarz, A., C. Gonzalez, S. Esplugas, and M. Vicente. 1992. Rheology of clarified juices. I. Peach juices. *J Food Eng* 15: 49–61.

Ibarz, A., J. Pagan, and R. Miguelsanz. 1992. Rheology of clarified fruit juices. II: Blackcurrant juices. *J Food Eng* 15: 63–73.

Kokini, J.L. 1992. Rheological properties of foods. In *Handbook of food engineering*, eds. D.R. Heldman and D.B. Lund, 1–38. New York: Marcel Dekker.

Krishna, A.G.G. 1993. Influence of viscosity on wax settling and refining loss in rice bran oil. *J Am Oil Chem Soc* 70: 895–898.

Lang, W., S. Sokhansanj, and F.W. Sosulski. 1992. Modelling the temperature dependence of kinematic viscosity for refined canola oil. *J Am Oil Chem Soc* 69: 1054–1055.

Lopez, A., A. Ibarz, J. Pagan, and M. Vilavella. 1989. Rheology of wine musts during fermentation. *J Food Eng* 10: 155–161.

Manohar, B., P. Ramakrishna, and K. Udayashankar. 1991a. Some physical properties of mango pulp concentrate. *J Texture Studies* 21: 179–190.

Manohar, B., P. Ramakrishna, and K. Udayashankar. 1991b. Some physical properties of tamarind Tamarindus indica L. juice concentrates. *J Food Eng* 13: 241–258.

McCarthy, K.L., and J.D. Seymour. 1994. A fundamental approach for the relationship between the Bostwick measurement and Newtonian fluid viscosity. *J Texture Studies* 24: 1–10.

Mizrahi, S. 1979. A review of the physicochemical approach to the analysis of the structural viscosity of fluid food products. *J Texture Studies* 10: 67–82.

Munro, J.A. 1943. The viscosity and thixotropy of honey. *J Econ Entomol* 36: 769–777.

Noureddini, H., B.C. Teoh, and L.D. Clements. 1992. Viscosities of vegetable oils and fatty acids. *J Am Oil Chem Soc* 69: 1189–1191.

Paredes, M.D.C., M.A. Rao, and M.C. Bourne. 1989. Rheological characterization of salad dressings. 2. Effect of storage. *J Texture Studies* 20: 235–250.

Perry, J.H. 1950. *Chemical engineer's handbook.* 3d ed. New York: McGraw-Hill.

Qiu, C-G., and M.A. Rao. 1988. Role of pulp content and particle size in yield stress of apple sauce. *J Food Sci* 53: 1165–1170.

Ramaswamy, H.S., and S. Basak. 1991. Rheology of stirred yoghurts. *J Texture Studies* 22: 231–241.

Rao, K.L., W.E. Eipeson, P.N.S. Rao, M.V. Patwardhan, and P.K. Ramanathan. 1985. Rheological properties of mango pulp and concentrate. *J Food Sci Technol India* 22: 30–33.

Rao, M.A. 1977. Rheology of liquid foods—A review. *J Texture Studies* 8: 135–168.

Rao, M.A., and H.J. Cooley. 1984. Determination of effective shear rates of complex geometries. *J Texture Studies* 15: 327–335.

Rao, M.A., and H.J. Cooley. 1992. Rheology of tomato pastes in steady and dynamic shear. *J Texture Studies* 23: 415–425.

Rao, M.A. H.J. Cooley, C. Ortloff, K. Chang, and S.C. Wijts. 1993. Influence of rheological properties of fluid and semi solid foods on the performance of a filler. *J Food Process Eng* 16: 289–304.

Roscane, S., and D.E. Smith. 1993. Flow behavior properties and density of whole milk retenates as affected by temperature. *Michwissenschaft* 48: 6–10.

Rovedo, C.O., P.E. Viollaz, and C. Suarez. 1991. The effect of pH and temperature on the rheological behavior of Dulce de Leche: A typical dairy Argentine product. *J Dairy Sci* 74: 1497–1502.

Rozema, H., and W.A. Beverloo. 1974. Laminar isothermal flow of non Newtonian fluids in a circular pipe. *Lebensmittel Wissenschaft und Technologie* 7: 223.

Saravacos, G.D. 1968. Tube viscometry of fruit purees and juices. *Food Technol* 22: 1585–1968.

Saravacos, G.D. 1970. Effect of temperature on viscosity of fruit juices and purees. *J Food Sci* 35: 122–125.

Saravacos, G.D., and J.C. Moyer. 1967. Heating rates of fruit products in an agitated kettle. *Food Technol* 27: 372–376.

Sherman, P. 1970. *Industrial rheology*. New York: Academic Press.

Sterling, C., and J.J. Wuhrmann. 1960. Rheology of cocoa butter. I. Effect of contained fat crystals on flow properties. *Food Res* 25: 460–463.

Timms, R.E. 1985. Physical properties of oils and mixtures of oils. *J Am Oil Chem Soc 62,* no. 2: 241–248.

Varshney, N.N., and B.K. Kumbhar. 1978. Effect of temperature and concentration on rheological properties of pineapple and orange juice. *J Food Sci Technol India 15,* no. 2: 53–55.

Vitali, A.A., and M.A. Rao. 1984a. Flow properties of low-pulp concentrated orange juice: Serum viscosity and effect of pulp content. *J Food Sci* 49: 876–880.

Vitali, A.A., and M.A. Rao. 1984b. Flow properties of low-pulp concentrated orange juice: Effect of temperature and concentration. *J Food Sci* 49: 882–888.

Wayne, J.E.B., and C.F. Shoemaker. 1988. Rheological characterization of commercially processed fluid milks. *J Texture Studies* 19: 143–152.

Yoo, B., and M.A. Rao. 1994. Effect of unimodal particle size and pulp content on rheological properties of tomato puree. *J Texture Studies* 25: 421–436.

Yoo, B., and M.A. Rao. 1996. Creep and dynamic rheological behavior of tomato concentrates: Effect of concentration and finisher screen size. *J Texture Studies* 27: 451–459.

CHAPTER 6

Rheological Behavior of Food Gel Systems

José Antonio Lopes da Silva and M. Anandha Rao

Rheological studies can provide much useful information on sol-gel and gel-sol transition, as well as on the characteristics of a gel. There are several definitions of what a gel is, based on phenomenological and/or molecular criteria. Flory (1953) defined a gel to consist of polymeric molecules cross-linked to form a tangled interconnected network immersed in a liquid medium. Hermans (1949) defined it as a two-component system (e.g., gelling polymer and the solvent, water or aqueous solution in foods) formed by a solid finely dispersed or dissolved in a liquid phase, exhibiting solid-like behavior under deformation; in addition, both components extend continuously throughout the entire system, each phase being interconnected. At the molecular level, gelation is the formation of a continuous network of polymer molecules, in which the stress-resisting bulk properties (solid-like behavior) are imparted by a framework of polymer chains that extends throughout the gel phase. Further, gel setting involves formation of cross-links, while softening or liquefaction (often called melting) involves their destruction.

In many food products, gelation of polysaccharides is critical to the formation of desired texture. Most biopolymers form physical gels, structured by weak interactions (hydrogen, electrostatic, hydrophobic), and thus at a gross level belong to Flory's third mechanism (Flory, 1974), described as polymer networks formed through physical aggregation which are predominantly disordered, but with regions of local order. They are often thermoreversible and almost invariably occur in the presence of excess of solvent, usually water or aqueous electrolyte. Therefore, they are called solvated networks. In most biopolymer gels, the polymer chains form extended junction zones by means of side-by-side associations of a physical nature, in contrast to the typical single covalent bonds found in chemically cross-linked networks. Consequently, in physical gelation the formation and breakdown of the junction zones are usually reversible, the cross-link functionality is very high, and the junction zones have a finite lifetime. The formation of these kinds of transient networks is determined by the chemical composition of both the polymer and the solvent that constitute the gelling system, and by temperature and time. In this chapter, the formation of gels and their softening (melting) are reviewed. Further, much of the contents of this chapter was discussed by the authors in a recent review of the subject (Lopes da Silva et al., 1998).

DYNAMIC RHEOLOGICAL TESTS TO EVALUATE PROPERTIES OF GEL SYSTEMS

As mentioned in Chapter 3, dynamic (oscillatory) rheological tests provide valuable information on the viscoelastic nature of foods. Because gels are viscoelastic materials, dynamic rheological tests are well suited for studying the characteristics of gels as well as of gelation and melting. From dynamic rheological tests in the linear viscoelastic range, the storage modulus, G', and the loss modulus, G'', and $\tan \delta = (G''/G')$, the loss tangent, can be obtained (Rao, 1992). In addition, the complex viscosity $\eta^* = (G^*/\omega)$ is another useful parameter, where ω is the frequency of oscillation (rad s^{-1}) and $G^* = \sqrt{(G')^2 + (G'')^2}$. Three types of dynamic tests can be conducted to obtain useful properties of gels and of gelation and melting. (1) Frequency sweep studies in which G' and G'' are determined as a function of frequency (ω) at fixed temperatures. (2) Temperature sweep in which G' and G'' are determined as a function of temperature at fixed ω. (3) Time sweep in which G' and G'' are determined as a function of time at fixed ω and temperature.

MECHANISMS OF GELATION

Because of the large number of gel systems, several gelation mechanisms have been encountered among the various biopolymer gels. Because of variations in the number and nature of the cross-links, in framework flexibility, in attractions and repulsions between framework elements, and in interactions with solvent, different properties of the formed gels are to be expected (Rees, 1972). The range of systems is quite extensive, from systems formed purely by topological entanglements to the complex networks formed by ordered fibrous assemblies, such as by actin and fibrin filaments *in vivo*. Some of the characteristics of the gelation processes found in food polymers are given in Table 6–1 (Lopes da Silva et al., 1998).

Polysaccharides of the galactan group (carrageenans, agarose) form thermoreversible gels through a transition from a disordered to a helical macromolecular conformation (Rees, 1969; Morris et al., 1980a). Alginates and low methoxyl pectins form gels by chain-stacking through the interaction between divalent ions and the carboxyl groups of the polysaccharide macromolecule (Grant et al., 1973; Gidley et al., 1979). In gels of methyl and hydroxypropyl derivatives of cellulose, the gel network is formed by hydrophobic association of chains, showing very distinct properties, like gel setting when heated and gel melting when cooled (Haque and Morris, 1993). A phase separation process was suggested to interpret the gelation of amylose (Miles et al., 1985; Doublier and Choplin, 1989). Gels of bovine serum albumin are fibrillar in nature, first unfolding by heating and then by aggregation, leading to gels of different properties depending on pH and ionic strength (Richardson and Ross-Murphy, 1981a). Gelatin networks are formed by triple helices randomly distributed in space and separated by chain segments of random-coil conformation (Pezron et al., 1990).

CLASSIFICATION OF GELS

Clark and Ross-Murphy (1987) classified biopolymer gels based on the level of order of the macromolecule both before and during the network formation: (1) gels formed from disordered biopolymers, such as carrageenans, pectins, starch, gelatin, and (2) gel networks that involve specific interactions between denser and less flexible particles, such as thermally denatured globular proteins and aggregated proteins from enzymatic or chemical action.

More information than given here is available for specific gelling systems. For example, in globular protein gelation, the activation phase occurs (Foegeding et al., 1998) when (1) the solvent quality is changed (e.g., salting out), (2) molecules are linked covalently by enzymatic action (e.g., transglutaminase) and protein unfolding or hydrolyzed. In foods, the common method for globular protein gelation is heat-induced protein denaturation that causes changes in quaternary, tertiary, and secondary structures. It takes one globular protein molecule to denature, two or more to aggregate, and many more to form a gel matrix. Many denatured protein gels are considered to be particulate in nature (Clark and Ross-Murphy, 1987).

Gel forming biopolymers can also be divided into "cold-setting" or "heat-setting," based on the two main gelation mechanisms. In the former, gelation is induced by cooling and includes those biopolymer gels that occur in nature and provide structures in biological systems. In the latter, gelation occurs by heating and includes those systems where gelation involves extensive denaturation of the biopolymer, such as thermally unfolded globular proteins.

Based on the macroscopic behavior of these systems, a practical and useful distinction can be made between those systems that are free-standing as a consequence of the development of the three-dimensional network, called "true gels," and those characterized by a tenuous gel-like network that is easily broken when submitted to a high-enough stress, called "weak gels" (Doublier et al., 1992). Clark and Ross-Murphy (1987) distinguished between these two kinds of gelled systems based on the operational definition of a gel from mechanical spectroscopy, designating strong gels as those networks that have "finite energy," and weak gels as those systems that

Table 6–1 Gelation Characteristics of Important Food Polymers

Polymer	*Important Structural Features Affecting Gelation*	*Main Factors Promoting Gelation*	*Mechanism of Gelation*
Carrageenans	Alternating β-1,3- and α-1,4-linked galactose residues; presence of 3,6-anhydride residues; presence of sulfated residues	Cooling; presence of potassium or other gel-promoting cations	Association of the molecular chains into double helices followed by aggregation of the ordered "domains"
Agarose	Alternating 1,3-linked β-D-galactopyranose and 1,4-linked 3,6-anhydro-α-L-galactopyranose	Cooling	Helix formation followed by aggregation
Alginates	Alternating blocks of β-1,4-linked D-mannuronic acid and α-1,4-linked L-guluronic acid residues	Presence of divalent cations (usually Ca^{2+})	Specific site binding of calcium with the carboxyl groups of the polyuronic acid residues (mainly the polyguluronic), "egg-box" model
Pectins	α(1,4)-D-galacturonic acid residues, partially methyl esterified; presence of 1,2-linked L-rhamnosyl residues		
• high-methoxyl pectins	• degree of methylation higher than 50%	• Cooling; low pH and low water activity	• Junction zones stabilized by hydrogen bonds and hydrophobic interactions between the ester methyl groups
• low-methoxyl pectins	• degree of methylation lower than 50%	• Presence of divalent cations (usually Ca^{2+})	• Specific site binding of calcium with the carboxyl groups of the polyuronic acid residues, "egg-box" model
Gellan gum	Glucose, glucuronic acid and rhamnose residues (2:1:1); presence of *O*-acetyl and *O*-L-glyceryl substituents	Cooling; presence of gel-promoting cations	Ion-mediated aggregation of double helices

continues

Starch	Linear fraction (amylose): α-1,4-linked D-glucose residues Branched fraction (amylopectin): α-1,4- and α-1,6-linked D-glucose residues; higher DP	Cooling	Composite gels: amylopectin granules threaded by a gelled amylose matrix
Gelatin	High content of pyrolidine residues (pro+hypro); presence of glycine as every third residue; isoelectric point affected by acid or alkali treatment	Heating solutions followed by cooling	Formation of triple helices randomly distributed in space and separated by chain segments of random-coil conformation
Bovine serum albumin	Globular protein (M.W. ~66500 Da); isoelectric point ~ 5.1; characteristic secondary structure, with a specific content of α-helix, β-sheet, and disordered peptide chain conformations; presence of disulphide bonds and one free sulphydryl group	Heating solutions followed by cooling	Denaturation followed by aggregation, likely by formation of β-sheet regions and other less specific protein-protein associations
β-lactoglobulin	Globular protein (M.W. ~18500 Da); characteristic secondary structure, with a specific content of α-helix, β-sheet, and disordered peptide chain conformations; isoelectric point ~ 5.4–5.5; presence of two disulphide bridges and a free thiol group	Heating solutions followed by cooling	Denaturation followed by aggregation
Caseins	Different proteins: variation in the degree of phosphorylation, glycosylation, hydrophobicity, and amphipathic structures Presence of disulphide groups Small molecules, c. 20000–24000 Da Isoelectric points: pH 4.5–4.9	• Renneted gels: Enzymatic action followed by precipitation by Ca^{2+} • Acid coagulation: Acidification and instability of the casein micelles	• *Renneted gels:* k-casein hydrolysis, instability of the casein micelles followed by coagulation, mainly by electrostatic interactions involving Ca^{2+} • *Acid coagulation:* Solubilization of the colloidal calcium phosphate and aggregation when the i.p. is approached (charge neutralization)

are transient in time and that even well above c_0 show some of the properties usually attributed to the presence of hyperentanglements. A traditional example of weak gel behavior is the viscoelastic behavior of xanthan gum dispersions.

The difference between the entanglement network, the strong (true) gel (diagrams (A) in Figures 6–1 and 6–2) and the weak gel (diagrams (B) in Figures 6–1 and 6–2), can be easily established using mechanical spectroscopy or by a strain dependence of the moduli (Clark and Ross-Murphy, 1987). In the former, the molecular rearrangements within the network are very reduced over the time scales analyzed, G' is higher than G'' throughout the frequency range, and G' is almost independent of frequency (ω). In weak gels there is a higher dependence on frequency for the dynamic moduli, suggesting the existence of relaxation processes occurring even at short time scales, and lower difference between moduli values, indicating that a lower percentage of the stored energy is recovered.

THEORETICAL TREATMENT OF GELS

Theoretical and experimental treatments of gels go hand-in-hand. The former are covered first because they are the basis for an understanding of gel point and other concepts. Two main theories have been used to interpret results of experimental studies on gels: the classical theory developed by Flory and Stockmayer, and the percolation model credited to de Gennes. Gelation theories predict a critical point at which an infinite cluster first appears. As with other critical points, the sol-gel transition can be characterized, in general, in terms of a set of generally applicable (universal) critical exponents.

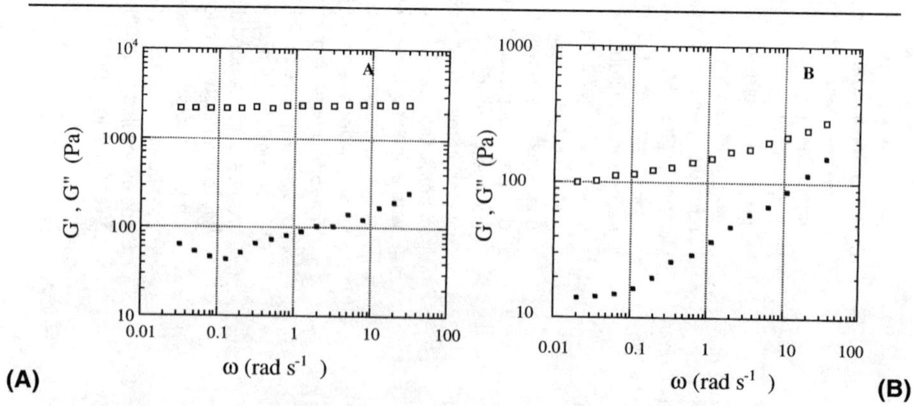

Figure 6–1 Mechanical Spectra (□ G′, ■ G″) of a 1% High-Methoxyl Pectin Gel, pH 3.0, 60% (w/w) Sucrose Measured after Aging for 36 Hours at Two Different Temperatures: (A) 5°C, (B) 30°C. *Source*: Reprinted from *Carbohydrate Polymers*, Vol. 24, J.A.L. da Silva and M.P. Goncalves, Rheological Study into the Aging Process of High Methoxyl Pectin/Sucrose Aqueous Gels, pp. 235–245.© 1994, with permission from Elsevier Science.

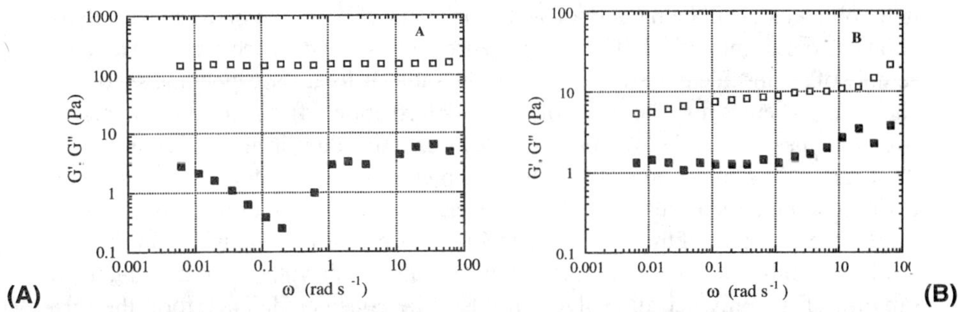

Figure 6–2 Mechanical Spectra (□ G', ■ G'') of Low-Methoxyl Pectin/Calcium Gels: 8.1 g L^{-1}, $[Ca^{2+}] = 6.2 \times 10^{-3}$ mol L^{-1}, 0.1 mol L^{-1} NaCl, Measured at 20°C after a Quenching Time of 21 Hours, for Two Different pH Values: (A) pH 3.0, (B) pH 6.8. *Source*: Reprinted from J.A.Lopes da Silva, M.A. Rao, and J-T. Fu, Rheology of Structure Development and Loss During Gelatin and Melting, in *Phase/State Transitions in Foods: Chemical, Rheological and Structural Changes*, M.A. Rao and R.W. Hartel, eds., pp. 111–156, by courtesy of Marcel Dekker, Inc.

Assuming high gel homogeneity for single biopolymer systems and reversibility of the cross-linking process to explain the critical concentration aspect, the classical theory and its extension and modifications, including the cascade theory (Gordon and Ross-Murphy, 1975), were used to describe the gelling process of several biopolymers. It was used to describe the gel growth process after the gel point and to explain the role of important parameters in the gelation of physical gels, such as the critical gel concentration and the concentration dependence of the gel modulus. The classical theory was used to describe bovine serum albumin (Richardson and Ross-Murphy, 1981a), gelatin (Peniche-Covas et al., 1974), agarose (Watase et al., 1989), egg albumen and ovalbumin (Hsieh and Regenstein, 1992a), and maltodextrins (Kasapis et al., 1993).

Rubber Elasticity

An equation for the modulus of ideal rubber was derived from statistical theory that can be credited to several scientists, including Flory, and Guth and James (Sperling, 1986). A key assumption in derivation of the equation is that the networks are Gaussian. For the same Gaussian networks, Treloar (1975) derived an equation for the shear modulus, G:

$$G = \left(\frac{gcRT}{M_c}\right) \tag{6.1}$$

where T is temperature (K), R is the gas constant, and M_c means molecular weight of chains joining adjacent cross-links. Equation 6.1 can be valid only under the following assumptions (Treloar, 1975; Fu, 1998): (1) The network contains N chains per unit volume, a chain being defined as the segment of molecule between successive points of cross-linkage. (2) The mean-square end-to-end distance for the whole assembly of chains in the unstrained state is the same as for a corresponding set of free chains. (3) There is no change of volume of deformation. (4) The junction points between chains move on deformation as if they were embedded in an elastic continuum. As a result, the components of length of each chain change in the same ratio as the corresponding dimensions of the bulk rubber. (5) The entropy of the network is the sum of the entropies of the individual chains. This means the chains are Gaussian.

In a solvent-swollen polymer, the volume fraction of swollen polymer, v_2, in the mixture of the polymer and solvent needs to be considered. Therefore, the relationship between G' and M_c is given by the following (Treloar, 1975):

$$G = \left(\frac{\rho RT}{M_c}\right)(v_2)^{1/3} \tag{6.2}$$

where ρ (g cm^{-3}) is the density of the polymer in the unswollen state. Most biopolymer networks have structures that do not conform with the main assumptions of the theory of rubber elasticity, because the crosslinks are of low energy physical interactions and they are not limited to single points on the chains but correspond to more or less extended junction zones. In addition, the polymer chains between junction zones can be characterized by several degrees of rigidity and deviate from Gaussian behavior. Stiff chains increase their flexibility with increasing temperature due to the increased number of allowed torsion angles. Consequently, changes in the rigidity of the molecular chains can also explain the decrease in the gel modulus when the temperature increases. Therefore, the theory is not applicable to most biopolymer gels, since the assumption of a purely entropic contribution to the elasticity of a network chain is likely invalid in most cases. For thermoreversible gels the decrease in the number of junction zones with temperature is usually more pronounced than any increase from the entropic contribution.

In spite of the above drawbacks, Equations 6.1 and 6.2 are popular, mainly due to their simple form, and have been used to estimate M_c of protein (e.g., egg, gelatin) and other gels from shear modulus–concentration data (Table 6–2) (Fu, 1998). The equations' successful application to protein gels initially has been attributed to the greater flexibility of polypeptide chains in comparison to polysaccharide chains, which are relatively stiff. In addition, the inability to employ low strain rates in early experimental studies on polysaccharide gels could be another reason.

Oakenfull's Modification

Oakenfull (1984) developed an extension of Equation 6.1 for estimating the size of junction zones in noncovalently cross-linked gels subject to the following as-

Table 6–2 Examples of Application of the Rubber Elasticity Theory to Biopolymer Gels

System	*Measurement and Results*	*Reference*
Collagen from epimysial connective tissues	Using stress-strain tests to calculate molecular weight between covalent cross-links (M_c)	McClain et al., 1969
Wheat gluten and dough	Using extension tests to calculate the number of cross-links per unit volume	Muller, 1969
Kamaboko	Using extension and compression tests to develop the relationship between stress and strain and then to conclude that orientation of segments of kamaboko's network was negligible and similar to the behavior of the vulcanized rubber in small deformation	Takagi and Simidu, 1972
Wheat flour doughs	Using shear modulus obtained at different temperatures to conclude that model of rubber cannot be applied to dough elasticity	Bloksma and Nieman, 1975
Calcium, potassium, and sodium iota-carrageenate	Using rigidity modulus to calculate molecular weight of polymer chains between the network junctions (M_c)	Morris and Chilvers, 1981
HMP/sucrose and gelatin	Using shear modulus obtained from compression tests for noncovalent gels to calculate the association constant for the formation of junction zones, the number average molecular weight of the junction zones, and the number of cross-linking loci per junction zone	Oakenfull, 1984
Surimi	Using elastic moduli obtained from stress relaxation tests to calculate the number of cross-links per unit volume	Iso et al., 1984
Fish meat	Using elastic modulus (E_o) obtained from compression tests to calculate the number of chains per unit volume	Mochizuki et al., 1987
Egg white	Using stress relaxation tests to calculate the density of permanent cross-links	Ziegler and Rizvi, 1989

continues

Table 6–2 *continued*

System	*Measurement and Results*	*Reference*
Egg white and ovalbumin	Using compression tests to calculate molecular weight of the interchain (M_c)	Hsieh and Regenstein, 1992b
Soy protein isolate	Using storage modulus (G') obtained from dynamic oscillatory tests to calculate the number of disulfide bridges per molecule	Chronakis et al., 1996
Xanthan/galactomannans	Using shear modulus (G) obtained from creep tests to calculate molecular weight between cross-links (M_c)	Kloek et al., 1996
Waxy maize starches	Using plateau storage modulus (G') obtained from small amplitude oscillatory tests to calculate molecular weight between cross-links (M_c)	Gluck-Hirsch and Kokini, 1997

sumptions (Oakenfull, 1987): (1) The shear modulus can be obtained for very weak gels whose polymer concentration is very low and close to the gel threshold; that is, the polymer chains are at or near to maximum Gaussian behavior. (2) The formation of junction zones is an equilibrium process that is subject to the law of mass action. Oakenfull's expression for the modulus is as follows (Oakenfull, 1984):

$$G = -\frac{RT_c}{M}\frac{M[J] - c}{M_J[J] - c} \tag{6.3}$$

where M is the number average molecular weight of the polymer, R is the gas constant, T is the absolute temperature, c is the weight concentration of the polymer, M_J is the number average molecular weight of the junction zones, and $[J]$ is the molar concentration of junction zones. From assumption 2, it follows:

$$K_J = [J]\, M_J^n \{n(c - M_J[J])\}^{-n} \tag{6.4}$$

where K_J is the association constant for the formation of junction zones, and n is the number of segments from the different polymer molecules involved. Numerical methods have been used to solve for M, M_J, K_J, and n after substituting for $[J]$ to obtain a relationship between shear modulus and concentration (Oakenfull, 1984, 1987).

Cascade Theory

The cascade theory, developed by Gordon, is an extension of the classical Flory-Stockmayer concepts of gelation. An expression for the shear modulus, *G,* can be written as:

$$G = Ne\, aRT \tag{6.5}$$

where *Ne* is the number of elastically active network chains per primary chain ($\approx$moles V^{-1}), and *a* is generalized front factor. From cascade theory

$$G = \{Nf\alpha(1 - \nu^2)(1 - \beta)/2\}aRT \tag{6.6}$$

$N = (c/M)$ initially, f is the number of sites (functionalities) along each molecule's length, α is the fraction of such sites that have reacted, ν and β are derived from cascade approach to network formation:

$$\nu = (1 + \alpha + \alpha\nu)^{f-1} \tag{6.7}$$

$$\beta = (f - 1)\alpha\nu/(1 - \alpha + \alpha\nu) \tag{6.8}$$

The influence of temperature is taken into account for the cross-linking equilibrium constant *K*:

$$K = \exp(\Delta S°/R)\exp(-\Delta H°/RT) \tag{6.9}$$

The critical gelling concentration (c_c) (the symbol c_c is used here to denote a concentration different than c_o at initiation of gelling) of polymer at the melting temperature T_m can be calculated from

$$c_c = M(f - 1)/2K(T_m)f(f - 2)^2 \tag{6.10}$$

$$T_m = \Delta H°/\{\Delta S° + R\ln(2cf(f - 2)^2/M(f - 1))\} \tag{6.11}$$

Clark (1994) modeled the variation of the modulus of κ-carrageenan gels as a function of polymer's molecular weight using the cascade approach. From modulus-temperature data (Clark et al., 1994), thermodynamic and functional parameters were determined for gelatin: high $\Delta H°$ (-600 to 1,600 kJ mol^{-1}), $T_m \approx 30°$C, $a \approx 1, f \approx 3$–100, and for pectin: low $\Delta H°$ (-0.8 kJ mol^{-1}), $T_m \approx 130°$C, $a \approx$ 1000–5000, f $\approx$ 3–100. The relative insensitivity of the parameters *a* and *f,* and the high value of T_m ($\approx$130°C) noted for pectin gels point out the need for additional investigations on the applicability of the cascade theory to other gels.

Percolation Theory

Percolation describes the geometrical transition between disconnected and connected phases as the concentration of bonds in a lattice increases. It is the foundation for the physical properties of many disordered systems and has been applied to gelation phenomena (de Gennes, 1979; Stauffer et al., 1982). At just above gelation threshold, denoting the fraction of reacted bonds as p and $p = p_c + \Delta p$, p_c the critical concentration (infinite cluster), the scaling laws (critical exponents) for gel fraction (S_∞) and modulus (E) are

$$S_\infty \cong (p - p_c)^\beta;\ E \cong (p - p_c)^t \tag{6.12}$$

Theoretical values for the exponents from the percolation theory are $\beta = 0.39$, $t \sim 1.7$–1.9; $t > \beta$ because of dangling chains. In contrast, the classical theory predictions for the same quantities are

$$\beta_{cl} = 1,\ t_{cl} = 3 \tag{6.13}$$

According to the percolation model, the chains in a swollen gel need not be Gaussian. For a true gel ($\sigma \gg E$), stress (σ)-strain (λ) relationships from the percolation model (Equation 6.14) and the classical theory (Equation 6.15) are

$$\sigma = E\lambda^{5/2} \tag{6.14}$$

$$\sigma = E\lambda^2 \tag{6.15}$$

Verification of the values of the exponents in Equations 6.14 and 6.15 continues to be the subject of considerable debate between proponents of the two theories. The viscoelastic properties of the system that characterize the sol-gel transition are also important features of the percolation model.

$$\eta_o \sim \varepsilon^{-s} \tag{6.16}$$

$$G_o \sim \varepsilon^t \tag{6.17}$$

$$G'(\omega) \sim G''(\omega) \sim \omega^\Delta \tag{6.18}$$

$$\eta^*(\omega) \sim \omega^{\Delta - 1} \tag{6.19}$$

where ε is the relative distance to the gel point, $\varepsilon = (|p - p_c|/p_c)$ and $\Delta = t/(s + t)$. From Equation 6.18, it follows that tan δ is independent of frequency at the gel point:

$$\tan\delta = G''/G' = \text{constant} = \tan(\Delta\pi/2) \tag{6.20}$$

The predictions of the scalar percolation theory have been experimentally confirmed for chemical gelation (Durand et al., 1989; Adam, 1991) and physical gelation (Tokita et al., 1984; Axelos and Kolb, 1990; Wang et al., 1991).

GEL POINT AND SOL-GEL TRANSITION BY RHEOLOGICAL MEASUREMENTS

During gelation, a polymer undergoes a phase transition from a liquid to a gel. The sol-gel transition is a critical point where one characteristic length scale (i.e., the size of the largest molecule) diverges. It is a transition in connectivity between a sol, where the "monomers" are not connected, to a gel, where they are connected. It is not a thermodynamic transition, nothing special is happening to the free energy, and it is not associated with any singularity in concentration fluctuations (Joanny, 1989). Thus, independently of the system studied or the mechanism involved, gelation is a critical phenomenon where the transition variable will be the connectivity of the physical or chemical bonds linking the basic structural units of the material.

Gel Point

The idea of gel point has received much attention in synthetic and biopolymer gels. One can talk about gel point as an instant in time or as a specific temperature. In this chapter, the symbols t_c and T_g will be used for the gel time and gel temperature (Tobitani and Ross-Murphy, 1997a, 1997b; Foegeding et al., 1998), respectively. One can imagine many qualitative definitions of gel point, especially based on visual observation of individuals. Here, methods based on rheological measurements are emphasized.

Before the gel point, the connectivity is small and the material typically relaxes rapidly. Near the gel point, the relaxation time rises sharply, and at the gel point it diverges to infinity (or at least to very long times for a finite sample); in addition, the relaxation spectrum does not contain a characteristic time anymore. After the gel point, if the network has reached a high degree of development, the maximum relaxation time of the final network is also very short.

Classical Concept

Experimental detection of the gel point is not always easy, since the equilibrium shear modulus is technically zero at the gel point and any applied stress will eventually relax, but only at infinite time. From the classical theory, the attributes of the gel point are an infinite steady-shear viscosity and a zero equilibrium modulus at zero frequency limit (Figure 6–3) (Flory, 1953). These criteria have been widely employed to detect the gel point of chemical gels. However, because continuous shearing affects gel formation, accurate information from viscosity measurement is not possible in the close vicinity of the gel point. Further, information regarding the transition itself could only be obtained by extrapolation, thereby introducing uncertainties in the determination of the gelation moment.

Recently, focus has been placed on small amplitude oscillatory shear technique to measure the dynamic moduli during the gelation process to identify the gel point. By this method, the continuous evolution of the viscoelastic properties throughout the gelation process can be followed. Because the strain is kept small, modification of molecular structure caused by shear is minimized, which is an im-

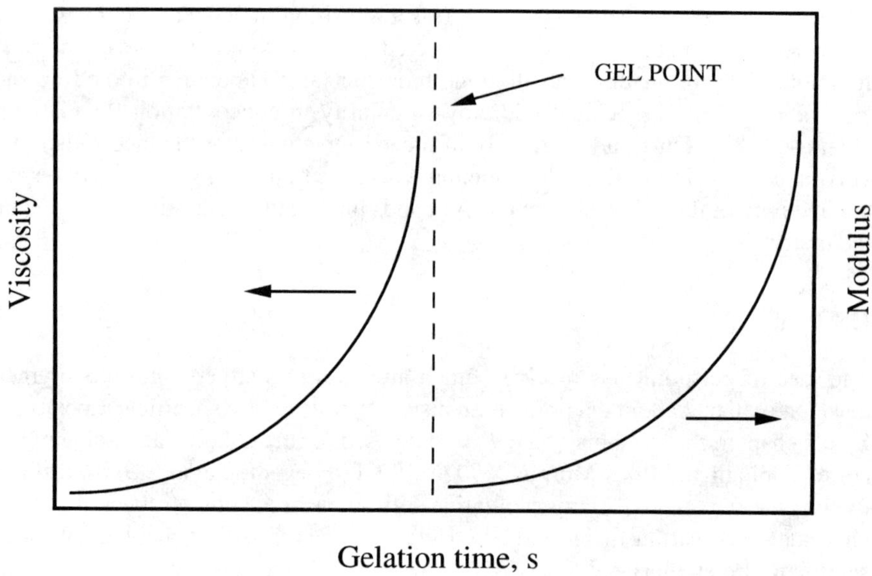

Figure 6–3 Schematic Diagram of the Classical Definition of Changes in Viscosity and Modulus at the Gel Point.

portant advantage over the classical methods based on the diverging viscosity. Some of the magnitudes of the critical exponents obtained for the steady-state viscoelastic properties of different biopolymers are shown in Table 6–3 (Lopes da Silva et al., 1998), and they are in agreement with different approaches to the percolation problem. More important is the fact that irrespective of the mechanism or complexity of the gelation process, the changes in the viscoelastic parameters near the gel point obey similar scaling laws.

G′ and G″ Crossover

Studies on thermosetting resins (Tung and Dynes, 1982) suggested that the gel point might occur at the time at which G' and G'' cross each other at a given frequency. This criterion was applied to the gelation of some biopolymers, such as gelatin (Djabourov et al., 1988b), β-lactoglobulin (Stading and Hermansson, 1990), maltodextrins (Kasapis et al., 1993), and κ-carrageenan gels (Stading and Hermansson, 1993). However, the time of the G'-G'' crossover was found to be dependent on the oscillatory frequency in the case of both synthetic polymers (Winter et al., 1988; Muller et al., 1991) and biopolymers (Djabourov et al., 1988b). As the gel time is an intrinsic property of the material, it cannot be de-

Table 6–3 Some Values of the Critical Exponents (*s* and *t*) for Biopolymer Gelation

Polymer Systems	s	t	*Reference*
Casein		1.8–2.6*	Tokita et al., 1984
	1.48	1.82	Djabourov et al., 1988b
Gelatin	1.1	1.7–1.9	Borchard and Burg, 1989
	0.95	—	Dumas and Bacri, 1980
LMP/calcium†	0.82	1.93	Axelos and Kolb, 1990
Alginate/calcium	0.88	—	Wang et al., 1991
Alginate/cobalt	1.14	—	

* Depending on frequency.
† $c_p = 14.9$ g/L ($c[\eta] = 4.5$), ionic strength 0.1 mol/L, pH 7, 20°C, ratio $R = [2(Ca^{2+})/(COO^-)]$ varied from 0.03 to 0.13.

pendent on the frequency of the dynamic rheological experiment. Another concern is that in low-concentration gelling systems, the viscoelastic moduli may be too low to give a measurable signal by a conventional rheometer. However, it is likely that the G'-G'' crossover time might be close to the sol/gel transition time.

Relaxation Modulus

Winter and coworkers (Winter and Chambon, 1986; Chambon and Winter, 1987; Winter et al., 1988) gave the first experimental demonstration of the power law behavior of the relaxation modulus at gel point and that the scaling of the dynamic moduli and dynamic viscosity behave according to power laws in oscillatory frequency. Power law relaxation at gel point seems to be a general property of both chemical gelation and physical gelation involving either synthetic polymers or biopolymers, respectively. However, the different values of the relaxation exponent (Δ) (Table 6–4) reflect the fact that it is not a universal parameter for gelation. Even for the same polymeric material, a relatively wide range of values have been obtained for this critical exponent, and for some physical gels it was found to be dependent on polymer concentration and thermal history (Michon et al., 1993; Lopes da Silva and Gonçalves, 1994) as well as chain stiffness (Matsumoto et al., 1992). As the exponent Δ depends on molecular structure, it may be useful to distinguish between different gel structures (Lopes da Silva et al., 1998).

Equation 6.20 is the most generally valid gel point criterion and it has been successfully applied for the detection of the gel point of a wide variety of polymers (te Nijenhuis and Winter, 1989; Lin et al., 1991; Muller et al., 1991; Scanlan and Winter, 1991; Michon et al., 1993).

Table 6–4 Some Values of the Critical Relaxation Exponent (Δ) Obtained for Physical Gels

Physical Gels	Δ	*Reference*
Poly(vinyl chloride) in plasticizers	0.8	te Nijenhuis and Winter, 1989
Thermoplastic elastomeric poly(propylene)	0.125	Lin et al., 1991
LM pectin/calcium systems	0.71	Axelos and Kolb, 1990
Xanthan/carob and iota-carrageenan	0.5	Cuvelier et al., 1990
Xanthan/chromium	0.35	Nolte et al., 1992
Sodium alginates	0.53–0.84	Matsumoto et al., 1992
Gelatin	0.22–0.71	Michon et al., 1993
HM pectin/sucrose systems	0.41–0.65	Lopes da Silva and Gonçalves, 1994

Threshold G′ *Value*

However, for many biopolymer systems, gelation begins from a sol state characterized by a very low viscosity, which is often below the instrument resolution in small strain oscillatory instruments. Therefore, other criteria have been used for the gel point. Many authors have assessed the gel point as when G' has increased to a value greater than the experimental noise level, an approach that is not very rigorous. Nevertheless, this criterion was used for amylose (Doublier and Choplin, 1989), β-lactoglobulin (Stading and Hermansson, 1990), gelatin (Ross-Murphy, 1991b), and BSA (Richardson and Ross-Murphy, 1981a; Tobitani and Ross-Murphy, 1997a). In the latter, it was shown that this criterion systematically indicated a gel time less than that measured on the basis of viscosity changes.

Extrapolation of G′ *Value*

The extrapolation of the rapidly rising values of G' to the time axis was also suggested as an accurate estimate of the gel point, t_c (Clark, 1991). Hsieh et al. (1993) determined T_g of whey and egg proteins by extrapolation of the rapidly rising values of G' to the temperature axis. In addition, the gel time was calculated for the gelation of bovine serum albumin by monitoring the increasing viscosity as a function of time (Richardson and Ross-Murphy, 1981a) and taking the time corresponding to the asymptotic value of viscosity. In this method, although small shear rates were used, the high strains necessarily imposed probably affected the gel critical point.

Critical Behavior at Sol-Gel Transition

For high-methoxyl pectin (HM pectin), the gel point changed with the aging temperature of the system (Lopes da Silva and Gonçalves, 1994) probably due to changes in the mobility of the pectin chains, and consequently in the lifetime of junction zones;

weaker pectin networks are formed under thermal conditions unfavorable to the development of hydrophobic interactions. Figure 6–4 shows the mechanical spectra recorded at different stages of the pectin network formation at 15°C. At this aging temperature, the HM pectin/sucrose system seems to be characterized by a relaxation exponent at sol/gel transition around 0.55. At higher aging temperatures, the viscoelastic behavior through the gel point was qualitatively and quantitatively different: The evolution of the dynamic moduli was quicker and occurred with a first tendency to an elastic plateau at low frequencies, with $G' > G''$ at low frequencies and G' and G'' superimposed at higher frequencies. For aging temperatures in the range 20°–60°C, the viscoelastic behavior near the gel point was qualitatively similar.

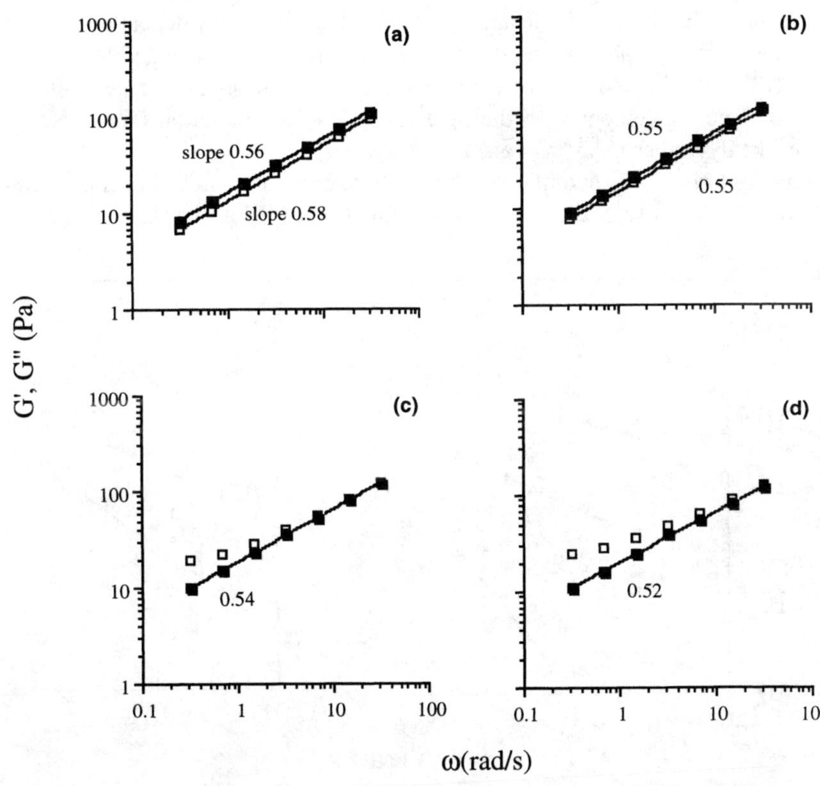

Figure 6–4 Mechanical Spectra (□ G', ■ G'') Recorded at 15°C and at Several Aging Times in the Initial Gelation Stages of a 1% (w/w) High-Methoxyl Pectin Dispersion (60% w/w sucrose, pH 3): (a) 34 min, (b) 41 min, (c) 68 min, (d) 84 min; values are slopes of the straight lines. *Source*: Reprinted from *Carbohydrate Polymers*, Vol. 24, J.A.L. da Silva and M.P. Goncalves, Rheological Study into the Aging Process of High Methoxyl Pectin/Sucrose Aqueous Gels, pp. 235–245, © 1994, with permission from Elsevier Science.

Power law dependence upon oscillatory frequency of both moduli, with the same relaxation exponent (Δ), was found to occur for values of Δ in the range 0.60–0.65.

Dynamic rheological experiments have been used to follow the aggregation process before the gel point and throughout the sol-gel transition for calcium-induced low-methoxyl pectin (LM pectin) gelation as a function of the calcium added to the system and as a function of the aging time (Lopes da Silva et al., 1996). Figure 6–5 shows G' and G'' at 20°C plotted against angular frequency (ω) at various calcium concentrations added to a pectin solution ($c_p = 6.1$ g L^{-1}, pH 7, 0.1 mol L^{-1} NaCl). The LM pectin dispersion without added calcium shows the classical viscoelastic behavior of a macromolecular solution in the terminal zone at low frequencies with $G'(\omega) \propto \omega^2$ and $G''(\omega) \propto \omega$. From this low-frequency behavior the system smoothly evolves toward a plateau value with increasing calcium concentration, with both moduli increasing in magnitude and decreasing their frequency dependence. It was very close to the gel point when the calcium concentration equaled 0.083 g/L, for which $G'(\omega) \propto \omega^{0.73}$ and $G''(\omega) \propto \omega^{0.71}$. Thus, this critical behavior at gel point was in satisfactory agreement with the percolation model prediction (2/3) based on Rouse-like dynamics (Martin et al., 1989).

Figure 6–6 shows the evolution of the dynamic moduli for an LM pectin/calcium system in the vicinity of the gel point as a function of the aging time. The evolu-

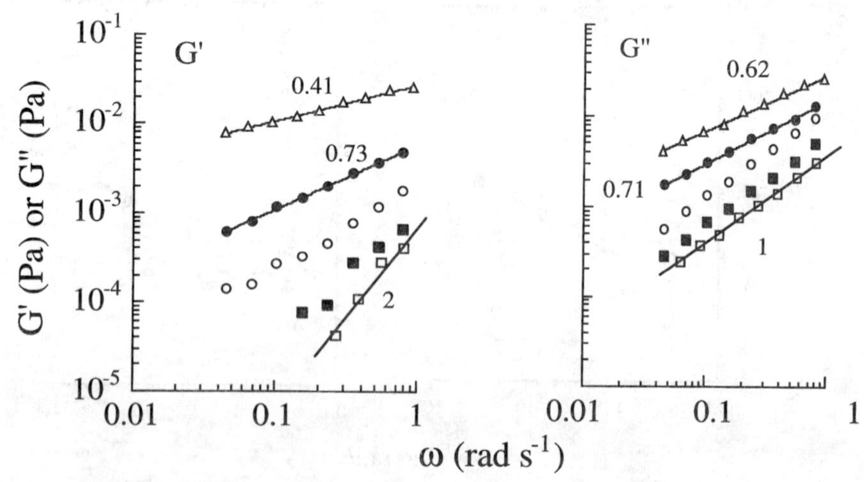

Figure 6–5 Mechanical Spectra Recorded at 20°C of 6.1 g L^{-1} Low-Methoxyl Pectin (pH 7, 0.1 mol L^{-1} NaCl) with Different Calcium Contents Aged for 20 Hours at 20°C: □ no added calcium, ■ 1.4×10^{-3} mol L^{-1}, ○ 1.9×10^{-3} mol L^{-1}, ● 2.1×10^{-3} mol L^{-1}, Δ 2.4×10^{-3} mol L^{-1}; numbers are values of the slopes of lines. *Source*: Reprinted from *Polymer Gels and Networks*, Vol. 4, J.A.L. da Silva, Effect of Galactomannan on the Viscoelastic behavior of Pectin/Calcium Networks, pp. 65–83, © 1996, with permission from Elsevier Science.

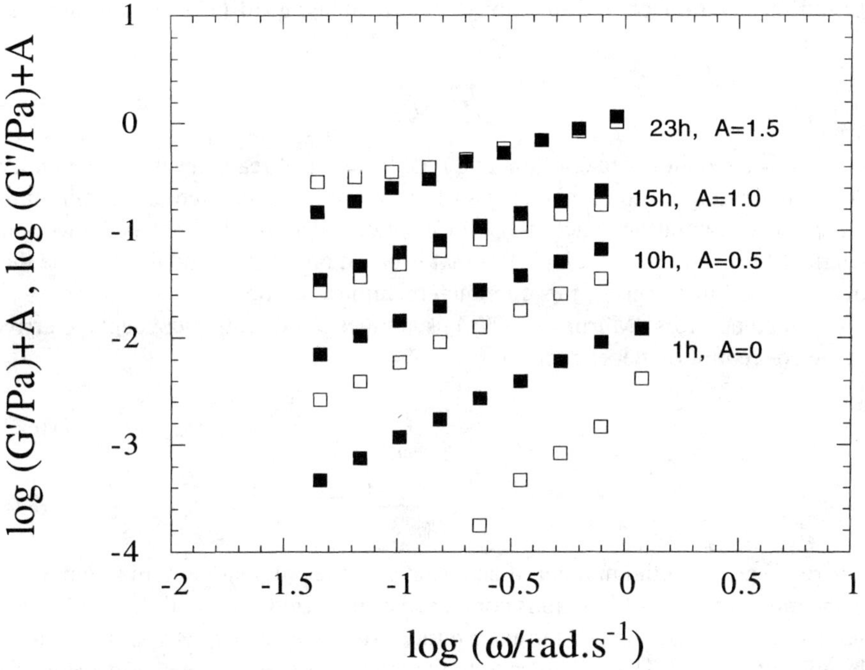

Figure 6–6 Dynamic Moduli (□ G', ■ G'') as a Function of Frequency at 20°C, Measured at Different Aging Times ([low-methoxyl pectin] = 6.1 g L^{-1}, [Ca^{2+}] = 2.4 × 10^{-3} mol L^{-1}, pH 7, 0.1 mol L^{-1} NaCl). To avoid superposition of data, values (A) shown were added to G'. *Source*: Reprinted from *Polymer Gels and Networks*, Vol. 4, J.A.L. da Silva, Effect of Galactomannan on the Viscoelastic Behavior of Pectin/Calcium Networks, pp. 65–83, © 1996, with permission from Elsevier Science.

tion of the dynamic moduli was similar to that observed as a function of the calcium concentration. In the initial period of aging, the system showed the typical liquid-like behavior. Then both moduli increased with time, G' increasing more rapidly than G'' and with lower dependence on frequency. After 15 hours of aging, the system was just above the critical gel point, with a viscoelastic exponent Δ in the range of 0.65–0.68. After the gel point, G' passed beyond G'', first in the lower frequency range where one can observe the initial formation of the elastic plateau.

Influence of Concentration and Temperature on Gel Time

The influence of concentration on gel time of gelatin gels was described (te Nijenhuis, 1981) by the equation

$$\ln(t_c) = a_1 + a_2 c \tag{6.21}$$

Based on gelation of high-methoxyl pectins, Oakenfull (1984) suggested the equation

$$t_c = \frac{1}{kc^n} \tag{6.22}$$

where k is a reaction rate constant and n is the order of reaction. It is to be understood that reaction orders higher than four are not encountered and a minimum (critical) concentration is necessary for gelation. Also of relevance is the work of Clark (1994) and Clark et al. (1994) that is based on kinetics and the cascade theory with the limitations of this approach mentioned earlier.

Tobitani and Ross-Murphy (1997a) used kinetic and scaling concepts to arrive at the concentration dependence as

$$t_c \approx \frac{nN_{b0}}{kN_{a0}^n} \tag{6.23}$$

$$t_c \sim \frac{1}{N_{a0}^n} \sim \frac{1}{c^n} \tag{6.24}$$

where N_{a0} is the initial number of active sites of the polymer per unit volume, N_{b0} is the initial number of junctions composed of n reacted sites, and the symbol ~ is used to indicate that the two terms on either side can be equated by scaling but not by absolute values. The influence of temperature, T (K), was taken into account in the reaction rate constant, k:

$$k = A \exp\left(\frac{-E_a}{RT}\right) \tag{6.25}$$

where A is the pre-exponential factor, E_a is the apparent activation energy, and R is the gas constant. Lopes da Silva et al. (1995) also found that the effect of temperature on gelation rates followed the Arrhenius equation. Tobitani and Ross-Murphy (1997a) combined the above equations and then modified the resulting expression to take into account the need for the cross term of (ln c) and temperature:

$$\ln(t_c) = a_1 + a_2 \ln(c) + a_3 T + a_4 T \ln(c) \tag{6.26}$$

where a_1, a_2, a_3, and a_4 are coefficients to be determined from experimental data.

Experimental gel threshold times (symbols) of bovine serum albumin (Tobitani and Ross-Murphy, 1997a) are shown as a function of temperature and the threshold times 100, 1000, and 10,000 s in Figure 6–7. The threshold values of time were determined as the time at which G' of a sample reached a threshold value of between 2 and 3 Pa. Further, values calculated using the above equation are shown as lines in the figure. In the diagram, the upper region represents the gel state and the lower region the sol state of a sample. For a given set of concentration and temperature values, the threshold time (e.g., 100 s) for gel formation can be found

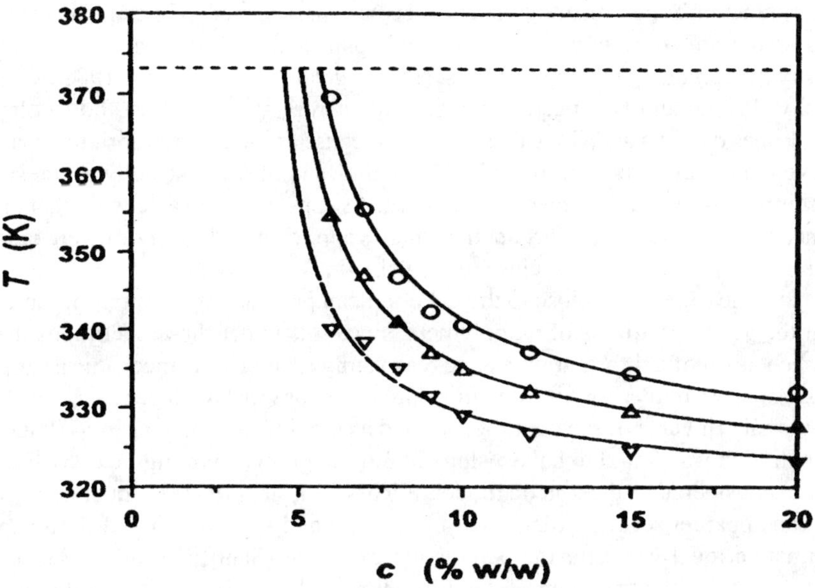

Figure 6–7 Temperature versus Concentration State Diagram Illustrating the Sol-Gel Boundary for BSA and Showing Threshold Times for Gelation: ○ 100 s, Δ 1000 s, ∇ 10,000 s. Horizontal broken line is boiling point of solvent. Curves shown are based on Equation 6.26. *Source:* Reprinted with permission from A. Tobitani and S. B. Ross-Murphy, "Heat-Induced Gelation of Globular Proteins. 1. Model for the Effects of Time and Temperature on the Gelation Time of BSA," *Macromolecules,* Vol. 30, pp. 4845–4854, © 1997, American Chemical Society.

from the figure. The diagram can also be used to predict the possibility of gel formation for any given concentration and temperature.

Evaluation of Structure Development during Biopolymer Gelation

Aqueous biopolymer gels exhibit a general characteristic behavior when the structure development is evaluated by dynamic rheological measurements (Clark and Ross-Murphy, 1987). The gradual development of the network structure is reflected by a progressive increase in storage modulus (G'). The beginning of the gelation process is dominated by the viscous behavior of the system ($G'' > G'$), and the elastic behavior dominates the final stages of the experiment ($G' \gg G''$). Both moduli increase as a result of the increasing density of junction zones, but G' rises more sharply until it intercepts and then exceeds G''. After an initial rapid increase, resulting from the rapid formation of junction zones into the polymer network, G' keeps increasing continuously as a result of the slower formation and rearrangement of junction zones, resulting in a net decrease in the length of elastically active chains; values of G' eventually reach a pseudo plateau region. It is emphasized that in many biopolymer gelation processes, such as gelatin (te Nijenhuis, 1981; Ross-Murphy, 1991b) and HM pectin/fructose and sucrose systems (Rao and Cooley, 1993; Lopes da Silva and Gonçalves, 1994), because of the nonequilibrium behavior, even after aging for long periods, G' continues to increase steadily as a result of a continuous reorganization of the network. As expected from a network that has attained a state close to equilibrium, the kinetics applicable at longer times are quite different from those observed close to the gel point.

Kinetic measurements close to the gel point can provide important information about the gelation process of biopolymers, since it is under these conditions that the mechanism of gel formation can be reasonably related to its mechanical properties. As an example, for HM pectin gelation, the power law dependence of the gelation rate (inverse of the gel time, defined as the time at which $G' = G''$) upon concentration was found to have a slope of 3.6 (diagram (A) in Figure 6–8). Taking into consideration the kinetic model of Ross-Murphy (1991a), Equation 6.27, using an apparent second-order reaction ($n = 2$), and assuming $p = 1.8$, the approximate critical value for the percolation exponent (Stauffer et al., 1982), the limiting value of the exponent of a log-log plot of $(1/t_c)$ versus c at high c/c_o was

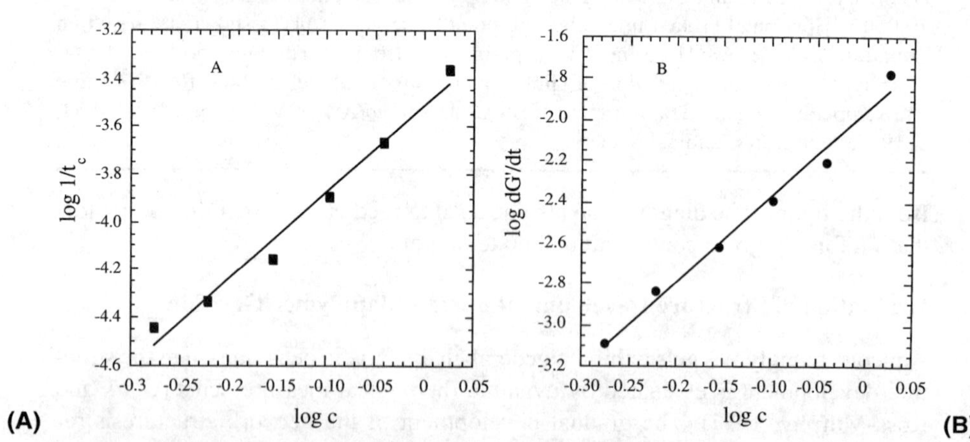

Figure 6–8 Influence of High-Methoxyl Pectin Concentration on the Gelation Rate (inverse of the apparent gel time, t_c) (A) and on the Rate of Change in G' at the Gelation Threshold (B). The slope of the straight line in (A) was 3.6 ($R^2 = 0.981$); least squares fit of data in (B) gave a slope of 4.1 and $R^2 = 0.986$. *Source*: Reprinted from *International Journal of Biological Macromolecules*, Vol. 17, J.A. Lopes da Silva et al., Kinetics and Thermal Behavior of the Structure Formation Process in HMP/Sucrose Gelation, pp. 25–32, © 1995, with permission from Elsevier Science.

equal to *np,* in close agreement with the slope of 3.6 obtained in Figure 6–8, diagram (A) (Lopes da Silva and Gonçalves, 1994).

$$t_c \approx K/[((c/c_o)^n - 1)^p] \tag{6.27}$$

The rate of change in G', dG'/dt, and the modulus G' at the gelation threshold are also sensitive to changes in pectin concentration. The rate of change in G' increased with pectin concentration (Figure 6–8, diagram (B)). The usual high power dependence obtained for dG'/dt with polymer concentration was probably related to the complex behavior typical at low biopolymer concentration close to the gel threshold. Pronounced concentration dependence has been reported for a wide range of systems. For BSA gelation, the slope of the modulus growth near the gel point was found to be proportional to c^{27} for low concentrations, and even at higher concentrations dG'/dt is still proportional to c^6 (Richardson and Ross-Murphy, 1981a). For physical gels of poly(acrylonitrile) in dimethylformamide close to the critical gel concentration, Bisschops (1955) found $dG'/dt \sim c^{22}$. In contrast, Nolte et al. (1992) found a much lower power dependence of dG'/dt on polymer concentration, $dG'/dt \sim c^{1.25}$, for the gelation of xanthan with chromium ions, even close to the critical gelling concentration. Close to the critical concentration, the interpretation of these exponents is complex. It is likely that the different mechanisms of macromolecular association involved account for the wide range of values observed.

When the storage modulus is plotted against log aging time, typically the curves obtained are approximately linear after some time. The slope of these curves (dG'/dt) can be interpreted as a rate of aging in this long time range, a criterion used by te Nijenhuis (1981) to analyze the rate of aging of gelatin and polyvinyl chloride gels. Examples of the application of this approach are shown in Figure 6–9 for LM pectin/Ca^{2+} gelation. Magnitudes of the rates of aging increased when the pectin concentration was increased (Figure 6–9, diagram (A)) or when the calcium/pectin ratio was increased (Figure 6–9, diagram (B)), suggesting that the annealing processes, such as rearrangement of junction zones and lateral chain aggregation resulting in the net increase in the length of elastically active chains, became more prevalent at higher pectin concentration and/or higher calcium/pectin ratio.

Effect of Temperature on the Rate of Structure Formation and Kinetic Data

Most gelation processes involving biopolymers can be seen as thermal effects, because the gelation mechanism basically is related to the change of only one thermodynamic parameter: the temperature. Depending on the molecular organization of the polymer network, temperature can have different effects on the viscoelasticity of the gel. The study of the influence of temperature on structure formation in biopolymer systems used as gelling agents in food formulations could be useful in understanding

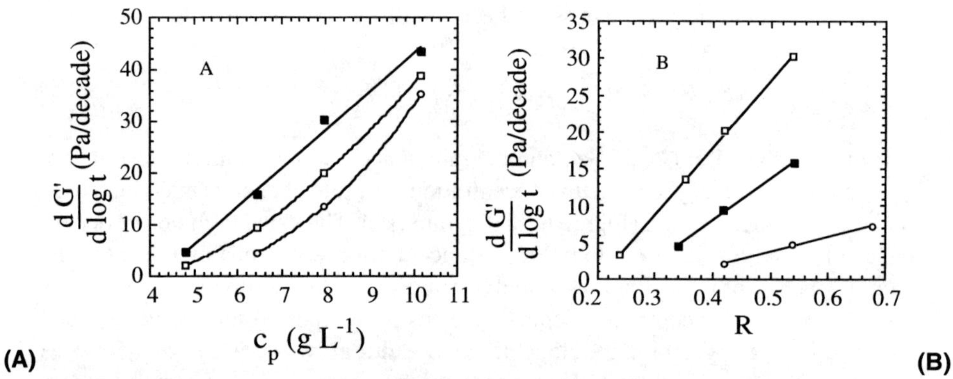

Figure 6–9 Rate of Aging for Low-Methoxyl Pectin Gelled Systems Studied Near the Pseudo Plateau Region. (A) dG'/d log t, against pectin concentration (0.1 mol L^{-1} NaCl, pH 7) for different calcium/pectin ratios: ○ $R = 0.34$, ■ $R = 0.42$, □ $R = 0.54$; (B) dG'/d log t, against the stoichiometric calcium/pectin ratio, R, for different concentrations: ○ 4.78 g L^{-1}, ■ 6.45 g L^{-1}, □ 7.96 g L^{-1}. *Source*: Reprinted from J.A.Lopes da Silva, M.A. Rao, and J-T. Fu, Rheology of Structure Development and Loss During Gelation and Melting, in *Phase/State Transitions in Foods: Chemical, Rheological and Structural Changes*, M.A. Rao and R.W. Hartel, eds., pp. 111–156, by courtesy of Marcel Dekker, Inc.

the thermal behavior of food commodities where they are incorporated and can provide additional insights into the nature of the gelation process of these polymers.

Structure Development Rate

Structure development rate (SDR), defined as dG'/dt, can be measured under isothermal conditions, at different aging temperatures (cure curves), and also during the gelation process promoted by decreasing the temperature of the biopolymer dispersion (nonisothermal conditions) (Rao and Cooley, 1993). The analysis of SDRs under isothermal conditions, as opposed to nonisothermal conditions, can be very useful to avoid the nonequilibrium effects and the effect of thermal history of the system.

Structure Development under Isothermal Conditions

The evolution of dG'/dt with time during a cure experiment has been applied to the gelation of HM pectin/sucrose systems (Lopes da Silva et al., 1995). The gelation rate of HM pectin/sucrose systems was strongly dependent upon the temperature, showing a complex behavior due to the several kinds of intermolecular interactions involved. The influence of temperature on these intermolecular interactions, mainly hydrogen bonding and hydrophobic interactions, could be described by an Arrhenius relationship (Figure 6–10). Two different temperature ranges could be identified, one between 15°C and 30°C and the other between 30°C

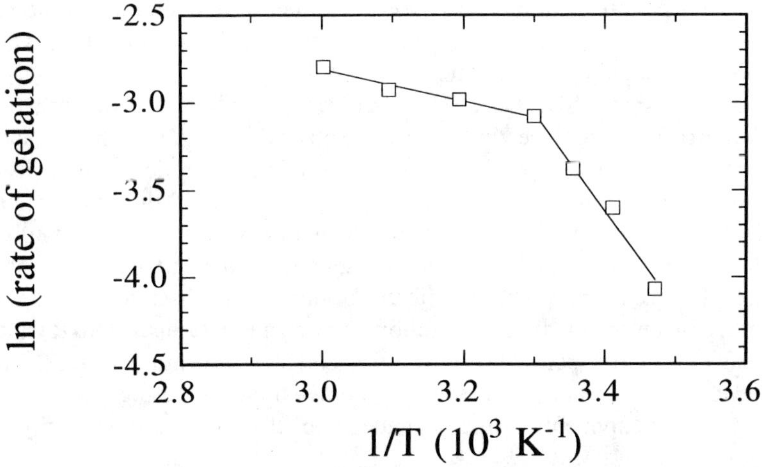

Figure 6–10 Arrhenius Treatment for the Influence of Temperature on the Gelation Rate of 1% High-Methoxyl Pectin Dispersion (60% sucrose, pH 3), Measured at Different Aging Temperatures (in the range 15°–60°C). *Source*: Reprinted from *International Journal of Biological Macromolecules*, Vol. 17, J.A. Lopes da Silva et al., Kinetics and Thermal Behavior of the Structure Formation Process in HMP/Sucrose Gelation, pp. 25–32, © 1995, with permission from Elsevier Science.

and 60°C, yielding apparent activation energies of 46.4 kJ mol^{-1} and 7.7 kJ mol^{-1}, respectively. It is clear that the gelation rate was energetically favored above 30°C.

Two different processes were proposed to explain the discontinuity observed, each one having rates with different temperature dependence. One of them was the initial intermolecular local contacts, essentially by hydrogen bonding and involving short chain segments. This stage can have similarities with the nucleation step found in the gelation of biopolymers that undergo a coil-to-helix transition. This process was considered the rate-determining process above 30°C. The second different rate process was attributed to the subsequent lateral aggregation of chains and enlargement of junction zones, essentially governed by hydrophobic interactions, favored by increasing temperature, and probably stabilized or strengthened by van der Waals forces. This was considered to be the rate-limiting process below 30°C. Hydrogen bonds are less dependent upon the separation of molecules and have enough energy to be stable even when they are few in number. Hydrogen bonds acting at larger intermolecular distances can be essential to the subsequent development of hydrophobic interactions and van der Waalsforces through clustering of nonpolar groups involving larger portions of the pectin chains.

The kinetics at longer times were quite different. As shown in Figure 6–11, the initial period of gelation was characterized by concave cure curves, after which they became convex and then almost linear. At 5°C, the convex region was not perceptible. This different behavior, and the effect of temperature on the rate of aging of the HM pectin gels, could be clearly illustrated when the SDRs were analyzed at different aging temperatures.

Figure 6–12 shows SDRs as a function of the aging time at different temperatures. Significant differences in SDRs were observed during the first 4 hours of aging. The general behavior is an initial increase in SDR, followed by a decrease in the SDR values, which continue to decrease steadily until the end of the aging period. The sharpest initial increase in the aging process was observed at 30°C; the maximum in SDR was reached at a lower aging time than at the other aging temperatures. For the aging at 5°C, a different behavior was observed, with an initial rapid decrease during the first hour, followed by an almost constant SDR value until the end of the aging period. Between 15°C and 50°C, the initial increase in SDR was due to the initial rapid increase in the storage modulus during the aging of the pectin gel. This exponential time dependence of the increase in elasticity of the

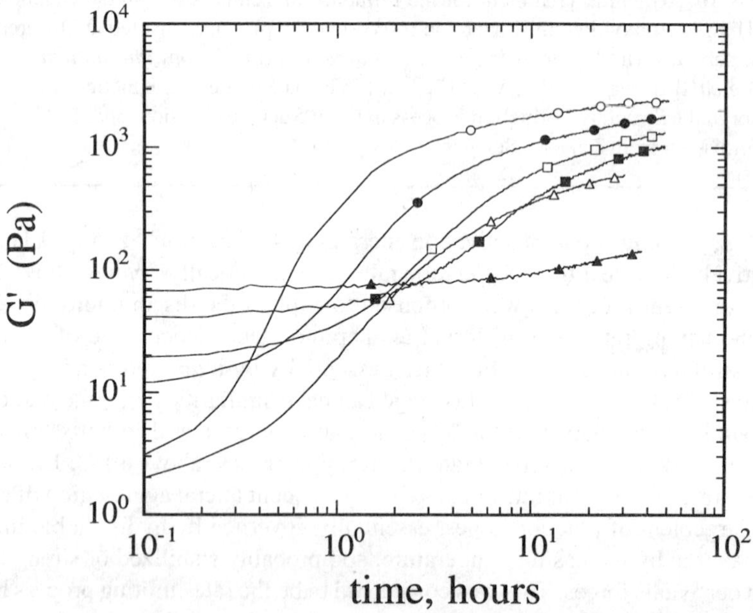

Figure 6–11 Storage Modulus (G') as a Function of the Aging Time for 1% High-Methoxyl Pectin (60% sucrose, pH 3), at Several Aging Temperatures: ▲ 5°C, ■ 10°C, □ 15°C, ● 20°C, ○ 30°C, Δ 50°C. Data were obtained at 3% strain and 0.5 Hz every 3 min during first 3 hours, then every 10 min. For clarity, only some of the experimental points are shown. *Source*: Reprinted from *International Journal of Biological Macromolecules*, Vol. 17, J.A. Lopes da Silva et al., Kinetics and Thermal Behavior of the Structure Formation Process in HMP/Sucrose Gelation, pp. 25–32, © 1995, with permission from Elsevier Science.

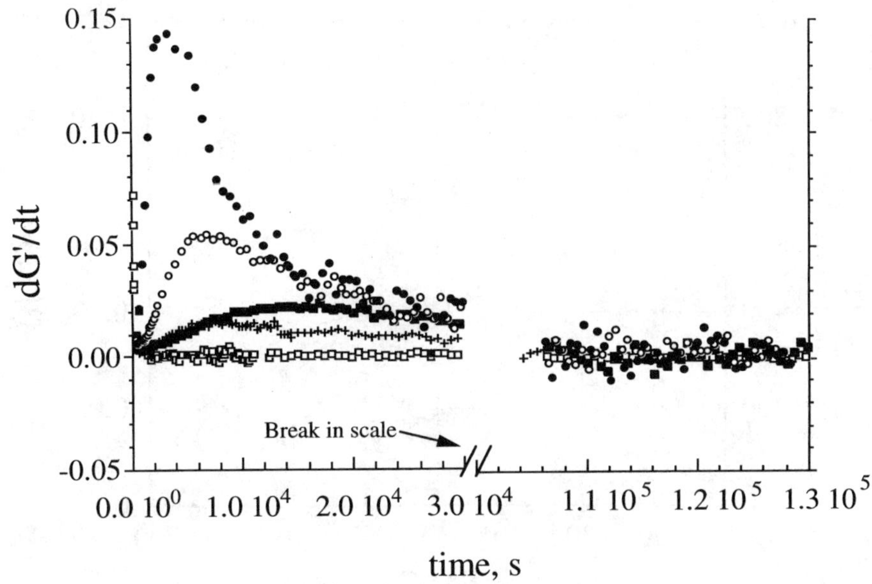

Figure 6–12 Structure Development Rates during Aging of a 1% High-Methoxyl Pectin/60% Sucrose (pH 3.0) Dispersion under Isothermal Conditions at Different Temperatures: □ 5°C, ■ 15°C, ○ 20°C, ● 30°C, and + 50°C. SDR was calculated using numerical differentiation followed by smoothing of the derivative values. *Source*: Reprinted from *International Journal of Biological Macromolecules*, Vol. 17, J.A. Lopes da Silva et al., Kinetics and Thermal Behavior of the Structure Formation Process in HMP/Sucrose Gelation, pp. 25–32, © 1995, with permission from Elsevier Science.

system should correspond to the rapid formation of junction zones between pectin chains. After this stage, the SDR decreased until it reached a value, always higher than zero but almost independent of the aging time or of the aging temperature, that corresponded to a slow increase in storage modulus and to the almost linear region of the cure curves. This latter stage should correspond to a slow reorganization of the network, allowing for creation of new junction zones or an increase in the extension of the junctions between molecular chains.

Structure Development Rate under Nonisothermal Conditions

Based on rheological data obtained from temperature sweep experiments, SDRs can also be measured during the gelation process promoted by decreasing the temperature of the biopolymer dispersion (nonisothermal conditions), as illustrated for the HM pectin/sucrose system in Figure 6–13. During gelation induced by cooling, the SDR increased with decreasing temperature, showing a higher variation at the higher and lower temperature ranges. A temperature range where the

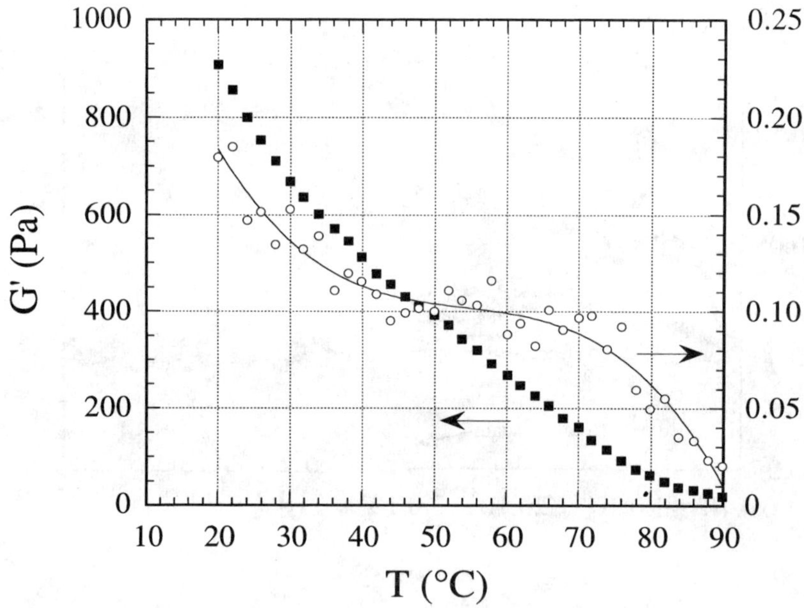

Figure 6–13 Influence of Temperature on Storage Modulus (G') and dG'/dt (SDR) for 1% High-Methoxyl Pectin/60% Sucrose Dispersion during Cooling at 0.5°C min^{-1} from 90°C to 20°C at 3% Strain and 0.5 Hz: ■ G'; ○ dG'/dt calculated after cubic spline fit; —— dG'/dt calculated after polynomial fit. *Source*: Reprinted from *International Journal of Biological Macromolecules*, Vol. 17, J.A. Lopes da Silva et al., Kinetics and Thermal Behavior of the Structure Formation Process in HMP/Sucrose Gelation, pp. 25–32, © 1995, with permission from Elsevier Science.

SDR increases with decrease in temperature can be found during the gelation of biopolymers that undergo coil-helix transitions, such as gelatin (Djabourov et al., 1988a) or agarose (Braudo et al., 1991). However, in these cases the effect of temperature on the formation of ordered structures typically follows a sigmoidal shape, instead of the logarithmic shape observed, and consequently the effect of temperature on the rate of structure development shows a bell-shaped curve, with a maximum value at an intermediate temperature.

Kinetics of Gelation

Kinetic data can also be obtained for the structure development process during cooling of biopolymer dispersions. This kind of analysis is difficult due to the combined effects of time and temperature. Usually, the data are obtained using time as an independent variable and changing the temperature and test sample for each run.

A nonisothermal kinetic model was used to describe the gelation process of the high-methoxyl pectin/sucrose system during cooling (Lopes da Silva et al., 1995), taking into consideration kinetic effects as well as thermal effects on biopolymer gelation. It was adapted from the "linear increasing temperature method" (Rhim et al., 1989) to changes in structure under nonisothermal conditions (temperature sweeps) by linearly decreasing the temperature at a constant rate. This nonisothermal kinetic model was based on the combination of the classical rate equation, the Arrhenius equation, and the time-temperature relationship, yielding:

$$\ln\left(\frac{1}{G'^n}\frac{dG'}{dt}\right) = \ln k_o - \left(\frac{E_a}{RT}\right) \tag{6.28}$$

where G' is storage modulus, taken as a measure of the number of junction zones in the network; n is reaction rate order; t is time, k_o is the Arrhenius frequency factor; E_a is the activation energy of the process; R is the universal gas constant; and T is the absolute temperature.

An important benefit in using this method to describe the gelation process of biopolymers is that order of the reaction rate can be evaluated, a parameter that can be difficult to determine for this kind of process. It was assumed that at least in the incipient gelation process of the HM pectin/sucrose aqueous system, macromolecular aggregation could be described by a bimolecular association equilibrium, that is, $n = 2$. This was also supported by the results obtained for the concentration dependence of the gelation rate. However, that a more complex mechanism due to the change in temperature or in the number of preformed junction zones during the cooling-induced gelation of the system can also occur should not be excluded.

Assuming the reaction order $n = 2$, Equation 6.28 was used to estimate the activation energy of the process by using a linear least-squares routine (Figure 6–14). The process could be satisfactorily described by assuming a two-step process, corresponding to two different temperature ranges. Between 90°C and 62°C, the least squares linear regression yielded $E_a = 132.0$ kJ mol^{-1} and $k_o = 5.09 \times 10^{14}$ Pa^{-1}s^{-1}, and between 50°C and 20°C, $E_a = 27.6$ kJ mol^{-1} and $k_o = 1.77 \times 10^{-2}$ Pa^{-1}s^{-1}.

Since data points were estimated in a continuous manner during the initial period of gelation induced by cooling, the kinetic parameters obtained were likely affected by the previous thermal history of the material. The higher activation energy in the higher temperature range may reflect the higher energy barrier between the unaggregated and aggregated macromolecular states, under temperature conditions not favorable to the development of intermolecular interactions that could promote the formation of the pectin network. Under thermal conditions that favor interchain aggregation, as was the case below 50°C, the structure development process was energetically easier, and the activation energy was lower.

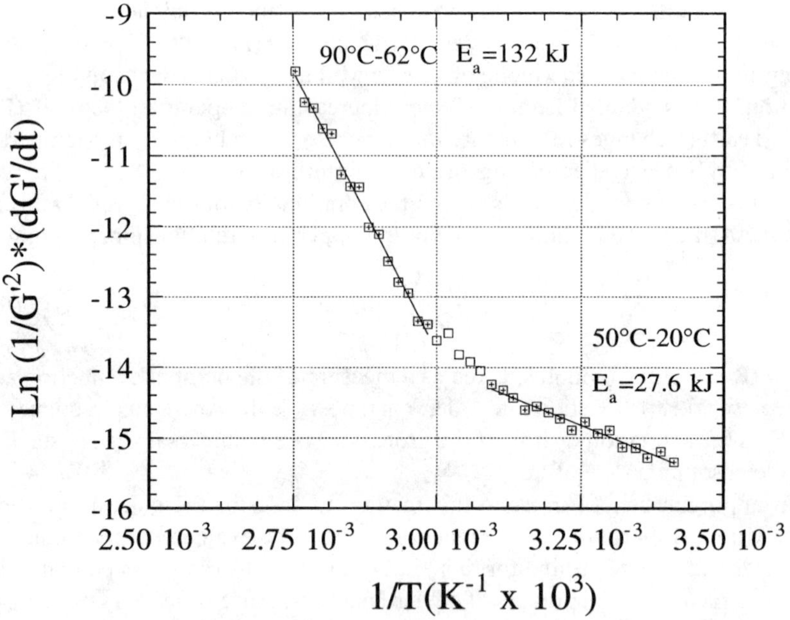

Figure 6–14 Arrhenius Plot for the Change in the Storage Modulus (G') during Cooling of a 1% High-Methoxyl Pectin/60% Sucrose Dispersion (pH 3.0). Apparent second-order kinetics and a two-stage gelation process were assumed. *Source*: Reprinted from *International Journal of Biological Macromolecules*, Vol. 17, J.A. Lopes da Silva et al., Kinetics and Thermal Behavior of the Structure Formation Process in HMP/Sucrose Gelation, pp. 25–32, © 1995, with permission from Elsevier Science.

Evaluation of Structure Loss During Melting/Softening of Biopolymer Gels

Many biopolymer gels follow the general behavior of viscoelastic amorphous polymers: a decrease in viscoelastic parameters with increasing temperature (Ferry, 1980). Such biopolymer gels include alginates (Andresen and Smidsrød, 1977), κ-carrageenan (Watase and Nishinari, 1987b), agarose/κ-carrageenan mixed gels (Zhang and Rochas, 1990), globular proteins (Richardson and Ross-Murphy, 1981a, 1981b), and LM pectin/calcium gels (Garnier, 1992). This behavior is typical of polysaccharide gels where hydrogen bonding or electrostatic interactions are the only significant interactions that stabilize the polymer network. Consequently, for most biopolymer gels the rupture of junction zones stabilized by thermolabile interactions may explain the decrease in modulus with increasing temperature, in addition to the increase in flexibility of the polymer chains. Nishinari et al. (1985) introduced the "reel-chain" model for the molecular mechanism underlying the temperature dependence of the viscoelastic behavior of thermoreversible biopolymer gels, which was further theoretically treated by Higgs and Ball (1990). Structure loss rates of only a few binary gels, such as HM pectin/sodium alginate (Rao and Cooley, 1995), have been studied.

Clearly, temperature plays a crucial role in the formation and disruption of the typical physical cross-links found in biopolymer gels. Activation energies associated with the breakage of junction zones in biopolymer gels have been estimated from the temperature dependence of different rheological parameters. For both protein and polysaccharide gels, activation energies ranging from 10 to 273 kJ mol^{-1} have been reported (Table 6–5).

Eldridge-Ferry Plot

Eldridge and Ferry (1954) studied the relationship between the melting point (T_m) and concentration (c) and molecular weight (M) of gelatin gels. They found linear relationships between ln (c) and $1/T_m$ and between ln (M) and T_m. Assuming an equilibrium binary association of polymer chains and using the van't Hoff law, they suggested the following relationship from which it is possible to calculate the enthalpy (ΔH_m) of the cross-linking process and to get some information about the stability of the junction zones into the gel:

$$-\left[\frac{\mathrm{d}\ln c}{\mathrm{d}T_m}\right]_{\mathrm{MW}} = \frac{\Delta H_m}{RT^2} \tag{6.29}$$

This approach was satisfactorily applied to other biopolymer gelled systems, including agarose (Watase and Nishinari, 1987a), HM pectin/dimethyl sulfoxide (DMSO) gels (Watase and Nishinari, 1993), κ-carrageenan (Watase and Nishinari, 1987b), and LM pectin/Ca^{2+} and LM pectin/Ca^{2+}/sucrose systems.

Melting temperatures (T_m) were determined for several LM pectin/calcium gels at different pectin and calcium concentrations. During the heating process of the gel, the storage modulus G' decreased with increasing temperature. T_m was selected to be either the temperature at which G' became too small to be accurately measured or that corresponding to the inflection point of the plot of G' versus temperature. T_m increased as either the pectin or the calcium concentration increased, due to the increase in cross-linking junctions between the pectin chains. For the LM pectin + calcium and the LM pectin + calcium + sucrose systems, plots of ln c_p versus T_m^{-1} (Figure 6–15) showed good linearity.

The dissociation enthalpies of junction zones (ΔH_m) for melting of the LM pectin/Ca^{2+} and LM pectin/Ca^{2+}/sucrose gels that reflect the thermal stability of the junction zones were estimated from the slopes of the straight lines of the Eldridge-Ferry plots shown in Figure 6–15. No significant effects of the pectin and

Table 6–5 Magnitudes of Activation Energy for Viscoelastic Properties of Biopolymer Gels

Sample	E_a *(kJ/mol)*	*Technique*	*Reference*
κ-carrageenan gels (0.1 mol/L KCl) 15 g/L	87	Temperature dependence of the shift factor required for superposition of creep curves (9–50°C)	Braudo et al., 1984
κ-carrageenan gels 15 g/L	10.0	Arrhenius equation applied to the elastic modulus measured at different temperatures by compression tests	Zhang and Rochas, 1990
xanthan/chromium gels	42	Temperature dependence of the gelation rate measured at different temperatures	Nolte et al., 1992
1.5% HM pectin gels (70% sucrose, pH 3)	155–160	Temperature dependence of the shift factor required for superposition of creep curves (25–50°C)	Plashchina et al., 1979
2% HM pectin gels (65% sucrose, pH 3)	130	Temperature dependence of the shift factor required for superposition of creep curves (11–49°C)	Kawabata, 1977
2.5% LM pectin gels (0.05% calcium)	273	Temperature dependence of the shift factor required for superposition of creep curves (5.5–24.5°C)	Kawabata, 1977
LM pectin/calcium gels (c_p = 4–8.5 g/L; R = 0.3–0.5)	11–86	Temperature dependence of the shift factor required for time-temperature superposition of mechanical spectra (5–40°C).	Garnier, 1992
Whey protein gels (protein 14%, pH 8)	87–143	Time-temperature superposition of creep curves and WLF equation (15–75°C)	Katsuta and Kinsella, 1990

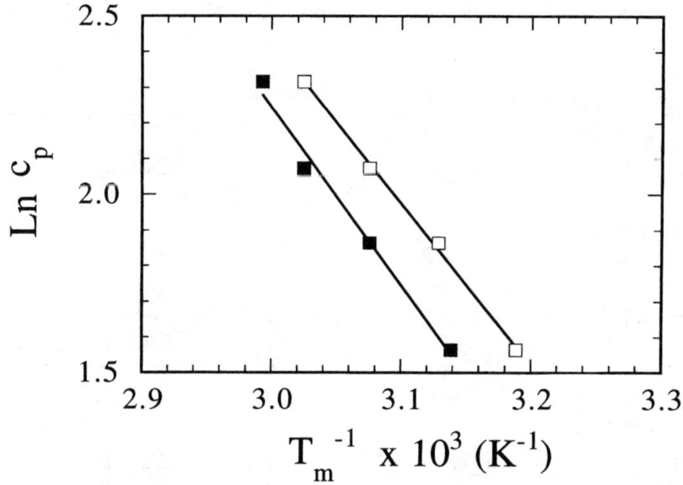

Figure 6–15 Eldridge-Ferry Plots in ln c_p versus $1/T_m$ for Low-Methoxyl Pectin + Calcium Gels: no sugar, 0.1 mol L^{-1} NaCl, pH 7.0, for two different *R* ratios: □ $R = 0.42$ and ■ $R = 0.54$. *Source*: Reprinted from J.A. Lopes da Silva, M.A. Rao, and J-T. Fu, Rheology of Structure Development and Loss During Gelation and Melting, in *Phase/State Transitions in Foods: Chemical, Rheological and Structural Changes*, M.A. Rao and R.W. Hartel, eds., pp. 111–156, by courtesy of Marcel Dekker, Inc.

---

calcium/pectin ratio (*R*) on the ΔH_m values were observed. The enthalpy change obtained in this way characterizes the enthalpy associated with the strongest network junctions averaged over the concentration range, that is, the junction zones that remain stable until the melting temperature is reached (Braudo et al., 1984). The temperature dependence of the viscoelastic behavior of the gel is another method to measure the breakdown enthalpy of network junctions (ΔH_{bT}). The enthalpy change obtained in this way characterizes gels of a certain concentration and relates to the temperature interval for which the viscoelastic properties were measured. Its magnitude can be regarded as a measure of the variation in the cross-link strength (i.e., in the number of cross-links) within the gel (Braudo et al., 1984). Both parameters can serve as measures of stability of the gel junction zones, but each of them reflects a different manifestation of the breakdown process, and thus their magnitudes are not necessarily the same for all gelled systems. In fact, as can be seen in Table 6–6, it seems that the difference between ΔH_{bT} and ΔH_m is larger for gels with low water activity, for which an important contribution from hydrophobic interactions to the stability of the gel may be expected. Therefore, junction zones with different stability might be present, and the hydrophobic interactions may contribute to the relatively high values of ΔH_m.

Table 6–6 Magnitudes of Enthalpy (ΔH_m) of Melting of Network Junctions of Biopolymer Gels

Sample	ΔH_m *(kJ/mol)*	*Technique**	*Reference*
Gelatin gels (*c* = 2–6%)			
chilled at 0°C	205–305†	1	Eldridge and Ferry, 1954
chilled at 15°C	502–920†		
HM pectin gels (*c* = 2%)			
(70% sucrose, pH 3.0)	20	2	Plashchina et al., 1979
HM pectin gels	126	1	Braudo et al., 1984
(50% glycerol, pH 2.7)	17	2	
Ca-salt of furcellaran	121	1	Braudo et al., 1984
(70% sucrose, pH 7.6)	19	2	
κ-Furcellaran gels	54	1	Braudo et al., 1984
(0.15M KCl, pH 7.6)	40	2	
Agarose gels (*c* =4–18%)	$1.5–1.6 \times 10^3$	1	Watase and Nishinari, 1987a
κ-carrageenan gels (*c* = 2–12%)	60	1	Watase and Nishinari, 1987b
LM pectin/calcium gels	38–45**	1	Lopes da Silva and Rao (unpublished results)
	50–75**	2	Garnier, 1992
LM pectin/calcium/ sucrose gels	43–50**	1	Fu (1998) (unpublished results)
HM pectin/DMSO gels	115–440††	1	Watase and Nishinari, 1993
pH 2.4, pH 6.9	45–125††		

* Method 1—Eldridge-Ferry approach (Eldridge and Ferry, 1954); Method 2—superposition of viscoelastic data at different temperatures, and plot of the slope of ln b_T versus $1/T$ (Schultz and Myers, 1969).

† Depending on molecular weight.

** Depending on *R* and pectin concentration.

†† Depending on DMSO content.

Values of ΔH_m obtained for the LM pectin/calcium gels were close to those found for κ-carrageenan and κ-furcellaran gels but much lower that those of either gelatin or agarose gels. The κ-carrageenan and κ-furcellaran gels are formed by helical structures aggregated by potassium ions (Morris et al., 1980a), a mechanism that energetically is similar to the gelation of LM pectin via cross-linking by calcium ions. In the case of gelatin and agarose, gelation is associated to helix formation, which results in a large entropy decrease that must be compensated by a large enthalpy decrease. The difference between agarose and κ-carrageenan gels may be due to the different sulfate content (Watase and Nishinari, 1987a). The smaller values of ΔH_m obtained for the LM pectin gels, if compared with these gels and even with the HM pectin/glycerol or HM pectin/DMSO gels, suggest that the junction zones in LM pectin/calcium are less heat-resistant. In contrast, the enthalpy ΔH_{bT} obtained for LM pectin/calcium gels is higher than that obtained for HM pectin gels, which suggests a cooperative process in the formation or disruption of effective junction zones in the pectin/calcium system (Garnier, 1992).

Application of Thermal Cycles

Application of thermal cycles is one of the methods that has been used to study the thermal behavior of biopolymer gels. For thermoreversible gels, important hysteresis effects related to the nonequilibrium nature of the gel phase can be identified by different methods, including dynamic rheological methods, for systems such as gelatin (Djabourov et al., 1988a), agarose (Braudo et al., 1991; Watase and Nishinari, 1987a), κ-carrageenan gels (Morris et al., 1980a), and methylcellulose gels (Haque and Morris, 1993). Thermal hysteresis has its origin in stabilization of the ordered form by further aggregation (Liang et al., 1979; Morris et al., 1980b): When the conformational ordering on cooling is accompanied by aggregation of the ordered structures, the melting of the aggregates usually occurs over a high temperature range.

Nishinari et al. (1990) described the thermoreversible gel-sol transition by a zipper model approach. The junction zones were assumed to be made up of association of molecular zippers standing for a rigid ordered molecular structure such as helices or extended molecules. The gel-sol transition for thermoreversible gels was compared to a zipper opening from both ends. In the case of heating, the gel-sol transition caused by the opening of molecular zippers will start as soon as the temperature arrives at the point where the segments begin to be released from a zipper. On cooling, in contrast, the pairwise coupling cannot start as easily because of the difficulty for a long molecule to find its partner in appropriate positions for zipper construction. Hence, a state that is like supercooling may occur. This explains why T_m is far higher than T_g and why the rise of G' in cooling is far steeper than the decay of G' on heating, as observed for several thermoreversible biopolymer gels.

Structure Loss Rate

Rates of structure disruption or loss rates (SLR), defined as $-dG'/dt$, can also be measured when the gels are subjected to an increasing temperature, in order to generate kinetic data for the melting process. The changes in the high-methoxyl pectin network structure when subjected to an increasing temperature are of a complex nature, as seen in plots of temperature versus dG'/dt during the heating of an HM pectin gel, immediately after a cooling scan (Figure 6–16), or after aging the gel at 20°C for 48 hours (Figure 6–17). In contrast to most polysaccharide gels, HM pectin gels are considered to be heat-irreversible (Rao and Cooley, 1993). Qualitatively similar behavior was obtained for the two experiments. Three different zones can be identified: an initial increase in rate of structure loss, then a decrease in SLR, and finally another temperature range, above 60°C, where SLR increased again. Some quantitative differences can be found between the behavior of these two networks. The gel analyzed in Figure 6–16 was far from an equilibrium state, and a higher mobility of the pectin chains in the network can be expected. For some temperatures it can be seen that values of SLR were negative, meaning that an increase in temperature even contributed to structure formation. In the gel previously aged (Figure 6–17), the higher number and/or the higher extension of the junction zones also led to higher SLR values.

A different behavior is expected with thermoreversible gels. Using the LM pectin/calcium network as an example, the influence of temperature on SLR was examined (Figure 6–18). The influence of temperature on the rate of structure disruption followed a bell-shaped curve, with a maximum value at an intermediate temperature.

Kinetic data can also be obtained for the structure loss process during heating of biopolymer gels. The nonisothermal kinetic model previously described (Equation 6.28) can be used to describe the melting process of thermoreversible gels and to obtain apparent activation energies for this nonisothermal process, and the

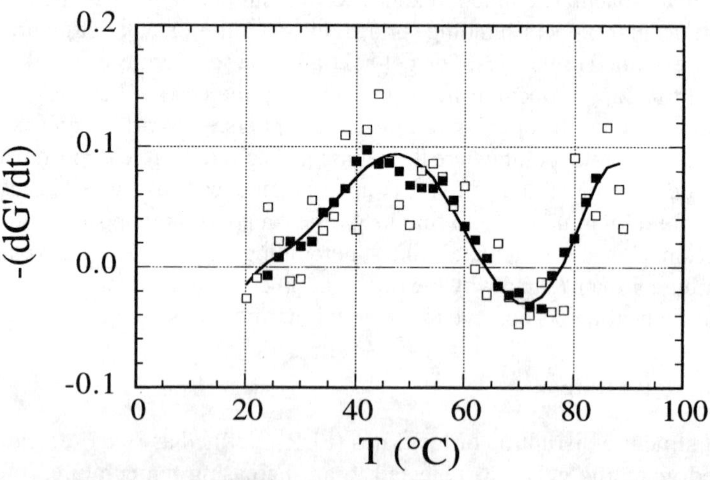

Figure 6–16 Effect of Temperature on SLR ($-\mathrm{d}G'/\mathrm{d}t$) of a 1% High-Methoxyl Pectin/60% Sucrose Gel during Heating from 20°C to 90°C, Right after a Cooling Scan from 90°C to 20°C: □ dG'/dt calculated after cubic spline fit; —— dG'/dt calculated after fitting a polynomial equation to the data; ■ smoothed dG'/dt values calculated after cubic spline fit. *Source*: Reprinted from J.A. Lopes da Silva, M.A. Rao, and J-T. Fu, Rheology of Structure Development and Loss During Gelation and Melting, in *Phase/State Transitions in Foods: Chemical, Rheological and Structural Changes*, M.A. Rao and R.W. Hartel, eds., pp. 111–156, by courtesy of Marcel Dekker, Inc.

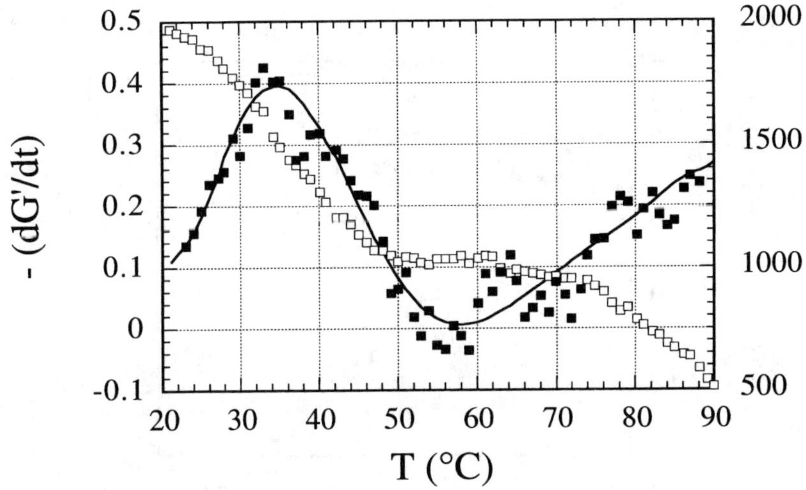

Figure 6–17 Effect of Temperature on SLR of a Cured Gel (20°C, 48 h) during Heating from 20°C to 90°C: dG'/dt values calculated by derivation of a polynomial equation fitted to the data (——), or after cubic spline fit and smoothed using a "moving average" procedure (■); □ denotes G' values. *Source*: Reprinted from J.A. Lopes da Silva, M.A. Rao, and J-T. Fu, Rheology of Structure Development and Loss During Gelation and Melting, in *Phase/State Transitions in Foods: Chemical, Rheological and Structural Changes*, M.A. Rao and R.W. Hartel, eds., pp. 111–156, by courtesy of Marcel Dekker, Inc.

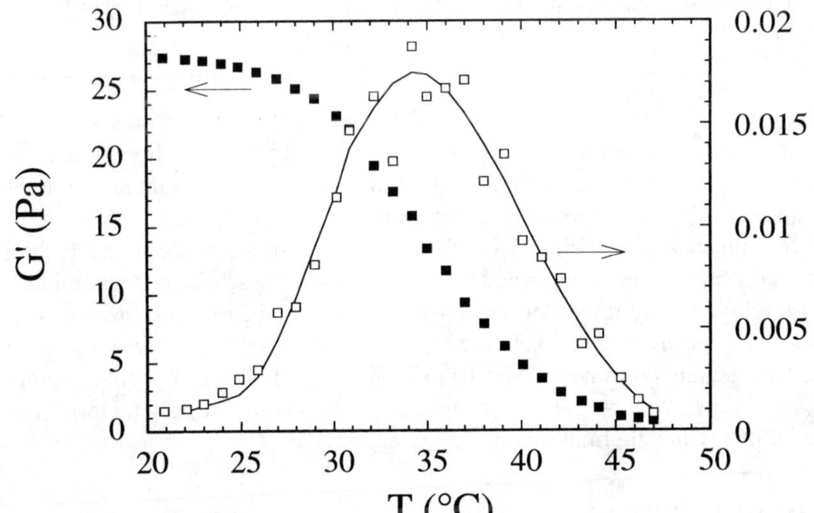

Figure 6–18 Dependence on Temperature of SLR for a Cured Gel (24 h) under Heating from 20°C to 50°C ([low-methoxyl pectin] = 6.4 g L^{-1}, [Ca^{2+}] = 0.18 g L^{-1}, 0.1 mol L^{-1} NaCl, pH 7.0). Values of dG'/dt calculated after cubic spline fit (G); also shown for comparison are the dG'/dt values calculated by derivation of a polynomial equation fitted to the data (line) and the G' (ω) values. *Source*: Reprinted from J.A. Lopes da Silva, M.A. Rao, and J-T. Fu, Rheology of Structure Development and Loss During Gelation and Melting, in *Phase/State Transitions in Foods: Chemical, Rheological and Structural Changes*, M.A. Rao and R.W. Hartel, eds., pp. 111–156, by courtesy of Marcel Dekker, Inc.

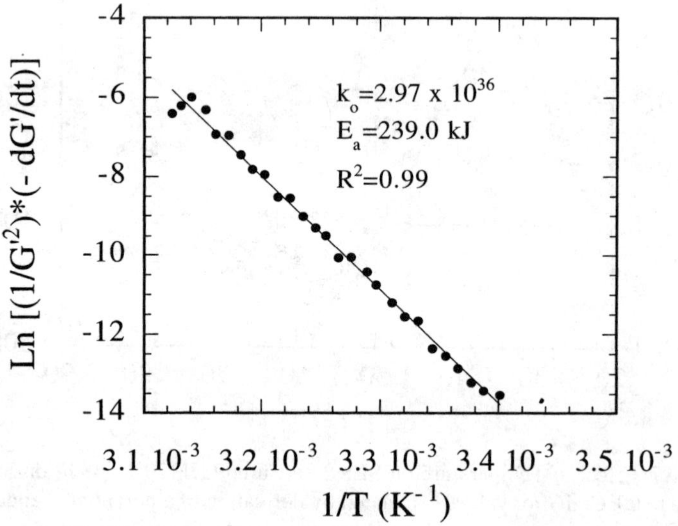

Figure 6–19 Arrhenius-Type Plot for Change in the Storage Modulus (G') during Heating for the Low-Methoxyl Pectin System in Figure 6–18. *Source*: Reprinted from J.A. Lopes da Silva, M.A. Rao, and J-T. Fu, Rheology of Structure Development and Loss During Gelation and Melting, in *Phase/State Transitions in Foods: Chemical, Rheological and Structural Changes*, M.A. Rao and R.W. Hartel, eds., pp. 111–156, by courtesy of Marcel Dekker, Inc.

model was applied to the melting of LM pectin/calcium gels. A multiple linear regression of ln ($-dG'/dt$) versus ln G' and $1/T$ was conducted to evaluate the order of reaction (n). This treatment showed invariably that the order of reaction (n) for the process under analysis should be close to 2. Exceptions were found for the higher pectin concentration analyzed, and for high R values, where a general tendency for a higher reaction order was observed. Taking the reaction order $n = 2$, Equation 6.28 was used to estimate the activation energy of the process, by using a linear least-squares routine. As shown in Figure 6–19 for an LM pectin/calcium gel ($c_p = 6.4$ g L^{-1}, $R = 0.42$), the melting process could be satisfactorily described by this nonisothermal kinetic approach.

MIXED POLYMER GELS

It was pointed out in Chapter 4 that in protein-polysaccharide systems, the asymetry of modulus versus concentration plots is due to phase separation and inversion, as first shown by Tolstoguzov (1985). The behavior in systems of two or more polymers is important because it differs from how each would behave in the absence of the other. They will interact with each other, either by sticking together or by separating apart, called associative or segregative effects, depending on whether heterotypic interactions between the two polymers are enthalpically more or less favorable than homotypic interactions (Morris, 1998).

One common mechanism of association between unlike biopolymers is electrostatic interaction between polyanions (e.g., acidic polysaccharides) and polycations (e.g., proteins below their isoelectric point) that may lead to the formation of a gel involving a "coupled network" or a "coprecipitate" of the two polymers as a "complex coacervate." In only a few systems, gelation occurs by association of two different polymers into conformationally ordered cooperative junction zones that are analogous to those in single-component polysaccharide ("synergistic") gels: konjac glucomannan or low-galactose content galactomannans in combination with xanthan or with the aggregating polysaccharides—carrageenan/agar and guluronate-rich alginate in combination with high-methoxyl pectin under high-acid conditions. In this section, segregative interactions that lead to heterogeneous gels are described. Schmitt et al. (1998) presented an extensive review of the structure and functional properties of protein-polysaccharide complexes.

Phase Diagram

Phase separation in solution can be detected by the immediate development of turbidity on mixing, eventually resulting in two separate layers whose compositions fall on a "cloud-point curve" or "binodal" that define the boundary between the single-phase and two-phase states of the system. In Figure 6–20, the curve represents the binodal of a polysaccharide (alginate, [A]) and protein (cassein, [C]). The thin straight lines are tie lines from which the concentrations of coexisting protein-rich and polysaccharide-rich phases can be determined, as shown at the filled circles. Below the binodal, the polymer concentrations of the single homogeneous phase solution can be determined. In particular, the open circle represents the critical point below which the system remains as a single homogeneous phase.

In Chapter 2, simple Takayanagi models for the modulus of two-component blends were discussed that are applicable to the shear modulus, G_c, of a composite formed from components (polymers) x and y with shear moduli G_x and G_y, respectively. If ϕ_x and $\phi_y = 1 - \phi_x$ are the respective volume fractions of the two components, the equations for the upper and lower bound values are

$$G_c = \phi_x G_x + \phi_y G_y \qquad \text{(upper bound, isostrain, if } G_x > G_y\text{)} \qquad (6.30)$$

$$(1/G_c) = (\phi_x/G_x) + (\phi_y/G_y) \qquad \text{(lower bound, isostress, if } G_x < G_y\text{)} \qquad (6.31)$$

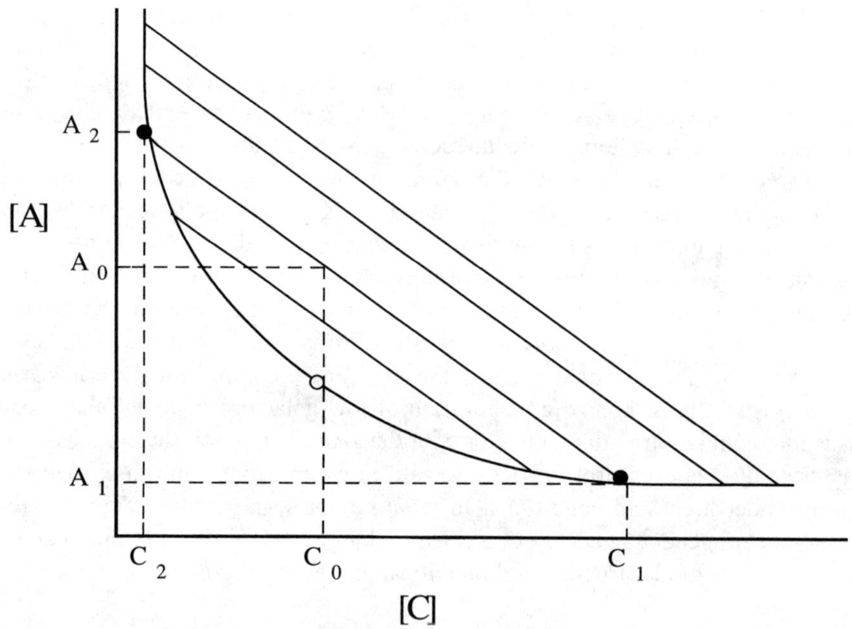

Figure 6–20 Phase Diagram of a Mixed Solution of Two Biopolymers (e.g., sodium alginate and casein). The curve represents the binodal, and the thin straight lines are tie lines from which the concentrations of coexisting protein-rich and polysaccharide-rich phases can be determined, as shown at the filled circles. The open circle represents the critical point below which the system remains as a single homogenous phase. Morris, 1998.

The former approximates the weighted-average of the individual moduli, while the latter approximates the weighted-average of the individual compliances. If the filler *x* is either much softer or much harder than the surrounding matrix, the above equations can be reduced, respectively, to

$$G_c \approx G_x\phi_x \text{ or } (G_c/G_x) = \phi_x, \text{ when } G_x \gg G_y \tag{6.32}$$

$$(1/G_c) \approx (\phi_x/G_x) \text{ or } (G_c/G_x) \approx (1/\phi_x), \text{ when } G_x \ll G_y \tag{6.33}$$

It is reasonable to expect the major component to form the continuous phase, that is, $\phi_x > 0.5$, so that, irrespective of the modulus of the filler particles (G_y), it is likely that the modulus of the composite (G_c) will not be different from that of the matrix (G_x) by more than a factor of ~2. More complex Takayanagi models have been reviewed by Manson and Sperling (1976).

Morris (1992) explored the manner in which values of G_x, G_y, and the upper ($G_u = G_c$) and lower bound values ($G_l = G_c$) vary with the fraction of the component

x. It was assumed that the two polymers x and y are confined entirely to their respective phases, the modulus varies as the square of the polymer concentration, and the volume of the polymer chains is negligible compared to the volume of the solvent, so that:

$$\phi_x = x \text{ and } \phi_y = 1 - x \tag{6.34}$$

Further, it was also assumed that $M_y = (n')^2 M_x$, where n' is an integer. The results obtained with $n' = 2$ (the symbol n' is used to avoid confusion with the symbol n used for the reaction rate order) are shown in Figure 6–21 where data reported in Morris (1992) were replotted. The values of the modulus G_x are high at low values of x and decrease with increasing x, reaching a final value $G_x = M_x$ at $x = 1$. Conversely, G_y is high at high values of x and decreases to $G_y = M_y$ at $x = 0$. At a critical point, $G_x = G_y$; that is, the moduli of both phases are identical. In addition, at the critical point, G_u and G_l reach minimum and maximum values, respectively. Also of interest are the general results for the critical concentration and modulus:

$$x_{crit} = (1/[n' + 1]) \text{ and } G_{crit} = M_x(n' + 1)^2 \tag{6.35}$$

Phase-separated polymer mixtures are important in many practical applications, such as confectionery and low-fat foods. Values of the fraction (x) of a component can be replaced by those of the fraction of solvent (S_x) associated with polymer (Morris, 1998), so Figure 6–21 may be used also to understand the role of solvent partitioning. A key assumption is that there is complete separation of the two polymers into their respective phases, so the relative phase volumes (hence, the polymer concentration in each phase) may be characterized by the "p-factor," defined as:

$$\text{p-factor} = \frac{\text{(water/polymer) in phase 1}}{\text{(water/polymer) in phase 2}} \tag{6.36}$$

Using calibration curves of modulus versus concentration of each polymer, experimental moduli of mixed gel networks, and the Takayanagi blending laws discussed above, one can calculate the best value of the p-factor by using numerical methods. Such data then provide information on the distribution of the two polymers in a mixed gel system. Additional examples of mixed gel systems that have been studied (Morris, 1998) are maltodextrin/milk protein and gelatin/maltodextrin.

Relatively small reductions in modulus result when one polymer is added to another gelling polymer, such as agar-bovine serum albumin and agar-gelatin (Clark et al., 1982, 1983). Another example (Morris, 1998) is on the addition of oxidized starch to low-methoxyl pectin + Ca^{2+} in which, at low concentrations of Ca^{2+} (up to ~23% stoichiometry equivalent), G' increased due to the addition of starch; however, at higher Ca^{2+} concentrations (e.g., ~100% stoichiometry equivalent), G' decreased by more than four orders of magnitude. The reduction may be due to

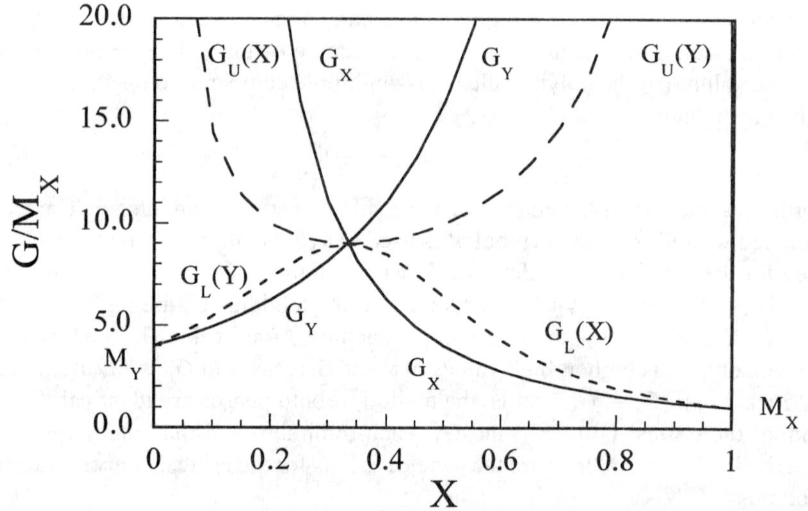

Figure 6–21 Distribution of the Moduli of Components *x, y*, Upper Bound (G_u), and Lower Bound (G_l) from the Takayanagi Isostress and Isostrain Blending Laws. The continuous phase is indicated in parentheses. Morris, 1998.

the starch forming the continuous phase that included only a small amount of the gelling polymer or due to phase inversion at high concentrations of the second polymer in some systems. From the discussion of rheological properties of starch dispersions in Chapter 4, the important role of the starch granules in mixed gel systems should be recognized.

Low-Fat Spreads

Low-fat spreads are mixed gel systems of fat, water, emulsifiers, and hydrocolloids. Because they often undergo emulsion breakdown, starch derivatives and gelatin are used to provide the aqueous phase with acceptable rheological and melting characteristics. In a recent study, creep-compliance studies were used to examine the rheological characteristics of butter, margarine, and a low-fat spread (Chronakis and Kasapis, 1995; Evageliou et al., 1997). It will be recalled from Chapter 4 that an experimental creep-compliance curve is described by a general equation of the form

$$J(t) = J_0 + J_m[1 - \exp(t/\tau_m)] + (t/\eta_N) \tag{6.37}$$

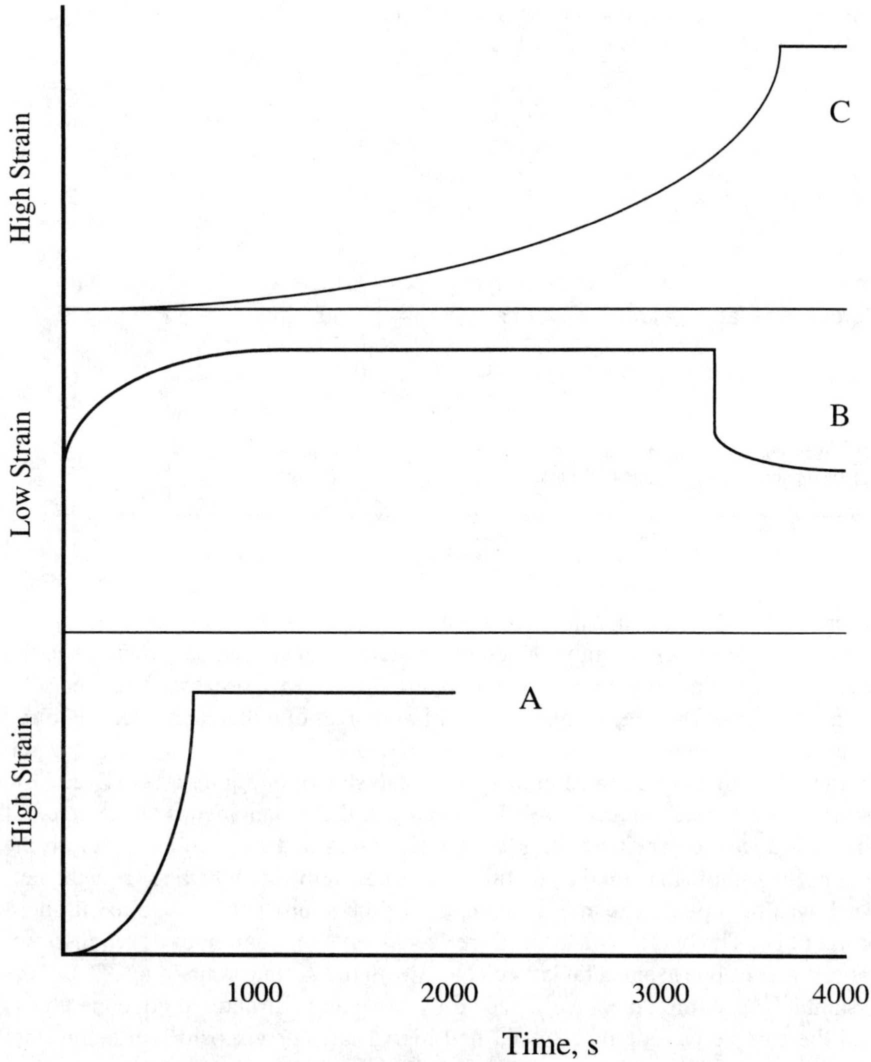

Figure 6–22 Creep-Compliance Curves at 5°C of Butter (A), a Commercial Low-Fat Spread Containing Gelatin (B), and an Experimental Product with 5.2% Fat Content and Containing Maltodextrin (C). Evageliou et al., 1997, and Stanley et al., 1998.

Table 6–7 Creep-Compliance Parameters of Low-Fat Spreads at 5°C. Evageliou et al., 1997, and Stanley et al., 1998.

Sample	*Applied Stress** *($Pa \times 10^3$)*	J_O *($Pa^{-1} \times 10^{-4}$)*	*Voigt Units*	η_N *($Pa\ s \times 10^6$)*
Gelatin gel, 3.6%	0.2	4.52	1	51
	3.0	4.25	3	62
Margarine, 80% fat	1.8	0.10	2	473
	2.2†	0.14	3	69
Spread, 3% fat + gelatin	0.3	0.69	2	172
	1.4	0.98	4	199
Spread, 22% fat + gelatin	0.53	0.35	2	173
	0.75	0.47	3	45

* Stress was applied for 1 hour.
† Disruption occurred after 44 minutes.

where J_0 describes the instantaneous compliance, η_N is the viscosity, and $J_m[1 - \exp(t/\tau_m)]$ describes the time-dependent retarded elastic region, where m is the number of Kelvin-Voigt units required to describe the experimental data precisely.

In Table 6–7, obvious differences can be seen among the parameters required to describe the creep-compliance data obtained during 1 hour on a gelatin gel, margarine, and an experimental low-fat spread shown in Figure 6–22 (Evageliou et al., 1997; Stanley et al., 1998). For example, the instantaneous compliance of margarine was lower, increasing the applied stress had more of an effect on the margarine sample than on the gelatin sample, and at the higher stress level the network was disrupted. The effect of increasing the applied stress resulted in more Voigt units. The two low-fat spreads reflected the combination of a gelatin-structured aqueous phase and a fat lattice phase; with the sample containing 3% fat, the instantaneous compliance was affected by the gelatin-structured aqueous phase; and the sample containing 22% fat had high values of viscosity, reflecting that only a small deformation caused terminal flow.

REFERENCES

Adam, M. 1991. Growth process of polymers near the gelation threshold. *Die Makromolekulare Chemie,* Macromolecular Symposia 45: 1–9.

Andresen, I.L., and O. Smidsrød. 1977. Temperature dependence of the elastic properties of alginate gels. *Carbohydrate Res* 58: 271–279.

Axelos, M.A.V., and M. Kolb. 1990. Crosslinked biopolymers: Experimental evidence for scalar percolation theory. *Phys Rev Lett* 64: 1457–1460.

Bisschops, J. 1955. Gelation of concentrated polyacrylonitrile. *J Polymer Sci* 17: 89–98.

Bloksma, A.H., and W. Nieman. 1975. The effect of temperature on some rheological properties of wheat flour doughs. *J Texture Studies* 6: 343–361.

Borchard, W., and B. Burg. 1989. Investigations of the complex shear modulus and the optical rotation in the system gelatin-water during the thermoreversible gelation process. In *Molecular basis of polymer networks,* eds. A. Baumgärtner and C.E. Picot, 162–168. Berlin, Germany: Springer-Verlag.

Braudo, E.E., I.R. Muratalieva, I.G. Plashchina, and V.B. Tolstoguzov. 1991. Correlation between the temperatures of formation/breakdown of the gel network and conformational transitions of agarose macromolecules. *Carbohydrate Polymers* 15: 317–321.

Braudo, E.E., I.G. Plashchina, and V.B. Tolstoguzov. 1984. Structural characterisation of thermoreversible anionic polysaccharide gels by their elastoviscous properties. *Carbohydrate Polymers* 4: 23–48.

Chambon, F., and H.H. Winter. 1987. Linear viscoelasticity at the gel point of a crosslinking PDMS with imbalanced stoichiometry. *J Rheol* 31: 683–697.

Chronakis, I.S., and S. Kasapis. 1995. A rheological study on the application of carbohydrate-protein incompatibility of the development of low fat commercial spreads. *Carbohydrate Polymers* 28: 367–373.

Chronakis, I.S., S. Kasapis, R.K. Richardson, and G. Doxastakis. 1996. Characterisation of a commercial soy isolate by physical techniques. *J Texture Studies* 26: 371–389.

Clark, A.H. 1991. Structural and mechanical properties of biopolymer gels. In *Food polymers, gels and colloids,* ed. E. Dickinson, 322–338. Cambridge, England: The Royal Society of Chemistry.

Clark, A.H. 1994. Rationalisation of the elastic modulus-molecular weight relationship for kappa-carrageenan gels using cascade theory. *Carbohydrate Polymers* 23: 247–251.

Clark, A.H., K.T. Evans, and D.B. Farrer. 1994. Shear modulus-temperature meltdown profiles of gelatin and pectin gels. *Int J Biol Macromolecules* 16: 125–130.

Clark, A.H., R.K. Richardson, G. Robinson, S.B. Ross-Murphy, and A.C. Weaver. 1982. Structural and mechanical properties of agar/BSA co-gels. *Prog Food Nutr Sci* 6: 149–160.

Clark, A.H., R.K. Richardson, S.B. Ross-Murphy, and J.M. Stubbs. 1983. Structural and mechanical properties of agar/gelatin co-gels. Small deformation studies. *Macromolecules* 16: 1367–1374.

Clark, A.H., and S.B. Ross-Murphy. 1987. Structural and mechanical properties of biopolymer gels. *Adv Polymer Sci* 83: 57–192.

Cuvelier, G., C. Peigney-Noury, and B. Launay. 1990. Viscoelastic properties of physical gels: Critical behaviour at the gel point. In *Gums and stabilisers for the food industry 5,* eds. G.O. Phillips, D.J. Wedlock, and P.A. Williams, 549–552. Oxford, England: IRL Press.

de Gennes, P.G. 1979. *Scaling concepts in polymer physics.* Ithaca, NY: Cornell University Press.

Djabourov, M., J. Leblond, and P. Papon. 1988a. Gelation of aqueous gelatin solutions. I. Structural investigation. *J Physique* 49: 319–332.

Djabourov, M., J. Leblond, and P. Papon. 1988b. Gelation of aqueous gelatin solutions. II. Rheology of the sol-gel transition. *J Physique* 49: 333–343.

Doublier, J.L., and L. Choplin. 1989. A rheological description of amylose gelation. *Carbohydrate Res* 193: 215–226.

Doublier, J.L., B. Launay, and G. Cuvelier. 1992. Viscoelastic properties of food gels. In *Viscoelastic properties of foods,* eds. M.A. Rao and J.F. Steffe, 371–434. Barking, England: Elsevier Science.

Dumas, J., and J.C. Bacri. 1980. New method of viscosity measurement near the gelatin sol-gel transition. *J Physique—Letters* 41: 279–282.

Durand, D., F. Naveau, J.P. Busnel, M. Delsanti, and M. Adam. 1989. Evolution of polyurethane gel fraction near the gelation threshold. *Macromolecules* 22: 2011–2012.

Eldridge, J.E., and J.D. Ferry. 1954. Studies of the cross-linking process in gelatin gels. III. Dependence of melting point on concentration and molecular weight. *J Phys Chem* 58: 992–995.

Evageliou, V., S. Alevisopolous, and S. Kasapis. 1997. Application of stress-controlled analysis to the development of low-fat spreads. *J Texture Studies* 28: 319–335.

Ferry, J.D. 1980. *Viscoelastic properties of polymers.* 3d ed. New York: John Wiley & Sons.

Flory, P.J. 1953. *Principles of polymer chemistry.* Ithaca, NY: Cornell University Press.

Flory, P.J. 1974. Introductory lecture. *Faraday Discussions of the Chemical Society* 57: 7–18.

Foegeding, E.A., H. Li, and S.R. Bottcher. 1998. Gelation of globular proteins. In *Phase/state transitions in foods: Chemical, rheological and structural changes,* eds. M.A. Rao and R.W. Hartel, 111–156. New York: Marcel Dekker.

Fu, J-T. 1998. Rheology of sol-gel and gel-sol transition of low-methoxyl pectin + Ca^{2+} gels: The effect of sweeteners. PhD diss., Cornell University, Ithaca, NY.

Garnier, C. 1992. Gelification des pectines en presence de calcium: Étude physico-chimique et rheologique. PhD diss., Université de Nantes, Nantes, France.

Gidley, M.J., E.R. Morris, E.J. Murray, D.A. Powell, and D.A. Rees. 1979. Spectroscopic and stoichiometric characterisation of the calcium-mediated association of pectate chains in gels and in the solid state. *J Chemical Society, Chemical Communications* 990–992.

Gluck-Hirsch, J.B., and J.L. Kokini. 1997. Determination of the molecular weight between crosslinks of waxy maize starches using the theory of rubber elasticity. *J Rheol* 41: 129–139.

Gordon, M., and S.B. Ross-Murphy. 1975. The structure and properties of molecular trees and networks. *Pure Appl Chem* 43: 1–26.

Grant, G.T., E.R. Morris, D.A. Rees, P.J.C. Smith, and D. Thom. 1973. Biological interactions between polysaccharides and divalent cations: The egg-box model. *FEBS Letters 32,* no. 1: 195–198.

Haque, A., and E.R. Morris. 1993. Thermogelation of methylcellulose. Part I: Molecular structures and processes. *Carbohydrate Polymers* 22: 161–173.

Hermans, P.H. 1949. Gels. In *Colloid science,* ed. H.R. Kruyt, Vol. 2, 483–651. Amsterdam: Elsevier Publishing.

Higgs, P.G., and R.C. Ball. 1990. A "reel-chain" model for the elasticity of biopolymer gels, and its relationship to slip-link treatments of entanglements. In *Physical networks: Polymers and gels,* eds. W. Burchard and S.B. Ross-Murphy, 185–194. Barking, England: Elsevier Applied Science.

Hsieh, Y.L., and J.M. Regenstein. 1992a. Modeling gelation of egg albumen and ovalbumin. *J Food Sci* 57: 856–861.

Hsieh, Y.L., and J.M. Regenstein. 1992b. Elastic attributes of heated egg protein gels. *J Food Sci 57,* no. 4: 862–868.

Hsieh, Y.L., J.M. Regenstein, and M.A. Rao. 1993. The gel point of whey and egg proteins from dynamic rheological data. *J Food Sci* 58: 116–119.

Iso, N., H. Mizuno, T. Saito, F. Ohzeki, and N. Kurihara. 1984. The change of the rheological properties of surimi (minced fish-meat) by heating. *Bull Jpn Soc Sci Fisheries* 50: 1045–1049.

Joanny, J.F. 1989. The sol-gel transition. *Physica B* 156 and 157: 381–385.

Kasapis, S., E.R. Morris, I.T. Norton, and A.H. Clark. 1993. Phase equilibria and gelation in gelatin/maltodextrin systems. Part I: Gelation of individual components. *Carbohydrate Polymers* 21: 243–248.

Katsuta, K., and J.E. Kinsella. 1990. Effects of temperature on viscoelastic properties and activation energies of whey protein gels. *J Food Sci* 55: 1296–1302.

Kawabata, A. 1977. Studies on chemical and physical properties of pectic substances from fruits. *Memoirs of the Tokyo University of Agriculture* 19: 115–200.

Kloek, W., H. Luyten, and T. van Vliet. 1996. Small and large deformation behaviour of mixtures of xanthan and enzyme modified galactomannans. *Food Hydrocolloids* 10: 123–129.

Liang, J.N., E.S. Stevens, E.R. Morris, and D.A. Rees. 1979. Spectroscopic origin of conformation-sensitive contributions to polysaccharide optical activity: Vacuum-ultraviolet circular dichroism. *Biopolymers* 18: 327–333.

Lin, Y.G., D.T. Mallin, J.C.W. Chien, and H.H. Winter. 1991. Dynamic mechanical measurement of crystallization-induced gelation in thermoplastic elastomeric poly(propylene). *Macromolecules* 24: 850–854.

Lopes da Silva, J.A.L., and M.P. Gonçalves. 1994. Rheological study into the ageing process of high methoxyl pectin/sucrose aqueous gels. *Carbohydrate Polymers* 24: 235–245.

Lopes da Silva, J.A.L., M.P. Gonçalves, J.L. Doublier, and M.A.V. Axelos. 1996. Effect of galactomannans on the viscoelastic behaviour of pectin/calcium networks. *Carbohydrate Polymers* 24: 235–245.

Lopes da Silva, J.A.L., M.P. Gonçalves, and M.A. Rao. 1995. Kinetics and thermal behaviour of the structure formation process in HMP/sucrose gelation. *Int J Biol Macromolecules* 17: 25–32.

Lopes da Silva, J.A.L., M.A. Rao, and J-T. Fu. 1998. Rheology of structure development and loss during gelation and melting. In *Phase/state transitions in foods: Chemical, rheological and structural changes,* eds. M.A. Rao and R.W. Hartel, 111–156. New York: Marcel Dekker.

Manson, J.A., and L.H. Sperling. 1976. *Polymer blends and composites.* New York: Plenum Press.

Martin, J.E., D. Adolf, and J.P. Wilcoxon. 1989. Viscoelasticity near the sol-gel transition. *Phys Rev A Gen Phys* 39: 1325–1332.

Matsumoto, T., M. Kawai, and T. Masuda. 1992. Influence of chain stiffness on the gelation and gel structure of alginate aqueous systems. *J Chem Soc, Faraday Trans 88,* no. 18: 2673–2676.

McClain, P.E., E. Kuntz, and A.M. Pearson. 1969. Application of stress-strain behaviour to thermally contracted collagen from epimysical connective tissues. *J Agric Food Chem* 17: 629–632.

Michon, C., G. Cuvelier, and B. Launay. 1993. Concentration dependence of the critical viscoelastic properties of gelatin at the gel point. *Rheologica Acta* 32: 94–103.

Miles, M.J., V.J. Morris, and S.G. Ring. 1985. Gelation of amylose. *Carbohydrate Res* 135: 257–269.

Mochizuki, Y., T. Saito, N. Iso, H. Mizuno, A. Aochi, and M. Noda. 1987. Effects of adding fat on rheological properties of fish meat gel. *Bull Jpn Soc Sci Fisheries* 53: 1471–1474.

Morris, E.R. 1992. The effect of solvent partition on the mechanical properties of biphasic biopolymer gels: An approximate theoretical treatment. *Carbohydrate Polymers* 17: 65–70.

Morris, E.R. 1998. Segregative interactions in biopolymer co-gels. In *Phase/state transitions in foods: Chemical, rheological and structural changes,* eds. M.A. Rao and R.W. Hartel, 111–156. New York: Marcel Dekker.

Morris, E.R., M.J. Gidley, E.J. Murray, D.A. Powell, and D.A. Rees. 1980. Characterization of pectin gelation under conditions of low water activity, by circular dichroism, competitive inhibition and mechanical properties. *Int J Biol Macromolecules* 2: 327–330.

Morris, E.R., D.A. Rees, I.T. Norton, and D.M. Goodall. 1980. Calorimetric and chiroptical evidence of aggregate-driven helix formation in carrageenan systems. *Carbohydrate Research* 80: 317–323.

Morris, V.J., and G.R. Chilvers. 1981. Rheological studies on specific ion forms of iota-carrageenate gels. *J Sci Food Agric* 32: 1235–1241.

Muller, H.G. 1969. Application of the statistical theory of rubber elasticity to gluten and dough. *Cereal Chem* 46: 443–446.

Muller, R., E. Gérard, P. Dugand, P. Rempp, and Y. Gnanou. 1991. Rheological characterization of the gel point: A new interpretation. *Macromolecules* 24: 1321–1326.

Nishinari, K., S. Koide, and K. Ogino. 1985. On the temperature dependence of elasticity of thermoreversible gels. *J Physique* 46: 793–797.

Nishinari, K., S. Koide, P.A. Williams, and G.O. Phillips. 1990. A zipper model approach to the thermoreversible gel-sol transition. *J Physique* 51: 1759–1768.

Nolte, H., S. John, O. Smidsrød, and B.T. Stokke. 1992. Gelation of xanthan with trivalent metal ions. *Carbohydrate Polymers* 18: 243–251.

Oakenfull, D.G. 1984. A method for using measurements of shear modulus to estimate the size and thermodynamic stability of junction zones in noncovalently cross-linked gels. *J Food Sci* 49: 1103–1104, 1110.

Oakenfull, D.G. 1987. The chemistry of high-methoxyl pectins. In *The chemistry and technology of pectin,* ed. R.H. Walter, 87–118. New York: Academic Press.

Peniche-Covas, C.A.L. et al. 1974. The critically branched state in a covalent synthetic system and in the reversible gelation of gelatin. *Faraday Discussions of the Chemical Society* 57: 165–180.

Pezron, I., T. Herning, M. Djabourov, and J. Leblond. 1990. Scattering from a biopolymer solution in the sol and gel states: The gelatin example. In *Physical networks: Polymers and gels,* eds. W. Burchard and S.B. Ross-Murphy, 231–252. Barking, England: Elsevier Applied Science.

Plashchina, I.G., O.A. Fomina, E.E. Braudo, and V.B. Tolstoguzov. 1979. Creep study of high-esterified pectin gels. I. The creep of saccharose-containing gels. *Colloid Polymer Sci* 257: 1180–1187.

Rao, M.A. 1992. Measurement of viscoelastic properties of fluid and semisolid foods. In *Viscoelastic properties of foods,* eds. M.A. Rao and J.F. Steffe, 207–231. Barking, England: Elsevier Applied Science.

Rao, M.A., and H.J. Cooley. 1993. Dynamic rheological measurement of structure development in high-methoxyl pectin/fructose gels. *J Food Sci* 58: 876–879.

Rao, M.A., and H.J. Cooley. 1995. Rates of structure development during gelation and softening of high-methoxyl pectin-sodium alginate-fructose mixtures. *Food Hydrocolloids* 9: 229–235.

Rees, D.A. 1969. Structure, conformation, and mechanism in the formation of polysaccharide gels and networks. *Adv Carbohydrate Chem Biochem* 24: 267–332.

Rees, D.A. 1972. Polysaccharide gels—A molecular view. *Chem Industry* 19: 630–636.

Rhim, J.W., R.V. Nunes, V.A. Jones, and K.R. Swartzel. 1989. Determination of kinetic parameters using linearly increasing temperature. *J Food Sci* 54: 446–450.

Richardson, R.K., and S.B. Ross-Murphy. 1981a. Mechanical properties of globular protein gels: 1. Incipient gelation behaviour. *Int J Biol Macromolecules* 3: 315–322.

Richardson, R.K., and S.B. Ross-Murphy. 1981b. Mechanical properties of globular protein gels. II: Concentration, pH and ionic strength dependence. *British Polymer J* 13: 11–16.

Ross-Murphy, S.B. 1991a. The estimation of junction zone size from geltime measurements. *Carbohydrate Polymers* 14: 281–294.

Ross-Murphy, S.B. 1991b. Incipient behaviour of gelatin gels. *Rheologica Acta* 30: 401–411.

Scanlan, J.C., and H.H. Winter. 1991. Composition dependence of the viscoelasticity of end-linking poly(dimethylsiloxane) at the gel point. *Macromolecules* 24: 47–54.

Schmitt, C., C. Sanchez, S. Desobry-Banon, and J. Hardy. 1998. Structure and technofunctional properties of protein-polysaccharide complexes: A review. *Critical Reviews in Food Science and Nutrition* 38: 689–753.

Schultz, R.K., and R.R. Myers. 1969. The chemorheology of poly(vinyl alcohol)-borate gels. *Macromolecules* 2: 281–285.

Sperling, L.H. 1986. *Introduction to physical polymer science.* New York: John Wiley & Sons.

Stading, M., and A.M. Hermansson. 1990. Viscoelastic behaviour of β-lactoglobulin gel structures. *Food Hydrocolloids* 4: 121–135.

Stading, M., and A.M. Hermansson. 1993. Rheological behaviour of mixed gels of κ-carrageenan-locust bean gum. *Carbohydrate Polymers* 22: 49–56.

Stanley, D.W., J.M. Aguilera, K.W. Baker, and R.L. Jackman. 1998. Structure/property relationships of foods as affected by processing and storage. In *Phase/state transitions in foods: Chemical, rheological and structural changes,* eds. M.A. Rao and R.W. Hartel, 1–56. New York: Marcel Dekker.

Stauffer, D., A. Coniglio, and M. Adam. 1982. Gelation and critical phenomena. *Adv Polymer Sci* 44: 103–158.

Takagi, I., and W. Simidu. 1972. On rheological properties and structure of kamaboko. I. Application of rubber elasticity theory to kamaboko. *Bull Jpn Soc Sci Fisheries* 38: 299–303.

te Nijenhuis, K. 1981. Investigation into the ageing process in gels of gelatin/water systems by the measurement of their dynamic moduli. Part I—Phenomenology. *Colloid Polymer Sci* 259: 522–535.

te Nijenhuis, K., and H.H. Winter. 1989. Mechanical properties at the gel point of a crystallizing poly(vinyl chloride) solution. *Macromolecules* 22: 411–414.

Tobitani, A., and S.B. Ross-Murphy. 1997a. Heat-induced gelation of globular proteins. 1. Model for the effects of time and temperature on the gelation time of BSA. *Macromolecules* 30: 4845–4854.

Tobitani, A., and S.B. Ross-Murphy. 1997b. Heat-induced gelation of globular proteins. 2. Effect of environmental factors on single component and mixed protein gels. *Macromolecules* 30: 4855–4862.

Tokita, M., R. Niki, and K. Hikichi. 1984. Percolation theory and elastic modulus of gel. *J Phys Soc Japan* 53: 480–482.

Tolstoguzov, V.B. 1985. Functional properties of protein-polysaccharide mixtures. In *Functional properties of food macromolecules,* eds. J. Mitchell and D.A. Ledward, 385–415. London: Elsevier Applied Science.

Treloar, L.R.G. 1975. *The physics of rubber elasticity.* 3d ed. Oxford, England: Clarendon Press.

Tung, C.Y.M., and P.J. Dynes. 1982. Relationship between viscoelastic properties and gelation in thermosetting systems. *J Appl Polymer Sci* 27: 569–574.

Wang, Z.Y., Q.-Z. Zhang, M. Konno, and S. Saito. 1991. Sol-gel transition of alginate solution by the additions of various divalent cations: Critical behavior of relative viscosity. *Chem Phys Lett 186,* nos. 4, 5: 463–466.

Watase, M., and K. Nishinari. 1987a. Dynamic viscoelasticity and anomalous thermal behaviour of concentrated agarose gels. *Die Makromolekulare Chemie* 188: 1177–1186.

Watase, M., and K. Nishinari. 1987b. Rheological and thermal properties of carrageenan gels—Effect of sulfate content. *Die Makromolekulare Chemie* 188: 2213–2220.

Watase, M., and K. Nishinari. 1993. Effects of pH and DMSO content on the thermal and rheological properties of high methoxyl pectin-water gels. *Carbohydrate Polymers* 20: 175–181.

Watase, M., K. Nishinari, A.H. Clark, and S.B. Ross-Murphy. 1989. Differential scanning calorimetry, rheology, X-ray, and NMR of very concentrated agarose gels. *Macromolecules* 22: 1196–1201.

Winter, H.H., and F. Chambon. 1986. Analysis of linear viscoelasticity of a crosslinking polymer at the gel point. *J Rheol* 30: 367–382.

Winter, H.H., P. Morganelli, and F. Chambon. 1988. Stoichiometry effects on rheology of model polyurethanes at the gel point. *Macromolecules* 21: 532–535.

Zhang, J., and C. Rochas. 1990. Interactions between agarose and κ-carrageenans in aqueous solutions. *Carbohydrate Polymers* 13: 257–271.

Ziegler, G.R., and S.S.H. Rizvi. 1989. Determination of cross-link density in egg white gels from stress relaxation data. *J Food Sci* 54: 218–219.

CHAPTER 7

Role of Rheological Behavior in Sensory Assessment of Fluid Foods

José Antonio Lopes da Silva and M. Anandha Rao

STIMULI FOR EVALUATION OF VISCOSITY

Szczesniak (1979) pointed out that to describe mouthfeel of beverages, 136 terms were generated from 5350 responses. Of these, 30.7% were viscosity-related terms. It was also pointed out that the term "mouthfeel" was better than "viscosity," which was applicable to liquid foods. The former has the same meaning as texture, which has been defined by Szczesniak (1963) as "the composite of the structural elements of the food and the manner in which it registers with the physiological senses" (p. 386) and by Sherman (1970) as "the composite of those properties which arise from the physical structural elements and the manner in which it registers with the physiological senses" (p. 372). The high frequency of mention, and often the sole mention, of terms such as "thin," "thick," "viscous," "low viscosity," and "high viscosity" indicates that the sensory perception of viscosity is the most important mouthfeel sensation (Szczesniak, 1979). Some of the terms for describing viscosity of fluid foods may be related to their shear rate versus shear stress data, often referred to as objective evaluation. For meaningful results, both the objective data and sensory assessment data must be obtained, and such studies are reviewed in this chapter.

Wood (1968) suggested that the stimulus associated with oral assessment was the apparent viscosity determined at 50 s^{-1}. Shama et al. (1973) and Shama and Sherman (1973) made an important contribution in identifying the stimuli associated with sensory assessment of foods by nonoral methods—tilting a container and stirring a fluid food—and oral methods. They showed that instead of a single shear rate as suggested by Wood (1968), the stimuli encompass a wide range of either shear rates or shear stresses. They matched the foods assessed sensorially with the shear rate versus shear stress data on the foods. Even after 25 years, the methodology described and the results reported have considerable merit. The results of Shama and Sherman (1973) on stimuli associated with sensory assessment of viscosity were confirmed in subsequent studies (Cutler et al., 1983; Kokini and Cussler, 1984).

The key step in the procedure used was in the sensory assessment, especially in selecting a set of four fluid foods that could be arranged in a matrix, shown in Figure 7–1. The arrows were marked along the diagonals joining the corners of the matrix that indicated the direction of viscosity ratings in each pair examined. Each arrow pointed to the less viscous sample. It is important to note that for a wholly consistent evaluation of all four samples in a group, none of the triangles within the matrix should have all three arrows pointing in a clockwise or anticlockwise direction. Further, one of the arrows in each triangle has to point in the direction opposite to the other two. Obviously, considerable preliminary work was done

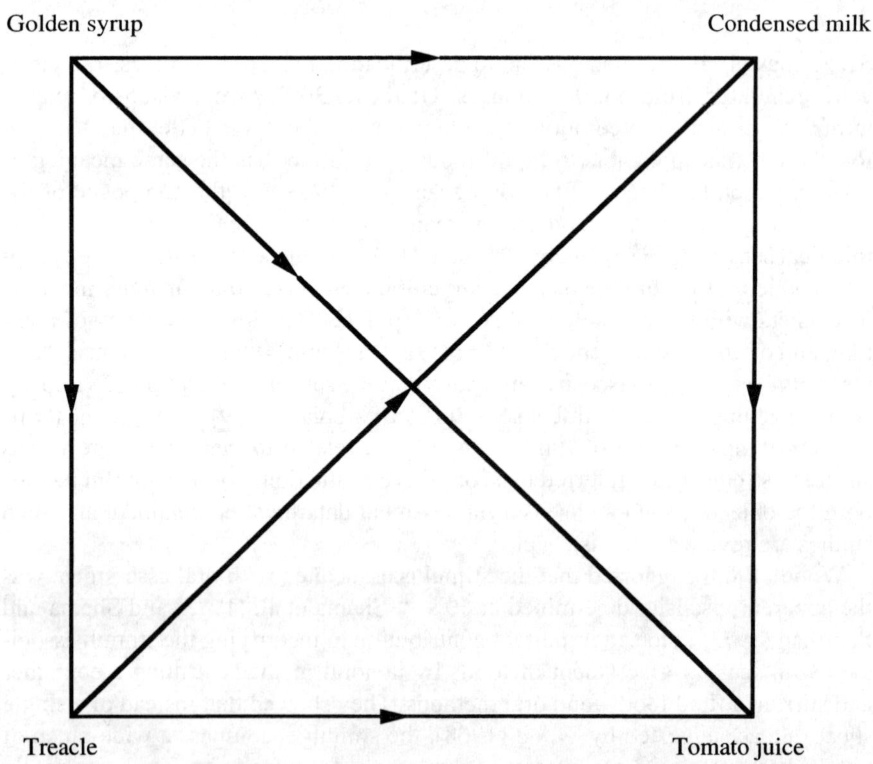

Figure 7–1 A Set of Four Fluid Foods That Could Be Arranged in a Matrix for Sensory Assessment. Each arrow points to the less viscous sample.

Table 7–1 Typical Viscosity Evaluation Data of a Sensory Panelist. Data from Shama et al., 1973

Product	*Code*	*Ranking*	*Code*	*Product*
1. Dil. golden syrup	W	>	B	Dil. treacle
2. Condensed milk	U	<	G	Dil. golden syrup
3. Condensed milk	H	<	R	Dil. treacle
4. Tomato juice	P	<	M	Golden syrup
5. Dil. treacle	V	>	L	Tomato juice
6. Tomato juice	D	<	O	Condensed milk

prior to choosing the four foods in a group. For the foods shown in Figure 7–1, the panelist's sensory evaluation of viscosity is shown in Table 7–1.

Stimuli Associated with Tilting Container and Stirring

In Figure 7–2, four sets of shear rate versus shear stress data are shown on a double logarithmic scale. The data of the soup intersected and crossed over the data of the three diluted treacles (molasses) at different points. According to the sensory evaluation, the soup was considered to have a viscosity similar to that of 80% treacle, and it was intermediate between the viscosities of the 77% and 84% treacle samples. In the sensory assessment of viscosity by stirring, the stimulus that controls the subject's evaluation of viscosity is the shear stress ($\sim 10^1$–10^3, Pa) developed at a shear rate (90–100 s^{-1}) that does not alter substantially with the rheological characteristics of the food (Figure 7–3). The subjects judged viscosity as the resistance offered by each sample to the stirring motion. The torque measured during stirring is given by:

Torque = Surface area of rotating body × radius of rotating body × shear stress

In the evaluation of viscosity by tilting the container, panel members indicated that their judgments were based on the rate at which each sample flowed down the side of the beaker; that is, viscosity was judged from the behavior of the apex of the film that flowed down the side of the container when it was first tilted. The major variable was the thickness of the film flowing down the side and its rate of flow; that is, viscosity evaluation by tilting the container depends on the shear rate of (10^{-1}–100 s^{-1}) developed at a shear stress (6–60 Pa) (Figure 7–4), which varies with the rheological properties of the sample.

Stimuli Associated with Oral Evaluation of Viscosity

Wood (1968) compared the sensory ratings provided by a panel of personnel with shear rate–shear stress data and concluded that the stimulus associated with the

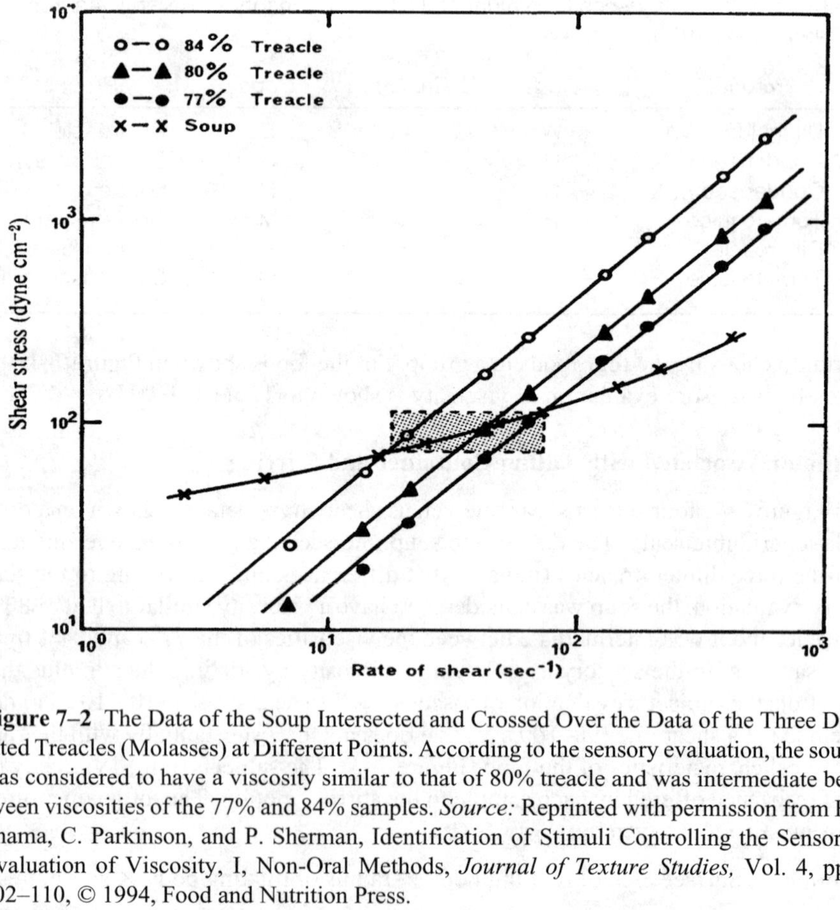

Figure 7–2 The Data of the Soup Intersected and Crossed Over the Data of the Three Diluted Treacles (Molasses) at Different Points. According to the sensory evaluation, the soup was considered to have a viscosity similar to that of 80% treacle and was intermediate between viscosities of the 77% and 84% samples. *Source:* Reprinted with permission from F. Shama, C. Parkinson, and P. Sherman, Identification of Stimuli Controlling the Sensory Evaluation of Viscosity, I, Non-Oral Methods, *Journal of Texture Studies,* Vol. 4, pp. 102–110, © 1994, Food and Nutrition Press.

oral evaluation of viscosity was the shear stress developed in the mouth at a constant shear rate of 50 s^{-1}. However, Shama and Sherman (1973) showed that the stimulus depends on whether the food is a low-viscosity or a high-viscosity food (Figure 7–5). For low-viscosity foods, the stimulus is the shear rate developed at an almost constant shear stress of 10 Pa. In contrast, for high viscosity foods, the stimulus is the shear stress developed at an approximately constant shear rate of 10 s^{-1}. Shama and Sherman (1973) also suggested that for foods that contain oil as the continuous medium, the foods may not spread spontaneously over the tongue and hard palate, and the spreading coefficient may be associated with the oral evaluation of viscosity. While Figure 7–5 contains constant viscos-

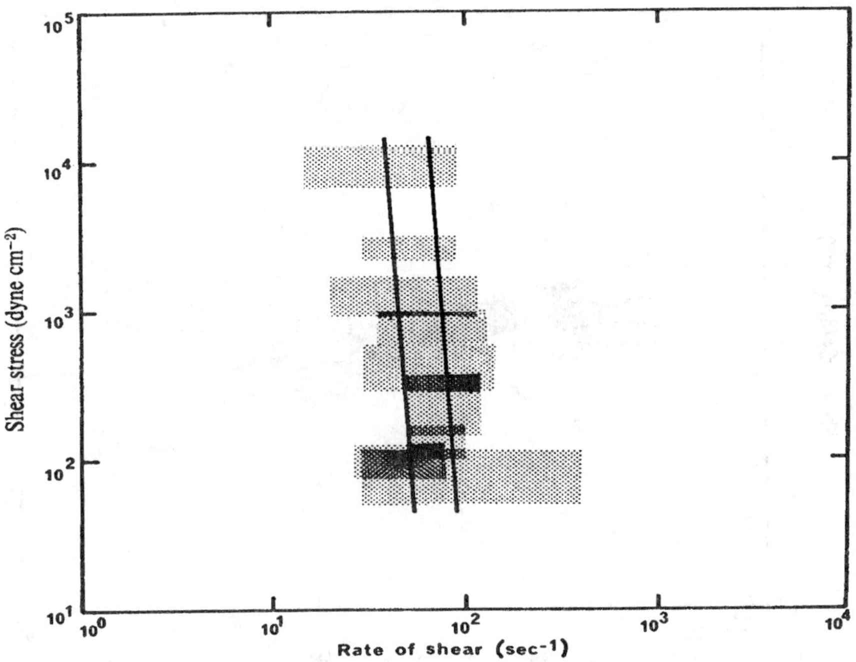

Figure 7–3 Stimuli Controlling the Sensory Assessment of Viscosity by the Nonoral Method of Stirring the Fluid. *Source:* Reprinted with permission from F. Shama, C. Parkinson, and P. Sherman, Identification of Stimuli Controlling the Sensory Evaluation of Viscosity, I, Non-Oral Methods, *Journal of Texture Studies,* Vol. 4, pp. 102–110, © 1994, Food and Nutrition Press.

---

ity lines, Figure 7–6 has the rheological data of the several foods that were employed.

An additional factor that may play a significant role is the dilution of fluid foods with saliva. The amount of saliva is related to the rate of production, which in turn depends on the composition of the fluid food. Therefore, matching relevant composition characteristics (e.g., pH) of the Newtonian and non-Newtonian foods used in the sensory evaluation may be desirable. The flow of low viscosity Newtonian fluids in the mouth may be turbulent, exhibiting higher sensory viscosity.

SENSORY ASSESSMENT OF VISCOSITY OF GUM DISPERSIONS

Gums are used extensively in foods, and their rheological behavior was discussed in Chapter 4. In general, gum dispersions exhibit shear-thinning behavior. Based

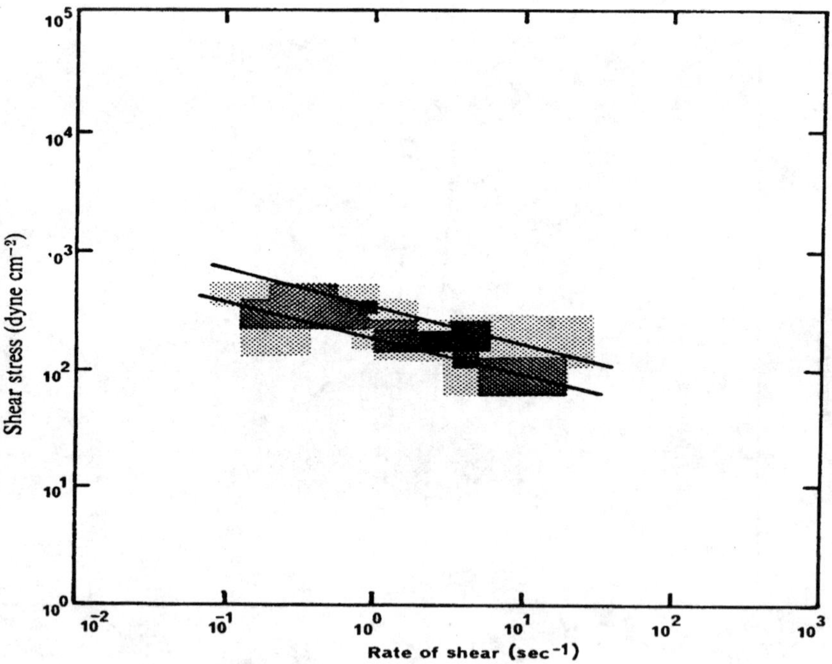

Figure 7–4 Stimuli Controlling the Sensory Assessment of Viscosity by the Nonoral Method of Tilting a Container. *Source:* Reprinted with permission from F. Shama, C. Parkinson, and P. Sherman, Identification of Stimuli Controlling the Sensory Evaluation of Viscosity, I, Non-Oral Methods, *Journal of Texture Studies,* Vol. 4, pp. 102–110, © 1994, Food and Nutrition Press.

on the shear-thinning characteristics with a Brookfield viscometer and sensory assessment by a trained panel, gum dispersions could be divided into three groups (Szczesniak and Farkas, 1962); however, true shear rates were not calculated in the objective measurement of flow behavior. Group A consisted of gums (e.g., 2% cooked, 0.75% phosphomannan, and 0.15% xanthan gum) and corn starch that exhibited a sharp decrease in viscosity with an increase in the shear rate; they were also judged by the trained panel to be nonslimy. Group C consisted of gums that showed a small to moderate change in viscosity and were considered by the trained panel to range from slimy to extremely slimy (e.g., 1% locust bean gum, 1.3% sodium alginate, 5% low-methoxyl pectin, and 2.6% methyl cellulose). Group B consisted of gums (e.g., 0.6% guar gum, 1% carrageenan, and 1% gum karaya) showing an intermediate dependence on the shear rate and described by the panel as at least somewhat slimy.

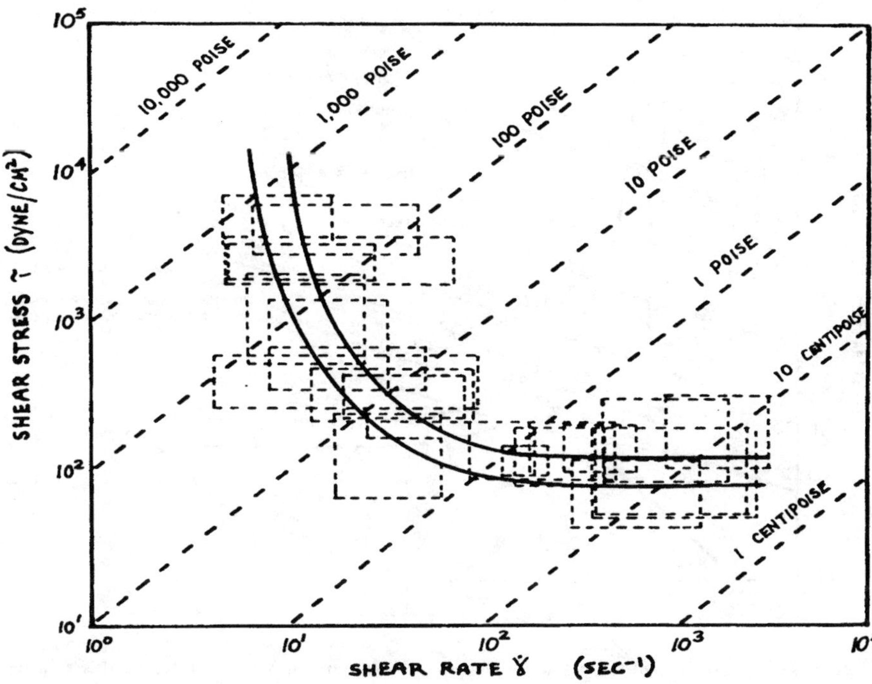

Figure 7–5 Stimuli Controlling the Sensory Assessment of Viscosity by Oral Methods: Lines of Constant Viscosity. *Source:* Reprinted with permission from F. Shama and P. Sherman, Identification of Stimuli Controlling the Sensory Evaluation of Viscosity, II, Oral Methods, *Journal of Texture Studies,* Vol. 4, pp. 111–118, © 1973, Food and Nutrition Press.

Perceived Thickness of Gum Dispersions

The results of Szczesniak and Farkas (1962) do not provide any quantitative relationships between rheological behavior and sensory assessment of viscosity in part because the mathematical form of rheological behavior of the gum dispersions was not determined. Further, it was shown in Chapter 4 that the reduced variables η/η_0 and $\dot{\gamma}/\dot{\gamma}_0$ can be used to describe rheological data of gum dispersions by a single master curve. Morris et al. (1984) correlated the perceived thickness (T) and stickiness (S) with the rheological parameters at 25°C of concentrated gum dispersions with $\eta_0 > 10$ mPa s: lambda carrageenan, high-methoxyl pectin, guar gum, locust bean gum, and high guluronate-alginate. The correlating equations obtained were

$$\log T = 0.25 \log(\eta_0) + 0.10 \log(\dot{\gamma}_{0.1}) + 1.31 \qquad (7.1)$$

$$\log S = 0.25 \log(\eta_0) + 0.12 \log(\dot{\gamma}_{0.1}) + 1.25 \qquad (7.2)$$

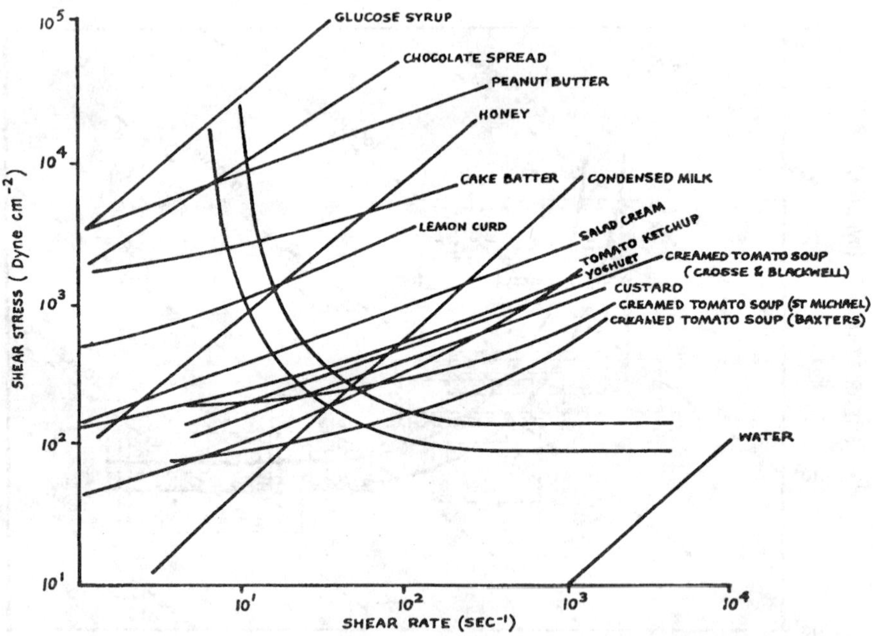

Figure 7–6 Stimuli Controlling the Sensory Assessment of Viscosity by Oral Methods: Shear Rate versus Shear Stress Data. *Source:* Reprinted with permission from F. Shama and P. Sherman, Identification of Stimuli Controlling the Sensory Evaluation of Viscosity, II, Oral Methods, *Journal of Texture Studies,* Vol. 4, pp. 111–118, © 1973, Food and Nutrition Press.

---

where $\dot{\gamma}_{0.1}$ is the shear rate when $\eta_a = 0.1\ \eta_0$. To facilitate comparison with other studies on sensory assessment, Morris et al. (1984) used poise as the unit of viscosity.

The 2.77 w/v % pectin solution (log η_0 in poise $= 1.64$ and log $\dot{\gamma}_{0.1} = 2.30$) was the standard and the blind control and was assigned a value of 100 for both thickness and stickiness, and a sample judged to be four times as "thick" as the standard was given a "thickness" score of $T = 400$. Thus, the ratio scaling approach of Stevens (1975) was used. Because the above equations were developed using the 2.77 w/v % pectin solution as the standard arbitrarily, panelists were also asked to assess the thickness of a range of Newtonian foods (mostly sugar syrups) against the standard pectin solution to obtain an equation for thickness:

$$\log T = 0.221 \log (\eta_N) + 1.691 \tag{7.3}$$

The above equation was also applicable to stickiness, *S*, substituted in place of the thickness, *T*. Because Cutler et al. (1983) have shown that perceived thickness and viscosity were linearly related, the correlations for perceived thickness can be used for perceived viscosity. By combining Equations 7.1 and 7.3, one can obtain an expression to calculate the viscosity η_N of a Newtonian liquid that would be perceived as having an identical thickness and stickiness to any polysaccharide dispersion with known η_0 and $\dot{\gamma}_{0.1}$:

$$\log \eta_N = 1.13 \log \eta_0 + 0.45 \log \dot{\gamma}_{0.1} - 1.72 \qquad (7.4)$$

Perceived Thickness of Foods

Cutler et al. (1983) extended the experimental techniques used by Morris et al. (1984) to fluid foods: syrups, chocolate spread, condensed milk, strawberry dessert sauce, lemon curd, tomato ketchup, rosehip syrup, milk, and sieved chicken soup. Again, the 2.77 w/v % pectin solution ($\log \eta_0$ (poise) $= 1.64$ and $\log \dot{\gamma}_{0.1} = 2.30$) was the standard and the blind control and was assigned a value of 100 for both thickness and stickiness; also, to facilitate comparison with earlier studies on sensory assessment, the viscosity of all samples was expressed in poise. Equation 7.3 can be rewritten to obtain an expression for the perceived thickness of Newtonian foods:

$$T = 49\,(\eta_N)^{0.22} \qquad (7.5)$$

The constant 49 is applicable to the selected standard fluid for scaling, but the exponent is independent of the standard chosen. Further, its low value means that to double the perceived thickness would require a 20-fold increase in the objective value of η_N. Also from Equation 7.3, one can derive an expression for the perceived thickness of Newtonian foods that can be used to compare equivalent Newtonian viscosity with observed flow curves for non-Newtonian foods:

$$\log(\eta_N) = 4.52 \log T - 7.65 \qquad (7.6)$$

The effective oral shear rates at which the objective viscosity was equal to the η_N (equivalent Newtonian viscosity) calculated from the above equation were in good agreement with the results of Shama and Sherman (1973) (Figure 7–7) (Cutler et al., 1983); the only exceptions to this observation were the highly shear-thinning xanthan gum dispersions that deviated slightly from the Shama and Sherman results.

SPREADABILITY USING FORCE AND UNDER NATURAL CONDITIONS

A master curve of the rheological conditions applicable during spreading of lipophilic preparations with force on the skin using a method similar to that of

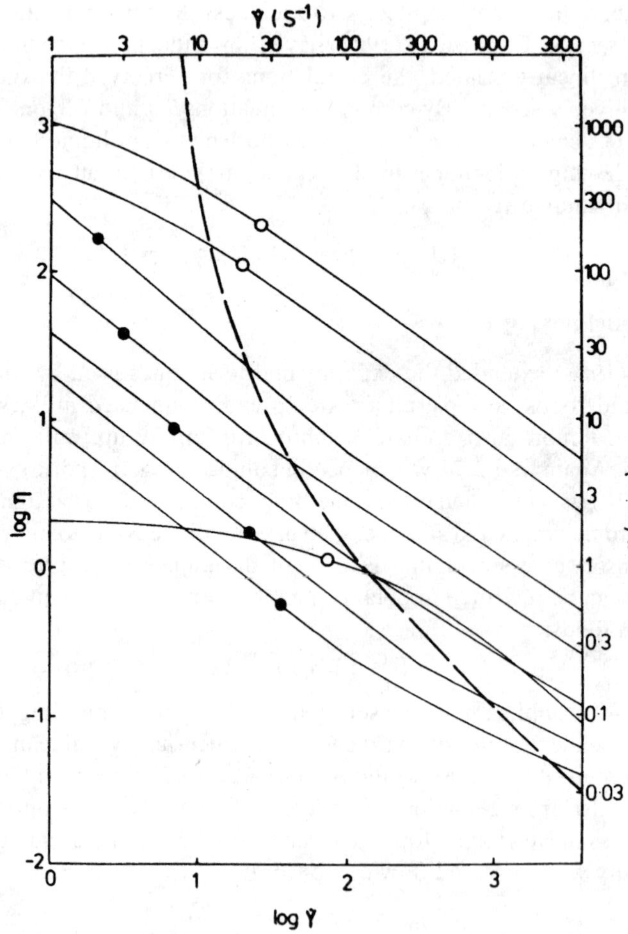

Figure 7–7 Estimated Oral Shear Rates by Cutler et al. (1983) Are Indicated (circles). Filled circles are of xanthan gum solutions; the dotted line represents the middle of the curve determined by Shama and Sherman (1973). *Source:* Reprinted with permission from A.N. Cutler, E.R. Morris, and L.J. Taylor, Oral Perception of Viscosity in Fluid Foods and Model Systems, *Journal of Texture Studies,* Vol. 14, pp. 377–395, © 1983, Food and Nutrition Press.

Wood (1968) showed that the range of acceptable apparent viscosity was about 3.9 poise to 11.8 poise, with an optimum of approximately 7.8 poise (Barry and Grace, 1972). The preferred region was approximately bounded by shear rates 400 to 700 s^{-1} and shear stress 2000 to 7000 dyne cm^{-2} (200–700 Pa, respectively).

The ranges of values of shear stress and shear rate associated with pourability of two commercial salad dressings under natural conditions were 190 to 430 dyne cm^{-2} (19–43 Pa, respectively) and 0.9 to 7.3 s^{-1}, respectively (Kiosseoglou and Sherman, 1983). The methodology used in this study was the same as that in Shama et al. (1973). These values agreed reasonably well with those prevailing when the viscosity of a food is judged by tilting the container and evaluating the rate of flow down the container wall. Also, spreadability evaluation was influenced by yield stress and shear-thinning characteristics; flow ceases when the operative shear stress falls below the yield stress.

APPLICATION OF FLUID MECHANICS

There are a large number of terms that are used to describe food texture. However, only a few studies have attempted to relate the descriptive words to physical data (Jowitt, 1974). Kokini et al. (1977) examined 12 textural terms and suggested that for liquid and semisolid foods the three terms thick, smooth, and slippery are adequate. Assuming that as a first approximation of the deformation process, a liquid food is squeezed between the tongue and the roof of the mouth, thickness is assumed to be related to the shear stress on the tongue; that is:

$$\text{thickness} \propto (\text{shear stress on the tongue}, \sigma)^{\beta} \tag{7.7}$$

Assuming that $\beta = 1$ for a Newtonian fluid, the expression for the thickness of a Newtonian fluid is

$$\text{thickness} \propto \text{shear stress on the tongue } \sigma = (\eta_{\mathrm{N}})^{1/2}\, F^{1/2}\, v \left(\frac{4t}{3\pi r_{\mathrm{o}}^{4}}\right)^{1/2} \tag{7.8}$$

where F is the normal force between the tongue and the roof of the mouth, r_{o} is the radius of the tongue, v is the average velocity of the tongue, and t is time for assessment. From Equation 7.8, to double the perceived thickness, the magnitude of the viscosity must be increased four-fold, and the perceived thickness is a function of time. The work of Cutler et al. (1983) (Equation 7.5) shows that perceived thickness is independent of time and also predicts an increase in the perceived thickness with increase in viscosity, but a 20-fold increase in viscosity is suggested to double the perceived thickness.

For a non-Newtonian power law fluid ($\sigma = K\dot{\gamma}^{n}$), the expression for the perceived thickness that is proportional to shear stress on the tongue is

$$\text{shear stress on the tongue } \sigma = Kv^{n}\left[\frac{1}{h_{0}^{n+1/n}} + \left(\frac{F}{r_{\mathrm{o}}^{n+3}}\,\frac{n+3}{2\pi K}\right)^{1/n} \times \left(\frac{(n+1)t}{2n+1}\right)\right]^{(n^{2}/n+1)} \tag{7.9}$$

Fluid Mechanics of Spreadability

For spreading foods, a torque is applied by the consumer during the spreading action that is equal to shear stress on knife × area of knife × length of knife (Kokini, 1987). Therefore, spreadability is inversely proportional to the shear stress on the knife:

$$\text{spreadability} \propto (\text{shear stress on knive})^{-1} \tag{7.10}$$

Smoothness and Slipperiness

Other sensory terms that can be related to various physical forces are smoothness and slipperiness:

$$\text{smoothness} \propto (\text{friction force})^{-1} = (fF)^{-1} \tag{7.11}$$

$$\text{slipperiness} \propto (\text{viscous force} + \text{friction force})^{-1} \propto \left[\frac{Kv^n}{h_s} (\pi r_o^2) + \eta F \right]^{-1} \tag{7.12}$$

where f is a friction coefficient and h_s is the average height of aspertites on the tongue's surface. In addition, creaminess is related to thickness and smoothness:

$$\text{creaminess} \propto (\text{thickness})^{0.54} \times (\text{smoothness})^{0.84} \tag{7.13}$$

ROLE OF RHEOLOGY ON THE PERCEPTION OF FLAVOR AND TASTE

The influence of the rheology of a particular food material on the perception of its taste or flavor can have two main origins: a physiological effect due to the proximity of the taste and olfactory receptors to the kinesthetic and thermal receptors in the mouth, since an alteration of the physical state of the material may have an influence on its sensory perception, and an effect related to the bulk properties of the material (e.g., texture, viscosity), since the physical properties of the material may affect the rate and the extent with which the sensory stimulus reaches the gustatory receptors.

For fluid systems, where mainly hydrocolloid solutions and dispersions have been used as model systems, it was demonstrated that, generally, an increase in viscosity of the system decreased the perception of sweetness. Different hydrocolloids were found to affect sweetness to differing extents (Vaisey et al., 1969; Pangborn et al., 1973). Vaisey et al. (1969) have also shown that hydrocolloid solutions with more pronounced shear-thinning behavior tended to decrease to a lesser extent the sweetness perception than did those that are less pseudoplastic. Although the sweetness was more difficult to detect when the viscosity was higher, for suprathreshold levels of sugar it was shown that the more viscous solutions were perceived as sweeter (Stone and Oliver, 1966).

Moskowitz and Arabie (1970) found that the taste intensity (sweetness, sourness, saltiness, and bitterness) was related to the apparent viscosity of carboxymethylcellulose solutions by a power function with a negative slope. Pangborn et al. (1973) observed that the influence of different hydrocolloids on the perception of some basic taste intensities (saltiness, bitterness, sourness) appeared to be more dependent on the nature of the hydrocolloid and the taste of the substance than on the viscosity level. In contrast, sweetness imparted by sucrose was found to be highly dependent on viscosity, that is, the hydrocolloid concentration; above a certain viscosity threshold, it was shown that the sweetness intensity of sucrose was significantly depressed. Saltiness was the taste attribute least affected; sourness, imparted by citric acid, was significantly reduced by all hydrocolloids tested; and for the other taste substances, the presence of a hydrocolloid generally enhanced the taste intensity of saccharin and depressed that of sucrose and caffeine (bitterness).

The effect of different kinds and levels of hydrocolloids on selected aromatic flavor compounds was also investigated (Pangborn and Szczesniak, 1974). The hydrocolloids were found to decrease both the odor and the flavor intensities, and once again, the overall effects seemed much more dependent on the nature of the hydrocolloid and of the taste substance than on the viscosity level imparted by the biopolymer. Similar results were found in a subsequent study conducted by Pangborn et al. (1978) about the effect of the same hydrocolloids on the aroma, flavor, and basic tastes of three distinct beverage systems.

Launay and Pasquet (1982) studied the relationship between sweetness and solution viscosity using sucrose solutions thickened with guar gum prepared to two constant viscosity levels. The results showed that the sweetness intensity decreased in the presence of the gum, but this reduction was not dependent on the viscosity level, a result that seems to be in conflict with the results previously reported by Moskowitz and Arabie (1970).

Some results from the several studies mentioned above are difficult to compare, and some apparent discrepancies are observed. The main causes for the differences are due to the different hydrocolloids and the taste substances that have been used, the different hydrocolloid concentration ranges that have been studied, and also, in some cases, the inaccurate evaluation of the effect of the added taste substance, such as sucrose, on the rheological behavior of the hydrocolloid.

Role of Hydrocolloid Concentration

The importance of the concentration regime of the hydrocolloid upon its effect on taste and flavor perception was clearly shown by Baines and Morris (1987). These authors studied the effect of guar gum concentration on flavor/taste perception in thickened systems, using solutions with constant concentration of sucrose and flavoring and a wide range of concentrations of three different samples of guar gum. The results obtained showed that the perceived intensity of both attributes was independent of polymer concentration up to c^* (the concentration threshold that marks the transition from a dilute solution of random-coil polymer molecules to an entangled network where the molecules are not free to move independently, described in Chapter 4), but decreased steeply at higher degrees of space occupancy by the polymer ($c[\eta]$), that is, above c^*. The interpretation that has been proposed for the observed behavior was that a restricted replenishment of surface depletion would occur with increasing coil overlap and entanglement of the hydrocolloid, thus decreasing the perception of the sensory attributes.

Engineering Approach

An interesting engineering approach was proposed by Kokini and coworkers to model viscosity-taste interactions. Kokini et al. (1982) have studied the perception of swcctness of sucrose and fructose in solutions with various tomato solids contents, and based on the observed decreasing of sweetness intensity as the percentage of tomato solids increased, they have proposed a more complex but rather comprehensive physical model relating viscosity and taste intensity, based on the physics and chemistry in the mouth. This model was further successfully applied (Cussler et al., 1979) to explain the effect of the presence of hydrocolloids at different levels on the subjective assessment of sweetness and sourness, giving a new and more consistent interpretation of the results previously reported by Moskovitz and Arabie (1970), and also to explain the interaction of butyric acid in solutions of carboxymethylcellulose reported by Pangborn and Szczesniak (1974). The main assumptions are rather simple and logical: Because the taste or odor attributes are assessed by very rapid chemical reactions at the human taste and olfactory receptors, the assessment process should be limited not by these chemical reactions but rather by a diffusion process in the mouth or in the nose. Thus, for example, variations in sweetness and sourness are proportional to the product of the solute concentration and the square root of the diffusion coefficient. The diffusion coefficients may be calculated from mass transfer considerations and are dependent on viscosity. The model seems to work quite well for those cases where mass transfer is the limiting factor. However, for attributes like saltiness and bitterness the above model does not work; the model also cannot explain why fructose and sucrose at the same concentration originate different taste intensities.

REFERENCES

Baines, Z.V., and E.R. Morris. 1987. Flavour/taste perception in thickened systems: The effect of guar gum above and below c*. *Food Hydrocolloids* 1: 197–205.

Barry, B.W., and A.J. Grace. 1972. Sensory testing of spreadability: Investigation of rheological conditions operative during application of topical preparations. *J Pharm Sci* 61: 335–341.

Cussler, E.L., J.L. Kokini, R.L. Weinheimer, and H.R. Moskowitz. 1979. Food texture in the mouth. *Food Technol* 33, no. 10: 89–92.

Cutler, A.N., E.R. Morris, and L.J. Taylor. 1983. Oral perception of viscosity in fluid foods and model systems. *J Texture Studies* 14: 377–395.

Jowitt, R. 1974. The terminology of food texture. *J Texture Studies* 5: 351–358.

Kiosseoglou, V.D., and P. Sherman. 1983. The rheological conditions associated with judgement of pourability and spreadability of salad dressings. *J Texture Studies* 14: 277–282.

Kokini, J.L. 1987. The physical basis of liquid food texture and texture-taste interactions. *J Food Eng* 6: 51–81.

Kokini, J.L., K. Bistany, M. Poole, and E. Stier. 1982. Use of mass transfer theory to predict viscosity–sweetness interactions of fructose and sucrose solutions containing tomato solids. *J Texture Studies* 13: 187–200.

Kokini, J.L., and E.L. Cussler. 1984. Predicting liquid food texture of liquid and melting semi-solid foods. *J Food Sci* 48: 1221–1225.

Kokini, J.L., J. Kadane, and E.L. Cussler. 1977. Liquid texture perceived in the mouth. *J Texture Studies* 8: 195–218.

Launay, B., and E. Pasquet. 1982. Sucrose solutions with and without guar gum: Rheological properties and relative sweetness intensity. *Prog Food Nutri Sci* 6: 247–258.

Morris, E.R., R.K. Richardson, and L.J. Taylor. 1984. Correlation of the perceived texture of random coil polysaccharide solutions with objective parameters. *Carbohydrate Polymers* 4: 175–191.

Moskowitz, H.R., and P. Arabie. 1970. Taste intensity as a function of stimulus concentration and solvent viscosity. *J Texture Studies* 1: 502–510.

Pangborn, R.M., Z.M. Gibbs, and C. Tassan. 1978. Effect of hydrocolloids on apparent viscosity and sensory properties of selected beverages. *J Texture Studies* 9: 415–436.

Pangborn, R.M., and A. Szczesniak. 1974. Effect of hydrocolloids and viscosity on flavor and odor intensities of aromatic flavor compounds. *J Texture Studies* 4: 467–482.

Pangborn, R.M., I.M. Tabue, and A. Szczesniak. 1973. Effect of hydrocolloids on oral viscosity and basic taste intensities. *J Texture Studies* 4: 224–241.

Shama, F., C. Parkinson, and P. Sherman. 1973. Identification of stimuli controlling the sensory evaluation of viscosity. I. Non-oral methods. *J Texture Studies* 4: 102–110.

Shama, F., and P. Sherman. 1973. Identification of stimuli controlling the sensory evaluation of viscosity. II. Oral methods. *J Texture Studies* 4: 111–118.

Sherman, P. 1970. *Industrial rheology.* New York: Academic Press.

Stevens, S.S. 1975. *Psychophysics—Introduction to its perceptual neural and social prospects.* New York: John Wiley & Sons.

Stone, H., and S. Oliver. 1966. Effect of viscosity on the detection of relative sweetness intensity of sucrose solutions. *J Food Science* 31: 129–134.

Szczesniak, A.S. 1963. Classification of textural characteristics. *J Food Sci* 28: 385–389.

Szczesniak, A.S. 1979. Classification of mouthfeel characteristics of beverages. In *Food texture and rheology,* ed. P. Sherman, 1–20. New York: Academic Press.

Szczesniak, A.S., and E. Farkas. 1962. Objective characterization of the mouthfeel of gum solutions. *J Food Sci* 27: 381–385.

Vaisey, M., R. Brunon, and J. Cooper. 1969. Some sensory effects of hydrocolloid sols on sweetness. *J Food Sci* 34: 397–400.

Wood, F.W. 1968. Psychophysical studies on the consistency of liquid foods. *S.C.I. monograph: Rheology and texture of foodstuffs,* 40–49. London: Society of Chemical Industry.

CHAPTER 8

Application of Rheology to Fluid Food Handling and Processing

M. Anandha Rao

VELOCITY PROFILES IN TUBES

Equations describing velocity profiles can be used, among other applications, to study the effect of different rheological models on the distribution of velocities and to understand the concept of residence-time distribution across the cross section of a pipe or a channel. The velocity profile of a fluid flowing in a tube can be derived from the relationship

$$v_z = \int_r^{r_o} \left(\frac{dv_z}{dr}\right) dr \tag{8.1}$$

where v_z is the velocity in the axial direction, r is the radial coordinate, and r_o is the radius of the tube (Figure 8–1). From a simple force balance for tube flow, one can obtain

$$\sigma_{rz} = \sigma_w \frac{r}{r_o} = -\frac{r}{2}\frac{dp}{dz} \tag{8.2}$$

where σ_{rz} is the shear stress at any radius r, σ_w is the magnitude of the shear stress at the wall ($r = r_o$), p is the pressure, and z is the axial coordinate. Using Equations 8.1 and 8.2 and noting that in the rheological equations described in Chapter 2, $\sigma = \sigma_{rz}$ and $\dot{\gamma} = (dv_z/dr)$, one can derive equations describing the velocity profiles for laminar flow in a tube.

The power law model ($\sigma = K\dot{\gamma}^n$, Equation 2.3) has been used extensively in handling applications. The relationship between the maximum velocity (v_{zm}) and the average velocity ($\bar{v}$) for the design of the length of a holding tube of a pasteurizing system is of special interest:

$$\frac{v_{zm}}{\bar{v}_z} = \frac{3n + 1}{n + 1} \tag{8.3}$$

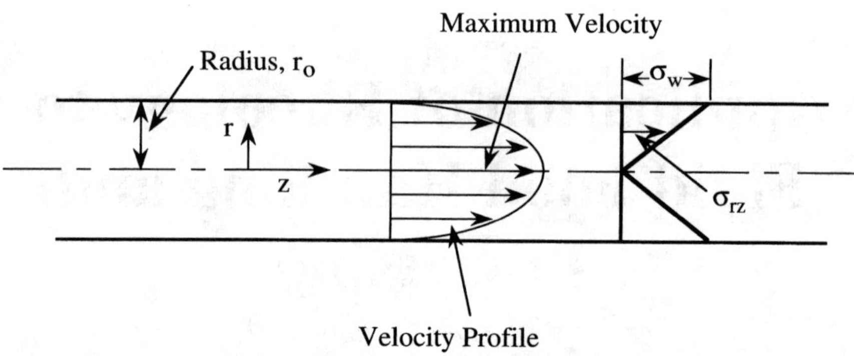

Figure 8–1 Schematic Diagram for Analysis of Laminar Flow in a Tube: v_z is the velocity in the axial direction, r is the radial coordinate, and r_o is the radius of the tube.

For Newtonian and pseudoplastic fluids in laminar flow, one can deduce from Equation 8.3 the popular relationship that maximum velocity is, at most, equal to twice the average velocity. In contrast, in the case of shear-thickening (dilatant) fluids, the maximum velocity would be more than twice the average velocity.

The volumetric flow rate Q is given by the following equation:

$$Q = \int_0^{r_o} 2\pi r v_z \, dr = \pi \int_0^{r_o} v_z(r) \, d(r^2) \tag{8.4}$$

Integrating by parts the second part of Equation 8.4, and using the boundary condition $v = 0$ at $r = r_o$, one can obtain the general equation for the volumetric flow of a fluid in a tube:

$$Q = \left(\frac{\pi r_o^3}{\sigma_w^3}\right) \int_0^{\sigma_w} \sigma_{rz}^2 \left(\frac{dv_z}{dr}\right) d\sigma_{rz} \tag{8.5}$$

Substituting for the shear rate appropriate expressions from different rheological models, one can derive equations relating Q and pressure drop Δp (Rao, 1995). Figure 8–2 contains expressions for the velocity profiles and the volumetric flow rates of the three rheological models: power law, Herschel-Bulkley, and the Bingham plastic models.

In the case of fluids obeying the power-law model, the pressure drop per unit length $\Delta p/L$ is related to Q and r_o by the relationship

$$\frac{\Delta p}{L} \propto \frac{Q^n}{r_o^{3n+1}} \tag{8.6}$$

From this relationship, we see that for Newtonian foods ($n = 1$) the pressure gradient is proportional to the $(r_o)^{-4}$ power. Therefore, a small increase in the radius of the tube will result in a major reduction in the magnitude of the pressure gradient. In contrast, for a highly pseudoplastic fluid (e.g., $n = 0.2$), increasing the pipe radius does not have such a profound effect on the pressure gradient.

Power Law Model: $\sigma = K\dot{\gamma}^n$

Velocity profile*: $v_z = \left(\frac{n}{n+1}\right)\left(\frac{\Delta p}{2KL}\right)^{(1/n)} [r_o^{(n/n+1)} - r^{(n/n+1)}]$

Volumetric flow rate: $\frac{Q}{\pi r_o^3} = \left(\frac{n}{n+1}\right)\left(\frac{\sigma_w}{K}\right)^{(1/n)}$

Herschel-Bulkley Model: $\sigma - \sigma_0 = K_H(\dot{\gamma})^{n_H}$

Velocity profile†: $v_z = \frac{2L}{\Delta p\,(m+1)\,K_H^m}\left[(\sigma_w - \sigma_0)^{(m+1)} - \left(\frac{r\Delta p}{2L} - \sigma_0\right)^{(m+1)}\right]$

when, $r_0 \leq r \leq r_o$

Volumetric flow rate: $\frac{Q}{\pi r_o^3} = \frac{(\sigma_w - \sigma_0)^{(m+1)}}{\sigma_w^3 K_H^m}\left[\frac{(\sigma_w - \sigma_0)^2}{m+3} + \frac{2\sigma_0(\sigma_w - \sigma_0)}{m+2}\frac{\sigma_0^2}{m+1}\right]$

Note that in the equations for the Herschel-Bulkley model, $m = (1/n_H)$

Bingham Plastic Model: $\sigma - \sigma_0 = \eta'\,\dot{\gamma}$

Velocity profile: $v_z = \frac{1}{\eta'}\left[\frac{\Delta p}{L}(r_o^2 - r^2) - \sigma_0(r_o - r)\right]$

when, $r_0 \leq r \leq r_o$

Volumetric flow rate: $\frac{4Q}{\pi r_o^3} = \frac{\sigma_w}{\eta'}\left[1 - \frac{4}{3}\left(\frac{\sigma_0}{\sigma_w}\right) + \frac{1}{3}\left(\frac{\sigma_0}{\sigma_w}\right)^4\right]$

The average velocity can be obtained by dividing the equation for volumetric flow rate by the area of cross section of the pipe.

* Maximum velocity occurs at the center line, $r = 0$.

† Velocity profiles are valid for $r_o \leq r \leq r_o$, where the radius of plug $r_o = (2\sigma_0 L/\Delta p)$; the maximum velocity occurs when $0 \leq r \leq r_o$ and is obtained by substituting r_o for r.

Figure 8–2 Velocity Profile and Volumetric Flow Rate Equations for Power Law, Herschel-Bulkley, and Bingham Plastic Fluids.

ENERGY REQUIREMENTS FOR PUMPING

Mechanical Energy Balance Equation

The energy required to pump a liquid food through a pipe line can be calculated from the mechanical energy balance (MEB) equation. The MEB equation can be used to analyze pipe flow systems. For the steady-state flow of an incompressible fluid, the mechanical energy balance can be written as follows (Brodkey, 1967):

$$gZ_1 + \frac{P_1}{\rho} + \frac{v_{z1}^2}{\alpha} - W = gZ_2 + \frac{P_2}{\rho} + \frac{v_{z2}^2}{\alpha} + E_f \quad (8.7)$$

where g is the acceleration due to gravity, Z is the height above a reference point, P is the pressure, v is the fluid velocity, W is the work output per unit mass, E_f is the energy loss per unit mass, α is the kinetic energy correction factor, and the subscripts 1 and 2 refer to two points in the pipe system (for example, Figure 8–3). To accurately estimate the energy required for pumping of frozen concentrated orange juice (FCOJ) in a specific piping and equipment system, the term $(-W)$ (J kg^{-1}) has to be estimated from Equation 8.7 after the other terms have been evaluated as described next.

The velocities at the entrance and exit of the system can be calculated from the respective diameters of the tanks or pipes and the volumetric flow rate of the food.

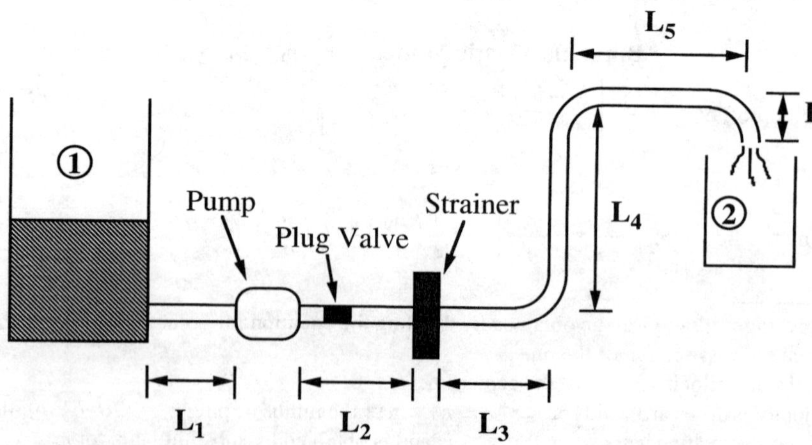

Figure 8–3 Schematic Diagram of Flow System for Application of Mechanical Energy Balance Equation. *Source*: Reprinted with permission from J.F. Steffe and R.G. Morgan, Pipeline Design and Pump Selection for Non-Newtonian Fluid Foods, *Food Technology*, Vol. 40, pp. 78–85, © 1986, institute of Food Technologists.

The energy loss term, E_f, consists of losses due to friction in the pipe and in valves and fittings:

$$E_f = \frac{2f\,v_z^2\,L}{D} + \sum_1^b \frac{k_f v_z^2}{2} \tag{8.8}$$

where f is the friction factor, v_z is the velocity, L is the length of straight pipe of diameter D, k_f is the friction coefficient for a fitting, and b is the number of valves or fittings. It is emphasized that k_f is unique to a particular fitting and that different values of u, k_f, and f may be required when the system contains pipes of different diameters. Further, losses due to special equipment, such as heat exchangers, must be added to E_f (Steffe and Morgan, 1986).

Friction Losses in Pipes

Because many fluid foods are non-Newtonian in nature, estimation of friction losses (pressure drop) for these fluids in straight pipes and in fittings is of interest. One can estimate the friction losses in straight pipes and tubes from the magnitude of the Fanning friction factor, f, defined as:

$$\frac{\Delta P_f}{\rho} = \frac{2\,f\,L\,v_z^2}{D} \tag{8.9}$$

where f is the friction factor. For laminar flow conditions, Garcia and Steffe (1987) suggested that based on the work of Hanks (1978), the friction factor can be calculated from a single general relationship for fluids that can be described by Newtonian ($\sigma = \eta\dot{\gamma}$, Equation 2.1), power law ($\sigma = K\dot{\gamma}^n$, Equation 2.3), and Herschel-Bulkley ($\sigma - \sigma_0 = K_H(\dot{\gamma})^{n_H}$, Equation 2.5) fluid models.

For fluids that can be described by the power law (Equation 2.3), such as FCOJ, the generalized Reynolds number (GRe) can be calculated from the equation

$$\text{G Re} = \frac{D^n v_z^{2-n}\,\rho}{8^{(n-1)}\,K}\left(\frac{4n}{3n+1}\right)^n \tag{8.10}$$

where D is pipe diameter, v_z is the average velocity of FCOJ, ρ is the density of FCOJ, and K and n are the power law parameters of FCOJ. When $n = 1$, Equation 8.10 reduces to the Reynolds number for Newtonian foods:

$$\text{Re} = \frac{D v_z \rho}{\eta} \tag{8.11}$$

In laminar tube flow, the Fanning friction factor can be calculated from the equation

$$f = \frac{16}{\text{G Re}} \tag{8.12}$$

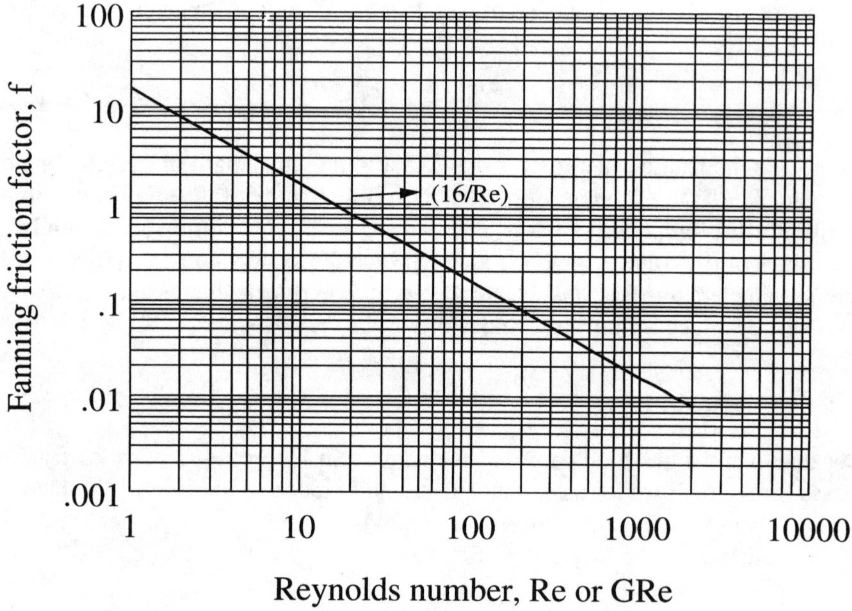

Figure 8–4 The Friction Factor–Generalized Reynolds Number Relationship for Power Law Fluids under Laminar Flow Conditions. It can also be used for Newtonian fluids in laminar flow.

Figure 8–4 illustrates the friction factor–GRe relationship for power law fluids under laminar flow conditions. It can also be used for Newtonian fluids in laminar flow with the Reynolds number being used in place of GRe. In fact, the Newtonian f versus Re relationship was established much earlier than extension to non-Newtonian fluids. Once the magnitude of the friction factor is known, the pressure drop in a pipe can be estimated from Equation 8.12.

For the laminar flow of a Herschel-Bulkley fluid ($\sigma - \sigma_0 = K_H(\dot{\gamma})^{n_H}$, Equation 2.5), the friction factor can be written as

$$f = \frac{16}{\Psi \text{G Re}} \tag{8.13}$$

where Ψ is related to the yield stress (σ_0) and the flow behavior index (n) (Garcia and Steffe, 1987):

$$\Psi = (3n + 1)^n (1 - \xi_0)^{1+n} \left[\frac{(1 - \xi_0)^2}{3n + 1} + \frac{2\xi_0(1 - \xi_0)}{(2n + 1)} + \frac{\xi_0^2}{(n + 1)} \right]^n \tag{8.14}$$

where,

$$\xi_0 = \frac{\sigma_0}{\sigma_w} = \frac{\sigma_0}{(D\Delta P/4L)} \tag{8.15}$$

Because of the highly viscous nature of non-Newtonian foods, laminar flow conditions are likely to be encountered more often than turbulent conditions. Nevertheless, it is important to be aware of developments with respect to prediction of friction factors in turbulent flow of non-Newtonian foods. For turbulent flow, except for Newtonian fluids, the predicted magnitudes of friction factors for non-Newtonian fluids may differ greatly depending on the relationship employed (Garcia and Steffe, 1987). However, for power law fluids, the relationships of Dodge and Metzner (1959) and Hanks and Ricks (1974) were found to predict similar magnitudes of the friction factors. For the Herschel-Bulkley model that is used when non-Newtonian foods exhibit yield stress, the analysis of Hanks (1978) was found to be the most comprehensive, but to use the derived relationship it is necessary to perform a numerical integration and several iterations.

Kinetic Energy Losses

In the MEB equation, kinetic energy losses can be calculated easily, provided that the kinetic energy correction factor α can be determined. In turbulent flow, often, the value of $\alpha = 2$ is used in the MEB equation. When the flow is laminar and the fluid is Newtonian, the value of $\alpha = 1$ is used. Osorio and Steffe (1984) showed that for fluids that follow the Herschel-Bulkley model, the value of α in laminar flow depends on both the flow behavior index (n) and the dimensionless yield stress (ξ_0) defined above. They developed an analytical expression and also presented their results in graphical form for α as a function of the flow behavior index (n) and the dimensionless yield stress (ξ_0). When possible, the values presented by Osorio and Steffe (1984) should be used. For FCOJ samples that do not exhibit yield stress and are mildly shear-thinning, it seems reasonable to use a value of $\alpha = 1$.

Friction Loss Coefficients for Fittings

Steffe et al. (1984) determined magnitudes of the coefficients for a fully open plug valve, a tee with flow from line to branch, and a 90-degree short elbow as a function of GRe using apple sauce as the test fluid. They found that, as for Newtonian fluids, k_f increases with decreasing values of GRe. The regression equations for the three fittings were as follows:

three-way plug valve $\qquad k_f = 30.3\ \mathrm{GRe}^{-0.492}$ (8.16) tee $$k_f = 29.4\ \mathrm{GRe}^{-0.504} \tag{8.17}$$

elbow $$k_f = 191.0\ \mathrm{GRe}^{-0.896} \tag{8.18}$$

In many instances, the practice is to employ values determined for Newtonian fluids, such as those in the *Chemical Engineers' Handbook.*

PUMP SELECTION AND PIPE SIZING

Steffe and Morgan (1986) discussed in detail the selection of pumps and the sizing of pipes for non-Newtonian fluids. Preliminary selection of a pump is based on the volumetric pumping capacity only from data provided by the manufacturers of pumps. Effective viscosity, η_e, was defined by Skelland (1967) as the viscosity that is obtained assuming that the Hagen-Poiseuille equation for laminar flow of Newtonian fluids is applicable:

$$\eta_e = \left\{\frac{(D\Delta P/4L)}{(32Q/\pi D^3)}\right\} \tag{8.19}$$

An alternate form of Equation 8.19 in terms of the mass flow rate, m, and the friction factor, f, is

$$\eta_e = \frac{fm}{4\pi D} \tag{8.20}$$

In calculating η_e from Equation 8.20, either the port size of a pump or the dimensions of the assumed pipe size can be used. Based on the magnitude of η_e, the suitability of the pump volumetric size must be verified from plots of effective viscosity versus volumetric flow rate. It is emphasized that a pump size is assumed based on the volumetric pumping requirements and the assumption is verified by performing detailed calculations.

A comprehensive example for sizing a pump and piping for a non-Newtonian fluid whose rheological behavior can be described by the Herschel-Bulkley model (Equation 2.5) was developed by Steffe and Morgan (1986) for the system shown in Figure 8–3, and it is summarized in the following. The Herschel-Bulkley parameters were as follows: yield stress = 157 Pa, flow behavior index = 0.45, consistency index = 5.20 Pa s^n.

Pump Discharge Pressure

The discharge pressure of the pump can be calculated by applying the MEB equation between the pump discharge and the exit point of the system so that the upper seal pressure limits are not exceeded. The MEB equation for this purpose can be written as

$$P_1 = \left[g(Z_2 - Z_1) + \frac{P_2}{\rho} + E_f \right] \rho \tag{8.21}$$

The energy loss due to friction in the pipe, valve, and fittings was estimated to be 329.0 J kg^{-1}, and the discharge pressure of the pump, P_1, was estimated to be 4.42 × 10^5 Pa (Steffe and Morgan, 1986).

Power Requirements for Pumping

The total power requirements for pumping are calculated by adding the hydraulic and the viscous power requirements. The former can be estimated from the MEB equation written for the work input, $-W$. Note that $P_1 = P_2$ and that E_f includes the friction losses not only on the discharge section but also on the inlet section. The former was, as stated earlier, estimated to be 329.0 J/kg, while the latter was estimated to be 24.7 J kg^{-1} by applying the MEB equation between the exit of the tank and the pump inlet. The work input ($-W$) was estimated to be 380.0 J kg^{-1}, and because the mass flow rate was 1.97 kg s^{-1}, the hydraulic power input was estimated to be 749.0 J s^{-1} or 0.749 kW. For estimating the viscous power requirements due to energy losses in the pump due to friction, the operating speed and the effective viscosity of the fluid food must be calculated. The former can be calculated from the displacement volume per revolution of the pump and the required volumetric flow, while the latter can be calculated from Equation 8.20. For a size 30 Waukesha pump, the volumetric displacement per revolution is 2.27 × 10^{-4} m^3 s^{-1}, and hence the pump speed was 417 rpm, while the equivalent viscosity for the fluid food under consideration was 0.703 Pa s (Steffe and Morgan, 1986). The energy losses in the pump were estimated from the manufacturer's data to be 0.835 kW. Therefore, the sum of the hydraulic and the viscous losses were 1.58 kW or 2.12 hp. These data allow selection of a suitable motor and drive system. In this example, pipe size was based on the pump port's diameter. It may also be based on plans for future expansion and ease of cleaning.

POWER CONSUMPTION IN AGITATION

Henry Rushton and coworkers developed the concept of the power number, Po, for studying mixing of fluids that for Newtonian fluids is defined as:

$$\text{Po} = \frac{p}{D_a^5 N_a^3 \rho} \tag{8.22}$$

where p is power ($p = 2\pi N_a \times T$), (Js^{-1}), D_a is agitator diameter (m), T is torque on agitator (Nm), N_a is agitator speed (s^{-1}), and ρ is the density of orange juice ($kg\ m^{-3}$). Several types of agitators, some with proprietary designs, are used in the food industry. For the purpose of illustration, the commonly used anchor agitator is shown in Figure 8–5.

In laminar mixing conditions of Newtonian fluids, Po is linearly related to agitator rotational Reynolds number, Re_a:

$$Po = \frac{A}{Re_a} \tag{8.23}$$

where A is a constant, and

$$Re_a = \frac{D_a^2 N_a \rho}{\eta} \tag{8.24}$$

The value of the constant A depends on the type of agitator. Laminar mixing conditions are encountered as long as Re_a is less than about 10.

For non-Newtonian foods, such as FCOJ, the viscosity is not constant but depends on the shear due to agitation. Therefore, for FCOJ and other non-Newtonian fluids we would like to define an agitator Reynolds number that can be used in place of Re_a to estimate Po from data presented in Figure 8–6. For fluids that follow the power law flow model, Metzner and Otto (1957), suggested a generalized

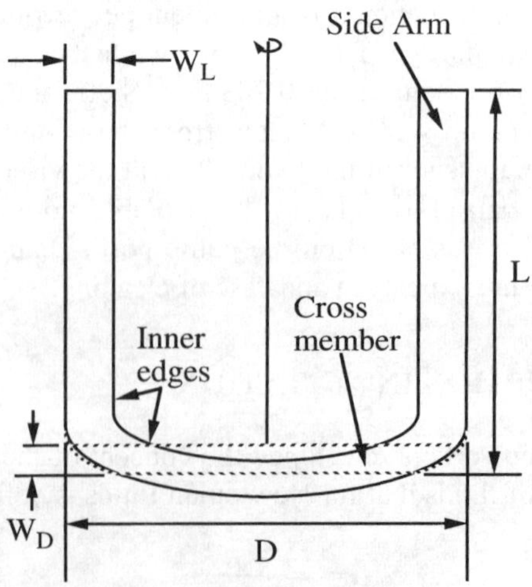

Figure 8–5 Schematic Diagram of an Anchor Agitator.

Reynolds number for mixing:

$$\mathrm{Re}_{\mathrm{mo}} = \frac{D_a^2 N^{2-n} \rho}{K} 8 \left(\frac{n}{6n+2} \right)^n \tag{8.25}$$

However, $\mathrm{Re}_{\mathrm{mo}}$ has not found wide acceptance because of deviations between predicted and experimental values of Po. A widely accepted procedure assumes that the average shear rate during mixing is directly proportional to the agitator rotational speed, N_a, that is,

$$\dot{\gamma} = k_s N_a \tag{8.26}$$

and the apparent viscosity is

$$\eta_a = K(k_s N_a)^{n-1} \tag{8.27}$$

Substituting the expression for apparent viscosity in the rotational Reynolds number for Newtonian fluids, $\mathrm{Re}_a = (D_a^2 N_a \rho / \eta)$, the power law agitation Reynolds number is

$$\mathrm{Re}_{\mathrm{pl}} = \frac{D_a^2 N_a^{2-n} \rho}{K(k_s N_a)^{n-1}} \tag{8.28}$$

The curves of $\mathrm{Re}_{\mathrm{pl}}$ versus Po for several agitators, adapted from Skelland (1967), are shown in Figure 8–6. It is emphasized that each line in Figure 8–6 is valid for a specific agitator, its orientation and dimensions, and the mixing tank dimensions, as well as the configuration of the tank's baffles. Given the strong likelihood of proprietary agitator designs and mixing tanks in food processing plants, it would be advisable to develop $\mathrm{Re}_{\mathrm{pl}}$ versus Po data for agitation systems being used for a specific food.

Once the magnitude of the power law agitation Reynolds number ($\mathrm{Re}_{\mathrm{pl}}$) is known, assuming that it is equal to Re_a, the corresponding value of the power number (Po) can be determined from the applicable curve for the specific agitator, such as Figure 8–6 or similar data. From the known values of Po, the diameter of the agitator (D_a), agitator rotational speed (N_a), and the density (ρ) of FCOJ, the power required (p) for agitation can be calculated.

Estimation of k_s

Procedures for determining k_s of a specific agitator and mixing tank can be found in Rao and Cooley (1984). In one procedure (Rieger and Novak, 1973; Rao and Cooley, 1984), the constant k_s can be determined from a plot of log $[p/KN_a^{n+1}D_a^3]$ versus $(1 - n)$; the slope of the line is equal to $-\log k_s$. For a given agitator, tests must be conducted such that the following data are obtained: p, the power ($p = 2\pi N_a \times T$)(Js^{-1}); D_a, agitator diameter (m); T, torque on agitator (Nm); N_a,

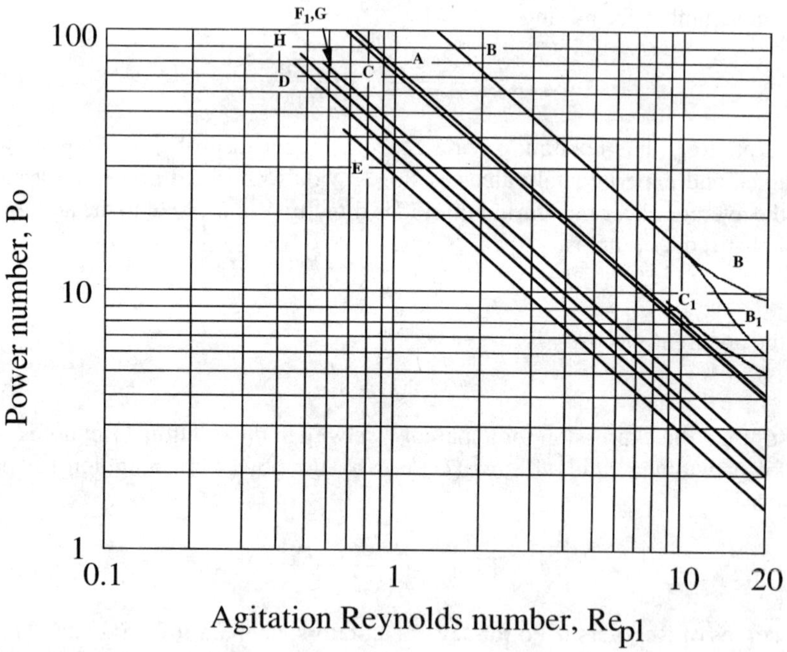

Figure 8–6 Curves of Power Law Agitation Reynolds Number (Re_{pl}) versus Power Number for Several Agitators. A—single turbine with 6 flat blades; B—two turbines with 6 flat blades; C—fan turbine with 6 blades at 45°; D and E—square-pitch marine propeller with 3 blades with shaft vertical and shaft 10° from vertical, respectively; F and G—square-pitch marine propeller with shaft vertical and with 3 blades and 4 blades respectively; and H—anchor agitator. *Source*: Reprinted with permission from A.H.P. Skelland, *Non-Newtonian Flow and Heat Transfer*, © 1967, John Wiley.

agitator speed (s^{-1}); and the power law rheological parameters of test fluids so that a wide range of $(1 - n)$ values are obtained. Typical values of the proportionality constant k_s for chemical industry impellers range from about 10 to 13 (Skelland, 1967).

RESIDENCE TIME DISTRIBUTION IN ASEPTIC PROCESSING SYSTEMS

In flow systems, not all fluid and solid food particles remain for the same time periods; that is, the particles have a distribution of residence times. Danckwerts (1953) proposed the concept of residence time distribution (RTD), and the theoretical and experimental principles of RTD have been well reviewed by Levenspiel (1972) and Himmelblau and Bischoff (1968). The RTD functions and the role of RTD in continuous pasteurization systems were reviewed (Rao and Loncin, 1974a, 1974b). Here only the necessary principles of RTD are discussed. For additional information the above references must be consulted.

With few exceptions, RTD studies represent lumped parameter analysis. Only in a few instances, such as that of fully developed laminar flow of fluids in tubes, can one attempt distributed parameter analysis in that the expressions for the RTD functions C and F can be derived from known velocity profiles.

Experimental RTD data on any equipment can be obtained by imposing at the inlet a pulse, step, or sinusoidal impulse and monitoring the response at the outlet. The tracer used must have physical properties such as density and viscosity similar to the test fluid food, and it should not in any manner alter the properties of the test fluid food. The sinusoidal impulse and response require more care and sophisticated equipment than the pulse and the step inputs. The response to a pulse impulse is called a C-curve, while the response to a step change is called an F-curve. For a closed vessel, that is, in which material passes in and out by bulk flow (i.e., no diffusion at inlet and outlet), the C and F functions are related as:

$$C(\theta) = \frac{\mathrm{d}F}{\mathrm{d}\theta} \tag{8.29}$$

One measure of the distribution of residence times (ages) of the fluid within a reactor is the E-function, defined so that $E\,\mathrm{d}\theta$ is the fraction of material in the exit stream with age between θ and $\theta + \mathrm{d}\theta$ (Levenspiel, 1972). It can be shown (Levenspiel, 1972) that the C and E functions are identical, and that for an isothermal process the ratio of the final (C) to initial (C_0) concentrations of either microorganisms or nutrients can be determined from the expression

$$\frac{C}{C_0} = \int_0^{\infty} \exp\left(\frac{-2.303\,t}{D}\right) E(t)\,\mathrm{d}t \tag{8.30}$$

where $D(T)$ is the D-value of the microorganism or a nutrient at a temperature T. For a nonisothermal process, (C/C_0) can be determined by recognizing that the D-value is dependent on temperature and taking this dependence into consideration. Note that $E(t) = [E(\theta)/t]$ is used in estimating the number of survivors or percent nutrient retention according to Equation 8.30.

Interpretation of RTD Data

Experimental RTD studies are usually interpreted in terms of the dispersion model (Equation 8.31) or the equal sized tanks in series model (Equation 8.33).

$$\frac{\partial C}{\partial \theta} = \left(\frac{D}{vL}\right)\frac{\partial^2 C}{\partial X^2} - \frac{\partial C}{\partial X} \tag{8.31}$$

where

$$X = \frac{z}{L}, \text{ and } \theta = \frac{t}{\bar{t}} = \frac{tv}{L} \tag{8.32}$$

$$C(\theta) = \frac{N\,(N\,\theta)^{N-1}}{(N-1)!}\,e^{-N\,\theta} \tag{8.33}$$

The former model is usually employed for flows in tubes and thus should be applicable to the interpretation of RTD in holding tubes. The theoretical RTD for laminar flow of Newtonian fluids can be derived as

$$C(\theta) = E(\theta) = \frac{1}{2\,\theta^3} \tag{8.34}$$

For non-Newtonian fluids that can be described by the power law model, Lin (1979) showed that the RTD is given by the expression

$$E(\theta) = \frac{\mathrm{d}\,F(\theta)}{\mathrm{d}\theta} = \frac{2n}{3n+1}\,\frac{1}{\theta^3}\left(1 - \frac{n+1}{3n+1}\,\frac{1}{\theta}\right)^{(n-1)/(n+1)} \tag{8.35}$$

This expression derived by Lin (1979) appears to be correct and needs to be tested with experimental data. There are a number of models for nonideal flows, that is, flows that fall between the ideal conditions of a perfect mixer and plug flow. Some of the models for nonideal flow were discussed in Levenspiel (1972) and Rao and Loncin (1974a).

Sizing Holding Tube Length

Holding tubes play an important role in thermal processing of fluid foods in high temperature–short time (HTST) or aseptic flow systems. One simple but important relationship for determining the length of a holding tube (L) is based on the residence time of the fastest particle, that is, $L = v_{zm} \times$ holding time. Because the average velocity ($\overline{v}_z$) of a fluid food can be readily calculated in many instances, a relationship between $\overline{v}_z$ and v_{zm} will be useful. Such a relationship applicable for the fully developed laminar flow of fluid foods obeying the power law model was given earlier:

$$\frac{\overline{v}_z}{v_{zm}} = \left(\frac{n+1}{3n+1}\right) \tag{8.3}$$

For the special case of a Newtonian fluid ($n = 1$), $v_{zm} = 2 \times (\overline{v}_z)$; this relationship provides a safe design criterion for Newtonian fluids and shear-thinning fluids in laminar and turbulent flow, but it will not provide a safe design for shear-thickening (dilatant) fluids. Even though there have been few reliable reports on shear-thickening (dilatant) behavior of fluid foods, the absence of such behavior must be confirmed by proper rheological tests in order to avoid underprocessing of fluid foods in holding tubes.

In practical terms, the Food and Drug Administration requires that holding tubes have positive slope of at least 0.25 inch per foot to avoid dead spaces. Also, to conserve plant space, holding tubes are coiled. Therefore, one can expect the flow conditions in holding tubes to be far from those in fully developed laminar flow in straight tubes with the result that the maximum velocities are less than twice the average velocities. The experimental data of Sancho and Rao (1992), discussed later, confirm the deviations from ideal fully developed laminar flow.

Experimental RTD Data

Most of the RTD data on thermal processing flow systems found in the literature were obtained using water as the test fluid. For a plate heat exchanger, Roig et al. (1976) found that the RTD data could be fitted with five tanks-in-series model. Veerkamp et al. (1974) also found that the RTD data in a holding tube could be fitted with the tanks-in-series model.

Heppell (1985) demonstrated that milk has a broader RTD than water in an infusion type ultra high temperature (UHT) sterilizer. Therefore, it is necessary to perform RTD studies on holding tubes and heat exchangers with either fluid foods that are to be processed or with model solutions that possess physical and rheological properties similar to the fluid foods. Sancho and Rao (1992) found that the maximum velocities in a holding tube were less than those estimated from fully developed flow assumption for both Newtonian and non-Newtonian fluids. Magnitudes of the dispersion number ($D/v2r_o$), shown in Figure 8–7 as a function of the Reynolds number for the Newtonian liquids and the generalized Reynolds number for the non-Newtonian liquids, followed three distinct trends depending on the magnitude of Re or GRe (Sancho and Rao, 1992): (1) Over the range 10–100, the dispersion number was nearly constant; (2) over the range 100–2000, the dispersion number increased with increase in Reynolds number; and (3) over the range 2000–10,000, the dispersion number decreased with increase in Reynolds number. Because the rheological behavior of the studied non-Newtonian guar gum solutions was known, interpretation of the data in terms of the generalized Reynolds number and grouping together with the Newtonian fluid data were possible.

Helical Flow

It is known that plug flow conditions are approached in laminar helical flow, and for this reason Rao and Loncin (1974a) suggested that this type of flow could be used advantageously in continuous sterilization of foods. It is interesting to note

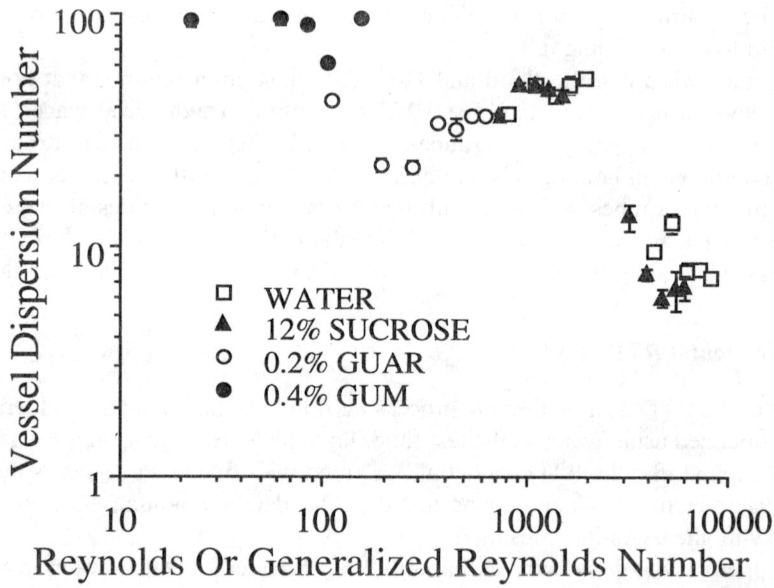

Figure 8–7 Magnitudes of the Dispersion Number ($D/v2r_0$) as a Function of the Reynolds Number for Newtonian Liquids and the Generalized Reynolds Number for Non-Newtonian Fluids. *Source:* Reprinted from *Journal of Food Engineering*, Vol. 15, M.F. Sancho and M.A. Rao, Residence Time Distribution in a Holding Tube, pp. 1–19, Copyright 1992, with permission from Elsevier Science.

that one continuous thermal processing system in which plug flow conditions are claimed to exist is based on helical flow.

HEAT TRANSFER TO FLUID FOODS

Thermorheological Models

To understand or study heat transfer phenomenon, the rheological behavior of a fluid food must be known as a function of both temperature and shear rate. Thermorheological models may be defined as those that have been derived from rheological data obtained as a function of both shear rate and temperature. Such models can be used to calculate the apparent viscosity at different shear rates and temperatures in computer simulation and food engineering applications. For a simple Newtonian fluid, because the viscosity (η) is independent of shear rate, one may consider only the influence of temperature on the viscosity. For many foods, the Arrhenius equation (Equation 2.41) is suitable for describing the effect of temperature on η:

$$\eta = \eta_{\infty A} \exp(E_a/RT) \tag{2.41}$$

where $\eta_{\infty A}$ is the frequency factor, E_a is the activation energy (J mol^{-1}), R is the gas constant (J mol^{-1} K^{-1}), and T is the absolute temperature (K).

As stated previously, for non-Newtonian foods, the simple power law model (Equation 2.3) can be used to describe shear rate ($\dot{\gamma}$) versus shear stress (σ) data at a fixed temperature:

$$\sigma = K\dot{\gamma}^n \tag{2.3}$$

where n is the flow behavior index (dimensionless) and K (Pa s^n) is the consistency index. Two very similar thermorheological equations have been obtained by combining the power law and the Arrhenius equations. The equation obtained by Christiansen and Craig (1962) was

$$\sigma = K_{TC}(\dot{\gamma} \exp\{E_{ac}/RT\})^{\bar{n}} \tag{8.36}$$

An equation commonly encountered in the food engineering literature and that has been used in several studies (Harper and El-Sahrigi, 1965; Rao et al., 1981b; Vitali and Rao, 1984a, 1984b) is

$$\sigma = K_{TH} \exp(E_{aH}/RT)\dot{\gamma}^{\bar{n}} \tag{8.37}$$

In both Equations 8.36 and 8.37, $\bar{n}$ is the average value of the flow behavior index for data at all the studied temperatures. Vitali and Rao (1984a, 1984b) showed that the activation energy terms in Equations 8.36 and 8.37 are related according to

$$E_{aH} = \bar{n}(E_{ac}) \tag{8.38}$$

Thermorheology of Starch Dispersions

In many foods, such as soups, salad dressings, gravies, and sauces, starch is present in excess water conditions. Because of the drastic increase in magnitude of the apparent viscosity (η_a) after the initial temperature of gelatinization is achieved, the rheological behavior of starch in excess water during the transition from fluid-like to viscoelastic behavior affects heat transfer during food thermal processing. Most models of starch gelatinization were developed under isothermal conditions, based on apparent first-order kinetics and the Arrhenius equation to describe the effect of temperature on the gelatinization rate (Kubota et al., 1979; Dolan and Steffe, 1990; Kokini et al., 1992; Okechukwu and Rao, 1995). Because of the temperature history imposed during thermal processing, data obtained under isothermal conditions may not be suitable to describe changes in η_a during gelatinization. In addition to the temperature of the starch sample, shear rate or dynamic frequency has a significant effect on η_a or complex viscosity (η^*), respectively.

Model of Dolan and Colleagues

A comprehensive thermorheological (TR) model taking into consideration the effect of time (t)-temperature (T) history was developed by Dolan et al. (1989). The model for constant shear rate and starch concentration is

$$\eta_{\text{dim}} = \frac{\eta_a - \eta_{ug}}{\eta_\infty - \eta_{ug}} = [1 - \exp(-k\psi)]^\alpha \tag{8.39}$$

where

$$\psi = \int T(t) \exp\left(\frac{-E_g}{RT(t)}\right) dt \tag{8.40}$$

where η_{dim} is dimensionless apparent viscosity defined in Equation 8.39, η_a is apparent viscosity at a specific time during heating, η_{ug} is apparent viscosity of the ungelatinized dispersion, and η_∞ is the highest magnitude of η_a during gelatinization. The model was extended to include the influence of shear rate, temperature, concentration, and strain history (Dolan and Steffe, 1990). Data obtained using back extrusion and mixer viscometers were used to evaluate the models. The activation energy of gelatinization (E_g) depended on the heating temperature (Dolan et al., 1989), and some of the factors affecting viscosity were negligible (Dolan and Steffe, 1990). Applicability of Equation 8.40 to η^* data during gelatinization of a 8% corn starch dispersion (STD) was demonstrated in Chapter 4.

Model of Yang and Rao

Because η^* data are obtained at low strains with minimal alteration of the STD structure, they provide unique opportunities for studying applicable models. Further, empirically obtained frequency shift factor (Ferry, 1980) has been used successfully in time-temperature superposition studies on food polymer dispersions (Lopes da Silva et al., 1994), and the applicability of similar, if not identical, scaling of frequency was explored for STDs (Yang and Rao, 1998a).

For fluid dynamics and heat transfer investigations related to food processing, the necessary η_a data may be obtained from models developed for η^* data using relationships based on the Cox-Merz rule (Rao, 1992; Rao and Cooley, 1992). These results were described in detail elsewhere (Yang, 1997; Yang and Rao, 1998a, 1998b, 1998c) and will be reviewed in brief here.

Experimental TR data were averaged and the results used to fit submodels for simulation as shown in Figure 8–8. The increasing segment of the η^* curve that was sigmoid shaped was fitted to a modified Oswin model (Equation 8.41):

$$\eta^*\left(\frac{\omega}{\omega_r}\right) = 7.4 \times 10^{-6.0}\left(\frac{T}{100 - T}\right)^{6.208}; \quad R^2 = 0.99 \tag{8.41}$$

The segment from 89.5°C to 92.5°C that contained the end of increase in viscosity section, the peak viscosity value, and beginning of the decreasing part of the viscosity curve was well described by a third-order polynomial (Equation 8.42)

$$\eta^*\left(\frac{\omega}{\omega_r}\right) = -69122.86 + 2244.36T - 24.28T^2 + 0.088T^3; \quad R^2 = 0.99 \tag{8.42}$$

It is emphasized that rheological data described by Equations 8.41 and 8.42 of an STD cannot be predicted a priori and must be obtained experimentally. Because viscosities at ≥95°C were severely affected by water evaporation in the rheometer (Yang and Rao, 1998a), for temperatures from 95°C to 121°C (retort temperature) (C-D), the decrease in magnitudes of the η^* with increase in temperature was assumed to follow an Arrhenius equation (Equation 8.43).

$$\eta^*\left(\frac{\omega}{\omega_r}\right) = 4.11 \exp\left[\frac{227.1 \times 10^3}{R}\left(\frac{1}{T} - \frac{1}{366.1}\right)\right]; \qquad R^2 = 0.97 \tag{8.43}$$

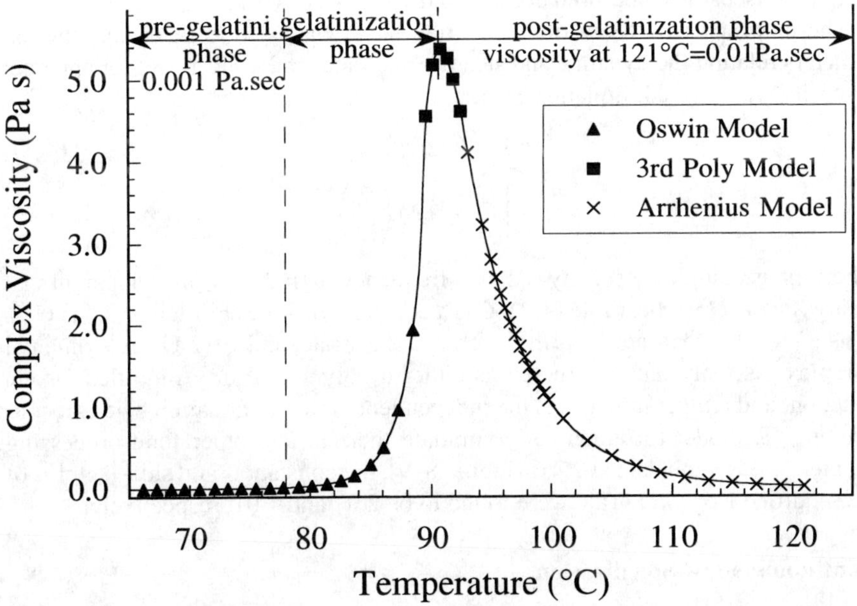

Figure 8–8 Thermorheological Data during Gelatinization of a Corn Starch Dispersion. *Source:* Reprinted with permission from W.H. Yang, *Rheological Behavior and Heat Transfer to a Canned Starch Dispersion: Computer Simulation and Experiment,* PhD Thesis, © 1997, Cornell University.

Because of the availability of reliable TR data from 65°C to 95°C during gelatinization (Equations 8.41 and 8.42), as well as of both the experimental heat penetration data and computer simulation–based temperatures, it was considered safe to assume the temperature profile described by Equation 8.43 from 95°C to 121°C. Comparison of the numerically calculated temperatures with the experimental profiles to be discussed strongly support the assumption of an Arrhenius model (Equation 8.43) for the magnitudes of η_a from 95°C to 121°C.

In Equation 8.43, the reference temperature, 366.1K, was the mean value of temperatures of the decreasing segment of η^*. The value of E_a in Equation 8.43, determined to be 227.1 kJ mol^{-1}, is higher compared to those reported for gelatinized corn starch at lower temperatures. Given the large decrease in viscosity, the value of E_a is reasonable and it is in good agreement with the value of E_g = 240 kJ mol^{-1} in the model of Dolan et al. (1989) for the increasing viscosity segment (Yang and Rao, 1998a). The decrease in viscosity with increasing temperature that occurs just after the peak viscosity has been reached is due to the decrease in starch granule size as a result of granule disruption and with E_g = 240 kJ mol^{-1}. In contrast, for the same STD sample after it was heated up to 95°C and cooled to 25°C, that is, after granule rupture has been completed, the value of E_a determined from the viscositics measured at 35°C and 75°C was 14.5 kJ mol^{-1}.

The modified Cox-Merz rule (Equation 8.44) was used to determine the parameters relating the dynamic and steady shear data, and a TR model for apparent viscosity was derived, Equation 8.45.

$$\eta^*(\omega) = C[\eta(\dot{\gamma})]^{\alpha}\Big|_{\omega=\dot{\gamma}} \tag{8.44}$$

$$\eta_a = \left[C'\ \eta^*(T)\left(\frac{\dot{\gamma}_r}{\dot{\gamma}}\right)\right]^{\alpha'} \tag{8.45}$$

where η^* is complex viscosity (Pa s) at frequency ω (rad s^{-1}), η_a is apparent viscosity (Pa s) at ($\dot{\gamma}$) shear rate (s^{-1}), C is a constant, α is the shift factor, C' is C^{-1}, α' is α^{-1}, $\dot{\gamma}$ is shear rate, $\dot{\gamma}_r$ is the reference shear rate, and $\eta^*(T)$ is experimental complex viscosity data expressed as either a polynomial or a modified Oswin function, and temperature (T) is the independent variable. Equation 8.45 is a convenient TR model to numerically simulate thermal and other food processing problems. Based on data at 25°C (Figure 8–9), the constant C and shift factor α of the modified Cox-Merz rule were found to be 2.07 and 1.01, respectively.

Continuous Flow Sterilization

The earlier works on heat transfer applicable to continuous flow sterilization (Graetz, 1883; Lyche and Bird, 1956; Wissler and Schechter, 1959), where a fluid was heated or cooled at constant wall temperature along the length of the heat exchanger, have been based on the assumption of temperature-independent physical properties. The fluid gains or loses heat along the length of the heat exchanger, and

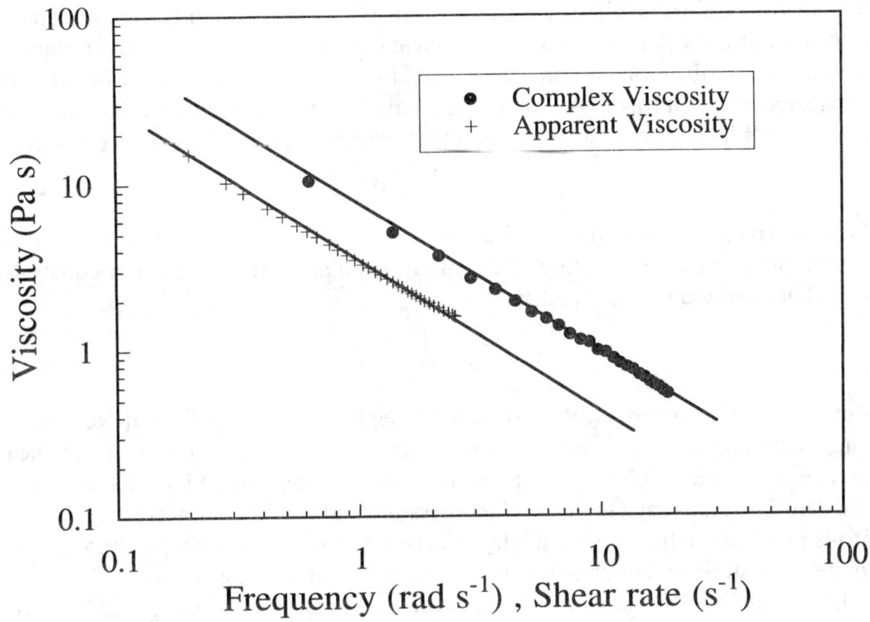

Figure 8–9 Modified Cox-Merz Rule for a Gelatinized Starch Dispersion. *Source*: Reprinted with permission from W.H. Yang, *Rheological Behavior and Heat Transfer to a Canned Starch Dispersion: Computer Simulation and Experiment*, PhD Thesis, © 1997, Cornell University.

therefore different parts of the fluid receive different thermal treatments. This causes a variation in temperature in the radial and axial directions and significant changes in the physical properties of the fluids, especially the rheological properties. The assumption of temperature-independent rheology appears to be of relatively minimal use for practical applications. Two expressions in terms of thermorheology have been introduced (Kwant et al., 1973): isoviscous and nonisoviscous flows. The isoviscous flow is defined as flow of a fluid with a temperature-independent rheological behavior, while the nonisoviscous flow is defined as flow of a fluid with temperature-dependent rheological behavior.

In isoviscous flow, the physical properties are independent of temperature, and therefore the velocity fields remain constant throughout the entire heat transfer section of the heat exchanger. A substantial improvement upon this simplification was first made by Christiansen and Craig (1962), who obtained more realistic velocity profiles by means of a modified power law equation to describe the flow properties, which incorporated the effect of temperature upon the rheological properties of the fluid being heated (Skelland, 1967). Simpson and Williams (1974) pointed out that the most important effect of the nonisothermal nature of the process is that radial temperature gradients will cause radial viscosity gradients, leading to considerable distortion of the velocity profiles. They have developed methods for the design of continuous sterilization equipment for pseudoplastic fluids with and without temperature-dependent rheology and recommended a total dimensionless length (ζ) of 1.2 with 0.8 for the heating section and the remaining 0.4 for the cooling section:

$$\zeta = (z\alpha/r_o^2\,\bar{v}_z) \tag{8.46}$$

This dimensionless length is valid within $\pm 2\%$ for $0.3 \leq n \leq 1.0$ and $-4 < \psi < 4$, and the strength of coupling of the momentum and energy transport equations was characterized by the equation

$$\psi = \frac{E_a}{R}\left(\frac{1}{T_0} - \frac{1}{T_w}\right) \tag{8.47}$$

Kumar and Bhattacharya (1991) simulated aseptic processing of a non-Newtonian fluid with temperature-dependent and shear-thinning viscosity in a tubular heat exchanger. Their results were similar to those of Simpson and Williams (1974) and demonstrated that a lower wall temperature and smaller diameter tube can improve product quality as well as the effective contribution of the heating section to the fluid sterility, which has to be accounted for in process design.

During thermal processing, the viscosity of a starch-containing fluid can increase drastically with temperature because of the amylose released and the starch granule swelling after the initial temperature of gelatinization has been reached. After the dispersion has attained its maximum viscosity, the granules start to rupture, so the viscosity decreases with temperature (Eliasson, 1986; Okechukwu and Rao, 1995). Therefore, as stated earlier, the viscosity model obtained by combining the power law model with the Arrhenius equation fails to describe the rheological behavior of starch-containing fluids during the course of gelatinization. Therefore, Liao (1998) recently examined the influence of thermorheology on the temperature and velocity profiles in a single straight tubular heat exchanger in which the starch-containing fluid food flowed.

Nauman (1977) has proposed that when fluids are heated in a continuous system, there is, in addition to a residence time distribution, a distribution of thermal time because of differences in heat treatment received by different fluid elements at different spatial locations within the reactor. The concept of thermal time distribution (TTD) has been studied since by several authors (Paulsson and Trägårdh, 1984; Trägårdh and Paulsson, 1985; Datta, 1989; Datta and Liu, 1992; Elborg, 1989; Bhamidipati and Singh, 1994) and is a way to provide a comprehensive description of time-temperature history in the nonisothermal process systems. Thermal time distribution can be defined as the distribution of F_0 values. In a manner similar to RTD, TTD was obtained as

$$E(F_0) = \frac{(\mathrm{d}Q/Q)}{\mathrm{d}F_0} \tag{8.48}$$

where dQ/Q is the fraction of fluid having lethality value between F_0 and dF_0. The integrated curve for the F_0 distributions was obtained as

$$F(F_0) = \int_0^{F_0} E(F_0)\, dF_0 \tag{8.49}$$

Figure 8–10 illustrates the analogy between thermal time distribution and resident time distribution (Liao, 1998).

Paulsson and Trägårdh (1984) simulated TTD in a tubular heat exchanger with turbulent flow using a stochastic combined flow reactor model. Datta (1989) has applied the concept of TTD to processing of a fluid with constant thermophysical properties in a can. Datta and Liu (1992) studied TTD in microwave and conventional heating of solids and liquids. Elborg (1989) has experimentally determined the TTD in a turbulent flow reactor utilizing the polymerization characteristics of dextran. Bhamidipati and Singh (1994) employed the concept of TTD to simulation of heat transfer to a pseudoplastic fluid in a tubular sterilizer. The axial and radial variation in velocities and temperature-dependent physical properties were considered in their study.

Role of Rheology in Broken Heating Phenomena

In thermal processing of canned foods, the formula methods provide a good approximation of the lethality when the heat penetration data can be described by a single straight heating curve (Equation 8.50). The parameters of Equation 8.50, particularly f_h, permit a formula method to evaluate processes under conditions different from those of experimental data.

$$\log \frac{RT - T}{j(RT - IT)} = -\frac{t}{f_{\mathrm{h}}} \tag{8.50}$$

The parameter j is a measure of the lag to achieve a uniform heating rate and is associated with the position of the cold spot or slowest heating point, the can size, and the IT (Ball and Olson, 1957); basically, these three factors determine the time to achieve a uniform heating rate. Although f_{h} is usually defined as the time required to traverse one logarithmic cycle on the temperature scale, the physical meaning of f_{h} is more complex and is associated with the mode of heat transfer. Ball and Olson (1957) derived analytical solutions for f_{h} in both ideal thermal convection (Equation 8.51) and conduction (Equation 8.52), respectively:

$$f_{\mathrm{h}} = 2.303\,\frac{c_{\mathrm{p}}\rho V}{UA} \tag{8.51}$$

$$f_{\mathrm{h}} = \frac{2.303}{\left[\dfrac{(2.405)^2}{r_0^2} + \dfrac{\pi^2}{L^2}\right]\alpha} \tag{8.52}$$

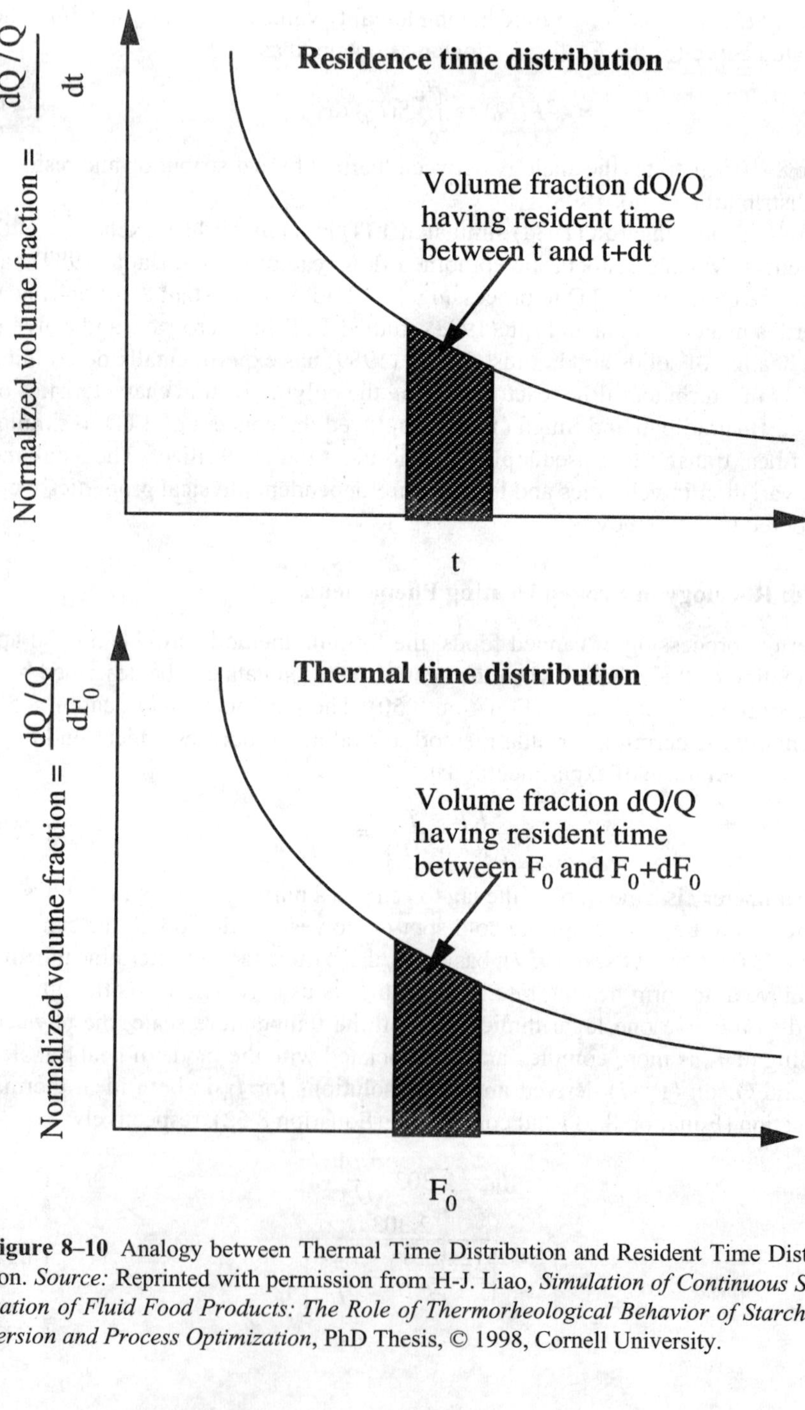

Figure 8–10 Analogy between Thermal Time Distribution and Resident Time Distribution. *Source:* Reprinted with permission from H-J. Liao, *Simulation of Continuous Sterilization of Fluid Food Products: The Role of Thermorheological Behavior of Starch Dispersion and Process Optimization*, PhD Thesis, © 1998, Cornell University.

Generally, Equation 8.50 is applicable to foods where the heat transfer mode is well defined, such as clarified fruit juices and thin soups that heat by convection mode, and corned beef and chicken loaf that heat by conduction mode. Broken heating (BH) heat penetration data usually can be fitted by two straight lines with different slopes on a semilog plot; in some cases, the heating curve is represented by more than two straight lines. Since there is a significant change in the heating rate, one can interpret that the break is the result of an increased resistance to heat transfer.

Ball and Olson (1957) associated viscosity data of corn starch dispersions with broken heating heat penetration data, but a direct correlation with starch gelatinization was not established because the data were obtained at a constant 60°C (140°F) with a Scott orifice viscometer. Stevens (1972) investigated broken heating phenomena, but due to limited computer capabilities and the lack of suitable thermorheological data, complete numerical simulation was not conducted.

Ball (1923, 1927, 1928) studied BH curves and proposed Equations 8.53 and 8.54 as extensions of the single heating curve method.

$$\log(g_{\mathrm{bh}}) = \log(jI) - \frac{x_{\mathrm{bh}}}{f_{\mathrm{h}}} \tag{8.53}$$

$$\log(g_{\mathrm{h2}}) = \frac{f_{\mathrm{h}} \log(jI) + (f_2 - f_{\mathrm{h}}) \log(g_{\mathrm{bh}}) - B_{\mathrm{b}}}{f_2} \tag{8.54}$$

Although Equations 8.53 and 8.54 are known to be not very precise (Ball and Olson, 1957; Stumbo, 1973), they are still used in the food industry. The break point (x_{bh}) is determined as the intersection of the two curves.

Transport Equations

Yang (1997) solved the governing partial differential equations for heat transfer to an STD in a 303 × 406 can with $R = 4.2$ cm and L = 10.7 cm, and no head space:

The energy equation:

$$\rho c_{\mathrm{p}}\left(\frac{\partial T}{\partial t} + v_r \frac{\partial T}{\partial r} + v_z \frac{\partial T}{\partial z}\right) = \frac{1}{r}\frac{\partial}{\partial r}\left(kr\frac{\partial T}{\partial r}\right) + \frac{\partial}{\partial z}\left(k\frac{\partial T}{\partial z}\right) \tag{8.55}$$

Because the STD exhibits viscoelastic behavior only after the initial gelatinization temperature is reached, the problem cannot be approximated to simple heat transfer by conduction. To simplify the buoyancy force caused by density variation with temperature, Boussinesq approximation (Equation 8.56) was used in the body force term of the momentum (Equation 8.57) in the vertical direction:

$$\rho_{\mathrm{o}} - \rho = \rho\beta(T - T_{\mathrm{o}}) \tag{8.56}$$

$$\rho\left(\frac{\partial v_z}{\partial t}+v_z\frac{\partial v_z}{\partial z}+v_r\frac{\partial v_z}{\partial z}\right)=-\frac{\partial \mathrm{p}}{\partial z}+\frac{1}{r}\frac{\partial}{\partial r}\left[r\,\eta\left(\frac{\partial v_z}{\partial r}+\frac{\partial v_r}{\partial z}\right)\right]$$
$$+2\frac{\partial}{\partial z}\left(\eta\frac{\partial v_z}{\partial z}\right)+\rho g[1-\beta(T-T_r)] \quad (8.57)$$

The momentum equation in the radial direction is

$$\rho\left(\frac{\partial v_r}{\partial t}+v_z\frac{\partial v_r}{\partial z}+v_r\frac{\partial v_r}{\partial r}\right)=-\frac{\partial \mathrm{p}}{\partial z}+\frac{\partial}{\partial z}\left[\eta\left(\frac{\partial v_z}{\partial r}+\frac{\partial v_r}{\partial z}\right)\right]$$
$$+\frac{2}{r}\frac{\partial}{\partial r}\left(r\,\eta\frac{\partial v_r}{\partial r}\right)-2\eta\frac{v_r}{r^2} \quad (8.58)$$

The continuity equation is:

$$\frac{1}{r}\frac{\partial(r\,\rho\,v_r)}{\partial r}+\frac{\partial(\rho\,v_z)}{\partial z}=0 \quad (8.59)$$

The boundary conditions for solving the equations were (Figure 8–11) (1) wall at $r = r_o$, $0 \le z \le L$, $T = RT$, $v_z = 0$, and $v_r = 0$; (2) top wall at $z = L$, $0 \le r \le r_o$, $T = RT$, $v_z = 0$, and $v_r = 0$; (3) bottom wall at $z = 0$, $0 \le r \le r_o$, $T = RT$, $v_z = 0$, and $v_r = 0$; and (4) axisymmetric line at $r = 0$, $0 \le z \le L$, $(\partial T/\partial r) = 0$, $(\partial v_z/\partial r) = 0$, and $v_r = 0$. The initial conditions were $0 \le r \le r_o$, $0 \le z \le L$, $T = IT$, $v_z = 0$, and $v_r = 0$. Since the magnitude of η_a during gelatinization could be from 10 to 2000 times higher than that of ungelatinized STD, the changes in the other fluid properties with the temperature were neglected and were assumed to be equal to those of water: $\rho = 1000$ kg m^{-3}, $c_p = 4.18$ kJ kg^{-1} K^{-1}, $k = 0.66$ J s^{-1} m^{-1} K^{-1}, and $\beta = 0.0005$ K^{-1} at 70°C.

The governing equations with boundary and initial conditions described above were solved with FIDAP fluid dynamics analysis program (Version 7.5, Fluid Dynamics International, Inc., Evanston, IL) on a Gateway 2000 computer platform (P5-100, Intel Pentium, 32 MB RAM, North Sioux City, SD). FIDAP is based on the finite element method for simulating fluid flow, and heat and mass transfer problems (Engelman and Sani, 1983; Dhatt and Touzot, 1984; Engelman, 1993). In the finite element method, the continuum region of study is divided into a number of shaped elements, and the method of weighted residuals with Galerkin's criteria is used to reduce the errors resulting from the approximation of the velocity, pressure, and temperature fields within each element. The penalty approach (Equation 8.60) was used to replace the continuity equation, eliminating the unknown pressure and thus simplifying solution of the problem.

$$\frac{1}{r}\frac{\partial(r\,\rho\,v_r)}{\partial r}+\frac{\partial(\rho v_z)}{\partial z}=-\frac{1}{\lambda}p \quad (8.60)$$

where λ is the penalty parameter with a magnitude range of 10^5–10^8. Because λ is

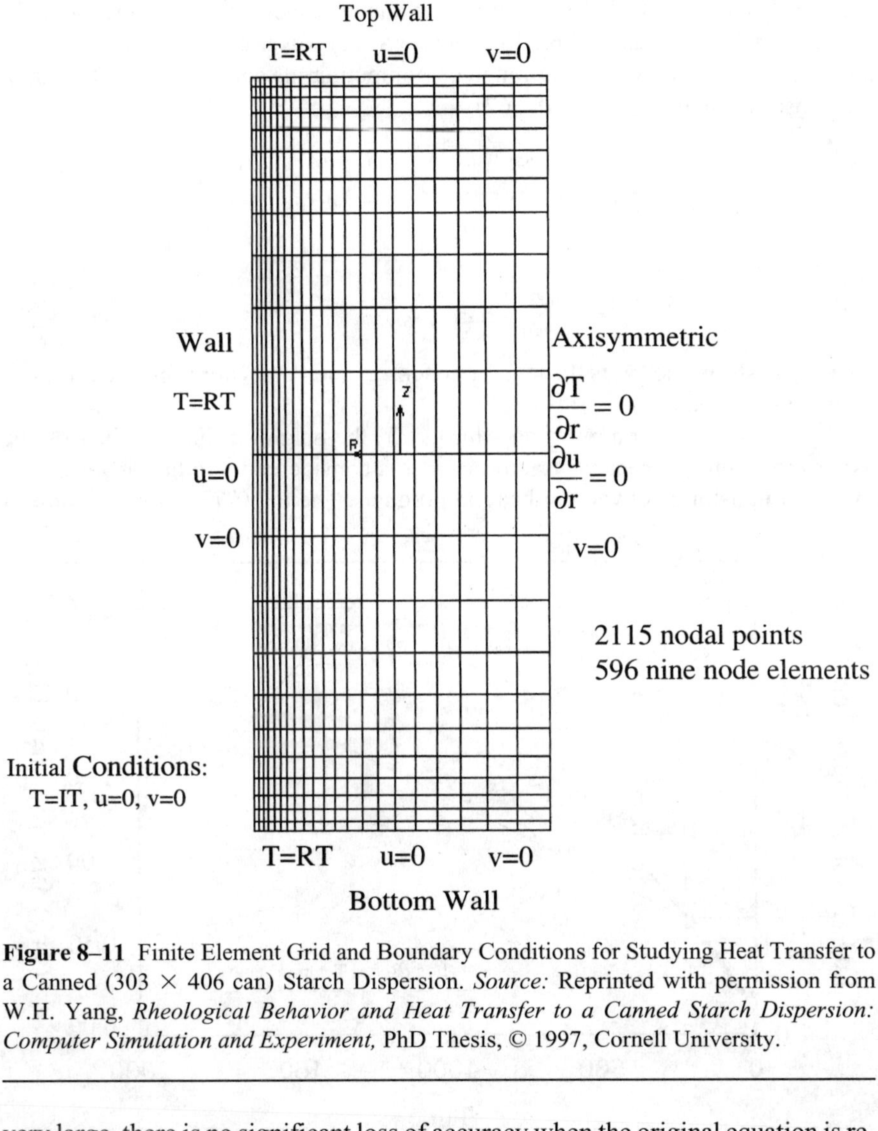

Figure 8–11 Finite Element Grid and Boundary Conditions for Studying Heat Transfer to a Canned (303 × 406 can) Starch Dispersion. *Source:* Reprinted with permission from W.H. Yang, *Rheological Behavior and Heat Transfer to a Canned Starch Dispersion: Computer Simulation and Experiment,* PhD Thesis, © 1997, Cornell University.

very large, there is no significant loss of accuracy when the original equation is replaced by the Equation 8.60. An implicit time integration (backward Euler) with variable time increment was used to study the transient behavior. At each time step, the nonlinear system was solved by the quasi-Newton procedure. This method was chosen over the other Newton-type algorithms because the form of Jacobian matrix that is recomputed and updated reduces considerably the computer time (Engelman and Sani, 1983).

A mesh made up of 2115 nine-node isoparametric elements, 45 in the radial direction and 47 in the axial direction (Figure 8–11) was used (Yang, 1997). Magnitudes of shear rate were determined by computing the components of the strain rate tensor (Equations 8.61 and 8.62).

$$\dot{\gamma}_{ij} = \begin{pmatrix} \dfrac{\partial v_z}{\partial x} & \dfrac{1}{2}\left(\dfrac{\partial v_z}{\partial y} + \dfrac{\partial v_r}{\partial x}\right) \\ \dfrac{1}{2}\left(\dfrac{\partial v_r}{\partial x} + \dfrac{\partial v_z}{\partial y}\right) & \dfrac{\partial v_r}{\partial y} \end{pmatrix} \tag{8.61}$$

$$\dot{\gamma} = (\dot{\gamma}_{ij}\dot{\gamma}_{ij})^{1/2} \tag{8.62}$$

where $\dot{\gamma}$ is shear rate, $\dot{\gamma}_{ij}$ is the strain rate tensor, and i and j are x and y coordinate directions, respectively.

The temperatures at a point one-third (1/3) the height of the container on the axis were examined because traditionally it is considered to be the slowest heating point in natural convection-heating products. The log ($RT-T$) versus time at

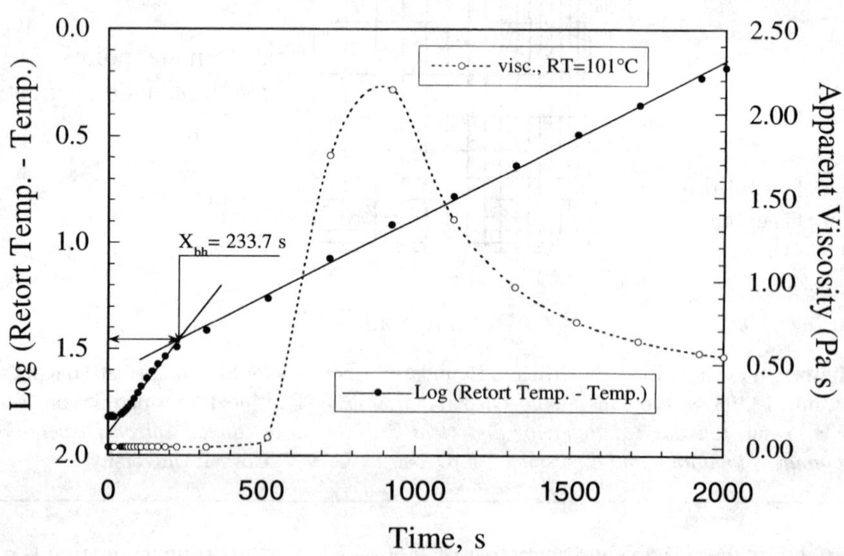

Figure 8–12 Broken Heating Curve from Simulation of Heat Transfer to a Canned Starch Dispersion at *RT* 101°C. The log ($RT-T$) versus time at the assumed slowest heating point and the corresponding fluid apparent viscosity (η_a) versus time profiles are shown. *Source:* Reprinted with permission from W.H. Yang, *Rheological Behavior and Heat Transfer to a Canned Starch Dispersion: Computer Simulation and Experiment,* PhD Thesis, © 1997, Cornell University.

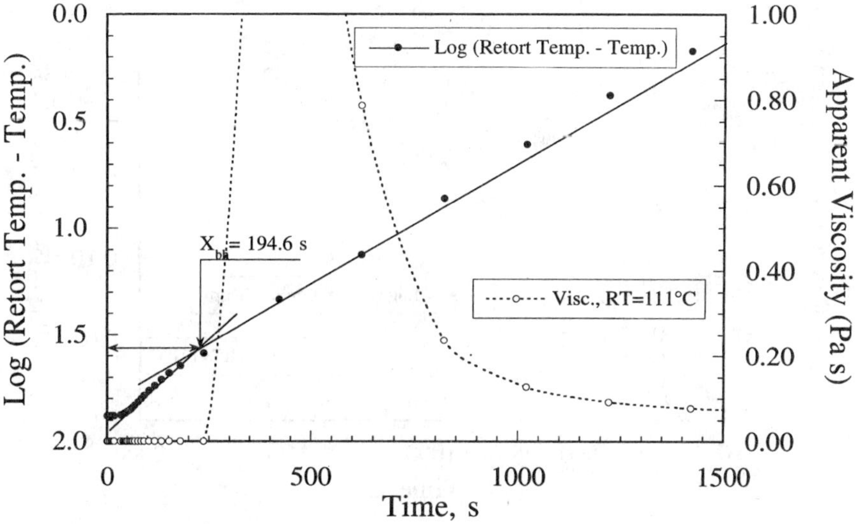

Figure 8–13 Broken Heating Curve from Simulation of Heat Transfer to a Canned Starch Dispersion at *RT* 111°C. The log $(RT-T)$ versus time at the assumed slowest heating point and the corresponding fluid apparent viscosity (η_a) versus time profiles are shown. *Source:* Reprinted with permission from W.H. Yang, *Rheological Behavior and Heat Transfer to a Canned Starch Dispersion: Computer Simulation and Experiment,* PhD Thesis, © 1997, Cornell University.

the assumed slowest heating point and the corresponding η_a versus time profiles are shown in Figures 8–12 through 8–15 for different heating conditions.

Estimated Heat Penetration Parameters

The following salient features of the calculated heat penetration parameters were noted:

1. The j values did not change significantly because of the short lag phase and use of the corrected zero time.
2. The f_h values were nearly the same except for the simulations at *RT* 131°C and can size 211 × 400.
3. The x_{bh} values could be separated into two distinct groups: one for simulations at *RT*s 101°C and 111°C, and the other for simulations at *RT*s 121°C and 131°C.
4. The f_2 values decreased as the *RT* increased. A considerable decrease in f_2 values occurred when the *RT* was increased from 101°C to 111°C.
5. The temperatures at which x_{bh} occurred were approximately 70°C for simulations with *RT*s at 101°C and 111°C, and close to the *RT* for simulations with *RT*s at 121°C and 131°C.

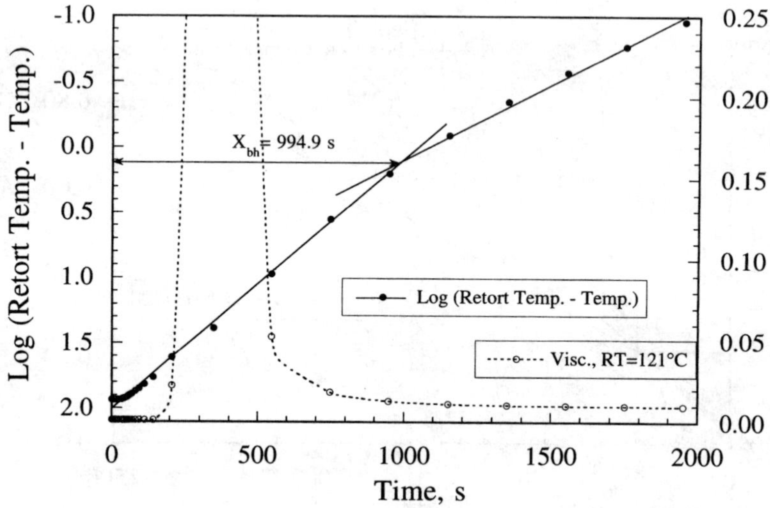

Figure 8–14 Broken Heating Curve from Simulation of Heat Transfer to a Canned Starch Dispersion at *RT* 121°C. The log (*RT*−*T*) versus time at the assumed slowest heating point and the corresponding fluid apparent viscosity (η_a) versus time profiles are shown. *Source:* Reprinted with permission from W.H. Yang, *Rheological Behavior and Heat Transfer to a Canned Starch Dispersion: Computer Simulation and Experiment,* PhD Thesis, © 1997, Cornell University.

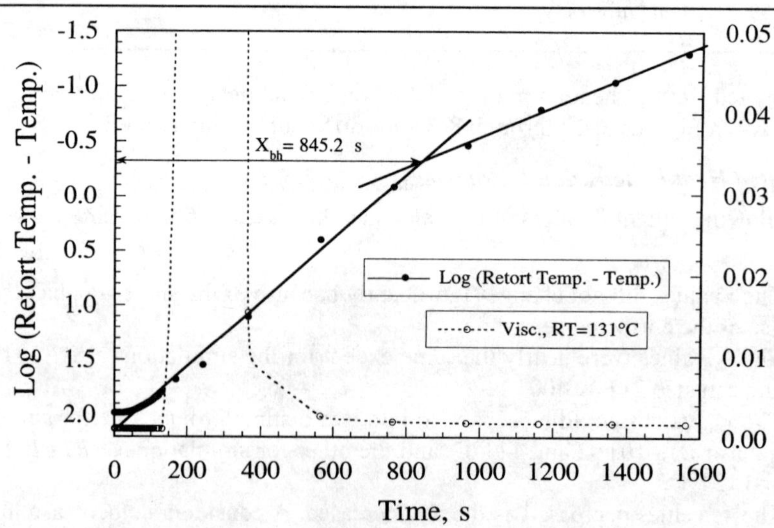

Figure 8–15 Broken Heating Curve from Simulation of Heat Transfer to a Canned Starch Dispersion at *RT* 131°C. The log (*RT*−*T*) versus time at the assumed slowest heating point and the corresponding fluid apparent viscosity (η_a) versus time profiles are shown. *Source:* Reprinted with permission from W.H. Yang, *Rheological Behavior and Heat Transfer to a Canned Starch Dispersion: Computer Simulation and Experiment,* PhD Thesis, © 1997, Cornell University.

The two process conditions: RT = 101°C and 111°C, IT = 35°C that had the shortest x_{bh} values, 3.9 and 3.2 minutes, respectively, also were characterized by the longest heating times during which the magnitudes of η_a were greater than 0.55 Pa s and also the highest magnitudes of η_{final}, 0.55 Pa s and 0.06 Pa s, respectively; in addition, the temperature at which the x_{bh} occurred was close to the initial temperature of gelatinization. The significant difference between the f_2 value (f_2 = 22.7 min) at RT = 101°C and at RT = 111°C (14.8 min) was due to the large differences in the magnitudes of η_a over the entire heating process after initiation of gelatinization (Figures 8–12 and 8–13). At RTs 121°C and 131°C, the x_{bh}s were approximately four times higher than those at RTs 101°C and 111°C, and the temperatures at which the x_{bh}s occurred were close to the corresponding retort temperatures. The x_{bh}s for these conditions were mostly due to the significant decrease in the heat transfer driving force (i.e., $RT-T$).

Empirical Correlations for Heat Transfer to Fluids Flowing in Tubes

Empirical correlations are based on dimensionless groups, so they are applicable for a wide range of physical and thermal properties and process conditions. In turn, the dimensionless groups are obtained from dimensional analysis. Dimensional analysis is based on the assumption that the only fundamental units are those of mass, length, time, temperature, and heat, and that the quantity under consideration is a function of several known variables. As an example, the heat transfer coefficient in convection heat transfer in containers can be expected to depend on the characteristic dimension of the container, the physical and thermal properties of the fluid food, and rotational speed of can agitation. The pertinent dimensionless groups for an application can be determined by the application of Buckingham's Pi theorem, which states that the relationship among q quantities whose units may be expressed in terms of u fundamental units may be written as a function of $q-u$ dimensionless groups (Bird et al., 1960).

There are many well-known dimensionless groups that are used in transport phenomena. Earlier, the Reynolds number was used to correlate data on pressure drop in pipe flow. For correlating data on heat transfer, often the dimensionless groups Nusselt (Nu), Reynolds (Re), Prandtl (Pr), and Grashof (Gr) are used. They are defined as follows:

$$\mathrm{Nu} = \frac{h_c D}{k} \tag{8.63}$$

$$\mathrm{Pr} = \frac{c_p \eta}{k} \tag{8.64}$$

$$\mathrm{Re} = \frac{D v \rho}{\eta} \tag{8.65}$$

$$\mathrm{Gr} = \frac{gL^3\rho^2\beta\Delta T}{\eta^2} \tag{8.66}$$

For non-Newtonian fluids, the viscosity in the above equations must be replaced by an appropriate apparent viscosity. In pipe-flow applications, as shown in Chapter 3, the pipe-flow apparent viscosity (η_{ap}) based on the diameter (D) and the average axial velocity in the tube ($\overline{v}_z$) can be used in deriving the appropriate dimensionless numbers:

$$\eta_{\mathrm{ap}} = K'\left(\frac{8\,\overline{v}_z}{D}\right)^{n'-1} \tag{8.67}$$

For example, one can obtain the generalized Reynolds number (GRe) for pipe flow:

$$\mathrm{G\,Re} = \frac{D^n\,\overline{v}_z^{2-n}\,\rho}{8^{(n-1)}K}\left(\frac{4n}{3n+1}\right)^n \tag{8.68}$$

In addition to the dimensionless groups mentioned above, the ratio of the viscosity at the wall, η_{w}, to that of the bulk fluid, η_{b}, is used to account for significant temperature gradients between the wall and the bulk fluid. It is helpful to think of dimensionless groups as ratios of forces or effects in a system (Bird et al., 1960; Foust et al., 1980). For example, the Reynolds number can be written as the ratio of inertial forces to viscous forces; thus, a low value of the Reynolds number means that the viscous forces are large in comparison to the inertial forces. The Prandtl number, which is used extensively in studies on convective heat transfer, is the ratio of molecular diffusivity of momentum to molecular diffusivity of heat.

The Nusselt number contains the heat transfer coefficient, h_{c}(W m^{-2} K), defined as:

$$q = h_{\mathrm{c}}A(\Delta T) \tag{8.69}$$

where q is the rate of heat transfer (W), A is the area (m^2) of heat transfer, and ΔT is the temperature difference (K) and is the driving force to heat transfer. Equation 8.69 is often referred to as Newton's law of cooling.

There have been several empirical heat transfer relationships for laminar flow in tubes proposed that attempt to allow for variation in rheological properties. Empirical equations developed by several authors for the fluids with different rheological behaviors have been well reviewed by Skelland (1967). Here, a few equations containing dimensionless groups are given to illustrate such relationships.

Newtonian Fluid

An empirical expression, credited to Sieder and Tate by Skelland (1967), for heat transfer to highly viscous Newtonian fluids in laminar flow through tubes has found extensive application:

$$\frac{h_m D}{k} = 2.0\left(\frac{wC_p}{kL}\right)^{1/3}\left(\frac{\eta_b}{\eta_w}\right)^{0.14} \tag{8.70}$$

where η_b is the viscosity at the bulk temperature and η_w is the viscosity at the wall.

Power Law Fluid

The equation proposed by Charm and Merrill (Charm, 1971) for heat transfer to power law fluid is

$$\frac{h_m D}{k} = 2.0\left(\frac{wC_p}{kL}\right)^{1/3}\left[\frac{K_b\,(3n+1)}{K_w\,2(3n+1)}\right]^{0.14} \tag{8.71}$$

where h_m is the mean heat transfer coefficient (W m^{-2} K), K_b is the consistency index (Pa s^n) at the bulk average temperature, and K_w is the consistency index (Pa s^n) at the wall temperature.

Empirical Correlations for Heat Transfer to Canned Fluids

Rao and Anantheswaran (1988) reviewed studies on convective heat transfer to canned fluids in detail. Here, the dimensionless groups and relationships applicable to both Newtonian and non-Newtonian fluids are reviewed in brief.

The rotational Reynolds number ($\mathrm{Re_a}$) is defined as

$$\mathrm{Re_a} = \frac{D^2\,N\,\rho}{\eta} \tag{8.72}$$

where D is the characteristic dimension, such as the diameter of the can, and N is the rotational speed. For a non-Newtonian fluid being heated in a can, Anantheswaran and Rao (1985b) suggested that the pipe apparent viscosity can be used in the various dimensionless groups. The expressions for the generalized Reynolds ($\mathrm{GRe_a}$), Prandtl (GPr), and Grashof (GGr) numbers are

$$\mathrm{G\,Re_a} = \frac{D^2\,N^{2-n}\,\rho}{8^{(n-1)}\,K}\left(\frac{4n}{3n+1}\right)^n \tag{8.73}$$

$$\mathrm{G\,Pr} = \frac{c_p\,K 8^{n-1}}{k\,N^{1-n}}\left(\frac{3n+1}{4n}\right)^n \tag{8.74}$$

$$\mathrm{G\,Gr} = \frac{gL^3\,\beta\,\Delta T\,\rho^2\,N^{2-2n}}{K^2\,8^{2n-2}}\left(\frac{4n}{3n+1}\right)^{2n} \tag{8.75}$$

Anantheswaran and Rao (1985a, 1985b) developed the following respective correlations for heat transfer to cans under end-over-end agitation for Newtonian and non-Newtonian fluids:

$$\mathrm{Nu} = 2.9\,\mathrm{Re_a^{0.436}}\,\mathrm{Pr^{0.287}} \tag{8.76}$$

$$\mathrm{Nu} = 1.3\ (\mathrm{GRe_a})^{0.485}\ (\mathrm{GPr})^{0.367} \tag{8.77}$$

The correlations were for data obtained with model fluids in copper cans with headspace volume in the range 3–9%, and the rate of heat transfer was independent of the radius of rotation in the range 0–14.9 cm.

Heat transfer to canned liquids being heated in a Steritort was examined by Rao et al. (1985). Both Newtonian (water, and 30%, 40%, 50%, and 60% w/v aqueous sucrose solutions) and non-Newtonian fluids (0.3%, 0.4%, 0.5%, and 0.75% aqueous guar gum solutions) were used. The heat transfer coefficient for both Newtonian and non-Newtonian fluids was correlated by an equation that was the sum of the contributions due to natural and forced convections:

$$\mathrm{Nu} = A[(\mathrm{Gr})\ (\mathrm{Pr})]^{B} + C[(\mathrm{Re_a})\ (\mathrm{Pr})\ (D/L)]^{D} \tag{8.78}$$

The contribution of natural convection heat transfer was significant with the Newtonian fluids. The coefficients A, B, C, and D for Newtonian fluids were 0.135, 0.323, 3.91×10^{-3}, and 1.369, respectively. For the non-Newtonian dispersions, the generalized dimensionless numbers $\mathrm{GRe_a}$, GPr, and GGr were used in place of $\mathrm{Re_a}$, Pr, and Gr, respectively. The magnitudes of the coefficients A, B, C, and D for the non-Newtonian guar gum dispersions were 2.319, 0.218, 4.10×10^{-7}, and 1.836, respectively. With respect to using dispersions of biopolymers in heat transfer experiments, it should be noted that when subjected to extensive heat treatment, most dispersions suffer loss in viscosity. The kinetics of decrease in viscosity of guar gum and sodium carboxymethylcellulose dispersions during heating were studied by Rao et al. (1981a), and additional studies are needed for selecting biopolymers that can be used in heat transfer studies.

REFERENCES

Anantheswaran, R.C., and M.A. Rao. 1985a. Heat transfer to model Newtonian liquid foods in cans during end-over-end agitation. *J Food Eng* 4: 1–19.

Anantheswaran, R.C., and M.A. Rao. 1985b. Heat transfer to non-Newtonian liquid foods in cans during end-over-end rotation. *J Food Eng* 4: 21–35.

Ball, C.O. 1923. Thermal process time for canned food. *Bulletin 37.* Washington, DC: National Research Council.

Ball, C.O. 1927. Theory and practice in processing. *Canner* 64: 27–32.

Ball, C.O. 1928. Mathematical solution of problems on thermal processing of canned food. *Univ Calif (Berkeley) Publ Public Health 1,* no. 2: 15–245.

Ball, C.O., and F.C.W. Olson. 1957. *Sterilization in food technology*. 1st ed. New York: McGraw Hill.

Bhamidipati, S., and R.K. Singh. 1994. Thermal time distribution in tubular heat exchangers during aseptic processing of fluid foods. *Biotechnol Prog* 10: 230–236.

Bird, R.B., W.E. Stewart, and E.N. Lightfoot. 1960. *Transport phenomena*. New York: John Wiley & Sons.

Brodkey, R.S. 1967. *The phenomena of fluid motions*. Reading, MA: Addison-Wesley.

Charm, S.E. 1971. *Fundamentals of food engineering*. 2d ed. Westport, CT: AVI Publishing Co.

Christiansen, E.B., and S.E. Craig. 1962. Heat transfer to pseudoplastic fluids in laminar flow. *AIChE J* 8: 154–160.

Danckwerts, P.V. 1953. Continuous flow systems. *Chem Eng Sci* 2: 1–13.

Datta, A.K. 1989. Integrated thermokinetic modeling of processed liquid food quality. In *Engineering and food. Vol 1. Physical Properties and Process Control,* eds. W.E.L. Spiess and H. Schubert, 838–847. New York: Elsevier Applied Science.

Datta, A.K., and J. Liu. 1992. Thermal time distributions for microwave and conventional heating of food. *Trans IChemE 70,* Part C: 83–90.

Dhatt, G., and G. Touzot. 1984. *The finite element method displayed.* 1st ed. New York: John Wiley & Sons.

Dodge, D.W., and A.B. Metzner. 1959. Turbulent flow of non-Newtonian systems. *AIChE J* 5: 189–204.

Dolan, K.D., and J.F. Steffe. 1990. Modeling rheological behavior of gelatinizing starch solutions using mixer viscometry data. *J Texture Studies* 21: 265–294.

Dolan, K.D., J.F. Steffe, and R.G. Morgan. 1989. Back extrusion and simulation of viscosity development during starch gelatinization. *J Food Process Eng* 11: 79–101.

Elborg, A. 1989. Thermal time distribution in a tube heat exchanger. PhD Diss., University of Lund, Sweden.

Eliasson, A.C. 1986. Viscoelastic behavior during the gelatinization of starch. *J Texture Studies* 17: 253–265.

Engelman, M.S. 1993. *FIDAP*. Evanston, IL: Fluid Dynamics International, Inc.

Engelman, M.S., and R.L. Sani. 1983. Finite-element simulation of an in-package pasteurization process. *Numerical Heat Transfer* 6: 41–54.

Ferry, J.D. 1980. *Viscoelastic properties of polymers.* 3d ed. New York: John Wiley & Sons.

Foust, A.S., L.A. Wenxel, C.W. Clump, L. Maus, and B. Andersen. 1980. *Principles of unit operations.* 2d ed. New York: John Wiley & Sons.

Garcia, E.J., and J.F. Steffe. 1987. Comparison of friction factor equations for non-Newtonian fluids in pipe flow. *J Food Process Eng* 9: 93–120.

Graetz, L. 1883. On the heat transfer in liquids. *Annal Physik* 18: 79.

Hanks, R.W. 1978. Low Reynolds number turbulent pipeline flow of pseudohomogeneous slurries. In *Proceedings of the Fifth International Conference on the Hydraulic Transport of Solids in Pipes Hydrotransport,* C2-23 to C2-34, Hanover, West Germany.

Hanks, R.W., and B.L. Ricks. 1974. Laminar-turbulent transition in flow of pseudoplastic fluids with yield stress. *J Hydronautics* 84: 163–166.

Harper, J.C., and A.F. El-Sahrigi. 1965. Viscometric behavior of tomato concentrates. *J Food Sci* 30: 470–476.

Heppell, N.J. 1985. Comparison of the residence time distributions of water and milk in an experimental UHT sterilizer. *J Food Eng* 4: 71–84.

Himmelblau, D.M., and K.B. Bischoff. 1968. *Process analysis and simulation*. New York: John Wiley & Sons.

Kokini, J.L., L-S. Lai, and L.L. Chedid. 1992. Effect of starch structure on starch rheological properties. *Food Technol 46,* no. 6: 124–139.

Kubota, K., Y. Hosakawa, K. Suzuki, and H. Hosaka. 1979. Studies on the gelatinization rate of rice and potato starches. *J Food Sci* 44: 1394–1397.

Kumar, A., and M. Bhattacharya. 1991. Numerical analysis of aseptic processing of a non-Newtonian liquid food in a tubular heat exchanger. *Chem Eng Comm* 103: 27–51.

Kwant, P.B., R.H.E. Fierens, and A. Van Der Lee. 1973. Non-isothermal laminar pipe flow—I. Theoretical. *Chem Eng Sci* 28: 1303–1316.

Levenspiel, O. 1972. *Chemical reaction engineering*. 2d ed. New York: John Wiley & Sons.

Liao, H.J. 1998. Simulation of continuous sterilization of fluid food products: The role of thermorheological behavior of starch dispersion and process. Ph D diss., Cornell University, Ithaca, NY.

Lin, S.H. 1979. Residence time distribution in continuous sterilization process. *Process Biochem 14,* no. 7: 23–25.

Lopes da Silva, J.A., M.P. Gonçalves, and M.A. Rao. 1994. Influence of temperature on dynamic and steady shear rheology of pectin dispersions. *Carbohydrate Polymers* 23: 77–87.

Lyche, B.C., and R.B. Bird. 1956. The Graetz-Nusselt problem for a power-law non-Newtonian fluid. *Chem Eng Sci* 6: 35–41.

Metzner, A.B., and R.E. Otto. 1957. Agitation of non-Newtonian fluids. *AIChE J* 3: 3–10.

Nauman, E.B. 1977. Non-isothermal reactors: Theory and applications of thermal time distributions. *Chem Eng Sci* 32: 359–367.

Okechukwu, P.E., and M.A. Rao. 1995. Influence of granule size on viscosity of corn starch suspension. *J Texture Studies* 26: 501–516.

Osorio, F.A., and J.F. Steffe. 1984. Kinetic energy calculations for non-Newtonian fluids in circular tubes. *J Food Sci* 49: 1295–1296, 1315.

Paulsson, B., and C. Trägårdh. 1984. Calculation of sterilization effect distribution in heat exchangers in turbulent flow. In *Engineering and food. Vol. 1. Engineering Sciences in the Food Industry,* ed. B.M. McKenna, 151–164. London and New York: Elsevier Applied Science.

Rao, M.A. 1992. Measurement of viscoelastic properties of fluid and semisolid foods. In *Viscoelastic properties of food,* eds. M.A. Rao and J.F. Steffe, 207–232. London: Elsevier Applied Science.

Rao, M.A. 1995. Rheological properties of fluid foods. In *Engineering properties of foods,* 2d ed., eds. M.A. Rao and S.S.H. Rizvi, 1–53. New York: Marcel Dekker.

Rao, M.A., and R.C. Anantheswaran. 1988. Convective heat transfer to fluid foods in cans. *Adv Food Res* 32: 39–84.

Rao, M.A., M.C. Bourne, and H.J. Cooley. 1981. Flow properties of tomato concentrates. *J Texture Studies* 12: 521–538.

Rao, M.A., and H.J. Cooley. 1984. Determination of effective shear rates of complex geometries. *J Texture Studies* 15: 327–335.

Rao, M.A., and H.J. Cooley. 1992. Rheology of tomato pastes in steady and dynamic shear. *J Texture Studies* 23: 415–425.

Rao, M.A., H.J. Cooley, R.C. Anantheswaran, and R.W. Ennis. 1985. Convective heat transfer to canned liquid foods in a Steritort. *J Food Sci* 50: 150–154.

Rao, M.A., and M. Loncin. 1974a. Residence time distribution and its role in continuous pasteurization. Part I. *Lebensm - Wiss Technol* 7: 5–13.

Rao, M.A., and M. Loncin. 1974b. Residence time distribution and its role in continuous pasteurization. Part II. *Lebensm - Wiss Technol* 7: 14–17.

Rao, M.A., R.H. Walter, and H.J. Cooley. 1981. Effect of heat treatment on the flow properties of aqueous guar gum and sodium carboxymethylcellulose CMC solutions. *J Food Sci* 46: 896–899, 902.

Rieger, F., and V. Novak. 1973. Power consumption of agitators in highly viscous non-Newtonian liquids. *Trans IChem E* 51: 105–111.

Roig, S.M., A.A. Vitali, E. Ortega Rodriguez, and M.A. Rao. 1976. Residence time distribution in the holding section of a plate heat exchanger. *Lebensm - Wiss Technol* 9: 255–256.

Sancho, M.F., and M.A. Rao. 1992. Residence time distribution in a holding tube. *J Food Eng* 15: 1–19.

Simpson, S.G., and M.C. Williams. 1974. An analysis of high temperature/short time sterilization during laminar flow. *J Food Sci* 39: 1047–1054.

Skelland, A.H.P. 1967. *Non-Newtonian flow and heat transfer.* New York: John Wiley & Sons.

Steffe, J.F., I.O. Mohamed, and E.W. Ford. 1984. Pressure drop across valves and fittings for pseudoplastic fluids in laminar flow. *Trans Am Soc Agric Engineers* 27: 616–619.

Steffe, J.F., and R.G. Morgan. 1986. Pipeline design and pump selection for non-Newtonian fluid foods. *Food Technol 40,* no. 12: 78–85.

Stevens, P.M. 1972. Lethality calculations, including effects of product movement, for convection heating and broken heating foods in still-cook retorts. PhD diss., University of Massachusetts, Amherst.

Stumbo, C.R. 1973. *Thermobacteriology in food processing.* 2d ed. New York: Academic Press.

Trägårdh, C., and B. Paulsson. 1985. Heat transfer and sterilization in continuous flow heat exchangers. In *Developments in food preservation,* Vol. 3, ed. S. Thorne, 245–272. London: Elsevier Applied Science.

Veerkamp, C.H., A.J.M. Romijn, and J.C. Pol. 1974. Influence of varying residence time distribution on inactivation of microorganisms during pasteurization of egg products. *Lebensm - Wiss Technol* 7: 306–310.

Vitali, A.A., and M.A. Rao. 1984a. Flow properties of low-pulp concentrated orange juice: Serum viscosity and effect of pulp content. *J Food Sci* 49: 876–881.

Vitali, A.A., and M.A. Rao. 1984b. Flow properties of low-pulp concentrated orange juice: Effect of temperature and concentration. *J Food Sci* 49: 882–888.

Wissler, E.H., and R.S. Schechter. 1959. The Graetz-Nusselt problem with extension for a Bingham plastic. *Chem Eng Prog Symp Ser 55,* no. 29, 203–208.

Yang, W.H. 1997. Rheological behavior and heat transfer to a canned starch dispersion: Computer simulation and experiment. PhD diss., Cornell University, Ithaca, NY.

Yang, W.H., and M.A. Rao. 1998a. Complex viscosity-temperature master curve of cornstarch dispersion during gelatinization. *J Food Proc Eng* 21: 191–207.

Yang, W.H., and M.A. Rao. 1998b. Numerical study of parameters affecting broken heating curve. *J Food Eng* 37: 43–61.

Yang, W.H., and M.A. Rao. 1998c. Transient natural convection heat transfer to starch dispersion in a cylindrical container: Numerical solution and experiment *J Food Eng* 36: 395–415.

Appendix A
Nomenclature

The principal symbols used are defined in this section. Other symbols have been defined in the text immediately after they have been used.

a	Generalized front factor, cascade theory
c	Polymer or solute concentration, w w^{-1}
c^*	Coil overlap concentration, w w^{-1}
c_o	Concentration when viscosity extrapolated to zero, w w^{-1}; critical gelling concentration, g L^{-1}
c_p	Heat capacity
$C(t)$, $C(\theta)$	Residence time distribution function for impulse input
cQ	Starch granule mass fraction
D	Starch granule, can, and pipe diameter; D-value; axial dispersion coefficient
D_a	Diameter of agitator, m
D_0	Initial average starch granule diameter
D_v	Diameter of the vane, m
E	Elongation and compression modulus, Pa
E_a	Activation energy of flow, J $mole^{-1}$
E_f	Energy loss per unit mass, J kg^{-1}
$E_g/\Delta E_g$	Activation energy of gelatinization, J $mole^{-1}$
E_i	Elastic modulus, Pa
$E(t)$, $E(\theta)$	Residence time distribution function for impulse input
E_{ac}, E_{ak}, $E_{a\eta}$	Activation energy for viscous flow
E_{aR}	Activation energy for reaction rate constant
f	Fanning friction factor, dimensionless; f number of sites (functionalities) along each molecule's length
f_h	Heat penetration parameter
F	Residence time distribution function for step change input
F_0	Lethality at 121.1°C and for z = 10°C
g	Rubber elasticity front factor
G	Shear modulus, Pa
G'	Storage modulus, Pa
G''	Loss modulus, Pa

G_0	Plateau storage modulus, G' value at high frequencies, Pa
G_p	Shear modulus of material of spherical particle
$G^* = \sqrt{(G')^2 + (G'')^2}$	Complex modulus, Pa
GSID	Starch granule size distribution
H	Height of vane, m
$J(t, \sigma_c) = \gamma/\sigma_c$	Shear compliance, Pa^{-1}
$[J]$	Molar concentration of junction zones, mol L^{-1}
k	Thermal conductivity
K, K_H	Consistency index, Pa s^n; equilibrium constant
k_f	Friction coefficient of a fitting
K_J	Association constant
k_s	Proportionality constant
L	Length of can and tube/pipe, characteristic dimension m
M	Number average molecular weight, Da
M_c	Mean molecular weight of chains, Da
M_J	Number average molecular weight of the junction zones, Da
n, n_H	Flow behavior index, dimensionless; number of polymer segments
$\boldsymbol{n}$	Unit normal vector
N	Equals (c/M) initially, cascade theory, mol L^{-1}
N_a	Agitator speed, revolutions s^{-1}
Ne	Number of elastically active network chains per primary chain ($\approx$mol L^{-1})
p	Pressure, Pa; power, J s^{-1}; fraction of reacted bonds; critical exponent
$\mathbf{q}$	Heat flux
Q	Mass of heated starch granules per unit weight of dry starch
r_i	Radius of inner cylinder, m
r_o	Radius of pipe or tube
R	Gas constant, 8.314 J mol^{-1} K^{-1}
s	Critical exponent
S_∞	Gel fraction
SD	Standard deviation of gelatinized starch granules
SDo	Standard deviation of ungelatinized starch granules
SDR	Structure development rate, dG'/dt
SLR	Structure loss rate ($-dG'/dt$)
SGM	Starch granule mass fraction
STD	Starch dispersion
t	Time, s; critical exponent in gelation

$\bar{t}$	Average residence time, s
t_c	Gelation time, s
T	Temperature of fluid (°C, K), torque on an agitator, N m
T_m	Maximum torque, N m; melting temperature, K
T_{cc}	Torque per unit height, N
T_{pv}	Peak viscosity temperature from dynamic test, °C
TR	Thermorheological
v	Velocity, m s^{-1}
v_x, v_y, v_z	Components of velocity in Cartesian coordinates
v_r, v_θ, v_z	Components of velocity in radial coordinates
v_r, v_θ, v_ϕ	Components of velocity in spherical coordinates
V	Volume, m^3
W	Work per unit mass, J kg^{-1}
Z_1 and Z_2	Heights above a reference point, m

GREEK LETTERS

α	Constant, thermal diffusivity $m^2\ s^{-1}$; shift factor, $\alpha' = \alpha^{-1}$; fraction of reacted sites
α_c	Time constant, s
β	Coefficient of thermal expansion (K^{-1}); critical exponent
Δ	Critical exponent
$\Delta H°$	Standard enthalpy of cross-linking, J mol^{-1}
ΔH_{b_T}	Enthalpy of network junctions, J mol^{-1}
ΔH_m	Enthalpy of the cross-linking, J mol^{-1}
$\Delta S°$	Standard entropy of cross-linking, J $mol^{-1}\ K^{-1}$
ε	Deformation; relative distance to the gel point
δ	Phase angle, rad
λ	Strain; penalty parameter
λ_c	Time constant, s
$\gamma(t)$	Oscillating strain or deformation
$\dot{\gamma}$	Shear rate, s^{-1}
γ_0	Strain amplitude
$\dot{\gamma}_c$	Critical shear rate at which shear-thickening begins, s^{-1}
ε_g	Index of starch granule rupture
η_a	Apparent viscosity (Pa s)
$\eta_{Ca} = (K_c)^2$	Casson viscosity, Pa s
η_{dim}	Dimensionless apparent viscosity, Pa s
η_0	Viscosity of water (Pa s)

η_s	Viscosity of continuous phase (Pa s)
η_o	Zero shear viscosity, Pa s
$[\eta]$	Intrinsic viscosity, dL g^{-1}
η_{ug}	Ungelatinized apparent viscosity, Pa s
$\eta^* = (G^*/\omega)$	Complex viscosity, Pa s
μ_p	Poisson's ratio (dimensionless)
η_∞	Highest magnitude of η_a during gelatinization, infinite shear viscosity, Pa s
ϕ	Volume fraction, $\phi = cQ$ for starch dispersions
ϕ_m	Maximum volume fraction
ϕ_{cp}	Volume fraction for close-packing $= \phi_m$
ϕ_{red}	Equals ϕ/ϕ_{cp}
ρ	Density, kg m^{-3}
σ	Shear stress, Pa
σ_c	Critical shear stress at which shear-thickening begins, Pa
$\sigma_{0c} = (K_{0c})^2$	Casson yield stress, Pa
σ_{ij}	Tangential components of stress, Pa
τ	Stress tensor
θ	Dimensionless time, $t/\bar{t}$; $t\,V/L$
ψ	Integral of time-temperature history, t-K
ψ_1	First normal stress coefficient, Pa s^2
ψ_2	Second normal stress coefficient, Pa s^2
ω	Dynamic frequency, rad s^{-1}

DIMENSIONLESS NUMBERS

De — Deborah number compares the time scale of the deformation process to the observation time

$D/v\,2r_o$ — Dispersion number

GGr — Generalized Grashof number

$$\mathrm{GGr} = \frac{gL^3\,\beta\,\Delta T\,\rho^2\,N^{2-2n}}{K^2\,8^{2n-2}}\left(\frac{4n}{3n+1}\right)^{2n}$$

GPr — Generalized Prandtl number

$$\mathrm{G\,Pr} = \frac{c_p\,K8^{n-1}}{k\,N^{1-n}}\left(\frac{3n+1}{4n}\right)^{n}$$

Gr — Grashof number

$$\mathrm{GR} = \frac{g\mathrm{L}^3\,p^2\,\beta\,\Delta T}{N^2}$$

GRe — Pipe flow generalized Reynolds number

$$\text{G Re} = \frac{D^n\, \bar{v}_z^{2-n}\, \rho}{8^{(n-1)}\, K}\left(\frac{4n}{3n+1}\right)^n$$

GRe_a — Generalized Reynolds number

$$\text{G Re}_a = \frac{D^2\, N^{2-n}\, \rho}{8^{(n-1)}\, K}\left(\frac{4n}{3n+1}\right)^n$$

Nu — Nusselt number, hD/k

Po — Power number, dimensionless

$$\text{Po} = \frac{p}{D_a^5 N_a^3 \rho}$$

Pr — Prandtl number, $c_p\eta/k$

Re — Reynolds number, $Dv\rho/\eta$

Re_a — Agitator and rotational Reynolds number

$$\text{Re}_a = \frac{D_a^2 N_a \rho}{\eta}$$

Re_{mo} — Metzner-Otto Reynolds number

$$\text{Re}_{mo} = \frac{{D_a}^2 N^{2-n} \rho}{K}\, 8\left(\frac{n}{6n+2}\right)^n$$

Re_p — Particle Reynolds number

$$\text{Re}_p = \frac{D^2 \dot{\gamma} \rho}{\eta}$$

Re_{pl} — Power law agitator Reynolds number

$$\text{Re}_{pl} = \frac{D_a^2 N_a^{2-n} \rho}{K(k_s N_a)^{n-1}}$$

OPERATIONS

Symbols in bold indicate vectors.

Substantial derivative $= \dfrac{\partial}{\partial t} + \nabla \cdot \boldsymbol{V}$

$\dfrac{d}{dt}$ — Derivative with respect to time

$\dfrac{\partial}{\partial t}$ — Partial derivative with respect to time

$\exp x$ — e^x = exponential function of x

$\ln x$ — Logarithm of x to base e

$\log_{10} x$ — Logarithm of x to base 10

∇ — Del or nabla operator $= \dfrac{\partial}{\partial x} + \dfrac{\partial}{\partial y} + \dfrac{\partial}{\partial z}$

Index